INTERNATIONAL SERIES OF MONOGRAPHS IN PURE AND APPLIED BIOLOGY

Division: **ZOOLOGY**

GENERAL EDITOR: G. A. KERKUT

VOLUME 39

PRINCIPLES OF ZOOPHYSIOLOGY I

PRINCIPLES
OF ZOOPHYSIOLOGY

(IN TWO VOLUMES)

VOLUME I

BY

B. RYBAK

Faculté des Sciences de Caen

"C'est surtout par la physiologie générale que nous
tenons vraiment le secret de l'être, du monde, de
Dieu, comme on voudra l'appeler."

ERNEST RENAN *(Souvenirs de jeunesse)*.

PERGAMON PRESS

OXFORD · LONDON · EDINBURGH · NEW YORK

TORONTO · SYDNEY · PARIS · BRAUNSCHWEIG

Pergamon Press Ltd., Headington Hill Hall, Oxford
4 & 5 Fitzroy Square, London W.1
Pergamon Press (Scotland) Ltd., 2 & 3 Teviot Place, Edinburgh 1
Pergamon Press Inc., 44–01 21st Street, Long Island City, New York 11101
Pergamon of Canada Ltd., 207 Queens Quay West, Toronto I
Pergamon Press (Aust.) Pty. Ltd., 19a Boundary Street, Rushcutters Bay,
N.S.W. 2011, Australia
Pergamon Press S.A.R.L., 24 rue des Écoles, Paris 5e
Vieweg & Sohn GmbH, Burgplatz 1, Braunschweig

First English edition 1968

This is a translation of the original French *Cours de Zoophysiologie*
published by Gauthier-Villars, Paris, 1962, which has been brought
up to date for the English edition

Library of Congress Catalog Card No. 67–22826

PRINTED IN HUNGARY

08 012158 6

CONTENTS

GENERAL INTRODUCTION

PHYSIOLOGY is an exact experimental science. It is the physico-chemistry of life. However, it is clear that, being a natural science, physiology may not be removed from its morphological context. Morphology and physiology, structures and functions are inseparable. In fact, physiology—the science of living functions—is an entity but can be thought of more as a integral than a unique science. Within physiology one may distinguish a physiology in natural conditions which is *ecology* and a physiology in laboratory conditions which simplifies the examination of phenomena but removes them from their natural context. Furthermore, in seeking to promote a completely explanatory physiology, it is necessary to consider both *special physiology* which is concerned with functions related to morphological differentiation, resulting from evolution of the living creature, and *molecular physiology* which, in one way or another, constitutes causal physiology. To illustrate the way in which relationships are established between causal physiology and this "symptomatic" physiology, I would say that the lung represents a morphological differentiation which is related to an increase in complexity of respiratory processes as a function of evolution, while the cytochrome system represents fundamental respiration.

However, having taken evolutionary physiology as the basis we are dealing, in fact, with *comparative physiology*. I would emphasize that I understand it to be comparative and not, as is erroneously said, compared since it is not compared with anything. Within this discipline, in the steps towards understanding, one fact is confronted with another, one function with another, both in the case of the one organism and very different organisms. Thus, this physiology must be both phylogenic and ontological, that is to say, embryological.

However, one realizes that in a work which seeks to provide the basic ideas of physiology, this methodology cannot be extensively applied and my aim is to stimulate thought rather than merely to provide facts.

To learn a science is to learn the terminology and overall concepts. This is what we shall endeavour to do.

FUNCTIONS OF NUTRITION AND INTERNAL ENVIRONMENT

CHAPTER 1

GENERAL PRINCIPLES OF NUTRITION

"L'énergétique a ce privilège de conférer aux problèmes une haute généralité et de nous instruire sur les mécanismes les plus minutieux de la Vie."

RENÉ WURMSER

In this chapter we shall consider those physico-chemical concepts that are necessary for an understanding of food transformation in animals, with special reference to biological thermodynamics. It will not be necessary to consider entropy at this stage.

Oxidation and reduction

Oxidation and reduction are the basis of all elementary nutrition.

(a) *Inorganic reactions*

$$Fe^{+++} + e' \rightleftharpoons Fe^{++}. \tag{1}$$

In the forward reaction a positive charge is neutralized (reduction).

(b) *Organic reactions*

Let us consider the various reactions which can occur with organic molecules:

$$R + \tfrac{1}{2}O_2 \rightarrow RO \text{ (Example: } CO + \tfrac{1}{2}O_2 \rightarrow CO_2) \tag{2}$$

R—an organic compound—is oxidized directly by oxygen (for example, from the air). R is therefore the *oxygen acceptor*.

But an organic compound can be oxidized in another way:

$$RH_2 + \tfrac{1}{2}O_2 \rightarrow R + H_2O \tag{3}$$

The oxidation of RH_2 involves the loss of two hydrogen atoms and is a *dehydrogenation*. Two examples of this type of reaction can be given:

(1) when molecular oxygen (principally from the air) accepts the hydrogen atoms, aerobic oxidation occurs;

(2) when the acceptor is an organic compound R', the reaction can be written:

$$RH_2 + R' \rightarrow R + R'H_2 \tag{4}$$

3

In equation (2) R can be oxidized not only by free oxygen but also by oxygen combined in a molecule of water:

$$R + H_2O \rightarrow R \underset{\text{OH}}{\overset{\text{H}}{<}} \rightarrow RO + 2H \qquad (5)$$

hydrate which
is oxidized by
dehydrogenation

This reaction can proceed in the absence of oxygen—anaerobic oxidation.

Biochemistry, as we know it, is a particular aspect of the chemistry of water but it is possible to visualize a biochemical system where the biophase is, for example, methane. If one accepts that there are many inhabited worlds, then this biochemical system could be that of some extragalactic planet. But we must return to earth.

It will be seen that oxidations by dehydrogenation occur more generally than direct oxidation. In dehydrogenation the hydrogen acceptor can vary. This distinguishes *aerobiosis* or *respiration* (where O_2 acts as the acceptor) from *anaerobiosis* or *fermentation* (where a product of intermediate metabolism acts as the acceptor, for example, R′ in equation (4)). The origin of the hydrogen acceptor determines whether an organism is aerobic or anaerobic.

It must be remembered in using the term fermentation that ferments have been identified as *diastases* or *enzymes* but enzymes are involved also in aerobic processes.

The pathways of oxidation and reduction of biological compounds during metabolism lead us to consider the principles of oxidation–reduction reactions.

In general terms, oxidation occurs when an atom gains positive charge or loses electrons and reduction takes place when an atom gains electrons.

It is important, before proceeding to theoretical considerations, to note $pH = \log \dfrac{1}{(H^+)}$, that is to say, pH equals the inverse of the logarithm —the cologarithm—of the concentration of hydrogen ions. $rH_2 = \log \dfrac{1}{(H_2)}$ or the cologarithm of the concentration of hydrogen gas.

Free energy is a concept which has considerable importance in the understanding of bioenergetics.

The first law of Carnot: equivalence of heat and work

In a closed cyclic system which passes from equilibrium state 1 to equilibrium state 2, the sum of external work is equivalent to the sum of heat exchanged with the sources; or:

$$\Sigma W = J \Sigma Q \qquad (6)$$

where J, the mechanical equivalent of heat, depends on the choice of units. Normally units for heat and work are adopted such that $J = 1$ and equation (6) can be rewritten:

$$\Sigma W = \Sigma Q \tag{7}$$

whence

$$W - Q = 0 \tag{7'}$$

Thus, in a closed isolated system energy is conserved and can only be transformed.

Let us consider now a non-closed system (for example, boiling water):

$$JQ - W = \Delta U = \text{the variation in internal energy.} \tag{8}$$

The internal energy of a system in the initial state is U_1 and in the final state is U_2, whence

$$\Delta U = U_2 - U_1 \tag{9}$$

The variation in internal energy of a system is the difference between the heat absorbed (or more generally, between the energy absorbed) and the work produced. We shall now consider, reactions occurring at constant volume and those occurring at constant pressure.

I. *Constant volume:* Reactions at constant volume are used in the measurement of the energy values of food substances. $W = 0$ and the variation in internal energy is equal to the heat evolved. Examples of some values obtained are given in Table 1.

TABLE 1

Type of food substance	kcal per g		
	Oxidation in a bomb calorimeter	Oxidation in the organism	Oxidation in the organism (values corrected for digestibility)
carbohydrates	4·1	4·0	3·68
lipids	9·4	9·0	8·65
proteins	5·6	4·1	3·88

erg = dyne×cm	calorie = $4·185 \times 10^7$ ergs or 4·185 joules
joule = 10^7 ergs	Calorie or kcal = $4·185 \times 10^{10}$ ergs or 4185 joules

(1) The figures given in the table are rounded values.

(2) They are the average values. For example, the caloric value of polysaccharides would not be the same as that of monosaccharides.

(3) A comparison of the two sets of values shows that while the energy released from carbohydrates and lipids is of the same order in the organism and in the calorimeter, the values for proteins are different. In the organ-

ism, absorbed carbohydrates and lipids are oxidized completely to carbon dioxide and water. In the case of protein, however, the oxidation is incomplete in the organism. The principal end product of protein metabolism, in man, is urea which contains potential energy which is not utilized by the body. Other examples of end products of protein metabolism are ammonia, creatinine and uric acid.

II. *Constant pressure:* In reactions at constant pressure there is a change in volume, ΔV; thus:

$$\Delta H = \Delta U + P\Delta V. \tag{10}$$

H is termed the *enthalpy* or *heat content*.

However, the external work produced by a reaction where the system is always near to the equilibrium state, depends only on the final state and the initial state and constitutes the *maximum work*, positive from the reaction, W_{max}. Subtraction of work done against a constant pressure from maximum work gives a measurement of the change in free energy:

$$-\Delta F = W_{max} - P\Delta V. \tag{11}$$

Whence: free energy = the energy available at constant pressure and, if E = energy available at constant volume,

$$F = E + PV. \tag{12}$$

What is important in physiological systems is that under conditions of constant temperature and pressure a reaction does not proceed spontaneously, except when ΔF has a negative value (has to be considered for heart pacemaker). At equilibrium $\Delta F = 0$.

This aspect of free energy is of help in explaining the *affinities* of reactants. Wurmser wrote: "Precise knowledge of free energy, in many cases, allows one to understand more fully that which was originally termed specific action. The role of hydrogen ions, the difficulty of cyclic syntheses and perhaps specific isodynamic action can be explained by an understanding of affinities." Modern research based on a knowledge of energy levels in molecules has led to an understanding of the mechanisms of oxidative phosphorylation and of certain aspects of muscular contraction.

Free energy in terms of oxidation–reduction

We have seen that, in inorganic chemistry, oxidation and reduction reactions proceed by the transfer of electrons. However, there are some organic compounds of biological interest which behave in the same way; in particular the natural pigments, for example, the yellow enzyme of Warburg, a flavoprotein. The electron active group is lactoflavin.

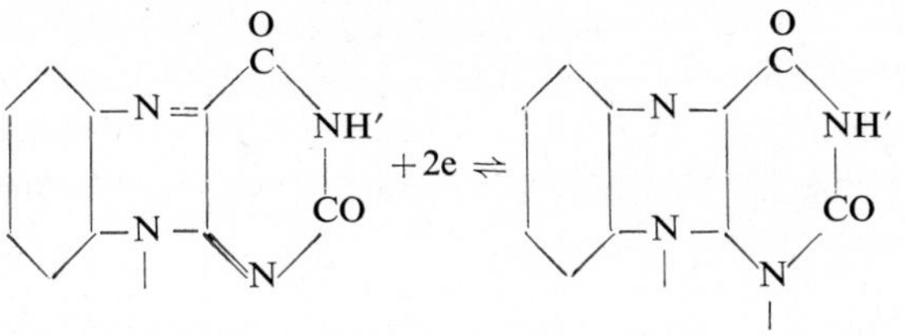

$$(13)$$

This electronic activity is not a property of all organic compounds. Most metabolites are inactive but natural catalysts exist, diastases called *dehydrogenases*, which allow these compounds to react with hydrogen acceptors. For example:

succinic ion + methylene blue $\rightleftharpoons$ fumaric ion + leucomethylene blue

(Quastel, Weetham, and Thunberg)

From these facts the following generalization concerning the importance of oxidation–reduction reactions in metabolism can be made. In metabolic oxidations the oxidized compound loses one or more electrons while another compound receives these electrons and is reduced. Thus, for each oxidation process there is an equivalent reduction process.

$$\text{Reductant} \rightleftharpoons \text{Oxidant} + n\,e' \tag{14}$$

This can be written:

$$\frac{[\text{oxidant}] \times [e']^n}{[\text{reductant}]} = k \tag{15}$$

$$[e'] = k \left(\frac{\text{reductant}}{\text{oxidant}} \right)^{1/n} \tag{16}$$

If R is the gas constant ($8 \cdot 32$ joules per mol per degree) and T is the absolute temperature, it follows, using natural logarithms, that:

$$\Delta F = RT \ln \frac{[\text{reductant}]}{[\text{oxidant}]} - RT \ln k \tag{17}$$

The direction of equation (14) depends on the number of electrons present. According to the law of mass action, when the number of electrons increases the system will produce more reductant. If the number of electrons decreases there will be an increase in the production of oxidant. This represents the most elementary form of chemical and biochemical control.

It follows that if the electronic state of a system is known its oxidizing and reducing ability can be determined. The ability of a compound to take up or give up electrons can be estimated by measuring the voltage produced by the oxidizing–reducing system, against that of a standard electrode, when two inert electrodes are placed in the system. The electrode potential is dependent on the transfer of electrons in the medium. The concentration of electrons in the medium is expressed by [e]. The concentration of electrons at the electrode is considered to be constant and is expressed by $[e_c]$. The work needed to move an electron from the medium (initial state) to the metal electrode (final state) is equal to that needed to transfer one faraday ($\mathcal{F} = 96{,}500$ coulombs) between a potential difference of E volts.

$$(-)\Delta F = E\mathcal{F} \tag{18}$$

thus

$$E = \frac{\Delta F}{\mathcal{F}} \tag{19}$$

but

$$\Delta F = RT \ln \frac{[\text{reductant}]}{[\text{oxidant}]} - RT \ln k \tag{17}$$

such that

$$\mathcal{F}E = RT \ln \frac{[e]}{[e_c]} - RT \ln k \tag{20}$$

therefore

$$E = \frac{RT}{\mathcal{F}} \ln [e] - \frac{RT}{\mathcal{F}} \ln [e_c] - \frac{RT}{\mathcal{F}} \ln k \tag{21}$$

since $[e_c]$ is constant, it follows that:

$$E = K - \frac{RT}{\mathcal{F}} \ln [e] - \frac{RT}{\mathcal{F}} \ln k \tag{22}$$

or

$$E = K - \frac{RT}{\mathcal{F}} \ln k - \frac{RT}{\mathcal{F}} \ln \frac{[\text{reductant}]}{[\text{oxidant}]} \tag{23}$$

or

$$E = k_1 - \frac{RT}{\mathcal{F}} \ln \frac{[\text{reductant}]}{[\text{oxidant}]}. \tag{24}$$

k_1 is a new constant since $K - (RT/\mathcal{F}) \ln k$ is a constant for all temperatures and reactions.

If E is measured with reference to a standard electrode (hydrogen electrode) then:

$$E_H = E - v \tag{25}$$

where v is the potential of the hydrogen electrode which by definition is equal to 0; then:

$$E_H = k_1 - \frac{RT}{\mathcal{F}} \ln \frac{[\text{reductant}]}{[\text{oxidant}]} \tag{26}$$

and for k_1 (= constant) = E_0, it follows that:

$$E_H = E_0 - \frac{RT}{\mathcal{F}} \ln \frac{[\text{reductant}]}{[\text{oxidant}]} \tag{27}$$

or

$$E_H = E_0 + \frac{RT}{\mathcal{F}} \ln \frac{[\text{oxidant}]}{[\text{reductant}]} \tag{28}$$

for one gram-equivalent. The equation for n gram-equivalents is the standard formula for E_H (Peter, 1898):

$$E_H = E_0 + \frac{RT}{n\mathcal{F}} \ln \frac{[\text{oxidant}]}{[\text{reductant}]} \tag{29}$$

From this equation it follows that:

(1) The greater the amount of reducing compound present the smaller will be the value of E_H and inversely for oxidizing compounds.

(2) If the concentration of reducing agent is equal to the concentration of oxidizing agent, [reductant]/[oxidant] = 1 and logarithm = 0. This is the *normal potential*. Each oxidation–reduction system possesses a value for the normal potential which is characteristic of the system.

The oxidation–reduction potential of many systems is dependent on pH. At a given pH the curve representing the ratio of reduced form to oxidized form is sigmoid (Fig. 1). The potential represented by a curve at 50 per cent reduction is the E_0' value of the system. This is the value used in practice together with rH_2 (defined by M. Clark) which is related to the ratio E_H. Experimentally the rH_2 value is measured by using a dye (of which the E_0' of decoloration is known) but this method is limited by the toxicity of the dyes and their ability to penetrate cells. rH_2 can be measured electrically. The following values were obtained by Aubel and Wurmser (1929) at pH 7 (Table 2).

TABLE 2

Physiological structure	rH_2	$E_0(V)$
Spleen, pancreas, intestine, stomach	16–20	
Muscle and kidney	14–16	–0·04
Liver and grey substance from brain	9	
Non-autolysing systems in anaerobic conditions	7	–0·20

(3) Thus, a tissue or biochemical systems may be characterized by the degree of positivity or negativity of the redox potentials.

For example:

(a) *In inorganic chemistry:* the system $F_2 + 2_e \rightleftharpoons 2F$ exhibits a normal potential of $+2·85$ V. The normal potential of $Li^+ + e = Li$ is $-2·96$ V.

2

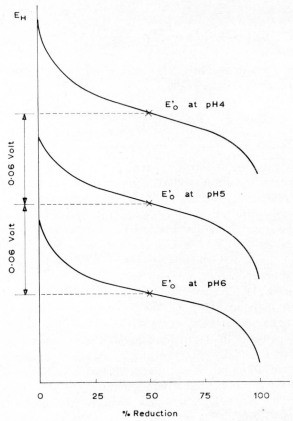

FIG. 1. Variation in the potential of a non-dissociated system at different pH as a function of percentage reduction. (WURMSER, 1930)

(b) *In biochemistry:* the system, succinic acid $\rightleftharpoons$ fumaric acid, shows an E_0' value of $+0\cdot024$ V at $25°C$ and pH 7. The E_0' value for the system lactic ion $\rightleftharpoons$ pyruvic ion is $-0\cdot176$ V at $36°C$ and pH 7.

Therefore, in principle but only in principle, an oxidizing agent can react with all compounds which have a lower E_0 value. Thus, apart from oxygen and hydrogen, the idea of an oxidizing agent is a relative one.

To determine whether a reaction is possible and in which direction it will proceed, a certain number of factors must be considered. These will be specific characteristics of each particular system; for example, solubility and the ability of the molecules to penetrate cells.

(4) The physiological importance of oxidation–reduction reactions is confirmed by the fact that synthesis of biological material is accompanied by an increase in available energy and is generally associated with reduction reactions. These reduction reactions are endothermic processes (or

more generally, endergonic, that is to say, they require energy). They cannot occur spontaneously and are associated with reactions which produce energy (exergonic) able to occur spontaneously. Thus, metabolism is always associated with coupled oxidation–reduction reactions.

Another fact which indicates the importance of oxidation–reduction phenomena in physiology is that biological oxidations and in particular, those under aerobic conditions, are essentially reactions involving the elimination of hydrogen. Whether with respect to aerobiosis or anaerobiosis, reactions involving the elimination of hydrogen are a form of excretion, the elimination of a toxic product. The non-excess hydrogen from anabolic processes is incorporated in syntheses. Thus, oxidation–reduction systems represent the most elementary form of nutrition and excretion, of incorporation and elimination.

Lavoisier considered the basis of life to be combustion which is supported by oxygen. But a more extended idea was developed from the discovery of anaerobic organisms by Pasteur and the generalized theory of Wieland (1912–1924) who visualized oxidations as successive dehydrogenations. From the two pathways of hydrogen utilization, it would appear that, during the course of evolution, organisms and organs associated with apparently distinct functions have been differentiated: syntheses by hydrogen fixation and respiration with the elimination of surplus hydrogen. Therefore, in life oxygen neither supports combustion nor is it a food. In my view it is a detoxicating agent and this function is associated with the liberation of energy which is necessary for syntheses. In general terms, an organic compound (R' on p. 3) could have the same function. Hydrogen must circulate continuously to avoid accumulation and the subsequent toxic effect on cells.

The concept of food

Food materials are required to build the structural constituents of cells, to provide reserves (plastic rôle) and to maintain specific functions by the provision of energy. If we could synthesize all the structural and energy components there would be no need to feed or at least we would require only simple compounds. Animals are dependent on plants which are able to synthesize complex food substances. For this reason, Pfeffer classified living organisms as autotrophes or plants and heterotrophes or animals.

Autotrophes — or autophytes — assimilate carbon dioxide and nitrogenous compounds: autotrophes must have been the first living organisms to evolve.

Heterotrophes, allotrophes or saprophytes require organic compounds.

Between these two classes are placed the *mixotrophes* which assimilate carbon dioxide and absorb organic compounds.

TABLE 3

Source of energy \ Ability to synthesize	Autotrophes Synthesis of all constituents. Reduction of nitrates and sulphates. No growth factors are required	Mesotrophes Inability to utilize nitrogen or sulphur or the oxidized forms, NO_3 and SO_4.	Metatrophes Require one or more growth factors.
Chemotrophes Energy is obtained by the oxidation of inorganic compounds. Carbon is provided by CO_2 assimilation	Chemo-autotrophes	Chemo-mesotrophes	Chemo-metatrophes (H)
Phototrophes Energy is obtained by photo synthesis. Carbon is provided by CO_2 assimilation.	Photo-autotrophes	Photo-mesotrophes	Photo-metatrophes (H)
Allotrophes Energy is obtained by the oxidation of organic compounds which constitute the source of carbon.	Allo-autotrophes (H)	Allo-mesotrophes (H)	Allo-metatrophes (H)

(H) : heterotrophe—requires organic compounds

Following a study of the nutritional requirements of Protozoa, A. Lwoff (1932) produced a more complex classification of physiological types (Table 3).

Lwoff wrote: "If a heterotrophe is defined by the need for organic compounds, it can be seen from the table that chemotrophes, phototrophes and allotrophes also come into this category. Even an autotrophe which synthesizes all its constituents requires organic carbon to provide energy (allo-autotrophe) and is therefore, heterotrophic". Lwoff agreed with the notion of Twort and Ingram (1911–1913) that the need for a growth factor is due to loss of the ability to synthesize it.

We have referred to foodstuffs, metabolites or substrates but, as yet, with the exception of dehydrogenases, we have not considered how physiological transformation of metabolites is maintained. Enzymes or diastases are present in the cell and these catalyse metabolic transformations. These catalysts are proteinous in nature. In order to understand this aspect of zoological physiology the following points should be remembered.

(1) The presence of an enzyme in an organism is controlled by a specific gene (Beadle). In the synthesis of an enzyme, a high molecular weight ribonucleic acid is synthesized at a desoxyribonucleic acid template in the nucleus (Gros, Ycas). This messenger ribonucleic acid transports the genetic information to the cytoplasmic ribonucleic acid where the processes of protein synthesis are directed towards the formation of the enzyme—a specific protein (Jacob and Monod, 1960).

(2) Karström (1930) divided bacterial enzymes into two groups, constitutive enzymes and inducible enzymes. Constitutive enzymes are formed normally by the cell and are dependent on genetic structure. The formation of inducible enzymes is dependent on the adaptation of the cell to a specific substrate or an inducer which is not necessarily a substrate. For example, in the formation of β-galactosidase by *E. coli*, methyl-β-D-galactoside

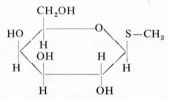

was found to act as a substrate and inducer, while methyl-β-D-thiogalacto-

CH₂OH structure

side was a more efficient inducer but did not act as a substrate (J. Monod, 1956). The inducible enzymes depend indirectly on genetic structure because constitutive enzymes are necessary for their manufacture.

(3) Although enzymes generally only catalyse a reaction in one direction, certain enzymes can act reversibly, for example, lipases.

(4) Some enzymes are produced as the inactive form or *zymogen* which in the presence of an activator, generally an enzyme or *kinase*, is converted to the active form. For example, trypsinogen from the pancreas is activated in the small intestine by enterokinase, produced by the duodenal mucosa. Pepsin is released into the stomach as the precursor, pepsinogen, which is activated by hydrochloric acid at pH 6. Therefore, zymogens can be activated by ionic activators.

(5) Many enzymes are composed of a protein fraction, the *apoenzyme* to which is attached a prosthetic group or *coenzyme*. One must not confuse coenzymes and kinases. The coenzyme is attached to the apoenzyme throughout the enzyme action while a kinase is not necessary for enzyme activity after it has transformed the precursor to the active form. Vitamins

act as coenzymes in many cases as will be seen for group B vitamins. Small amounts of certain inorganic ions are necessary in some enzyme actions.

(6) It is remarkable that the intestinal helminthes are not digested by the host (for example, the ascarid from dog resists the enzymic activity of pancreatic juice). It has been shown that extracts of ascarids contain compounds which inhibit the action of trypsin. These anti-enzymes are of great importance as they allow certain aspects of the specificity of parasites to be understood and thus contribute to an understanding of the ecology of parasites.

The presence of *anti-enzymes* can be demonstrated by the production of antibodies after parenteral injection of a protein—the antigen. If a foreign enzyme is injected into an organism, principally a mammal, antibodies acting as anti-enzymes are produced. J. Monod (1945) proposed that the antibodies are produced like inducible enzymes by the action of antigen.

(7) The activity of enzymes can be inhibited by many substances; specific antibodies produced after parenteral injection, salts of heavy metals (Hg^{++} and Pb^{++}), organic poisons; but the form of inhibitor which is of the greatest physiological interest is the *competitive inhibitor*. When a substance has a similar chemical configuration to that of the substrate it can compete with the substrate for the enzyme and prevent the normal action of the enzyme. These substances are called antimetabolites or *isosteres*:

for example, malonic acid $COOH \cdot CH_2 \cdot COOH$ is an isostere of succinic acid $COOH \cdot CH_2 \cdot CH_2 \cdot COOH$ and inhibits succinic acid dehydrogenase (Quastel and Wooldridge, 1928).

(8) The molecular weights within a group of enzymes—for example peroxidases, phosphatases—vary and thus the same enzyme activity can be supported by proteins which have different macromolecular configurations *(isozymes)* (Paul and Fottrell, 1961).

(9) In the organism, whether it be unicellular or multicellular, metabolism involves *multi-enzyme systems*, such that the products of one enzyme reaction become a substrate for the following reaction. These reactions proceed until ultimate-waste-products are formed which of necessity are eliminated by the organism. The final product of a metabolic chain must be eliminated because accumulation will result in inhibition of the enzyme which produced it (law of mass action) and a toxic effect will be produced in the animal as a whole. These considerations of molecular pathology provide another aspect of basic physiological regulation. Summarily, in the organism metabolic reactions are catenate and mutually interdependent. The significance of this will be appreciated when we consider metabolic cycles.

CHAPTER 2

ZOO-ENERGETICS

ZOO-ENERGETICS is the study of energy exchange in animals. Marcelin Berthelot emphasized that chemical reactions which are expressed by molecular formulae are incomplete unless quantitative data are provided as to the caloric processes involved in the reactions.

The first law of thermodynamics states that energy is conserved during transformations but this gives us no indication of the pathways of the reactions or the number of systems involved. The object of physiological chemistry is to provide precise information on these processes in the living organism.

Let us examine initially whether the first law of thermodynamics applies to animals. It is necessary to establish a balance sheet of that which enters the organism and that which leaves it, of food (or in more general terms—ingested material) and excreta.

Methods for measuring available energy

Several methods are available:

(1) A balance sheet can be drawn up by measuring the heat of combustion of foods and excreta in a bomb calorimeter. This is a useful method providing the subject undergoes no change in weight during the experimental period. If there is a change in weight, the values obtained must be corrected for the number of calories corresponding to the loss or gain in weight. By convention, this variation is taken always to be a variation in fat content and thus establishes a systematic error in the results obtained.

(2) An apparently simple method is used to measure the energy intake. The proportion of simple food materials in the diet is estimated together with their heat of utilization in the organism being considered. This method is not very accurate because the caloric values obtained, having taken account of digestibility, are mean values. Moreover, this method requires an involved analysis of the quantity and quality of the components in the diet.

(3) The amount of energy available to the organism can be estimated from the composition of the excreta by carrying out a continuous analysis of their caloric values. For this method to provide accurate data all the

excretory products must be considered (for example, sweat which contains lactic acid). Moreover, the time taken for the degradation of different food materials varies and there will be also variations in the latency according to the excretory pathway involved.

Measurements of energy output

I. Direct calorimetry

This is the best method available (Atwater; J. Lefèvre). The subject is placed in a thermally insulated calorimeter and the total amount of heat produced by the subject in a given period is measured directly. The heat released is absorbed either by water (Atwater) or by air (Lefèvre). During the experiment the subject can carry out a measured amount of work and the amount of oxygen consumed, carbon dioxide eliminated and nitrogen excreted can be determined.

By using thermo-electrically coupled micro-calorimeters, Calvet and Prat (1954) were able to follow small changes in the amount of heat produced by cockroach, *Drosophila* and young mice.

This method allows very accurate estimations of heat output to be made but involves the use of costly and complicated apparatus. Therefore, the method of indirect calorimetry is more often used.

II. Indirect calorimetry

The method of indirect calorimetry is based on *respiratory thermochemistry* and the heat output of the subject is calculated from his oxygen consumption or from the amount of water and carbon dioxide eliminated.

The basis of respiratory thermochemistry lies in the molecules of the respiratory chain. At each level of functional integration—enzyme–substrate systems, cells, tissues, organs, organisms—processes exist which are directed towards the same end and which show increasing complexity, such that more complex processes are dependent on those which are less complex.

In respiratory thermochemistry, the term *respiratory quotient* (R.Q.) is defined as the ratio of the volume of carbon dioxide produced to the volume of oxygen utilized. The value of R.Q. is different for different types of foodstuffs.

Carbohydrates—the complete oxidation of glucose is expressed by the equation:

$$C_6H_{12}O_6 + 6O_2 \rightarrow 6CO_2 + 6H_2O \tag{30}$$

and

$$R.Q. = \frac{CO_2}{O_2} = \frac{6}{6} = 1$$

Fats—the R.Q. of fats is < 1 because their oxygen content is low with respect to the carbon content of the molecules and thus, oxidation requires more external oxygen. Complete oxidation of tristearin is expressed by the following equation:

$$2C_{57}H_{110}O_6 + 163\ O_2 \rightarrow 114CO_2 + 110H_2O \tag{31}$$

and the R.Q. is 0·70 (114/163).

Proteins—oxidation of proteins cannot be expressed directly by formulae because their chemical structures are not yet known fully; by indirect methods the R.Q. is calculated as 0·8.

Abnormal respiratory quotients.

(1) In a mixed diet containing proteins, fats and carbohydrates the R.Q. is 0·85; in a diet rich in carbohydrates the R.Q. approaches 1 but during sugar diabetes, when there is poor utilization of sugars, the R.Q. decreases but may be increased by insulin treatment.

(2) It is possible to obtain values for R.Q. greater than 1·5. The metabolic significance of these values is that they result from interconversions in which one type of compound is transformed to another. For example, during the fattening of domestic animals, synthesis of lipids—low oxygen content—occurs at the expense of carbohydrate—high oxygen content—such that a relatively small amount of oxygen is utilized and the R.Q. can reach a value of 1·58.

Thermal coefficients of oxygen

The thermal coefficents of oxygen for fats and proteins are employed in respiratory thermochemistry in connection with the measurement of the amount of oxygen consumed. These coefficients are defined in the following way:

(1) *Carbohydrates.* Glucose

$$C_6H_{12}O_6 + 6O_2 \rightarrow 6CO_2 + 6H_2O + 675\ cal \tag{32}$$
$$but,\ 6O_2 = 6 \times 22\cdot4\ l. = 134\cdot4\ l.$$

and thus, the thermal coefficient of oxygen for glucose = $675/134\cdot4$ = 5·02 cal/l. For other sugars the coefficient is 5·048 since 1 g sugar releases 4·1 cal and requires 0·812 l. O_2 ($4\cdot1/0\cdot812 = 5\cdot048$).

(2) *Fats*

The combustion of 1 g fat releases 9·3 cal and uses 1·982 l. O_2; the coefficient of O_2 for fats = $9\cdot3/1\cdot98 = 4\cdot69$ cal/l.

(3) *Proteins*

The combustion of 1 g protein releases 4·1 cal and requires 0·915 l. O_2; the thermal coefficient of oxygen for proteins = $4\cdot1/0\cdot915 = 4\cdot48$ cal/l.

Thus, the thermal coefficient of oxygen

$$= \frac{\text{amount of heat produced} \, / \, \text{g simple food}}{\text{volume of } O_2 \text{ utilized in oxidation}}$$

COMMENTS: (1) The heat produced is related to units of oxidizing agent rather than to the weight of the compound oxidized.

(2) The thermal coefficient of CO_2 is not used because in short duration experiments errors can be introduced (for example, the elimination of carbon dioxide varies according to the emotional state of the subject) and in long term experiments acidosis can occur.

In choosing which thermal coefficient of oxygen to use in an experiment, it is necessary to know the nature of the subject's diet but this can be based on R.Q. If the R.Q. approaches 1, the thermal coefficient of oxygen for sugars is used (5·048); in the case of a R.Q. based on a mixed diet, the average thermal coefficient is used but this will decrease the accuracy of the experiment.

EXAMPLE: The R.Q. of a man who had been fasting for 16 h (post-absorptive period) was 0·83; the thermal coefficient of oxygen chosen was 4·8.

To calculate the number of calories produced by the subject, the number of litres of oxygen consumed is multiplied by the thermal coefficient of oxygen corresponding to the observed R.Q.

This calculation gives an approximate value which can be made more accurate by determining the amount of nitrogen excreted. Proteins contain on average 16 per cent nitrogen and from the weight of nitrogen excreted in unit time ($\times$ mg/mn), the weight of protein metabolized in the same unit of time can be calculated ($6·25 \times$ mg protein/mn). One can calculate also, the total carbon dioxide produced during oxidation of these proteins and the amount of oxygen needed in the oxidation. However, these latter values have more theoretical than practical significance.

Apparatus used in indirect calorimetry

(1) Techniques using gas analysis

Spirometers are used which consist principally of a gas holder which is counter-balanced such that the work needed to raise it is not supplied by the subject (Tissot's apparatus, Fig. 2). Portable spirometers can be used. These consist of a canvas bag, the Douglas bag, in which the expired gas is collected. A sample of this gas is analysed.

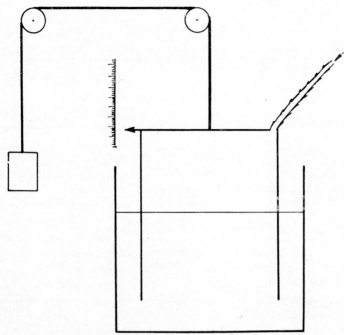

Fɪɢ. 2. Schema of a rigid water spirometer.

(2) Techniques not involving gas analysis

(1) Haldane's open circuit method (for use with small animals)

The principle of this method is the gravimetric analysis of oxygen consumed and carbon dioxide and water eliminated. The circuit consists of flasks containing concentrated sulphuric acid to collect water and soda lime to collect carbon dioxide. At the beginning of the experiment, time t_0, circuit B together with the animal and the cage are weighed. The experiment proceeds at constant temperature and at time t_x the circuit and the cage and animal are reweighed.

For example, an albino rat weighs 207 g; after 300 mn there is a decrease in weight of 1·13 g; 2·89 g carbon dioxide have been eliminated and 1·08 g water giving a total $(CO_2 + H_2O)$ of 3·97 g. Thus the weight of oxygen consumed = 1·13 − 3·97 = 2·84 g.

From this the R.Q. and heat formation in kcal/h are calculated.

(2) Closed circuit methods (Atwater and Benedict, 1904–1909)

This method is more complex but more accurate than the Haldane technique. It differs from the previous method, primarily, in that oxygen consumption can be measured directly and the circulating air can be analysed. Moreover, repeated measurements of water and carbon dioxide can

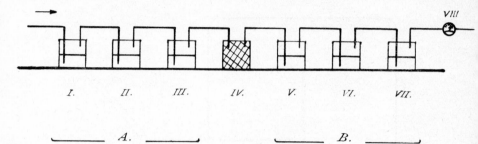

FIG. 3. Schema showing the arrangement of the Haldane apparatus. *I*, Concentrated H_2SO_4; *II*, soda lime; *III*, concentrated H_2SO_4; *IV*, air-tight cage immersed in a thermostat; *V*, concentrated $H_2SVI_4O_4$; soda lime; *VII*, concentrated H_2SO_4; *VIII*, oxygen or air flow meter. The direction of the current of air or oxygen is indicated by the arrow (*A* is the afferent circuit and *B* the efferent circuit).

be carried out as there is a double system of flasks containing concentrated sulphuric acid and soda lime. One set of flasks remains in the circuit while the other set are being weighed. The calorimetric chamber is sufficiently large to contain a man at rest or even undertaking physical excercise (for example, with the ergometric cycle of Fleisch).

From the results obtained from the various methods of direct and indirect calorimetry, it is apparent that the law of conservation of energy is applicable to living organisms (Table 4). J. Lefèvre (1911) wrote: "it has now been proved that the chemical energy taken in by the organism can be recovered completely as kinetic energy in the calorimeter".

TABLE 4. *Conservation of energy (mammals)*

Authors	Duration of the experiment (days)	Number of calories in the diet	Number of calories in calorimeter	% difference (experimental error)
Rübner	45	17,406	17,350	−0·32
Atwater– Benedict	93	1865	1869	−0·31
Benedict	24	95,075	95,689	0·65

Types of energy output

When an animal is fasting and completely at rest the minimum amount of energy is released. This energy corresponds to that necessary to ensure survival. It represents the "dépense de fond" (Magnus–Lévy) or basal

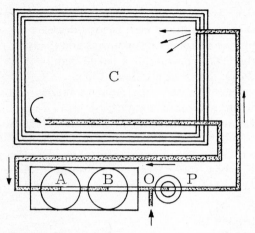

FIG. 4. Method of closed ventilation or method of determining total O_2
and CO_2. (From *Chaleur animale et bioénergétique*, LEFÈVRE, Masson,
1911.)
The apparatus is represented here without a spirometer and double series
of containers.
The air is either circulated always in the same direction or passes in and
out and the air that leaves the chamber containing the animal passes through
containers of potassium hydroxide in which it loses CO_2 and receives a new
supply of O_2 before returning purified to the chamber.
This principle is followed exactly in the apparatus of Atwater and Benedict.
The air circulated by pump P enters the upper part of the respiratory chamber; it leaves from the opposite side, charged with CO_2 and H_2O and deprived of O_2 consumed by the subject in the apparatus. The air is dried in con-
tainer A holding flasks of H_2SO_4; it then passes to B containing soda lime
where CO_2 is lost; finally it receives oxygen through O and is taken up by the
pump to be returned to the apparatus.
By this method the total respiratory exchange can be measured. By weighing
the containers A and B and the oxygen cylinder before and after the experiment, the weight of H_2O, CO_2 and O_2 of respiration is determined.

metabolism. Energy is produced to compensate for heat loss and to
maintain essential processes. Seguin and Lavoisier (1789) measured the
oxygen consumption corresponding to these metabolic states in man.

Basal metabolism

The mean caloric output of a man weighing 70 kg is of the order of
2400 cal per 24 h. This does not indicate, however, that the subject produces 100 cal/h. The amount of energy liberated varies according to the
amount of external and internal work (for example, digestion) and the
reactions of the subject to variations in ambient temperature. In order to
measure the basal metabolic rate, the subject must be in a state of muscular

relaxation, fasting and thermal neutrality. The temperature at which the subject does no work against an increase or a decrease in external temperature varies with the experimental conditions and the type of homoiotherm. The point of thermal neutrality for a naked man is at 30°C; for a lightly clothed man, 24°C; for a warmly clothed man, 16°C; for a dog, 25°C and for a mouse, 30°C. Therefore, the point of thermal neutrality depends, notably, on the presence of insulating coverings (clothes, fur or feathers).

Basal metabolism is defined by the number of calories produced per hour per square metre body surface by a fasting homoiotherm in a state of muscular relaxation and thermal neutrality. We shall examine in the following paragraphs, those terms which enter into the definition: (1) muscular relaxation; (2) temperature; (3) size; (4) body surface; (5) fasting.

(1) *Muscular relaxation*

Muscular work is carried out at the expense of glycogen and is associated with the production of heat. During muscular work the R.Q. approximates to 1. Atwater was the first to note this phenomenon. Table 5 demonstrates the changes in respiratory quotient with muscular activity (G. Schaeffer).

TABLE 5

Muscular activity	CO_2 eliminated per mn (in cm^3)	O_2 consumed per mn (in cm^3)	R.Q.	Frequency	
				radial pulse	ventilation
Subject lying down	204	243	0·83	65	20
Moderate muscular work	1780	1834	0·94	150	32
Intense muscular work	2227	3265	0·98	166	38

The daily energy output of man can be related to occupation:

Sedentary professions	2500 cal
Metallurgists	3200 cal
Woodmen, dockers	5000 cal

COMMENTS: During muscular activity in man, of 100 cal utilized by the skeletal musculature, 75 cal appear as heat and only 25 cal as work. The minimum muscular efficiency is 20 per cent and the maximum is 35 per cent. After a period of training muscular efficiency increases. This is not because there is an increase in the efficiency of individual muscle fibres but is due, essentially, to the suppression of superfluous movements. The initial training period is mainly a period of psychological change which allows the subject to perform more efficient movements. This gives an indication of the way in which the higher centres can influence processes which are essentially mechanical and chemical.

(2) *Temperature*

First of all, we must distinguish "cold-blooded" from "warm-blooded" animals. The former are called *poikilotherms*. These obey the law of Van't Hoff which states that for an increase in temperature of 10°C there is a two- or threefold increase in the intensity of chemical reactions.

The thermal coefficient or Q_{10} applies to most enzymic reactions within the limits of temperature which are compatible with life. However, these limits can be higher than those found normally for biological systems (for example, the thermophilic bacteria, or the H_2O_2–catalase system which is thermosensitive at 285°K). Genetic conditions modifed by "compensatory adaptation" can modify the Q_{10}. Specimens of the crab, *Uca pugilator*, from the region of the Atlantic around Woods Hole, exhibit increased respiratory gaseous exchange compared with *Uca* from Florida. In addition, *Uca* from the population at Woods Hole have a lower Q_{10}. From these metabolic characteristics two sub-species of *Uca pugilator* have been identified (Demeusy, 1957).

Homoiotherms are very different from poikilotherms. The central body temperature of an homoiotherm is not constant but shows only slight variations from the normal characteristic value. This value is generally higher in birds than in mammals. Within certain limits of environmental temperature the animal can compensate for changes in external temperature and maintain a constant body temperature (cf. thermoregulation, Volume II). Table 6 shows the variation in heat production as a function of ambient temperature in a lightly clothed man placed in current of air at a velocity of 3 m/s (Lefèvre, 1906).

TABLE 6

Environmental temp. in °C	Cal produced per 24 h
− 1	6650
+ 5	4700
+10	3700
+15	2750
+20	2050

In addition, the amount of carbon dioxide expired varies with external temperature (Table 7).

TABLE 7. *Experiments performed on the guinea pig by Rübner, 1880*

Environmental temp. in °C	Rectal temp. in °C	g CO_2 expired/kg body weight/h
0	37	2905
20	37	1766
25·7	37	1540
30·3	37·6	1317
34·9	38·2	1273
40	39	1454

As the environmental temperature increases there is a decrease in the volume of carbon dioxide, except at a temperature of 40°C. This anomaly has been studied principally in the rat (Table 8).

TABLE 8

Environmental temp. in °C.	Cal/m² body surface/24 h	
	1st day of fasting	2nd day of fasting
5	1962	1648
10	1628	1571
20	1186	1089
25	976	834
28	845	789
30	896	908
33	1062	982

Therefore, at a certain temperature there is minimum production of calories: this is the *critical point* (Rübner) or *point of thermal neutrality* (Fig. 5). "Un milieu assure la neutralité thermique lorsqu'il ne provoque ni le refroidissement ni l'échauffement du corps et permet l'écoulement de sa chaleur au fur et à mesure qu'elle se produit" (J. Lefèvre).

The normal processes of reflex thermoregulation in response to stimulation of the temperature sensitive receptors in the skin are inactive. Above and below the critical point the ability of the homoiotherm to maintain a constant body temperature decreases and, finally, death occurs due to hypothermia or hyperthermia. The zone of thermal neutrality may be wide or limited to a few degrees (Table 9).

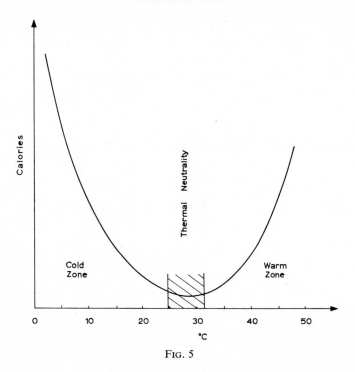

FIG. 5

TABLE 9. *Thermal neutrality*

Animals with a narrow zone	Animals with a wide zone
Rat ⎫ Mouse ⎬ 28–30°C Guinea pig ⎪ Pigeon ⎭ Dog (highly sensitive receptors in the skin) 25–29°C	Sheep (unshorn) 0–28°C Chicken 15–28°C Rabbit 10–21°C

(3) *Law of sizes*

A comparison of two homoiotherms of distinctly different weight shows that the smaller the animal or the lower its weight, the more heat it produces (Table 10). Treviranus (1832) was the first to establish this fact.

This law was confirmed by measurements made by Rübner on dogs (Table 11). Note that the smallest mammals have higher capillary densities (Krogh) but cold acclimatation, domestication, etc. interfere (Schmidt-Nielsen and Pennycuick).

TABLE 10. *Results from experiments carried out on rabbit by Schaeffer and Le Breton*

Wt. in g	Cal/kg/h
250	6
650	4
1500	3
3000	1·8

TABLE 11. *Metabolism of the dog per 24 h (Rübner)*

Wt. in kg	Cal/kg	Cal/m² body surface
30·4	34·8	984
11·0	57·3	1191
3·1	85·9	1099

Although there are differences in the values for cal/kg, the heat production per m² body surface shows little variation. The law of sizes leads to a consideration of a related factor (cf. p. 383).

(4) *Surface-area law (or Rübner's law (1883) or Richet's law (1899))*

This idea was originated by Rameaux and Sarrus of Strasbourg (1838). A small homoiotherm has a larger surface area relative to body weight than a large homoiotherm. The small animal loses more heat and must produce more heat to maintain the normal body temperature. Rameaux and Sarrus stated that the heat production per square metre of body surface was constant for homoiotherms of the same species. They wrote: "In stating that the body temperature of an animal is constant, it is recognized that there is a balance between heat production and heat loss. Heat loss is proportional to surface area or the square of homologous dimensions and it is important, therefore, that in comparing the quantity of oxygen consumed or the ratio of heat production to heat loss of two animals, these values are related to the square of the same measurements of size. This condition can be fulfilled in many ways."

If a homoiotherm is compared to a sphere with a central source of heat, then, Newton's law of cooling can be applied; if for two homoiotherms:

l^2 and L^2 represent the surface areas and q and Q represent the amount of heat produced, thus

$$\frac{Q}{q} = \frac{L^2}{l^2}$$

(33)

For animals of the same species and same density, surface area can be expressed by $\sqrt[3]{(W^2)}$ (where W = weight). There are variations between species and a species constant can be evaluated.

$$\frac{S_1}{S_2} = \frac{\sqrt[3]{(W_1^2)}}{\sqrt[3]{(W_2^2)}} \tag{34}$$

whence

$$\frac{S_1}{\sqrt[3]{(W_1^2)}} = \frac{S_2}{\sqrt[3]{(W_2^2)}} = \cdots \frac{S_n}{\sqrt[3]{(W_n^2)}} = K \tag{35}$$

thus

$$S = K\sqrt[3]{(W^2)} \qquad \text{(formula of Meeh, 1879)} \tag{36}$$

Species constants of various animals:

Man	12·3
Rabbit	12·0–12·9
Dog	10·3–11·2
White mouse	11·4
White rat	9·1

Several criticisms can be made of the formula of Meeh. For example, there can be as much as 36 per cent error in the calculation for obese subjects (reactions involving heat loss occur slowly in adipose tissue); expressed in another way: 100 kg body weight of an obese subject is not equivalent to 100 kg of an athlete (Harris and Benedict, 1919; Cobet, 1926; Bohnenkamp, 1946; Liberson, 1936). It should be noted that the species constant for man and rabbit are similar although the neuro-endocrine systems of these two groups are different. Moreover, age and sex should be taken into account.

Since the surface-area law was unreliable, bio-energeticians established empirical formulae:

(I) FORMULA OF BENEDICT AND HARRIS (1919)

If: W = weight of homoiotherm in kg

        H = height in cm

        A = age in years

        C = cal produced per 24 h

Benedict and Harris formulated the following equations which are related to *sex*:

$$C\,\male = 66.473 + 13.752\,W + 5.003\,H - 6.755\,A \tag{37}$$

$$C\,\female = 65.096 + 9.563\,W + 1.850\,H - 4.676\,A \tag{38}$$

Surface area which is a function of weight and height does not enter the formulae directly.

3*

These authors have established values for individuals aged 21–70 years. It is not certain, however, that the difference in caloric output of male and female has any biological significance. Moreover, these formulae do not apply to individuals with abnormal proportions.

(II) FORMULA OF DU BOIS (1916) The Du Bois brothers made very accurate measurements of the surface area of many subjects. They showed that the surface area in square metres could be estimated from the following formula (the same symbols are used as in the equation of Benedict and Harris):

$$\text{Area} = W^{0\cdot425} \times H^{0\cdot725} \times 71\cdot84 \tag{39}$$

The maximum error is 1·7 per cent. Tables of values obtained by this formula are available but even though it is the best formula to be proposed, it is not possible to verify the surface-area law.

There has been no progress in our understanding of the relationship between heat production and dimensions either from the mechanistic approach of Rameaux and Sarrus or the empirical approach. Although it is apparent that such a relationship exists, its precise nature remains unknown.

Returning to the factors to be considered in this problem, Rameaux and Sarrus stated that the ability to give off heat was constant for all surfaces of the body. However, this is incorrect: among homoiotherms there is a great variation in the heat released from various areas of the body and this is related to cutaneous circulation and position of the body (sitting, standing, etc.). The total surface area must not be confused with the area of the radiating surface. Measurements of these areas do not coincide. In the python (studied by Benedict) radiation is prevented by the scales and very little heat is lost from the skin. However, when Benedict applied the formula of Meeh to poikilotherms (alligator) he found that the surface area law was obeyed to a greater extent than when applied to homoiotherms for $K = 10$.

Has the law of surface-area any real significance? One tends to think that even in cases where the law is obeyed, factors other than body surface-area are involved. Lambert and Teissier believed that the surface-area law was a particular case of the law of *biological similarity*.

If for two animals which are geometrically similar, two homologous measurements will be in the ratio A, A can be estimated for all pairs of measurements of width and length, and thus, the size of the animals can be expressed as a function of one or more pairs of similar measurements. Cardiac and ventilatory frequency (cf. p. 312, Table 42) are inversely related to the weight of the animal (for example: the cardiac frequency of man is 72 beats/mn; of rat, 300 beats/mn and of humming bird, 1000 beats/mn). A comparison of these parameters and, in addition, basal metabolic rate, can be represented by a schematic graph of the type shown in Fig. 6 which consists of three parallel lines.

If W = weight

  K = species constant

it follows that:

$$\log \left(\begin{array}{c} \text{frequencies} \\ \text{or} \\ \text{metabolic rate} \\ \text{(cal)} \end{array} \right) = K - \frac{1}{3} \log W \qquad (40)$$

where $-1/3$ = slope of the line.

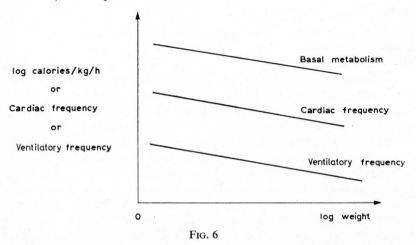

FIG. 6

Homologous time will have the same relationship as length, that is, proportional to the cubed root of body weight:

$$\tau = \alpha W^{1/3} \qquad (41)$$

Homologous frequencies will be inversely proportional to the cubed root of body weight (v. Hoeslin, Paul Bert)

$$f = \frac{1}{\alpha} W^{-1/3} \qquad (42)$$

There is no doubt that the relationship expressed in equation (40) establishes some interesting correlations but the law of similarity is static and cannot be determined by experimentation (tentatively suggested by Kayser). The problems arising from the surface-area law remain unsolved although, in practice, the surfacearea law is acceptable since values for basal metabolic rate are approximate in all cases.

Moreover, a living creature—in particular, a homoiotherm—cannot be compared with a machine producing heat and one must consider also the fact that heat is a degraded form of energy. A comparison of a factor of size with another energy factor—for example: oxygen consumption (in ml/h)—provides a valid interspecific ratio for both homoiotherms and poikilotherms (Benedict, 1938; Brody, 1945; Kleiber, 1961; Kayser, 1950, 1963; Zeuthen, 1947) which is expressed by the equation

$$V_{O_2} = aW^b \tag{43}$$

where b approximates to 0·70 for all vertebrates studied. The constancy of the value of b suggests that it has considerable biological significance.

(5) Fasting and specific dynamic action

It is easy to understand the importance of fasting in the measurement of basal metabolism. If for some experimental reason the animal has not been starved, a protein diet will result in an increase in gaseous exchange. For example, at thermal neutrality, if the basal metabolism is 100 cal during fasting, ingestion of a quantity of protein which theoretically should balance this heat output (100 cal), in fact, leads to the production of 130 cal, 30 cal being produced by the tissues. Under similar conditions, a carbohydrate diet results in the production of 105 cal and a lipid diet, 113 cal. Rübner (1902) called this action the *specific dynamic action* (S.D.A.) of proteins, carbohydrates and lipids.

The essential point is that outside thermal neutrality there is no production of extra heat and the three types of foodstuff appear to be isodynamic.

Rübner considered that the production of extra heat occurred during the transformation of proteins to glycogen. Thus, only the glycogenic amino acids would show this ability to produce extra heat. Glucose is the preferential food material of homoiotherms.

Max Rübner wrote in 1928: "Specific dynamic action is a loss of energy, a liberation of heat occurring during the processes of transformation of foodstuffs until they are converted to the true food material of cells, glucose; it is the amount of glucose which can be produced from albumin which gives a measure of the utilizable energy of albumin (18·6 cal)."

In his criticism of Rübner's idea, G. Lusk showed that during the maximum production of extra heat, approximately 2 h after the ingestion of glycogenic amino acids by phlorizinized animals (which are incapable of phosphorylating sugars), the elimination of sugar is proportional to the *quantity of glycogenic amino acids* (glycine, alanine) and not to the amount of sugar produced from them. This criticism is directed towards the explanation given by Rübner. However extra heat has been accounted for in another way. G. Schaeffer considered that extra heat could not be explained in metabolic terms alone but was the result of direct stimulation of the sympathetic nervous system by amino acids with the liberation of adrenaline; administration of ergotamine (a sympatholytic) inhibits the specific dynamic action of proteins. Tyramine (a liberator of catechol amines) and monoamine oxidase inhibitors should then be able to increase the heat production under those conditions which determine specific dynamic action. The theory of G. Schaeffer (1938) is an important contribution to the understanding of extra heat production.

Schaeffer's conception of heat production through adrenaline release is substantiated by the fact that adrenaline induces hyperglycemia (cf. p. 473) and I propose, therefore, the following scheme to explain the phenomena of specific dynamic action.

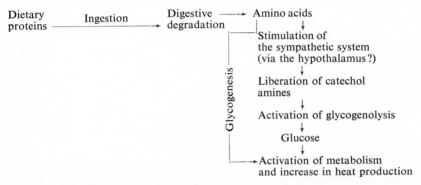

Some values of basal metabolic rate.

Man aged 20 − 50 years	39·7 cal/m²/h
Woman aged 20 − 50 years	36·9 cal/m²/h
New-born child	25 cal/m²/h

These values are for Europeans but the metabolic rate of individuals varies according to their race. For example, although the basal metabolic rate of Eskimos is 14–33 per cent higher than that of Europeans, the value for Chinese is approximately 12 per cent lower than that of Europeans. These anthropological differences have yet to be explained. During hyperthyroidism the basal metabolic rate increases by 20–150 per cent (Basedow's disease) while hypothyroidism decreases the metabolic rate. However, these facts do not explain the racial differences which may be related to differences in diet.

CHAPTER 3

EXCHANGE OF MATTER

Food and dietetics

General principles

The dietary compounds which, under normal conditions, are essential for the maintenance of animal life can be classed in two groups:

Metabolites (carbohydrates, lipids, proteins and certain mineral elements which constitute the basic material of part or the whole of the animal).

Catalysts for the transformation of these metabolites: (1) essential compounds: certain mineral elements and trace elements and vitamins; (2) additional compounds (inessential): for example, the *alkaloids* (nicotine, caffeine, etc.).

The latter group of compounds are related principally to pharmacology rather than to physiology and, therefore, will not be included here. Moreover:

(1) Dietary requirements are specific to the type of animal. Carnivores, herbivores and omnivores can be distinguished and within these groups certain species and sub-species show distinct requirements. These requirements depend on the enzyme characteristics of the organism which, in consequence, are of taxonomic importance.

(2) Increased metabolism can result from an increase in the amount of metabolites or of catalysts administered.

Intake of insufficient or superfluous amounts of catalysts can harm the animal and rapidly lead to a depressed state of metabolism.

In addition, the relative amount of a compound in the diet must be considered. For example, let us consider ethyl alcohol ingestion in man:

In low doses, alcohol favours digestion by pepsin.

In average doses, alcohol behaves as a foodstuff (98 per cent of the daily intake is oxidized, which represents 250–400 ml of 12–13 per cent ethyl alcohol taken at intervals). In the chick embryo 13 to 15 days after laying, two thirds of the ^{14}C-labelled ethyl alcohol appears as carbon dioxide and one third is utilized in the synthesis of lipids and proteins (Portet, 1961).

In high doses, alcohol acts as a *toxic compound* (*inhibitor* of metabolic transformations) and principally affects the liver and nervous system (cirrhosis of the liver, *delirium tremens*) (Fig. 7).

Thus, quantitative changes result in different qualitative effects.

(3) Food materials do not provide only calories. Their structural role is of equal importance to their energy role.

Deficiency symptoms can result from an unbalanced diet even though the lack of food is not apparent (partial hunger). Calcium deficiency provides a striking example of this. The symptoms of calcium deficiency are not apparent until after a long period.

We shall consider now what is meant by a balanced diet.

Metabolites

Nitrogenous foodstuffs

A diet must contain *assimilable nitrogen* (protein and amino acids) for the maintenance of life.

(1) The experimental minimum protein intake is "the smallest amount of albumin which will maintain nitrogen equilibrium". This minimum is dependent on other compounds present in the diet. Thus:

during fasting, 43·75 g of protein are degraded by the organism;
35·7 g of protein are degraded when the organism is receiving a lipid diet;
17·5 g of protein are degraded when the organism is receiving a carbohydrate diet.

Thus, carbohydrates minimize protein loss.

(2) The practical minimum protein intake is measured under normal conditions of a mixed diet. This value is 1 g protein per kg per day in man. However, foodstuffs contain various types of protein and, as a result, they have different nutritional values which depend mainly on the digestibility of the compound and the nature of the constituent amino acids.

Digestibility is defined either by the *coefficient of apparent digestibility*

which equals: $\dfrac{\text{ingested nitrogen} - \text{faecal nitrogen}}{\text{ingested nitrogen}}$ or by the *coefficient of true digestibility* which equals:

$$\frac{\text{ingested nitrogen} - \text{faecal nitrogen} - \text{faecal nitrogen of endogenous origin}}{\text{ingested nitrogen}}$$

A diet consisting of a mixture of amino acids can be equivalent to a protein diet. Amino acids can be classed in two groups—*essential* amino acids and *non-essential* amino acids. The essential amino acids must be present

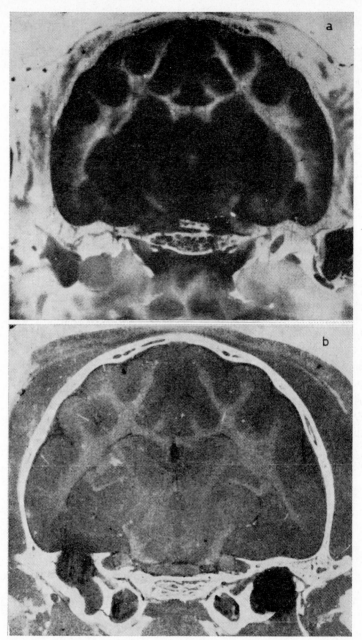

FIG. 7. Autoradiographs of frontal sections showing the distribution of
$1-{}^{14}C$-ethanol in the brain of cat; (a) 1 mn (cat 4 E 1) and (b) 76 mn (cat
2 E 76) after the start of intravenous injection. After 1 mn: concentration of
ethanol higher in gray matter than white matter, particularly high in the region of the geniculate bodies. After 76 mn: uniform distribution of ethanol
in gray matter. In white matter, concentration of isotope higher than after
1 mn but lower than that of gray matter; in the ventricles and sub-arachnoid
space, concentration greater than in cerebral tissue. Standards visible below
brain. Cat 4 E 1, 2·43 kg: injection in 30 s of 1·98 g ethanol with activity
of 250 μc. Cat 2 E 76, 2·70 kg: injection in 15 mn of 8·3 g ethanol with activity of 262 μc. (M. SCHERRER-ETIENNE and J.M. POSTERNAK, 1963.)

in the diet while the non-essential amino acids can be synthesized or, in more general terms, until we study mechanisms of formation, provided by the organism. The best method for determining to which group an amino acid belongs is to feed young animals on a protein-free diet containing a mixture of pure amino acids but excluding the amino acid under investigation. Table 12 (Rose, 1938) shows the results obtained for white rats and man.

TABLE 12

Essential amino acids	Non-essential amino acids
arginine	alanine
D-or L-histidine	aspartic acid
D-isoleucine	citrulline
L-leucine	cystine
D-lysine	glutamic acid
L-methionine	glycine
D-or L-phenylalanine	hydroxyglutamic acid
D-threonine	hydroxyproline
D-or L-tryptophan	norleucine
D-valine	proline
	serine
	tyrosine

The constituent proteins of an animal are composed of specific amino acids and are reconstructed from proteins in the diet. If an animal is fed on meat from the same species, the proteins lost by the animal are replaced more efficiently than if a non-specific diet is provided. The ingestion of *idiogenic* proteins (from the same species) is called *homophagia* and can be distinguished from the ingestion of *allogenic* proteins or *heterophagia*. The food value of plant proteins is generally lower than that of animal proteins. This can be seen easily by examining some values. Lambling wrote: "7·7 per cent of the glutamic acid of gliadin from wheat, containing 45 per cent glutamic acid, is incorporated into the serum albumin by the human organism and it is clear that the fraction of glutamic acid not utilized is greater than with casein which contains 11 per cent glutamic acid. In addition, this process of reconstruction is more costly when a given amino acid is less well represented in the protein being formed. The size of the fraction incorporated obeys the law of minimum which implies that the measure in which various constituents of protein can be incorporated into the reconstructed protein is dependent on the amount of the amino acid which is present in the lowest concentration. As a result, a certain number of amino fragments have no value in protein synthesis and are oxidized".

Therefore, both the quantitative and qualitative aspects of nutrition must be considered.

Zein from maize or gelatin are incomplete proteins with respect to their nutritional value. For example, zein which does not contain lysine or tryptophan is incapable of maintaining life in rat but zein provides an adequate diet for *survival* if it is supplemented by tryptophan. Addition of lysine also maintains *growth* processes.

The amino acid requirements of one part of an organism differ from that of another. M. Lafon (1939) compared the effect on rats and mice of lysine deficiency. Table 13 shows a summary of the results obtained.

TABLE 13

Animal	Organs showing growth	Organs showing regression	Organs showing no changes
Rat	heart lungs testes spleen salivary glands eyes	musculature thymus ovaries	liver digestive tract brain pituitary gland
Mice	liver kidney testes eyes	spleen thymus	heart lungs salivary glands

Thus lysine deficiency affected different organs in the two types of animals. For an exogenous protein to be incorporated in the protein of the organism it must be broken down to fragments of variable size. Therefore, both *proteolytic* and *proteosynthetic* enzymes are involved in the process of incorporation. One must not lose sight of the fact that an organism is in a dynamic state and degradation and resynthesis are continuous processes.

Hevesy and later Schoenheimer (1935–1939) decided to study the renewal of the "pool" of material in an organism using tracers, principally radioactive elements. After the administration of various ^{15}N-labelled amino acids to the organism, 10–12 per cent of administered nitrogen was retained as non-protein nitrogen; from 28 per cent (in the case of leucine) to 52 per cent (in the case of arginine) of ^{15}N was excreted within three days while 30 per cent (arginine nitrogen) to 66 per cent (lysine nitrogen) was incorporated into proteins. There was no change in the weight of the animals and thus, the ingested amino acids were continuously replacing the amino acids in tissue proteins by degradation and resynthesis of peptide linkages but without any change in the structure or amount of the proteins.

Using this method of radioactive tracers it is possible to measure the *turnover rate* of proteins. One finds that the rate of turnover is more rapid in liver and blood plasma and less rapid in skeletal muscle and skin.

Results obtained with bacterial and animal cells indicate that there is a balance between cell multiplication on the one hand, and protein turnover on the other (Harris and Watts, 1958). Protein turnover occurs when the cells are not in a state of multiplication but during multiplication, synthesis of proteins utilizes those amino acids which would serve normally to rebuild the tissues proteins.

It must be mentioned also that among the mammals at least, L *series* amino acids are metabolized more extensively than D *series* amino acids. The latter are said to be "unnatural" although they do exist in nature, for example, in lower plants. Table 14 summarizes the action of different D-amino acids on the growth of various organisms.

TABLE 14

D-amino acids	Rat	Mouse	Chicken
valine	−	−	−
leucine	−	−	+
isoleucine	−	−	−
threonine	−	−	−
lysine	−	−	−
histidine	+	+	
phenylalanine	+	+	
tryptophan	+	+	−
methionine	+	+	+

This difference in the behaviour of optical isomers is related principally to the fact that L-amino acids can be utilized in the synthesis of proteins while for D-amino acids to be used they must be transformed initially to the L-derivative.

I shall end this discussion of nitrogen metabolism by mentioning a special group of proteins—the *nucleoproteins*.

The nucleoprotein consists of a protein fraction of variable molecular weight (protamines, histones or true proteins) attached to a polynucleotide chain. Each nucleotide consists of a phosphate group, a five carbon sugar (ribose in ribonucleic acid and desoxyribose in desoxyribonucleic acid) and a purine or pyrimidine base.

Roes from fish, bone marrow, thymus—sold as sweetbreads by the butchers—and baker's yeast are particularly rich in nucleoproteins. These compounds are important in that they constitute chromosomes (DNA) and viruses and that the ribonucleic acids (of microsomes and cytoplasm) are considered, since Casperson and J. Brachet's early results, to have a direct function in protein synthesis (Zamecnik *et al.*, 1955). Moreover, it has been mentioned earlier that the synthesis of enzymes is controlled by a chromosomal gene, a nucleoprotein (cf. p. 13). From the point of view of nutrition, the protein fraction of a nucleoprotein is metabolized in the same way as any other protein while the nucleic fraction is destroyed by nucleases, nucleotidases and nucleosidases from the pancreatic and intestinal secretions and produces phosphoric acid, free bases and probably some free sugars. After further metabolic transformations (which will be considered in the chapter on excretion) these compounds are, in part, excreted. Again, I emphasize the fact that *nutrition serves to provide certain food elements which are incorporated into the substance of the organism or provide energy for this incorporation and for other forms of work.* The process of nutrition implies destruction, re-incorporation of certain fragments, elimination of non-utilized toxic fragments which can act in one way or another as enzyme inhibitors.

By the use of labelled precursors—for example, adenine labelled with [14]C—the incorporation of bases into nucleic acids can be measured. For example, if the value for the incorporation of labelled adenine into the bone marrow of rabbit is taken as 100, the following values were obtained for other tissues: spleen 40; liver 7; intestinal mucosa and thymus 32.

There is still much to be learnt as to the processes of metabolism of nucleoproteins in the alimentary tract; certain aspects of this problem will be examined on p. 146.

The ternary foodstuffs

These are composed of carbon, hydrogen and oxygen.

I. Carbohydrates

The absence of carbohydrates in the diet of mammals is incompatible with the balanced functioning of the organism.

The minimum carbohydrate intake for normal physiological function in man is of the order of 50 to 60 g per day. This minimum varies according to the proportion of lipids in the diet but the ratio sugars/lipids must not be less than 1/4.

Carbohydrates have an *energetic, structural* and *anti-ketogenic* role and finally, they have a *protective* function with respect to proteins.

The role of sugars in the provision of energy is part of the study of intermediate metabolism and muscular contraction (cf. Vol. II) and will not be dealt with in this chapter.

(1) STRUCTURAL ROLE. Sugars form notably part of the structure of the mucous membranes, supporting tissues and central nervous system.

(2) ANTI-KETOGENIC ROLE. *Carbohydrate starvation* rapidly leads to *acidosis* and *ketonuria*.

It was stated by Rosenfeld that fats are burnt in a carbohydrate fire; this is a Lavoisierian idea. However, this statement has some value as a mnemonic and there is no harm in retaining it for this reason. Thus, the explanation of the formation of ketone bodies during carbohydrate starvation or diabetes is to be found in lipid metabolism.

Under these conditions, sugar is not available for the provision of energy and the organism obtains energy by more extensive oxidation of lipids. This results in an increased production of products of incomplete catabolism of lipids (acetone, $CH_3 \cdot CO \cdot CH_3$; acetoacetic acid, $CH_3 \cdot CO \cdot CH_2 \cdot COOH$ and the related compound, β-hydroxybutyric acid $CH_3 \cdot CHOH \cdot CH_2 \cdot COOH$) such that the oxidative capacity of the tissues (mainly the liver) is saturated and these products accumulate in the blood *(ketonemia)* and are excreted in large amounts in the urine *(ketonuria)*. Ketonemia is associated with *acidosis* since, under normal conditions, the small amount of ketone bodies formed can be neutralized by the alkaline reserve of the blood but this becomes overloaded during carbohydrate deficiency.

Dietary ketogenesis can be prevented by the addition of sugars to the diet and *diabetic ketogenesis* is prevented by the administration of insulin.

During a diabetic coma or a coma due to sugar starvation, hallucinations can occur but these disturbances can be overcome simply by the administration of sugar.

CHO	CH$_2$OH	CHO
H$-$C$-$OH	C$=$O	H$-$C$-$OH
HO$-$C$-$H	HO$-$C$-$H	HO$-$C$-$H
H$-$C$-$OH	H$-$C$-$OH	HO$-$C$-$H
H$-$C$-$OH	H$-$C$-$OH	H$-$C$-$H
CH$_2$OH	CH$_2$OH	CH$_2$OH
D-Glucose	D-Fructose	D-Galactose

(3) PROTECTION OF PROTEINS. If the sugar content of the diet is replaced by an isodynamic quantity of lipids there is an increase in the nitrogen content of the urine which corresponds to the destruction of tissue proteins.

This excessive degradation of protein may be related to ketogenesis. Certain *amino acids* (for example, leucine, phenylalanine, tyrosine) are said to be ketogenic since they can form acetoacetic acid. Acetoacetic acid is involved, under normal conditions, in *lipogenesis* but as a result of the utilization of lipids during carbohydrate starvation this does not occur. Therefore, two types of compounds are concerned with ketonemia and acidosis during carbohydrate deficiency: lipids as source of energy and in addition, amino acids.

Not all sugars can be utilized by man who mainly uses *glucose, fructose* and *galactose*. Glucose is, by far, the most important sugar but galactose is required as a constituent of lactose, the disaccharide of milk, and of *cerebrosides* of the nervous system.

Finally, although starch can be assimilated by mammals other polysaccharides such as *cellulose* cannot be utilized. However, cellulose has a useful role. One reads quite often of the possibility of providing an adequate diet in pills containing necessary food materials. This is, however, a completely fictitious concept. It is not sufficient to provide the quantitatively and qualitatively correct chemical characteristics of the diet of an organism (principally mammals). The physical characteristics must be taken into consideration. Since the digestive tract consists of smooth muscle it is important for its functioning that it be kept under a certain amount of *mechanical tension*. Its digestive function is augmented by distension by the food contents. Therefore, mammals must be provided in the diet with substances which are refractory to the chemical processes but which favour mechanical processes. *Cellulose* is one of these compounds which have been called *"bulk" foodstuffs*. In herbivores, which have a very long digestive tract, these foodstuffs prevent constipation and intestinal atonia (there is greater muscular movement). For example, a rabbit which is maintained on a diet without cellulose will die of intestinal blockage. Moreover, among the ruminants, microbes contained in the rumen transform cellulose to simple sugars and in this case cellulose is functioning as a true foodstuff.

II. Lipids

Mammals can synthesize lecithins and cephalins, sterols and also glycerine and the principal fatty acids: oleic acid, palmitic acid and stearic acid which are contained in neutral fatty acids and which can be synthesized from sugars and proteins. As a result, it is considered that the diet of mammals need not contain neutral fatty acids. Nevertheless, mammals can not synthesize *unsaturated fatty acids* such as:

linoleic acid $C_{17}H_{31}COOH$, with two double bonds between carbon atoms 6–7 and 8–9;

linolenic acid $C_{17}H_{29}COOH$, with three double bonds between carbon atoms 3–4, 6–7 and 8–9.

In the absence of unsaturated fatty acids (a high concentration of which is found in linseed oil) growth and reproduction are inhibited (Burr, 1929, for linoleic acid) and hematuria is produced. These fatty acids are catalytic in action rather than metabolic, the double bonds containing high energy levels.

Water alimentation

Biochemistry is a particular aspect of the chemistry of water. Water is essential to ensure survival (the fakirs can deprive themselves of food but not of water). Water represents 70 per cent of the total body weight of man although this percentage decreases with aging. The amount of water in an animal can be measured either by desiccation or by studying the distribution of heavy water. Water can be derived either from *water in the diet* or from *water of oxidation* or *metabolic water*, which is produced during the oxidation of foodstuffs (Table 15). The fatty hump of the dromedary can provide a source of water for this animal which exists in hot desert regions and whose ability to go for long periods without drinking water is proverbial (Lombroso, 1891; Cauvet, 1925; Y. Charnot, 1958).

TABLE 15

Compound oxidized (100 g)	Water produced (g)
sugars	55
lipids	107
proteins	41

Water is eliminated from man by four pathways:
  skin: sensible and insensible perspiration (0·6 l. per day);
  lungs: water vapour in expired air (0·4 l. per day);
  kidneys: urine (1·4 l. per day);
  intestines: faeces (0·1 l. per day);
  (total, 2·5 l. per day).

Mineral alimentation

We shall consider in this section only those mineral elements which are taken in relatively large amounts. The mineral trace elements will be dealt with later when we consider the vitamins, that is to say, in the section on catalysts for the utilization of metabolites. Calcium, sodium, potassium, phosphorus, sulphur and chlorine constitute about 70 per cent of the total mineral content of mammals. Thus, although water is essential for life, the mineral ions are no less important for many reasons: basic function in the regulation of osmotic pressure, formation of skeleton (ratio of calcium ions to phosphate ions), maintenance of normal activity of nerves and muscle, etc. These ions are of *external origin*.

Calcium

The metabolism of calcium is associated with that of phosphorus and magnesium. This cation is that which is the most widely distributed in the organism and is important, principally, in the composition of bones and teeth; but it exists also in the body fluids and is of fundamental importance in blood coagulation, permeability of cell membranes and in the functioning of muscles (including the heart) and of nerves. In low concentrations, when it can be considered as a trace element, calcium activates numerous reactions which produce energy.

Calcium can be obtained from many sources but it is found notably in milk, cheese, egg yolk, kidney beans, cauliflowers and asparagus.

The human being requires 1 g of calcium per day. During gestation and childhood the requirements are increased to 2 g per day. Calcium can be assimilated in the form of the gluconate, carbonate, lactate and diphosphate. The ability to assimilate calcium depends on numerous factors:

(1) factors specific to certain individuals which are not fully understood;

(2) the nature of the diet; if the diet is rich in proteins approximately 15 per cent of Ca^{++} is assimilated while only 5 per cent of Ca^{++} is assimilated from a low-protein diet;

(3) the formation of insoluble compounds, phosphates, carbonates, oxalates (for example, in spinach), phytates formed from phytic acid (in cereal grains); free fatty acids react with calcium to form soaps.

(4) influence of vitamin D, parathormone and calcitonine.

TABLE 16. *The average distribution of calcium in man*

Type of structure (other than bone)	mg/l. (or/kg)
blood plasma	100
cerebrospinal fluid	50
muscular tissue	700
nervous tissue	150
milk	300

The major part of blood calcium is found in the plasma where it exists in two forms: *diffusible* (approximately 60 per cent) and *non-diffusible* which is probably in association with proteins. A decrease in the level of serum proteins demands an increase in diffusible calcium. A decrease in the ionized fraction of plasma calcium causes tetany. In man, only 200 mg/day, on average 20 per cent of total calcium eliminated, is excreted in the urine; most of the calcium is excreted in the faeces. Calcium excretion by carnivores is similar to that by herbivores.

The Madrepora which construct islands of coral in shallow waters of tropical seas (these are called *hermatypic* corals as opposed to *ahermatypic* corals living in deep waters and which do not produce reefs) have a skeleton composed of $CaCO_3$ in the form of aragonite (calcite is not present). The formation of the skeleton has been studied with the aid of ^{45}Ca (Goreau, 1959). In this case, as in the calcium deposit of the shells of the eggs of Galliformes or in the oyster, carbonic anhydrase has an important role.

Fig. 8. X-ray spectra of calcite (*below*) and aragonite (*above*). (Courtesy of Madame WALTER-LÉVY.)

However, in the case of corals, the rate of calcification is proportional to the intensity of illumination and is dependent on the presence of symbiotic unicellular algae, the zooxanthelles. According to Goreau, Ca^{++} from seawater is attached to a mucopolysaccharide matrix situated in the region of the mineral deposit and either under the influence of carbonic anhydrase alone in the absence of light, or during illumination, under the combined influence of this enzyme and photosynthetic activity—CO_2 and HCO_3-fixation—of the algae, calcium carbonate is precipitated.

Phosphorus

In a similar way to calcium, phosphorus is distributed throughout the organism and is found in particularly high concentrations in bone and teeth (80 per cent); 20 per cent is combined with sugars, proteins and lipids. The importance of phosphorus is derived from the fact that, on the one hand, it participates in structural processes (ossification) and on the other, it is concerned in high energy reactions, the phosphorylations and dephosphorylations.

The phosphorus requirements of an organism are similar to those for calcium; in man, 1 g per day.

Although the distribution of phosphorus is somewhat different from that of Ca^{++}, its metabolism is linked with that of calcium; in food the optimum ratio is 1 Ca/1 P; with respect to this it should be noted that cow's milk contains more phosphorus than calcium.

During hyperthyroidism and also after the absorption of sugars and certain lipids, there is a temporary hypophosphatemia. In certain renal

4*

TABLE 17. *Average distribution of phosphorus in man*

Type of structure	mg % (ml or g)
blood	40
muscular tissue	200
nervous tissue	350
bone and teeth	22000

diseases the retention of phosphorus leads to *acidosis* and the accompanying increase in serum phosphorus level tends to lower the blood calcium level. It is, therefore, not only ketonemia which can cause acidosis.

Magnesium

In bone approximately 70 per cent of the total magnesium is associated with calcium and phosphorus; the remaining 30 per cent is distributed throughout the tissue. Magnesium activates numerous enzymes, for example, ATPase, phosphoglucomutase $(G_1P \rightleftharpoons G_6P)$. The daily requirement for man is 300 mg.

TABLE 18. *Average distribution of magnesium in man*

Type of structure	mg/l (or/kg)
blood, cerebrospinal fluid	30
muscular tissue	210

While blood calcium is found mainly in the serum, more than half of the magnesium in blood is distributed in the cells.

The metabolism of magnesium is linked with that of calcium and phosphorus. Magnesium is eliminated by the urine and by the faeces in similar proportions; magnesium salts have a diuretic action. Mg^{++} can replace Ca^{++} in the salts of bone during Ca^{++} deficiency but excess Mg^{++} inhibits ossification.

A low magnesium diet (<20 mg/kg food) in rat induces cardiac arhythmia, hyperirritability, convulsions and tetany.

Intravenous injection of relatively high concentrations of magnesium (200 mg/l., or, in general, the concentration of the ion at the muscular level) induces *anaesthesia* with paralysis of voluntary muscle. Intravenous injection of a similar concentration of Ca^{++} leads immediately to suppression of anaesthesia and a return of muscular activity. It is thought that this antagonism is associated with the different actions of Ca^{++} and Mg^{++} at the level of the cell membrane. Mg^{++} is associated with K^+ inside the cell while Na^+, and perhaps Ca^{++}, are found principally outside the cell.

Sodium

Sodium is functionally associated with potassium, chlorine and water and is the most widely distributed in the extracellular spaces. It contributes in the regulation of acid–base equilibrium, participates in the maintenance of osmotic pressure, protects the organism against dehydration and has a function in the excitability of nerves and muscles.

Man requires on average 5 mg Na^+ per day, a requirement which is met easily as sodium is found in all food materials.

Almost all the sodium is excreted in the urine. Only a very small amount of sodium is found in the faeces as it is reabsorbed by the intestinal tract.

The regulation of sodium metabolism is controlled directly by steroids from the adrenal cortex; during hypofunction of the adrenal cortex there is a decrease in the blood level of sodium associated with an increase in sodium excretion. During Addison's disease when similar symptoms occur, this state is improved by gestation as a result of the production of steroid hormones by the placenta which leads to retention of sodium and water. An increased blood level of sodium is induced by adrenocorticotropin (ACTH), cortisone and desoxycortisone. Hypernatremia develops during hypothermia (Andjus).

TABLE 19a. *Average distribution of sodium in man*

Type of structure	g/l (or/kg)
blood	1·6
plasma	3·25
muscular tissue	1·1
nervous tissue	3·1

Deane and Smith (1952) have shown that the intracellular concentration of sodium is higher than was thought previously. Moreover, sodium can replace the potassium in the cell when sodium salts are administered to subjects deficient in potassium. ACTH and the adrenal cortex steroids can increase the *intracellular* concentration of sodium.

OEDEMA DUE TO WATER RETENTION. Diuresis is inhibited by sodium ions and augmented by Ca^{++} and K^+. Thus, sodium has a fundamental role in water metabolism. For this reason, "salt deficient" diets are recommended for certain cardiac and nervous conditions. "Salt deficient" is a misleading expression which derives from the use of the word "salt" to describe sodium chloride. In fact, a *sodium deficient* diet is administered which does not exclude the use of potassium chloride, for example, in the diet of these patients.

Potassium

Potassium is the principal low molecular weight cation of intracellular fluid. It is of importance in nervous and muscular activity and in the regulation of osmotic pressure and acid–base equilibrium.

TABLE 19b. *Average distribution of potassium in man*

Type of structure	g/l (or/kg)
blood	2·00
plasma	0·20
muscular tissue	3·00
nervous tissue	5·00

Man requires approximately 4 g/day. Potassium is found in all food-stuffs. Due to the presence of the isotope ^{40}K which emits β and γ radiation, natural potassium exhibits a certain amount of radioactivity which it confers to the organism, in particular mammals. Coursaget calculated that the rate of disintegration in man weighing 70 kg was 300,000 atoms of potassium per mn.

We shall consider now some abnormal states of general potassium metabolism.

(I) HYPERKALEMIA. This can result from several causes:

(1) When there is a decrease in water intake or an excessive loss of water, the rate of loss of water exceeds the rate of loss of electrolytes and the extracellular fluid becomes hypertonic compared with the intracellular fluid. As a result the cells are dehydrated and this results in vomiting, the production of concentrated urine and an increase in serum potassium.

(2) Shock: administration of adrenaline induces hepatic glycogenolysis together with a temporary hyperkalemia.

(3) Renal malfunction, acidosis.

(4) Injection of KCl or excessive cytolysis.

The symptoms of hyperkalemia are *cardiac* disorders (bradycardia and cardiac arrest) and *nervous* disorders (mental confusion, ventilatory weakness). The effects of hyperkalemia are always more serious than those of hypokalemia as potassium ion is localized mainly within the cell.

(II) HYPOKALEMIA. Hypokalemia develops during malnutrition, diarrhea, alkalosis and hypothermia (except in adrenalectomized mammals).

Potassium ions leave the cells and are eliminated by the kidney. Moreover, when glycogen is stored, a small but significant amount of K^+ is retained; in the treatment of *diabetic coma* with insulin and glucose, glycogenesis is rapid and the resultant hypokalemia may be fatal.

Hypokalemia is associated with tachycardia; the electrocardiogram (cf. p. 249) shows a flattened T wave which is inverted later; there is auriculoventricular block followed by cardiac arrest.

Chlorine

As a component of sodium chloride, this anion is essential in the regulation of osmotic pressure and the acid–base equilibrium. In the stomach, chlorine is of special importance in the production of HCl.

In the diet chlorine is associated with sodium and thus the requirements for the two are related. However:

(1) During vomiting, the loss of chlorine is higher than the loss of sodium. This leads to a decrease in plasma chloride with a compensatory increase in bicarbonate resulting in *hypochloremic alkalosis.*

(2) The levels of sodium and chlorine in blood plasma are not equal (cf. Table 19) and there is, therefore, no justification in expressing the concentration of chlorine in an organism in terms of sodium chloride. This is demonstrated by examining the average distribution of chlorine in man (Table 20).

TABLE 20

Type of structure	g/l (or/kg)
blood	2·50
plasma	3·65
cerebrospinal fluid	4·40
muscular tissue	0·40
nervous tissue	1·70

SOME IMPORTANT POINTS. Physiological fluids (Ringer and Tyrode solutions, etc.) have been developed to meet the main ionic requirements of an organ and to ensure its survival. The concentrations of calcium, potassium, sodium and chloride ions are of great importance in this respect.

A solution which is physiological for one type of animal may not be so for another and this must be taken into account when using the term "physiological". For example, a physiological solution which is used for isolated tissues from the frog (Ringer's solution) is unphysiological for the same tissues from a selacian. The latter is uremic and, therefore, for this animal a physiological saline must contain a considerable amount of urea.

Sulphur

Sulphur is distributed throughout the body of mammals, principally in the cell protein. As a result, it is associated with nitrogen metabolism. Sulphur is important in detoxication mechanisms and in tissue respiration (thiol group).

Higher animals cannot assimilate free sulphur or sulphates. The main sources of sulphur are the amino acids: *cysteine* or *cystine* and *methionine*. *Keratin*, the protein of hair, is rich in sulphur containing amino acids and thus, the sulphur requirements of hairy animals are higher than those of man.

Iron

The main role of iron is in cellular respiration; it is a component of hemoglobin, myoglobin and cytochrome (heme iron—linked to a porphyrin).

Iron can be derived from many sources but the main sources are liver, heart, kidney, spleen, yolk of egg, fish, oysters, spinach, asparagus, etc.

It is found also in sweat, skin and epidermal structures. *Growth of nails and hair thus constitutes a form of excretion, both of iron and sulphur.* During *hyposiderosis* the nails become brittle (this is seen also in vitamin B_2 and nicotinamide deficiency).

Most of the absorption of iron occurs in the mucous membrane of the duodenum and the proximal part of the jejunum. It is absorbed in the form of Fe^{++} (reduced). Absorption of iron from a diet rich in phosphate is reduced due to the formation of insoluble phosphates of iron. Granick showed that the formation of ferritin in guinea pig determines the rate of absorption of iron (ferritin = *apoferritin* (a protein) + basic oxide of iron or $[(FeO_2H)_3 - FeO - H_2PO_3]$).

TABLE 21. *Iron requirement in man*

Period	mg/day
early childhood	6
childhood	9
adolescence, gestation	
lactation	15
adults	
male	2–10 (?)
female	2–12 (?)

TABLE 22. *Average distribution of iron in man*
(S. Granick, 1954)

Type of substance	total in body (g)	iron content (g)
Heme compounds		
hemoglobin	900	3
myoglobin	40	0·13
cytochrome c	0·8	0·004
catalase	5·00	0·004
Non-heme compounds		
siderophilin	10	0·004
ferritin	2–4	0·4–0·8

With regard to these figures the following points should be made:

(a) The normal organic storage form of iron is *ferritin* (Schmiedeberg, 1894; Laufberger, 1935) containing approximately 23 per cent iron. If the capacity of the organism to store iron in the form of ferritin is saturated, iron accumulates in the liver in the form of *hemosiderin,* a protein containing approximately 35 per cent iron, which is visible under the light microscope (Neumann, 1888); hemosiderin appears after the injection of iron salts (hypersideremia). Bessis (1958) studied the distribution of ferritin in the reticular cells of the bone marrow using the electron microscope. Ferritin had a characteristic appearance, each particle of ferritin being in the form of a small square with sides measuring 50 Å and a mass of iron atoms at each corner. The amount of ferritin in horse, precipitated by anti-ferritin serum from rabbit, can be reduced to zero by pretreatment of the immuno-serum with horse hemosiderin. This reaction shows the immunological similarity of the two substances; hemosiderin is derived from ferritin by polymerization (B. Magnan de Bornier, 1963). But a rather surprising observation has been the multiplicity of transferrins that bind iron occurring in umbilical cord sera. It appears possible that the foetus may be relatively deficient in an enzyme system responsible for incorporating the prefabricated carbohydrate prosthetic groups into the transferrin molecules. Thus, at birth, although the majority of the transferrin molecules have their full complement of four carbohydrate prosthetic groups, some molecules have only three, two or one prosthetic groups attached (Bearn and Parker, 1963).

The rate of mobilization of iron from ferritin is low and, after hemorrhage, the iron content can remain low for several weeks.

(b) Ferritin and hemosiderin, which are *storage* forms of iron, must not be confused with *transport* forms of iron. A nonporphyrin iron compound was discovered in serum by Fontès and Thivolle (1925). This com-

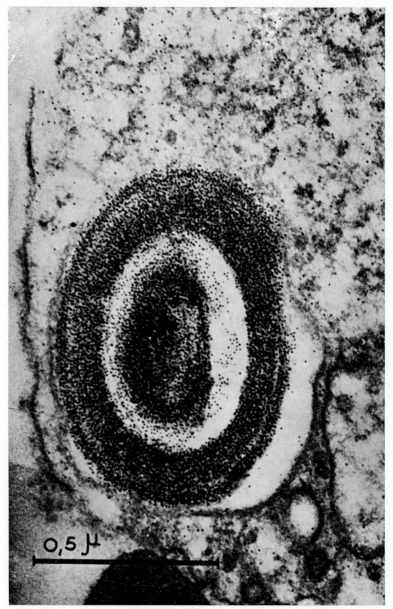

FIG. 9. Accumulation of hemosiderin and dispersed ferritin. (BESSIS, *Bull. Acad. Nat. Méd.*, 1958.)

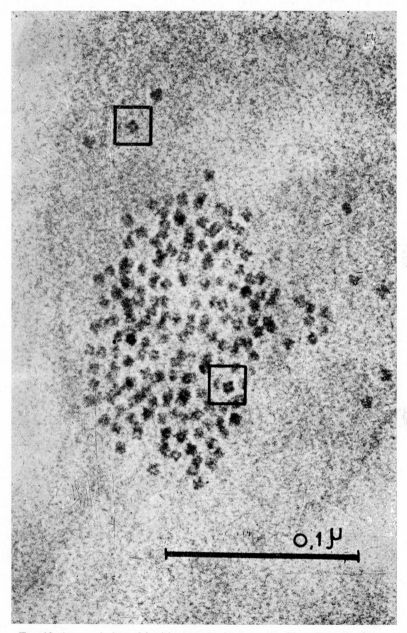

FIG. 10. Accumulation of ferritin granules in an erythroblast. In each of the two encircled areas one can distinguish clearly the ultra-structure of the ferritin molecule, consisting of four ferruginous nuclei. Each has a diameter of 15 Å. (BESSIS, *Bull. Acad. Nat. Méd.*, 1958.)

pound—*siderophilin*—is a complex of iron and a β-globulin. *Siderophilin* (or *transferrin*) is the normal carrier form of iron in the serum of higher animals. Thus, the iron content of blood exists in a heme form and in a non-heme form. In man, an average of 90 per cent of the iron atoms are eliminated from the plasma and fixed in the bone marrow, that is, in a hemopoietic organ. This estimation was carried out by G. de Hevesy (1963) using ^{59}Fe.

BRONZE (OR CIRRHOTIC) DIABETES. This is a *hereditary* condition: 40–50 g iron can be deposited in the skin in the form of *hemosiderin*; hepatic disorders develop and also pigmentation of the skin or *hemochromatosis*. Debré, Schapira and Dreyfus (1957) suggested that this condition was transmitted by a dominant gene or gene system.

This disorder must not be confused with *Bantu siderosis* (in peoples from the south of Africa) which is of *dietary* origin and results from a low-phosphate diet and the use of iron containers in the preparation of food.

CHAPTER 4

CATALYSTS FOR THE USE OF METABOLITES

The compounds used by the body in trace quantities are:

(I) inorganic ions;
(II) certain organic compounds.

I. Mineral trace elements

We shall consider copper, iodine, zinc, cobalt and fluorine. There is evidence to suggest that manganese, boron and molybdenum are important but, still, little is known of their action on the body.

Copper

The general function of copper may appear paradoxical, because, although it is an inhibitor of many enzymes (for example, ATPase) it is a constituent of:

(a) many enzymes such as *tyrosinase* (high levels in *Tenebrio molitor*) and ascorbic acid oxidase;

(b) oxygen carriers in certain invertebrates (*hemocyanin*, homologous with hemoglobin);

(c) a protein which contains copper from erythrocytes of mammals *(hemocuprin)* and a plasma α-globulin, *ceruloplasmin*, to which most of the copper in blood is bound.

TABLE 23. *Average distribution of copper in man*

Whole animal	100–150 mg
muscular tissue	65 mg
bone	25 mg
liver	20 mg
plasma	approx. $1 \mu g/ml$

TABLE 24. *Copper content (in mg/100 g dry weight) of some marine invertebrates (Vinogradov, 1953)*

Gorgonium flabellum	1·3
Bonellia viridis	0·265
Paracentrotus lividus (without test)	0·13
Marthasterias glacialis	0·47
Homarus americanus	3·88–16·7
Carcinus maenas	4·06

There is not doubt that copper is a food material. Man requires 2 mg copper per day which is easily obtainable from a normal diet. Copper favours the production of milk and in the rat, at least, which has been the subject of extensive experimentation (Dutt and Mills, 1960), copper deficiency inhibits reproduction. In general, copper deficiency leads to anemia, loss of weight and death. Disorders in ossification have been described in young dogs (Baxter) and cerebral lesions in sheep (Voisin, 1961). In mammals, copper is eliminated through the intestine (cf. volume II, Wilson's disease).

The presence of Cu^{++} appears to activate metamorphosis of planktonic larvae, for example, those of the tunicate *Stylea partita* (Bertholf and Mast, 1944). However, Cu^{++} in concentrations of 50 μg/l. seawater, inhibits fertilization in sea-urchins (F. A. Lillie, 1921) and growth of these larvae (P. Bougis, 1959) and of larvae of several other organisms, for example, *Bagula neritina* (Miller, 1946).

The following points should be considered in the interpretation of these results:

(1) The threshold dose for the action of a compound will be different for each type of animal and will depend also on the stage of the life cycle.

(2) The optimum concentration of a compound is important as this indicates the concentration which saturates the enzyme–substrate system and leads to inhibition. Thus, any large concentration of a compound, which acts normally in trace quantities, can lead to harmful effects. The action of a compound functioning in trace quantities is balanced, therefore, not only with the effects of sub-liminal doses which cause deficiency symptoms, but also with those of high doses.

Iodine

Iodine is essential for thyroid function and is, therefore, of great importance in the life of higher animals. We shall return to a study of this endocrine gland in a later chapter (cf. p. 398). The human organism requires 1 μg iodine per kg body weight per day, (this value was determined under conditions of basal metabolism). Iodine is found in fish, green vegetables, in crude sodium nitrate from Chile (which is used as a fertilizer and which will affect, therefore, the quantity and quality of plant and animal produce).

Zinc

In the animal as a whole, zinc deficiency inhibits growth of rats and results in sterility due to testicular atrophy. Zinc metabolism seems to be controlled by the adrenal glands (Voroshilovskja, 1963).

At the molecular level, zinc is important as a constituent of many enzymes which include *carbonic anhydrase* (cf. pp. 92, 386), essential in the regulation of acid–base equilibrium; alcohol dehydrogenase (cf. Volume II: Vision) and carboxypeptidase from the pancreas—which hydrolyses, for example, chloroacetyltyrosine to give tyrosine and monochloroacetic acid. The leucocytes contain zinc although they contain no carbonic anhydrase.

Man requires 0·3 mg zinc per kg per day which can be obtained from dietary compounds. In ox, horse, sheep and pig, zinc is eliminated by the mucosa of the tongue (Bhattacherje, 1934). In addition, as has been shown for copper, Zn^{++} delays the development of larvae of sea-urchins (P. Bougis, 1961), and it is most probable that this ion affects the distribution and abundance of planktonic organisms.

Cobalt

Cobalt activates numerous enzymes, notably *phosphatases;* it is a constituent of *vitamin B_{12}* which diminishes the toxic effects of excess thyroxine (thyrotoxicosis) in rat (Sure and Easterling, 1950) and in man (George and Chu Lai Haan, 1962). Cobalt affects the formation of erythrocytes and has been used in the treatment of certain anemias, notably in sheep pasturing on soils low in cobalt. Cobalt is eliminated in the urine.

Fluorine

Fluorine is found in bone and teeth in the form of fluorides. In low doses, fluorine favours the development of teeth and inhibits dental decay, an effect which is augmented by molybdenum. In higher doses fluorine is extremely toxic (sodium fluoride is an inhibitor of enolase which catalyses the conversion of phosphoglyceric acid to phosphopyruvic acid and which contains Mg^{++} (cf. glycolysis); it inhibits also acid phosphatase of liver and kidney). Fluorine is found in drinking water and in certain parts of the world, where there is a high level of fluorine in the water as, for example, in southern India, chronic poisoning results in man (fluorosis). This is characterized by hypercalcemia affecting the vertebral column and the ribs, such that the subjects exhibit abdominal ventilation (Shortt *et al.*, 1937); it would be of interest to study the action of fluorine on the parathyroid or on the thyroid which produces calcitonine.

II. Organic trace compounds in the diet and vitamins

The word "vitamin" was introduced by Funk. The history of vitamins is associated with that of certain diseases: scurvy, beriberi and pellagra.

Scurvy

This disease affected the early sailors and also decimated the army of Saint-Louis; it is characterized by loss of teeth, ulceration of the lungs and death following suffocation and syncope. This illness is caused by the absence in the diet of fresh vegetables and the active factor is ascorbic acid (vitamin C).

Beriberi

This disease was found among the populations of Asia which existed mainly on a diet of polished rice; it affects the nervous, gastric and cardiac systems. The active factor contained in husks of rice (Eijkmann, 1897) is thiamine (vitamin B_1).

Pellagra

This disease affected the poor classes of European and American societies who were badly fed and existed almost exclusively on a diet of maize. A balanced diet contains an anti-pellagra factor, *nicotinic acid* (vitamin P-P—pellagra preventive).

A vitamin can be described in the following way: vitamins are accessory food factors which are essential for some metabolic processes and which must be provided in the diet and which have an action in low doses. Among the mammals principally, the organism, with certain rare exceptions, cannot synthesize these compounds or, at least, not in sufficient quantities. Consequently, the essential amino acids could be placed in the category of vitamins; however, these amino acids function in relatively high concentrations and have a structural role and cannot be considered, therefore, as vitamins. In a similar way, all other factors of which the absence produces deficiency symptoms could be placed in the category of vitamins. But, there again, the concept of deficiency is misleading if it does not take into account the idea of biogenesis by the organism; in fact, *hormones* are also, to a certain extent, catalysts in the transformation of metabolites. Having understood the limitations of the definition of a vitamin with respect to dosage, function and autosynthesis, it remains now to examine the principal known vitamins. Vitamins are divided into two main classes:

(1) fat-soluble vitamins;
(2) water-soluble vitamins.

Most vitamins have been shown to be *cofactors for enzymes* and there appears to be little doubt that all vitamins enter this biochemical category.

In every case, they contain a high level of potential energy as a result of the presence in the molecule of labile hydrogen, double bonds or high energy phosphate bonds. In general terms, vitamins are involved in the appearance of hemorrhage syndromes, particularly in mammals.

This chapter should emphasize the dependance of manifestations in the whole animal on processes occurring at the intracellular level and the vitaminic interrelations.

Fat-soluble vitamins

VITAMIN A

Vitamin A is, in fact, two compounds; vitamin A_1 and vitamin A_2 which is 100 times less potent than vitamin A_1. We shall be concerned only with vitamin A_1.

Vitamin A_1 is formed by the hydrolysis of β-carotene (principal pro-vitamin of plant pigments) which occurs in the liver of man and in the intestinal wall in rat, pig, rabbit and chicken (equation 44). Although a normal rat can exist on a diet which contains either β-carotene or vitamin A, the thyrodectomized rat cannot exist on β-carotene alone. Vitamin A must be provided and it is, therefore, most probable that thyroid hormone is involved in the cleavage of β-carotene.

$$
\beta\text{-Carotene}
$$

$$
\begin{array}{c}
H_3C \quad CH_3 \\
\diagdown\diagup \\
C \\
\diagup\diagdown \\
H_2C \quad C-(CH=CH-\overset{CH_3}{\underset{|}{C}}=CH)_2-CH=CH-(CH=CH-\overset{CH_3}{\underset{|}{C}}=CH)_2-C \quad CH_2 \\
\underset{|}{H_2C} \quad \underset{\parallel}{C}-CH_3 \qquad\qquad\qquad\qquad\qquad\qquad H_3C-C \quad CH_2 \\
\diagdown\diagup \qquad\qquad\qquad\qquad\qquad\qquad\qquad\qquad \diagdown\diagup \\
C \qquad\qquad\qquad\qquad\qquad\qquad\qquad\qquad\qquad C \\
| \qquad\qquad\qquad\qquad\qquad\qquad\qquad\qquad\qquad | \\
H_2 \qquad\qquad\qquad\qquad\qquad\qquad\qquad\qquad\qquad H_2
\end{array}
$$

$$
\begin{array}{c}
H_3C \quad CH_3 \qquad\qquad\qquad\qquad\qquad\qquad\qquad\qquad H_3C \quad CH_3 \\
\diagdown\diagup \qquad\qquad\qquad\qquad\qquad\qquad\qquad\qquad\qquad \diagdown\diagup \\
C \qquad\qquad CH_3 \qquad\qquad\qquad\qquad\qquad\qquad CH_3 \qquad C \\
\diagup\diagdown \qquad |\qquad\qquad\qquad\qquad\qquad\qquad\qquad | \qquad\diagup\diagdown \\
H_2C \quad C-(CH=CH-C=CH)_2-CH_2OH+HOCH_2-(CH=CH-C=CH)_2-C \quad CH_2 \\
\underset{|}{H_2C} \quad \underset{\parallel}{C}-CH_3 \qquad\qquad\qquad\qquad\qquad\qquad\qquad\qquad H_3C-C \quad CH_2 \\
\diagdown\diagup \qquad\qquad\qquad\qquad\qquad\qquad\qquad\qquad\qquad\qquad \diagdown\diagup \\
C \qquad\qquad\qquad\qquad\qquad\qquad\qquad\qquad\qquad\qquad C \\
| \qquad\qquad\qquad\qquad\qquad\qquad\qquad\qquad\qquad\qquad | \\
H_2 \qquad\qquad\qquad\qquad\qquad\qquad\qquad\qquad\qquad\qquad H_2
\end{array}
$$

$$(44)$$

Vitamin A_1 is metabolized in the presence of lipoxidases by the oxidation of double bonds.

Functions

(1) Vitamin A_1 seems to act on lipases, which is of interest in view of the fact that this is a lipid-soluble vitamin.

(2) It maintains the integrity of epithelial tissue; in the absence of vitamin A_1 there is keratinization of epithelia.

(3) Vitamin A_1 is a constituent of visual purple (retina). In the absence of this vitamin night blindness occurs (hemeralopia). This must not be confused with day blindness (nyctalopia) which results from hypersensitivity of the retina to light. Moreover, xerophthalmia or keratinization of ocular tissue occurs in the absence of vitamin A and can lead to blindness. In vision, vitamins A and B_2 seem to be coupled (Theorell, 1935; Adler and v. Euler, 1938).

(4) Vitamin A_1 is a cofactor in the mechanisms of aerobic transfer of energy (Ernster).

Sources and requirements

Provitamin A is found in fruit (for example, apricots) and in green plants. Vitamin A is found in oils from fish liver, milk, butter and yolk of egg.

The international unit is equivalent to the activity of 0·6 μg of pure β-carotene. An adult requires 5000 I.U. per day and 7000 I.U. per day during gestation and lactation.

Hypervitaminosis leads to toxic effects, the principal symptoms being painful joints and loss of hair.

Hypovitaminosis A in man can occur as a result of certain liver diseases because, on the one hand, the liver is the site of hydrolysis of β-carotene, and on the other, all fat-soluble vitamins can be absorbed from the intestine only in the presence of bile and poor intestinal absorption will affect vitaminosis A.

The vitamins D

They are all *sterols* which occur principally in animals. Certain of these sterols—known as provitamins D—acquire antirachitic properties when irradiated with ultraviolet light; that is to say, administration of these compounds to vertebrates results in the prevention of disorders of calcification mechanisms. Although all of the vitamins D possess antirachitic activity, their potency varies according to the species studied (species specificity). Thus, irradiated ergosterol (vitamin D_2) is antirachitic in man and rat but not in chicken which is sensitive to vitamin D_3. Vitamin D_3 can be synthesized by mammals and differs from vitamin D_2 in the presence of three CH_2 groups in the place of CH=CH—CH— on

$$CH_3$$

the side chain.

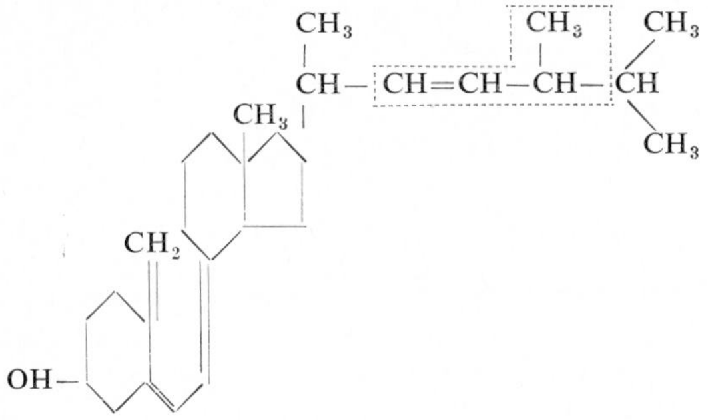

Vitamin D_2
(= calciferol or
activated ergosterol)

Functions

The general functions of vitamins D with antirachitic activity are:

(1) increase in the absorption of calcium and phosphorus from the intestine;

(2) increase in the elimination of phosphorus by the kidney—an action similar to that of parathormone; the ratio A/D_2 is to be considered (Gounelle *et al.*, 1951);

(3) according to Zetterström phosphorylated vitamin D_2 is water soluble and activates alkaline phosphatase of bone and also of the kidney and the intestinal mucosa.

Sources and requirements

The main sources of vitamins D are milk and viscera of fish, especially the liver.

An international unit corresponds to the activity of 0·025 μg calciferol. Man requires an average of 300 I.U. per day.

VITAMINS E

The different vitamins E are:

α-tocopherol:	5,7,8-trimethyltocol
β-tocopherol:	5,8-dimethyltocol
γ-tocopherol:	7,8-dimethyltocol
δ-tocopherol:	8-methyltocol
ε-tocopherol:	5-methyltocol
ζ-tocopherol:	5,7-dimethyltocol
η-tocopherol:	7-methyltocol

which are derived from the following compound — *tocol;*

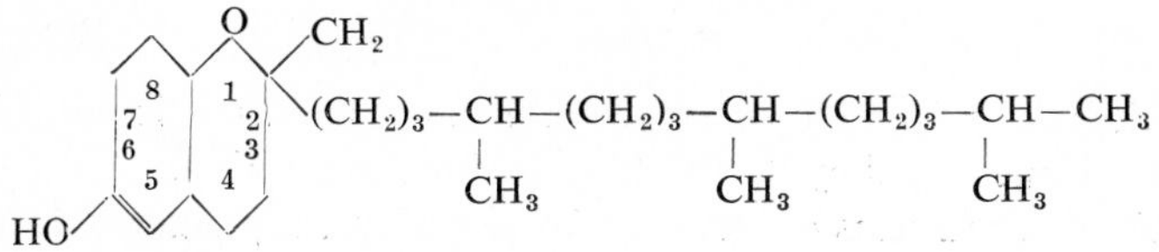

α-Tocopherol is the most widely distributed among animals.

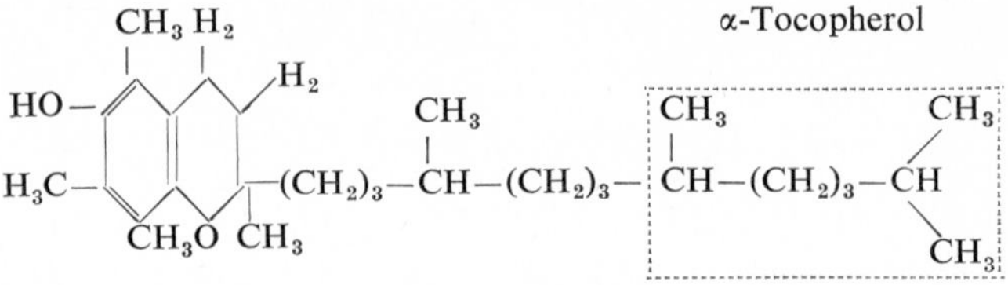

α-Tocopherol

The terminal fraction of the side chain of α-tocopherol is identical with that of vitamin D_3; in addition, oxidation of α-tocopherol yields α-tocopherylquinone

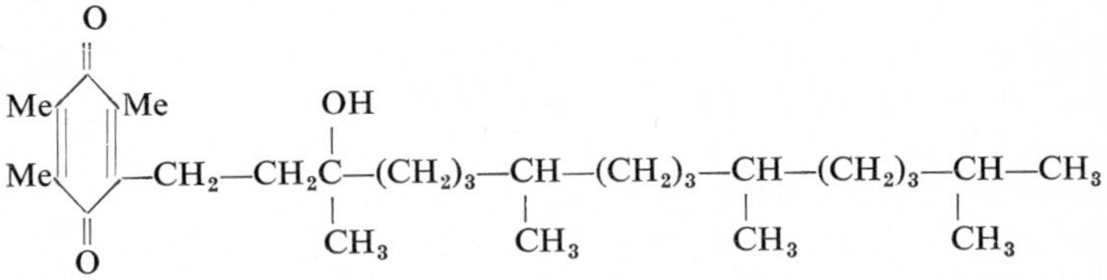

which possesses some similarity in structure with vitamin K_1, a fact which demonstrates, once again, the economical use of structural types in nature.

Functions

(1) According to the sex, deficiency in mammals results in:

    (a) atrophy of spermatogenic tissue with sterility;

    (b) resorption of the foetus during gestation (abortion).

(2) Vitamin E is a powerful *antioxidant* (it prevents the oxidation of vitamin A which results in loss of its vitamin properties).

(3) Transaminase activity (of the glutamic acid–oxaloacetic acid system) is reduced by half in the muscles of vitamin E-deficient guinea pigs.

(4) The rate of turnover of nucleic acids in muscle may be increased by vitamin E; this could explain the ability of vitamin E to potentiate gametogenesis.

(5) Certain cystine-deficient diets lead to necrosis of the liver in experimental animals. Infusion of vitamin E to the portal vein prevents this necrosis and lowers the rate of respiration of necrotic tissues.

(6) Vitamin E deficiency leads to uncoupling of oxidative phosphorylation (Pin, 1948); α-tocopherol is found in mitochondria (Slater, 1957).

Sources

This vitamin is found in wheat germ, fish, eggs and particularly in milk. Vitamin E is measured by its ability to maintain gestation.

The blood of humans contains approximately 4 mg vitamin E per l.

VITAMIN F

Evans proposed the general name of vitamin F to describe a group of unsaturated fatty acids (CH=CH, ethylenic linkage) as for example linoleic acid (two double bonds):

$$CH_3—(CH_2)_4—CH=CH—CH_2—CH=CH—(CH_2)_7—COOH$$

linolenic acid (three double bonds)

$$CH_3—CH_2—CH=CH—CH_2—CH=CH—CH_2—CH=CH—(CH_2)_7—COOH$$

Vitamin F deficiency in rat leads to arrest of growth, necrosis of the skin and death (cf. p. 40).

The vitamin activity of these compounds is questionable and they may be acting in a similar manner to the essential amino acids.

All compounds which contain one or more double bonds are potential sources of energy. For this reason, vegetable oils which are rich in unsaturated compounds are valuable not only in growth processes but also in maintaining general metabolism. In addition, numerous studies (notably in France, those of Enselme) have shown that, while a diet rich in saturated fatty acids can lead to the development of arteriosclerosis, a diet rich in unsaturated fatty acids (such as linolenic acid) lowers cholesterolemia and retards the development of arteriosclerosis. The following food oils are listed in order of decreasing content of double bonds: soya bean oil, sunflower oil, maize oil, colza oil and olive oil.

In fact, the effect of fatty acids is a controversial question since Pressman and Lardy (1956) have shown that unsaturated fatty acids—like oleic acid—can uncouple oxidative phosphorylation (cf. p. 318).

VITAMINS K

These compounds are derivatives of 2-methyl-1,4-naphthoquinone.

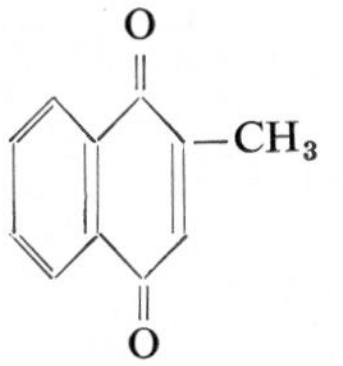

2-Methyl-1,4-naphthoquinone = vitamin K_3 (synthetic)

Vitamins K_1 and K_2 are of natural origin.

Vitamin K_1

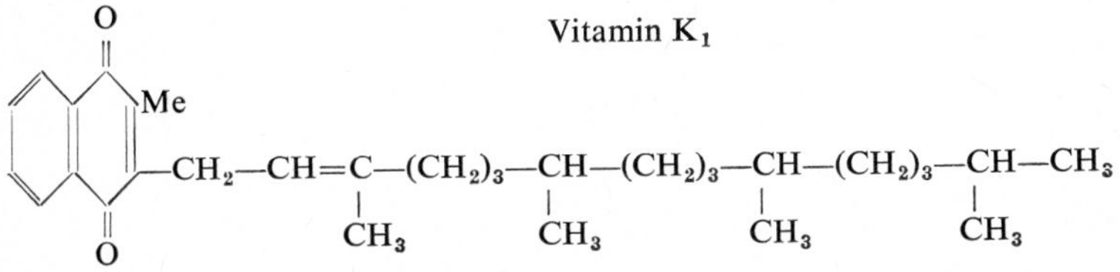

$$—CH_2—CH=C—(CH_2)_3—CH—(CH_2)_3—CH—(CH_2)_3—CH—CH_3$$
$$\quad\quad\quad\; CH_3 \quad\quad\; CH_3 \quad\quad\; CH_3 \quad\quad\; CH_3$$

Functions

(1) The vitamins K catalyse the synthesis of prothrombin in normal liver. (In cirrhosis the parenchyma cannot produce prothrombin and vitamin K is without effect). From this it can be seen that the results of hepatic disorders may be extremely serious. Vitamin K deficiency induces severe hemorrhages, principally in the new-born organism. This deficency, although rare, can occur as a result of biliary disorders as bile is required for the absorption of fat-soluble vitamins; however, water-soluble forms of vitamin K exist which can be absorbed in the absence of bile.

(2) Sulphonated vitamin K_3 (in concentrations of 3×10^{-5}M) inhibits choline acetylase (the enzyme for the synthesis of acetylcholine); however this action is related, probably, to the presence of sulphonic acid.

(3) Vitamin K is an electron acceptor and of importance in the cytochrome system of mitochondria (Colpa-Boonstra and Slater, 1958). Vitamin K is reduced by a specific enzyme—vitamin K reductase—possibly after oxidation by cytochrome b (Martius, 1956–1961). In particular, vitamin K deficiency decreases the oxidative phosphorylation by mitochondria and the phosphorylating ability is restored, *in vitro*, by the addition of vitamin K.

(4) Dihydroxycoumarin (or dicoumarol) obtained from two molecules of coumarin, is an isosteric antagonist of vitamin K and induces hypo-prothrombinemia after administration to an animal. This explains the powerful anti-coagulant activity of dicoumarol. Dicoumarol uncouples oxidative phosphorylation. It is called, therefore, an *anti-vitamin*. Another example of an anti-vitamin K can be found in the work of P. Meunier (1943) who obtained by coupling two molecules of phtiocol an anti-coagulant, methyleneoxynaphthoquinone. This compound, when administered to rabbit *(per os)* in concentrations of 20 mg/kg, decreased the level of prothrombin by 50 per cent within 24–48 h.

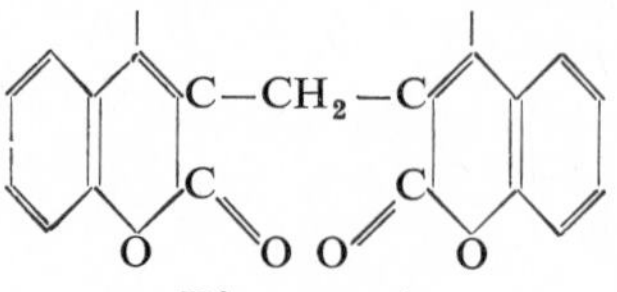

Dicoumarol 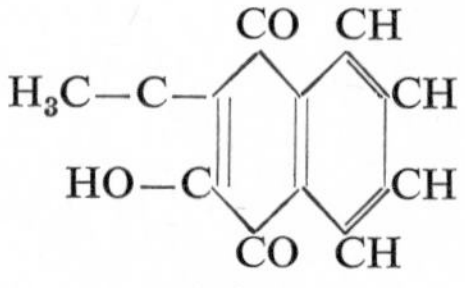   Phtiocol

Sources

These vitamins are found in most foodstuffs and they are synthesized by intestinal micro-organisms.

Water-soluble vitamins

VITAMIN C

This was the first vitamin to be chemically identified (Szent-Györgyi, 1932). The formula of vitamin C is similar to that of hexose (45):

$$
\begin{array}{ccc}
\begin{array}{l}
C=O \\
\;\;\diagdown \\
C-OH \\
\;\;\| \\
C-OH \\
\;\;| \\
H-C \\
\;\;| \\
HO-C-H \\
\;\;| \\
CH_2OH
\end{array}
&
\xrightleftharpoons[-2H]{+2H}
&
\begin{array}{l}
C=O \\
\;\;\diagdown \\
C=O \\
\;\;| \\
C=O \\
\;\;| \\
H-C \\
\;\;| \\
HO-C-H \\
\;\;| \\
CH_2OH
\end{array}
& \qquad (45)
\\[2pt]
\text{Reduced form} & & \text{Oxidized form} \\
\text{(ascorbic acid)} & & \text{(dehydroascorbic acid)}
\end{array}
$$

Both forms are physiologically active and are found in the organism. Ascorbic acid shows powerful *reducing activity*.

Functions

Although ascorbic acid is required by the animal organism, *certain mammals* (rat) do not require dietary vitamin C since they are *able to synthesize it* from glucose (Horovitz and King, 1953), an exception which confirms the definition of a vitamin given on p. 56. Aminopterin decreases the content of vitamin C in the liver (Williams, 1951).

However, ascorbic acid deficiency, whether of endogenous or exogenous origin, leads to the appearence of symptoms of scurvy, mainly:

(1) fragility of bones and loss of teeth;

(2) hemorrhages.

In high doses, (0·5–1 g per day *per os* in man) ascorbic acid shows an ability to prevent infection which is associated, without doubt, with activation of phagocytic activity.

Jonxis and Huisman (1954) described an increase in excretion of amino acids (glycine, tyrosine, lysine, etc.), in children deficient in ascorbic acid, which was probably due to increased production of amino acids in the kidney. The normal state can be reestablished within a few weeks by the administration of vitamin C.

Little is known as to the site of the biochemical activity of vitamin C: it could activate serum cholinesterase; a large amount of ascorbic acid is localized in the mitochondria of the adrenal cortex and this is depleted when the gland is stimulated by adrenocorticotropic hormone;

with the exception of cartilage cells, cells which have a high ascorbic acid content lose this at the time of division. Generally, the Golgi apparatus contains a large quantity of ascorbic acid.

Sources and requirements

Vitamin C is found in fruit (mainly the citrus fruits: lemons and oranges, etc.) and in green plants.

Is is destroyed easily by oxidation during cooking, particularly in an alkaline medium.

Man requires 30 to 150 mg per day depending on sex, age and conditions such as gestation and lactation.

When the level of vitamin C in the blood exceeds 10 mg/l it is eliminated in the urine.

VITAMIN C_2

This is also called vitamin P (affecting capillary permeability). Szent-Györgyi and Rusznyák, who extracted this compound in 1936, gave it the name *citrin* and believed that it was a diglucoside of a compound belonging to the flavone group. In fact, this compound has been the subject of much debate. According to different authors, many substances show vitamin P activity including *rutin* (a compound containing glucose, rhamnose and pentahydroxyflavone), *esculin* (a molecule containing dihydrocoumarin and glucose).

Flavone
(Phenyl-2-chromone)

This multiplicity of factors showing capillary permeability activity leads to the conclusion that these are precursors of the active substance. Lavollay and Parrot (1943) believe that catechin is the true vitamin P.

Function

The main physiological action of vitamin P would be in the reduction of hemorrhages through an increase in capillary permeability and a decrease in the resistance of the capillaries. These syndromes of hemorrhage are related to those of scurvy. Many conditions of avitaminosis are associated with symptoms of hemorrhage and it is most probable that the difficulties relating to the problem of vitamin P are due to the fact that the multiple causes of hemorrhage are little understood.

However, Lavollay (1940) considered that vitamin P acted at the molecular level by protecting adrenaline from destruction by auto-oxidation and thus, maintained the tone of the capillary walls. A weakness in this hypothesis is that, although many auto-oxidizable compounds can increase capillary resistance, vitamin C, which is a powerful reducing agent, has little vitamin P activity.

Sources and requirements

It is found in fruit (lemon juice, etc.). A daily dose of 1 g is given to man during capillary hemorrhages.

Vitamins of the B complex

Many compounds were found to have similar effects to those of the anti-beriberi factor, mainly on the nervous system, and therefore, the vitamins of the group B complex are chemically diverse compounds.

The elucidation of the structure of the constituents of the vitamin B group has clarified many of the problems relating to this group:

Thiamine, or aneurin, vitamin B_1.
Riboflavin, or lactoflavin, vitamin B_2.
Nicotinamide, or vitamin PP.
Pyridoxine, or rat antidermatitis factor, vitamin B_6.
Cyanocobalamin, or vitamin B_{12}.
Folic acid, or vitamin B_c or M.
Pantothenic acid or vitamin FF.
Inositol, or bios I or vitamin B_7.
Biotin, or bios II or vitamin H or B_8.
Lipoic acid.
We shall consider now the essential properties of these compounds.

VITAMIN B_1

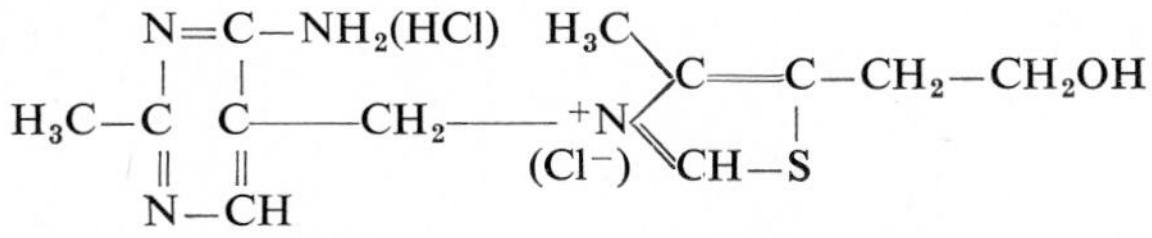

Pyrimidine fraction   Thiazole fraction
(found mainly in nucleic acids)

Functions

(1) Thiamine in the form of the diphosphate, is the coenzyme for the decarboxylation of α-keto acids—cocarboxylase (Lohmann and Schuster, 1937); it is of fundamental importance in basic respiratory processes.

For example:

$$CH_3-\overset{\overset{\displaystyle O}{\|}}{C}-COOH \xrightarrow[\text{thiamine diphosphate}]{\text{carboxylase}+} CH_3-\overset{\overset{\displaystyle O}{\|}}{C}-H+CO_2 \qquad (46)$$

Pyruvic acid　　　　　　　　　　　　　　Acetaldehyde

In yeast, where reduction of acetaldehyde gives ethanol, *direct decarboxylation* occurs.

In animal tissues, *oxidative decarboxylation* of pyruvic acid results in the formation of acetic acid which is an oxidation product of acetaldehyde. This reaction involves not only cocarboxylase but also lipoic acid, coenzyme A and diphosphopyridine nucleotide.

(2) Cocarboxylase can serve as a *phosphate carrier* (Kiessling and Lindahl, 1952).

(3) In mammals vitamin B_1 deficiency leads to nervous, cardiovascular, gastro-intestinal and growth disorders.

Vitamin B_1 is known mainly as an *anti-neuritic substance*. This property can be demonstrated in birds (avian polyneuritis) where neuritis is characterized principally by muscular spasms (Fig. 11). In the pigeon there

Fig. 11. Pigeon deprived of vitamin B_1. Polyneuritis is evident by contractures and convulsions. (HÉDON, 1950.)

is rigidity of the tail, extension of the limbs, etc. The cerebral tissue of these pigeons shows, *in vitro*, a decreased ability for oxygen consumption in the presence of glucose (Gavrilescu and Peters, 1931). Vitamin B_1 is a growth factor for certain ciliates, for example, *Glaucoma piriformis* (Lwoff); this property is used in the estimation of vitamin B_1 concentrations.

In the bovids, external provision of thiamine is not necessary because it can be synthesized by bacteria in the digestive tract. This does not constitute, however, an exception to the definition of a vitamin (p. 56) as the *animal* does not carry out the synthesis.

The recommended daily intake of thiamine in man is 0·5 mg per 1000 cal.

This vitamin is distributed in all plant foodstuffs (nuts, apricots, etc.) as well as in the liver of chicken.

VITAMIN B_2

Vitamin B_2 is a yellow-green, fluorescent pigment from milk; it is very sensitive to ultraviolet light and decomposes to give *lumiflavin* (cf. Volume II) in an alkaline medium and *lumichrome* in an acid medium, but is relatively thermostable.

This compound is related chemically to isoalloxazine but a *purine precursor* is involved in the biosynthesis of vitamin B_2.

The metabolic form of riboflavin is the phosphate derivative (in heart tissue the sugar attached to the flavin nucleus is lyxose and not ribose and this compound is called lyxoflavin).

Functions

(1) At the molecular level: vitamin B_2 exists in two forms:

(A) *Riboflavin phosphate* is a constituent of the Warburg *yellow enzyme* which catalyses the oxidation of TPN (triphosphopyridine nucleotide) or NADP; it is also a constituent of cytochrome c reductase and of L-amino acid oxidase.

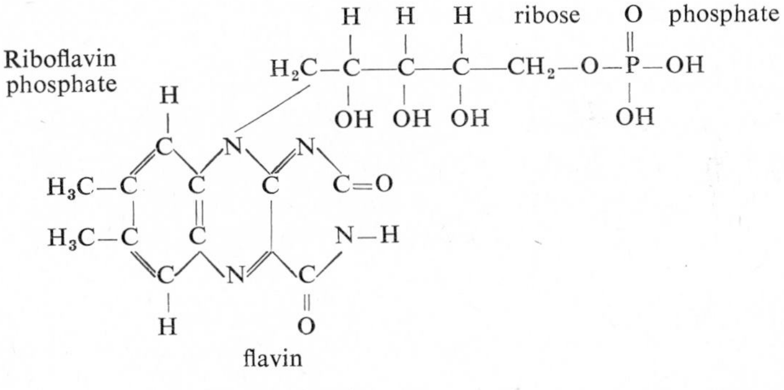

Riboflavin phosphate / flavin

(B) *Flavin adenine dinucleotide* (FAD) contains, in addition, adenylic acid, a potential high energy compound, whose role in this substance can be understood from the fact that FAD is a constituent of D-amino acid oxidase (and also of diaphorase which accepts hydrogen from DPN or NAD) the D-amino acids being metabolized less easily, at least, in the mammals.

O. Lindberg *et al.* (1958) suggested that FAD is a phosphate carrier during oxidative phosphorylation, a similar function being proposed for vitamin B_1.

Flavin adenine dinucleotide

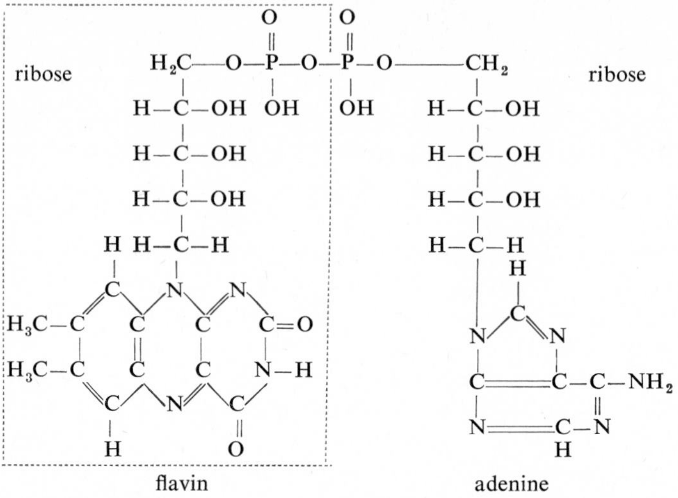

flavin   adenine

(2) In the animal as a whole: avitaminosis B_2 in young rats leads to arrest of growth, loss of weight and death. In rat, avitaminosis B_2 also leads to loss of hair (alopecia), and hypoproteinemia although there is no affect on the γ-globulin fraction (Jacquot-Armand *et al.*, 1963). In man, deficiency results in abnormal vascularization of the cornea.

This avitaminosis is the most prevalent in the U.S.A.

Sources and requirements

Vitamin B_2 is present mainly in milk, liver, heart and brewer's yeast. The normal concentration of vitamin B_2 in serum is 30 μg/l. The daily requirement for man is 2 mg. An international unit is equivalent to 5 μg.

VITAMIN PP

This compound is the amide of nicotinic acid—nicotinamide. *Nicotinic acid* ($=$ niacin) *is a provitamin.*

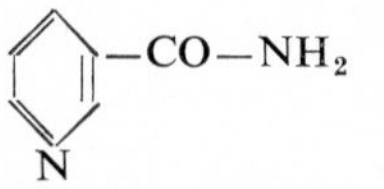

Nicotinamide

Functions

(1) At the molecular level, nicotinamide is a constituent of two coenzymes with considerable physiological importance:

coenzyme I or diphosphopyridine nucleotide (DPN) which exists also in the reduced form DPNH; following a decision at a recent international meeting on nomenclature, coenzyme I should be denoted by NAD, nicotinamide-adenine-dinucleotide, ($NADH_2$ for the reduced form);

coenzyme II or triphosphopyridine nucleotide (TPN) which is found also in the reduced form TPNH; it was decided also by the committee for chemical nomenclature to call coenzyme II NADP = nicotinamide-adenine-dinucleotide phosphate (and $NADPH_2$ for the reduced form).

The structure of coenzyme I is the same as that of flavin adenine dinucleotide except for the substitution of the flavin group by nicotinic acid. This is another example of the economical use of chemical units in nature (nucleotide coenzymes).

The coenzymes I and II are hydrogen and electron carriers in oxidation–reduction reactions (for example, coenzyme II is the cofactor for glucose monophosphate dehydrogenase which catalyses the oxidation of glucose-6-monophosphate (G6P) to phosphogluconic acid).

Ernster (1958) reported the existence of a cytoplasmic enzyme, *pyridine nucleotide diaphorase*, which is capable of reacting with numerous electron acceptors, and in particular with vitamin K_3. Addition of pyridine nucleotide diaphorase in the presence of vitamin K_3 increased the oxidation by mitochondria of $NADPH_2$ produced by the cytoplasmic system G6P–G6P dehydrogenase (cf. p. 125) (Conover and Ernster, 1960). This indicates the way in which distinct vitamins can be related at the functional level.

(2) In the animal as a whole: (a) Tryptophan is a precursor (provitamin) of nicotinamide and a diet consisting entirely of maize (deficient in tryptophan) results, therefore, in pellagra which is characterized by:

skin disorders: red and then brown pigmentation;
disorders of the digestive tract: gingivitis, inflammation of the tongue;
mental disorders: delirium, mental confusion.

Avitaminosis PP occurs widely in the U.S.A. where there are large areas of land under maize cultivation.

(b) Nicotinic acid is an arteriolar vasodilator.

Sources and requirements

Vitamin PP is found in yeast but liver, milk and tomatoes are also good sources for this vitamin. A recommended daily intake for man is 15 mg (for 3000 calories).

VITAMIN B_6

In vivo, the formation of pyridoxal from pyridoxamine is catalysed by pyridoxamine phosphate oxidase and also by transaminases through the reaction of pyridoxamine with oxaloacetate or α-ketoglutarate (Wada and Snell).

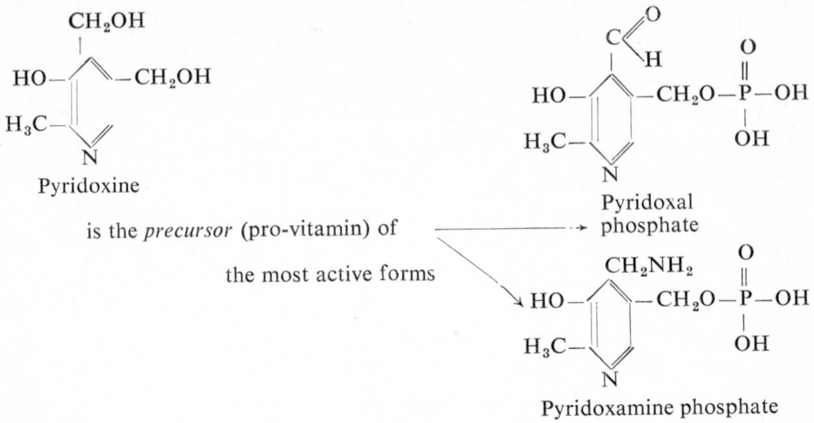

Pyridoxine

is the *precursor* (pro-vitamin) of   ⟶ Pyridoxal phosphate

the most active forms

Pyridoxamine phosphate

Pyridoxal can be transformed *in vivo* by phosphorylation to give pyridoxal-5-phosphate (catalysed by pyridoxal kinase), by oxidation to pyridoxic acid (catalysed by aldehyde oxidase), by the reverse transaminase reaction to give pyridoxamine or finally, by reduction (in baker's yeast) to pyridoxine. Pyridoxic acid is the excretory form of vitamin B_6 in man.

Functions

(1) *At the molecular level:* vitamin B_6 is an important coenzyme. Pyridoxal phosphate is the coenzyme for decarboxylase of many amino acids (tyrosine, arginine, etc.). It is also a codeaminase for serine and threonine. Therefore, as a coenzyme, pyridoxal phosphate has widespread importance.

Pyridoxamine phosphate acts as a cotransaminase. Moreover, the synthesis of NAD and NADP is affected by vitamin B_6 deficiency (it has a function in the transformation of tryptophan to nicotinamide); in some ways it can be considered to be a cofactor of enzymic cofactors. Pyridoxamine phosphate is the coenzyme of desulphydrases and in this way, functions in the processes of sulphur transfer and *in the removal of the elements of hydrogen sulphide* from cysteine, etc.

(2) *In the whole animal:* In rat, deficiency leads to hemorrhages, loss of hair, desquamation around the muzzle (characteristic) and anemia; in man, convulsions have been reported in deficient children. Vitamin B_6 appears to be essential in the metabolism of the central nervous system.

Sources and requirements

It is found principally in wheat grains and liver but is present also in milk and eggs.

Man requires 2 mg vitamin B_6 per day.

Vitamin B_{12}

Vitamin B_{12} is known also as cyanocobalamin or anti-pernicious anemia factor [extrinsic factor of Castle (cf. p. 89)]. Vitamin B_{12} contains a *tetrapyrrole nucleus* linked to cobalt (L. Smith, 1955); the pyrrole nucleus has great biological importance (it occurs in chlorophyll with Mg^{++} and in hemoglobin with Fe^{++}); thus, with each molecular change the tetra-pyrrole nucleus has a new function and this provides another example of economy in nature by the use of the same basic structure for different functions. Of special interest here is the presence of the $C \equiv N$ group.

Vitamin B_{12} has been isolated from liver (Rickes *et al.*, 1948). The vitamin is obtained industrially by the fermentation of *Streptomyces griseus*. It has been crystallized and appears as red crystals. This compound is thermostable in solution at pH 4 to 7 but above or below this range it is rapidly destroyed. This indicates the way in which thermostable or thermolabile properties are dependent on other factors.

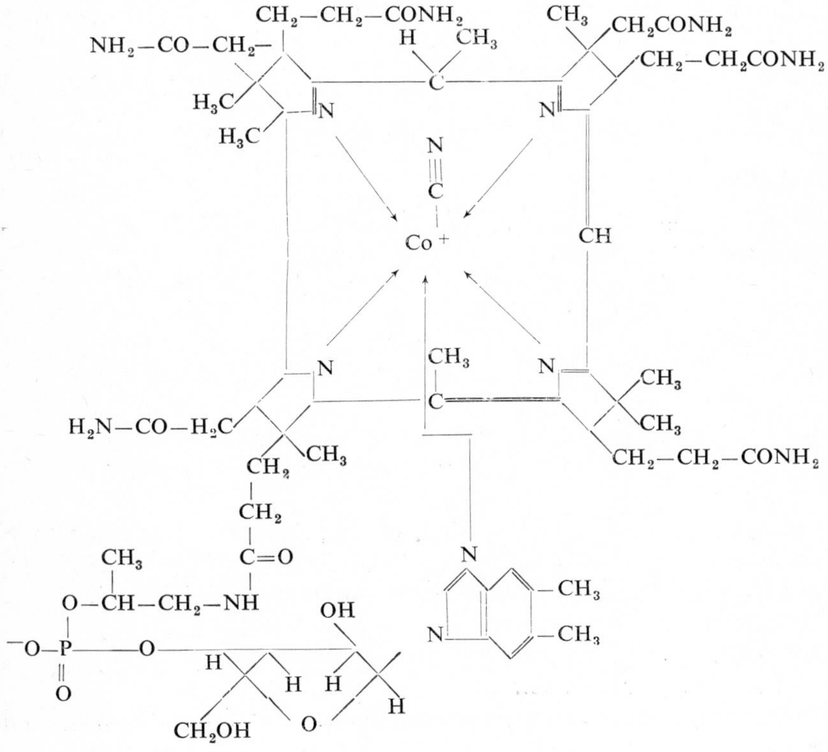

Functions

(1) *At the molecular level:*

(a) it favours the synthesis of nucleic acids in rats and monkeys;

(b) it is involved in processes of methylation (cf. p. 130).

(2) *In the animal as a whole:*

(a) 30-day-old rats can grow only if the diet contains methyl donors and also vitamin B_{12}. Poor growth results from a diet containing methionine but no vitamin B_{12}.

(b) the principal function of vitamin B_{12} is is the *prevention of pernicious anemia:*

The absorption of vitamin B_{12} by the intestine is dependent on the presence of hydrochloric acid and a compound in the gastric juice, called by Castle "intrinsic factor". This factor is probably a mucoprotein from the cardia and fundus region of the stomach. Pernicious anemia is associated with atrophy of the fundus and absence of hydrochloric acid secretion. In the blood vitamin B_{12} is attached to a carrier protein but this is an α-globulin not a mucoprotein.

Sources and requirements

Liver and kidney contain 50 μg vitamin per 100 g; milk, cheese and eggs contain 1–5 μg/100 g. Bacteria in the rumen of ruminants can synthesize vitamin B_{12} (50 μg vitamin B_{12}/100 g dry weight rumen; therefore the rumen contains an amount similar to that in liver or kidney).

Man requires 1–2 μg vitamin B_{12} per day.

VITAMINS B_c OR M

These vitamins constitute the folic acid group. Vitamin B_c is the monoglutamate derivative.

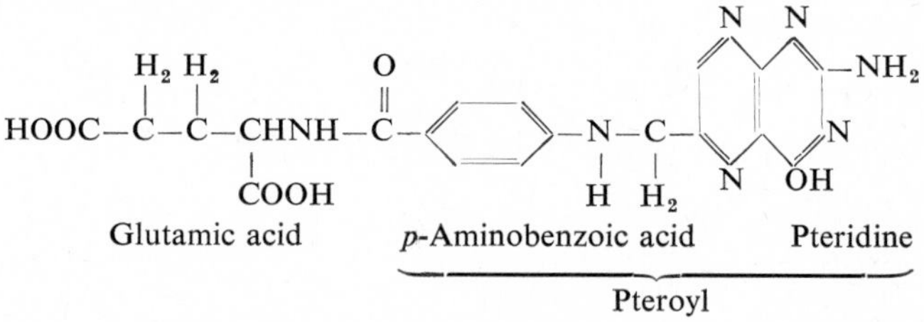

| Glutamic acid | *p*-Aminobenzoic acid | Pteridine |

Pteroyl

Folic acid is pteroylglutamic acid. In fact, there are three compounds which differ from one another only in the number of glutamic acid residues. In the organism folic acid can be converted to folinic acid.

Functions

(1) *At the molecular level:*

It participates in the incorporation of amino acids (for example, glycine) into purines

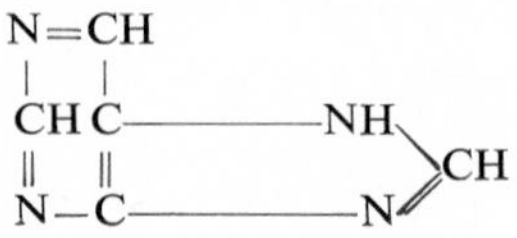

(2) *In the whole animal:*

Folic acid has a function in hemopoiesis and acts with vitamin B_{12} in the prevention of pernicious anemia; this hemopoietic property is based on the fact that folic acid is involved in the synthesis of nucleic acids via purines.

Deficiency in mammals leads to abnormal hemopoiesis, diarrhea, gingivitis and inhibits growth of young animals.

Aminopterin (L-aminofolic acid) which blocks the synthesis of nucleic acids is prescribed in the treatment of leukemia.

Sources and requirements

Folic acid is present in spinach, liver, kidney. Man requires 15 mg folic acid per day although different values have been quoted by certain authors.

PANTOTHENIC ACID

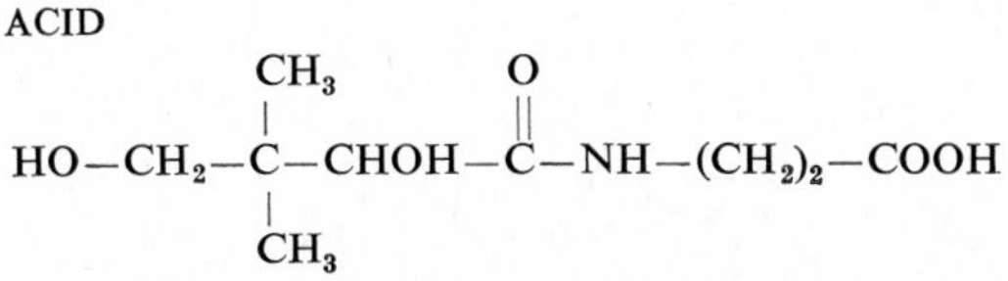

$$\underset{\underset{CH_3}{|}}{\overset{\overset{CH_3}{|}}{HO-CH_2-C-CHOH-}}\overset{O}{\overset{||}{C}}-NH-(CH_2)_2-COOH$$

The great importance of this compound in the life of the animal is derived from the fact that it is the principal constituent of *coenzyme A* or coacetylase (since it functions in acetylation reactions).

Biological synthesis of coenzyme A has been realized by Hoagland and Novelli (1954). Synthesis occurs in the presence of Mg^{++} and requires cysteine and ATP.

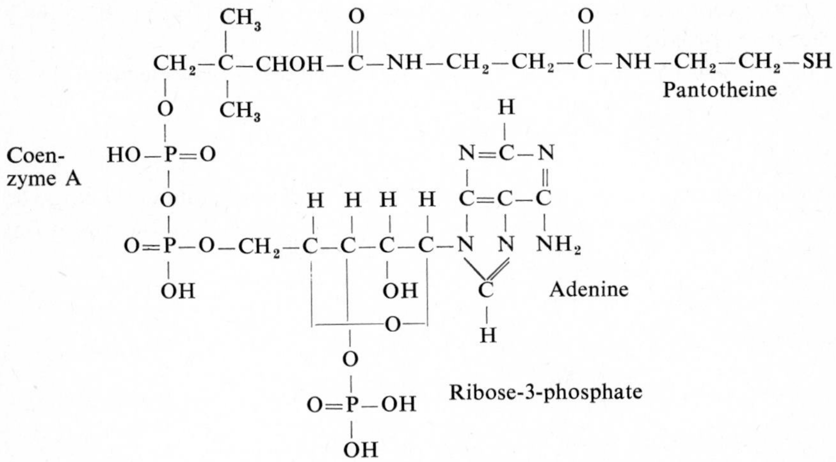

As in the case of FAD, NAD and NADP, this structure contains adenylic acid, associated here with pantothenyl phosphate. We have to emphasize that adenylic acid activates also the amino acids utilized in the biosynthesis of proteins (cf. p. 136).

Functions

(1) *At the molecular level:*

(a) in the form of acetyl-coenzyme A ("active acetate") acetic acid participates in numerous metabolic reactions. Acetic acid combines with oxaloacetic acid to form citric acid which enters the tricarboxylic acid cycle (cf. p. 124). Moreover, in the form of active acetate, acetic acid combines with choline to give acetylcholine, an equally important physiological reaction.

(b) Acetic acid is the metabolic precursor of cholesterol and, as a result, of steroid hormones. This reaction is catalysed by coenzyme A.

(c) Coenzyme A, in the form of acetyl-CoA, is required for the biosynthesis of fatty acids in the udder (Popják, 1955).

(2) *In the whole animal:*

(a) Pantothenic acid deficiency leads in chickens to whitening of the feathers and in mammals, including man, to depigmentation of hair. For this reason, this vitamin is known as the vitamin preventing greying of hair. One should not fail to note the disparity between the important molecular functions of pantothenic acid and this deficiency symptom.

(b) Deficiency leads also to diarrhea, cornification of the skin, hemorrhages and necrosis of the adrenal cortex associated with an increased appetite for salt (NaCl), which is related to the functioning of pantothenic acid in the biogenesis of sterols.

Sources and requirements

Spontaneous deficiency has not been shown in man; other mammals require approximately 1 mg per day.

It is found in yolk of egg, kidney, liver and heart in a concentration of 100–200 μg/g dry weight.

BIOS I

Bios I has been identified as inositol (Eascott). Nine isomers are known: the most important in nature is *meso-inositol* which has the following formula:

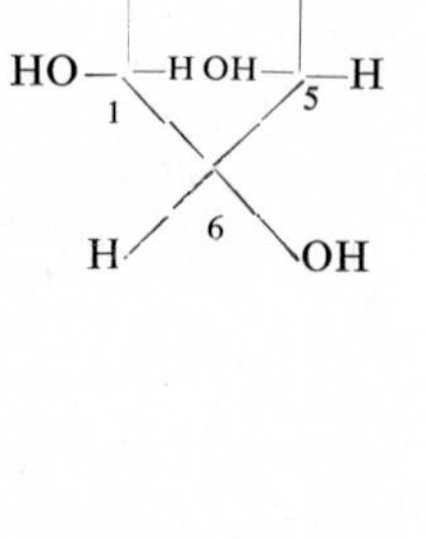

cis-1,2,3,5-*trans*-4,6-Cyclohexanehexol

Functions

(1) *At the molecular level.* Little is known as to the significance of this compound. Inositol could be a constituent of pancreatic amylase although this fact has been disputed. It has been isolated from heart (Scherer) and liver (Wooley, 1941).

(2) *In the whole animal.* The significance of inositol in human nutrition has not been established. It is known to be:

(a) a growth factor (Wildiers, 1901);

(b) a *lipotropic factor* (cf. p. 129); it forms *lipositols* — inositol-containing lipids — which are detoxication products; inositol prevents fatty degeneration of the liver; it shows anti-ketogenic properties;

(c) deficiency in mice leads to alopecia and failure of lactation;

(d) high concentrations are found in sperm of pig (Mann, 1951).

Requirements and sources

Man requires 500 mg per day. Inositol is found in plants (in the form of phytin or the hexaphosphoric ester of inositol) and in milk.

Bios II

It exists in an inactive form in combination with ovalbumin — *avidin*.

The naturally-occuring crystalline form of biotin is *biocytin*, a combination of biotin and lysine. Biotin is found in this form in cells and derives its name from this fact.

Functions

(1) *At the cellular level:*

(a) it is a growth factor: cancer cells are particularly rich in biotin; deficiency reduces the incorporation of ^{14}C-methionine to tissue proteins by 30 per cent (Dakshinamutri and Mistry, 1963);

(b) it activates numerous enzyme systems including succinate dehydrogenase;

(c) it functions in the fixation of carbon dioxide (Lynen) and in the biosynthesis of fatty acids (Wakil)

(2) *In the organism:*

If rats are fed on a diet in which the only source of protein is egg white, growth is retarded and there is loss of weight (Boas, 1924). Nausea, cutaneous lesions and modification of blood composition have been described during deficiency. Administration of biotin results in recovery.

Sources and requirements

Good sources of this vitamin are yolk of egg (Kögl and Tönnis, 1936), kidney, liver (György, Kuhn and Lederer, 1939), and tomatoes. The estimated daily requirement in man is 20 μg.

LIPOIC ACID

The vitamin action of this compound in higher animals has not been shown by experimental deficiency although it is important at the molecular level.

It is particularly important in the oxidative decarboxylation of pyruvate (cf. vitamin B_1); this reaction requires not only vitamin B_1 but another cofactor which is lipoic acid. For this reason lipoic acid is grouped with the vitamins B although, in the liver, it is associated with lipids and is lipid soluble.

$$CH_2\!-\!CH_2\!-\!CH\!-\!(CH_2)_4\!-\!CO_2H \quad \underset{-2H}{\overset{+2H}{\rightleftharpoons}} \quad CH_2\!-\!CH_2\!-\!CH\!-\!(CH_2)_4\!-\!CO_2H \quad (47)$$

$$\underset{SH\qquad\quad SH}{|\qquad\qquad\quad |} \qquad\qquad \underset{S\rule{1cm}{0.4pt}S}{|\qquad\qquad\quad |}$$

Reduced form   Oxidized form (disulphide bridge)

Functions

Lipoic acid is an essential factor for the nutrition of protozoa of the genus *Tetrahymena*. One of the main functions is in the decarboxylation of α-ketoglutarate to form "active succinate" (succinyl coenzyme A), which is the precursor of *porphobilinogen*, the first pyrrole synthesized in construction of the porphyrin nucleus utilized in hemoglobins, myoglobins, cytochromes and catalases.

Remark: we have seen in this chapter that each vitamin acts, in man for example, in different doses which are sometimes relatively large. As a result, the term "vitamin doses" to describe extremely small doses should be removed from scientific terminology.

CHAPTER 5

DIETARY BALANCE AND IMBALANCE

DURING starvation, energy is derived from the tissues of the animal; this is called *autophagia*. If no solid food materials (exogenous substrates) are available man can survive for several days provided that he obtains water (e.g. fakirs). Water allows the utilization of endogenous substrates. The *loss of weight* which accompanies prolonged dieting does not affect the essential organs: brain, heart, diaphragm; however, skeletal muscle loses up to 31 per cent of the wet weight, blood loses 27 per cent and adipose tissue 97 per cent. This loss in weight leads to the state of inanition.

During the initial period of metabolic exhaustion, *sugars* are utilized principally, followed by fats and finally proteins. In this last stage the animal is in a *critical condition* manifested by *starvation oedema* which was observed extensively after the 1939–1945 war in deportation victims.

It is interesting to note that, during the period of weight loss, the organism shows a *decreased ability to metabolize sugars*, such that when, for example, glucose is administered, the total glucose content of the blood increases *(hyperglycemia)* and there is *glycosuria*. This condition is described as *starvation diabetes*.

Examples of dietary imbalance

Inapparent deficiences associated with an *unbalanced diet* are manifested by *dysvitaminosis* principally, which, after a period of time, develops into typical *avitaminosis*. These disorders can be prevented by a varied but not excessive diet; in fact, "to eat well" is thought by many to mean "to eat large quantities".

In a diet, *excess lipids* produce fats; *excess sugars* produce liver and muscle glycogen and increased amounts form fats; finally, *excess proteins* are completely utilized with the elimination of nitrogen and storage of an equivalent amount of sugars and lipids. Therefore, contrary to what is commonly believed, meat can also cause one to put on weight.

The organism utilizes proteins immediately. This corresponds to the *"law of nitrogen balance"* which shows that *proteins cannot be stored* and as a result, in the raising of animals for meat production the important factor is the breed and not the diet. This provides another example of the relationship between *genetics* and *nutrition*, the gene–enzyme (or gene–protein) system (cf. p. 13).

Carnivorous chickens

La Mettrie, in 1747, stated that "raw meat made animals ferocious" and Diderot in his *Eléments de Physiologie* reported that "a man fed continously on meat would develop the characteristics of a carnivorous animal". Chickens normally eat grain, insects, bread, meat, etc. However, about fifty years ago F. Houssay fed chickens with meat alone through many generations. The line rapidly decreased in numbers but those which survived showed important morphological changes:

the intestine decreased in length,
the crop and gizzard were reduced,
the beak was curved and rapacions as in parrots,
the claws were longer and sharper,
the birds were more agressive.

It is essential, therefore, to provide a balanced diet and, as an example, a standard diet for man is given in Table 25.

TABLE 25. *Standard diet for adult man (Desgrez and Bierry)*

		in g		
		proteins	lipids	sugars
Substances of animal origin	milk (300 g)	10	10	15
	egg (55 g)	7	10	0
	meat, fish (120 g)	20	10	0
	animal fat, butter	0	40	0
	cheese (60 g)	15	15	0
	Total	52	75	15
Substances of plant origin	bread (240 g)	15	0	120
	potatoes (300 g)	7	0	60
	cereals (flour, rice biscuits, cakes (250 g))	7	0	120
	sugar (60 g)	0	0	60
	fresh vegetables, fruits	7	0	40
	Total	36	0	400
		88	75	415

For intensive physical work the proportion of *sugars* should be increased.

The concept of a balanced diet must take account of the type of animal being considered; for example, in a colony of bees the development of the queen is dependent on food (pantothenic acid is an essential factor, Weaver, 1956).

Cow during lactation

Firstly, a *unit of fodder* is defined classically in the following way: *if a kilogram of barley is taken as one unit, the nutritive value of food is expressed as the quantity of barley which produces the same amount of energy as one kilogram of this food* (Table 26).

TABLE 26

Food	Fodder unit
barley	1
oats	0·9
maize	0·95–1·1
beet	0·1

The *fodder equivalent* of a food is expressed, therefore, in the following way:

$$\text{F. Eq.} = \frac{1}{\text{number of fodder units}}$$

Examples:

$$\text{Hay} = \frac{1}{0\cdot4} = 2\cdot5$$

$$\text{Oats} = \frac{1}{0\cdot9} = 1\cdot11$$

$$\text{Beet} = \frac{1}{0\cdot1} = 10$$

Therefore, 2·5 kg hay or 1·11 kg oats or 10 kg beet is equivalent to 1 kg barley.

This is the *Danish method* (Fjord) of determining the nutritive value of food for animals. The results are *approximate* as the method does not take account of the *ability of the animal to utilize the foodstuff or the quality of the food.* For example, lipid deficiency is produced in animals by a diet containing a large proportion of beet while oats and oilcake are rich in fats.

A cow of 600 kg producing 15 kg milk per day during spring and 12 kg per day during autumn consumes 60 kg grass (or an amount equivalent approximately to one tenth the body weight). However, "grass" is an imprecise term. The nutritive value of grass varies throughout the year and, in particular, the consumption of dry grass in summer lowers milk production and, to prevent fluctuations in lactogenesis, it is necessary to supplement the food obtained by grazing. It is recommended that an added ration of 0·9 F.U. containing 100 g protein per F.U. be given for each additional litre of milk above the normal volume produced.

Dietetics is related to the species of the animal and therefore, to genetics and, in the production of meat, to specific productivity (cf. p. 11), the specific biosynthesis of proteins (cf. p. 135). "The pig is able to utilize the energy in food more efficiently than other animals: after 4 months gestation and 6 months rearing one obtains a litter of 8 pigs, each weighing 100 kg and a net production of approximately 600 kg meat in 10 months; however, during intensive rearing of beef cattle, it requires 9 months for gestation and 18 months rearing or a total of 27 months to produce an animal weighing 570 kg and giving 325 kg meat" (Craplet, 1955).

As physiological problems have a bearing on economic problems, it is appropriate to consider at this stage the acceleration of growth by certain antibiotics whose auxotropic properties have been demonstrated and utilized in domestic animals, notably pigs and chickens. Aureomycin (obtained from *Streptomyces aureofaciens*), penicillin and streptomycin exhibit these properties which were discovered by Stokstad and his co-workers (1949). For example, the addition of 11 mg penicillin per kg food to the diet of pigs accelerates growth and as a result, the calculations of Craplet could have greater significance although the mechanisms underlying this artificial acceleration of growth remain unknown.

The possibility that the relative increase in size in human young born since the advent of antibiotics is a result of the use of these drugs cannot be excluded.

The points made on p. 25 show that a small animal must metabolize more rapidly and here we find further evidence of the law of sizes and its evolutionary consequences. D.M.S. Watson (1950) wrote: "An animal requires as a maintenance diet a weight of food proportional to $W^{2/3}$ of the body weight W. When one takes into account the work done, the quantity of food required varies with the weight of the animal. Thus, the total amount of food required consists of two parts: one varying as the power $2/3$, the other as the weight of the animal.

"From this it can be seen that for two animals of the same species but of different weights the heavier animal is the more economic since, in the transformation of food into work, the efficiency is greater. The present day horse, two hundred times heavier than *Eohippus*, is more efficient. This fact must be considered in the interpretation of the constant increase in size in numerous groups, particularly in herbivorous mammals living in open country.

"Thus, it is an advantage to become large. However, there are certain problems, one of which is the strength of the skeletal structures. The alimentary system is completely dependent on the grinding surface of teeth and this determines the life of the animal. Furthermore, while the grinding surface would increase in square proportion, the corresponding body weight would increase in cubic proportion. The increase in dental surface is inadequate and will not allow work.

"Let us suppose that an animal doubles in length, the weight increases eight times but the dental surface only increases four times. It will be necessary for the animal to feed for twice as long to meet its requirements. Thus, an elephant can work for only four hours each day, the remaining time being utilized in feeding. As the hours of use of teeth increase, the rate of wear increases.

"In the South African veld, the horses do not generally live longer than ten years. At this age the teeth are worn down to the bone. For the teeth to last the normal life span, the problem is that of continuous elongation of the crown as the weight of the animal increases.

"In an animal feeding on leaves, the bunodont dentition allows complete utilization of food but grasses must be sheared and in this case, the important factor is the total height of the enamel crests, the height having increased considerably during evolution of the horse" (in *Paléontologie et Transformisme*, Albin Michel, Paris).

CHAPTER 6

DIGESTION AND ABSORPTION

Digestion

Nutrition consists of degradation of food materials and absorption of the constituents by the organism for incorporation into the basic structure or into energy producing cycles. Degradation of food requires the presence of analytical processes. These are both *mechanical* and *chemical* in nature. The first studies of digestion were carried out by Réaumur (on birds) and by Spallanzani (1780) whose remarkable experiments which are reported in his book *Sur la digestion* established our knowledge of digestion by demonstrating the more general action of gastric juices compared with that of trituration (he studied the products of digestion by artificially inducing vomiting in himself; he made the first attempts at artificial digestion, etc.).

To restrict the extent of this study, essentially I shall consider the morphology and physiological processes of the human digestive tract and, briefly, *rumination.*

The digestive tract consists of:

(a) *organs performing mainly mechanical work*;

(b) *glandular organs* (some of which form an integral part of mechanical organs) which are involved in *chemical processes.*

Chemical digestion becomes more important as one passes from buccal digestion to gastric digestion and then to intestinal digestion (numerous lytic enzymes) while mechanical digestion is of importance in the buccal cavity; then there is a *balance between mechanical and chemical processes.*

Buccal digestion

I. Mastication

Mastication is the essential mechanism in the buccal cavity; food is cut, torn, crushed by the teeth through the movements of the tongue and jaws. This is a voluntary and also reflex phenomenon (sensitive nerves pass to the bulbar masticatory centre which stimulates motor nerves, notably the trigeminal).

Bilateral section of the fifth pair of cranial nerves (trigeminal) in rabbit results in paralysis of the buccal cavity allowing the lower jaw to droop and the animal dies of starvation as it can no longer swallow or masticate food (Claude Bernard, 1858).

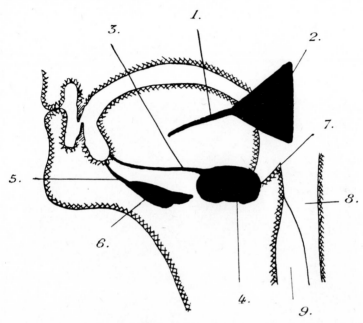

FIG. 12. Schematic representation of the relative position of salivary glands in man. *1*, duct of Stenon; *2*, parotid; *3*, duct of Wharton; *4*, submaxillary; *5*, duct of Bartholin; *6*, sublingual; *7*, epiglottis; *8*, oesophagus; *9*, larynx.

II. Salivation

Salivation occurs under the influence of peripheral and central stimuli: presence of food in the mouth, notably during mastication; stimuli originating in the stomach (increased salivation before vomiting); sight, touch and smell (cf. Conditioned reflexes, Volume II) and even psychological factors such as remembering certain foods.

Salivary secretion is discontinued as the result of adrenaline release during highly emotional states.

A total of 1000 to 1500 ml saliva is excreted each day by man. Cattle secrete 56 kg saliva per day (Colin, 1871) but, although the sublingual gland continuously produces saliva, the submaxillary glands are active during ingestion and not during rumination. Salivation varies quantitatively and qualitatively with the substance present in the mouth. Thus, fresh bread stimulates the production of more saliva from the parotid glands than dry bread or dry, powdered meat (School of Pavlov: Glinski, 1895; Wulfson, 1898); substances which have little flavour induce less abundant salivation, especially if they are insoluble in water, as, for example, $CaCO_3$ (experiments on dogs by Malloizel, 1905).

Innervation of the salivary glands

This is a problem which at the functional level has not yet been resolved satisfactorily.

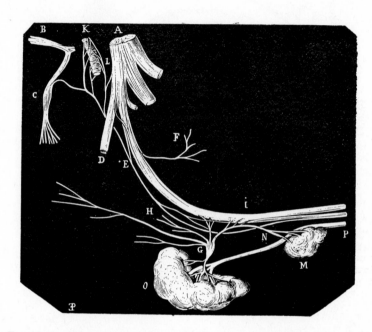

FIG. 13. Location of the chorda tympani in man. A, trunk of fifth pair; B, facial nerve giving rise at the bend to the petrosal nerves and from the ascending part to a gangliform area C, from which arises the chorda tympani; D, cut dentary nerve; E, chorda tympani separated by dissection from the lingual nerve; at E one sees a branch of the lingual nerve which appears to return at H; another part of the chorda tympani passes to the submaxillary ganglion G, while one of the branches follows the lingual nerve; F, buccal branch of lingual nerve; G, submaxillary ganglion receiving a branch of the chorda tympani and sending branches backward which are distributed to various parts of the buccal membrane, probably to the glands contained there; I was able to follow one very long branch to a glandular mass of the covering of the palate and pharynx; N, artery passing to sublingual gland M, I, lingual nerve; K, median meningeal artery surrounded by sympathetic branches connecting with the chorda tympani; L, branch connecting the inferior maxillary nerve with the chorda tympani; M, sublingual gland; N, nerve branch from the submaxillary ganglion to the sublingual gland; O, submaxillary gland; P, duct of submaxillary gland. (CLAUDE BERNARD, 1858.)

(a) SUBMAXILLARY AND SUBLINGUAL INNERVATION. Ludwig (1851) found that stimulation of the peripheral end of the lingual nerve induced secretion by the submaxillary gland.

Claude Bernard showed that this phenomenon was dependent on the *chorda tympani* which are derived from the facial nerves (VIIth cranial nerves) which also give rise to the lingual nerve; from this fact a historical controversy developed. Bernard noted that this excitation of the lingual was associated with vasodilation and thought that secretion was the result of increased blood circulation in the gland. Heidenhain, using atropine (the

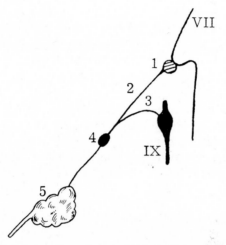

FIG. 14. *1*, geniculate ganglion; *2*, superficial small petrosal nerve; *3*, deep small petrosal nerve; *4*, otic ganglion; *5*, parotid gland; *VII*, facial; *IX*, glossopharyngeal.

parasympatholytic alkaloid from *Belladona* which inhibits the effects of stimulating the pneumogastric nerves), showed that with low doses of this inhibitor there was vasodilation without salivation after stimulation of the lingual nerve. Barbiturates also inhibit salivary secretion by the submaxillary gland (Guimarais *et al.*, 1955); these drugs act either on the ganglionic synapses or directly on the secretory cells. In addition, all parasympathomimetic compounds have a sialagogic action; for example, pilocarpine. It should be noted that sympathetic stimulation provokes, in addition to *vasoconstriction*, secretion of saliva which, in dog at least, is more viscous than that produced by excitation of the chorda tympani.

(b) PAROTID INNERVATION. The nerves to these glands are branches of the facial and glossopharyngeal nerves (IXth cranial nerves) via the otic ganglion.

Salivation

During activity the salivary glands produce electrical potentials which can be detected by external electrodes or intra-cellular microelectrodes (Bayliss and Bradford, 1886; Lundberg, 1955). Thus *biopotentials are not produced only by nerves and muscles.* Any isolated cell possesses a membrane potential and in certain eggs (for example, of the starfish) this potential changes at the moment of fertilization.

The function of saliva is essentially to moisten *food materials*, and in this way to facilitate their degradation by gastric enzymes. However, saliva from the posterior salivary glands of octopod cephalopods is proteolytic, rich in tyrosine and monophenolic amines (5-hydroxytryptamine, tyramine, etc.) and is toxic when injected to crab; when added to the perfusion fluid, it slows the rate of perfusion and induces the secretion of a clear fluid from the excretory duct (Bacq and Ghiretti, 1951, 1953).

The saliva of mammals contains a powerful stimulant of salivary secretion; it is active particularly when injected to the artery supplying the sub-maxillary or parotid salivary gland (Demoor, 1913; Guimarais, 1930). It is possible that there is a cyclic system whereby saliva activates salivation. Kerr (1961) stimulated the tongue by a mixture of citric acid and sugar and obtained intense reflex salivation which could be repeated in man every five minutes for three hours without apparent fatigue.

Composition of saliva

The pH of salivary secretions varies between 6·7 and 7·4; saliva contains notably vitamin C (approximately 0·5 per cent). There are differences in the composition of saliva from the three glands:

(1) *Parotid.* The saliva is clear, containing no mucin; in contact with air, $CaCO_3$ is precipitated; the presence of glucose in a concentration equivalent to 3–4 per cent of that in plasma has been shown in saliva from dog (L. L. Langley *et al.*, 1963).

(2) *Submaxillary.* The saliva is viscous, being rich in mucin and is rich also in acid phosphatase (Eggers–Lura, 1947).

(3) *Sublingual.* The saliva contains more mucin and is therefore more viscous and is alkaline.

In man saliva produced from each gland contains an enzyme, *ptyalin.* Ptyalin catalyses a reaction which would occur spontaneously although very slowly.

Chemical digestion in the buccal cavity

The importance of buccal digestion should not be overestimated. Ptyalin is an amylase (hydrolysis of starch to give dextrins and finally maltose, a reaction which can be followed by the colours given with iodine) but:

(1) removal of the salivary glands does not produce digestive disorders provided that abundant water is consumed.

(2) Cl ions are necessary for the activation of ptyalin but this enzyme is inactivated at pH ≤ 4, such that, when the food bolus reaches the stomach, ptyalin ceases to act. In fact, it is the *intestinal amylases* which are of fundamental importance in the digestion of starch. Moreover, the saliva of ruminants does not contain ptyalin and the presence of ptyalin in man appears to be a tentative evolutionary development.

However, human saliva contains other enzymes; in addition to acid and alkaline phosphatase, a lipase, proteinase, sulphatase and peroxidase are present. The ability of the buccal cavity to afford protection from bacterial infection may be derived, in part, from the presence in saliva of lysozyme and also "inhibines" (Dold, 1943) which digest many invading micro-organisms.

III. Deglutination

Deglutination of the food bolus is completed with the ingestion of the food by the oesophagus. It is a purely mechanical action.

In mammals deglutination takes place in three stages (Magendie):

(1) Swallowing is begun in the buccal cavity by the voluntary movement of the tongue against the roof of the palate which pushes the bolus towards the pharynx.

(2) Passage through the pharynx is rapid and automatic: the soft palate is pushed upward to block the nasal fossae. At the same time the Eustachian tube opens, the larynx is moved forwards (movement of the Adam's apple) and the epiglottis closes the larynx; thus for a short time, ventilation ceases.

(3) The oesophageal phase is automatic and characterized by contraction of the oesophagus; this ceases after vagotomy (Claude Bernard, 1857); with a *solid bolus* these contractions are rapid in the upper region of the oesophagus (presence of striated muscle) and slow in the two lower regions (presence of smooth muscle). However, with *fluids* the oesophagus is inactive (Arloing) and fluids are literally injected into the oesophagus by the pharyngeal movements, due to the rapidity of the pharyngeal phase.

The smooth muscle in the lower region of the oesophagus—the cardiac region—is associated with nervously controlled *expansion and contraction of the cardia.*

Two sounds are produced during deglutination:

a sound produced by movements during the pharyngeal phase (Meltzer);

a sound corresponding to the passage of the bolus to the dilated cardiac zone.

In seed-eating birds, the absence of the palatal septum and the isthmus of the gullet, coinciding with free connection between the anterior and posterior buccal cavity, leads to simplification of the act of deglutination which is accomplished in two stages, a buccopharyngeal stage and an oesophageal stage (S. Arloing, 1877).

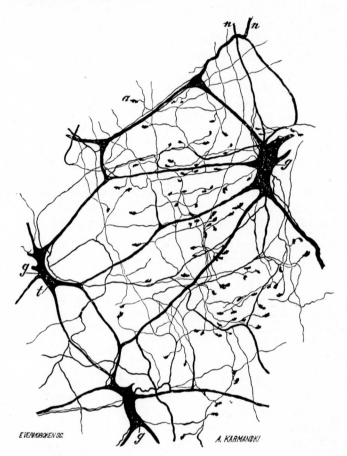

FIG. 15. Nerve plexus of the oesophagus of rabbit shown by gold staining method. *n n*, afferent nerve fibres; *g*, nerve ganglion; *t*, myelinated nerve running along the ganglion without penetrating it; *a*, terminal branching. (RANVIER, 1878.)

Nervous control of deglutination

Pharyngo-oesophageal deglutination is a reflex phenomenon, controlled by mechanical stimulation; swallowing is not possible when the mouth is empty and saliva must be present at least. *The reflex is abolished by cocaine anaesthesia* of the pharynx. The principal nerves involved are:
   the *superior laryngeal*;
   the *superior maxillary* branch of the trigeminal (sensory nerve from the

ganglion of Gasser as is the ophthalmic nerve; the inferior maxillary branch is sensory and motor).

The *central control* of this reflex originates in the *bulboprotuberalis region*. This centre stimulates the sequence of movements which constitutes deglutination through the motor nuclei of the following nerves:

hypoglossal, XIIth cranial nerves to the tongue;

trigeminal, Vth cranial nerves, the inferior maxillary branch of which functions in mastication and in salivation;

facial, VIIth cranial nerves to the submaxillary and sublingual salivary glands;

glossopharyngeal, IXth pair which functions in pharyngeal movements;

pneumogastric, Xth pair, which innervates the soft palate. (Classification of the cranial nerves by numbers according to Soemmering and Vicq–D'Azyr includes, in addition to those given above, I, olfactory (tract); II, optic (tract); III, common oculomotor; IV, trochlear; VI, external oculomotor; VIII, acoustic; XI, spinal).

The regulation of deglutination by the reflex centre has been compared to the playing of a keyboard by a pianist and from this analogy the term *"keyboard theory"* has been used to describe the processes of deglutination. Five nerves take part in these processes; it is a Chinese scale.

Gastric digestion

This is a *dispensable stage* of digestion as man can survive gastrectomy without serious effects provided that Castle's factors are administered. It may be mentioned that the development of the stomach in tadpoles of Anura does not reach the stage of production of pepsinogen and hydrochloric acid (Barrington, 1946). The stomach is essentially a silo. However, under normal conditions gastric digestion has some importance.

I. Chemical digestion

Gastric digestion is dependent on three types of stimuli (Ivy):

(1) *Cephalic*. Stimuli associated with inborn or conditioned reflexes (sight, smell, etc.) are transmitted by the pneumogastric nerves. Stimulation of the reticular formation (cf. Volume II) in the region of the pons of Varole does not excite gastric secretion of acid during fasting but does during feeding (Bakuradze, 1962); moreover, electrical stimulation of the hypothalamus and marginal cortex also produces marked secretion of acid (Bakuradze; E. C. Hoff *et al.*, 1962). Section (P. G. Bogarch *et al.*, 1962) or local inhibition of the vagi by an anaesthetic such as xylocaine (Bremer and Dubois, 1963) inhibits secretion. In the frog, stimulation of the pneumogastric nerves induces secretion by the oesophagus and stomach

7

but secretion of alkaline mucus by the oesophagus predominates such that any acidification is masked (Contejean, 1892); at the end of digestion, when the prey is reduced to chyme, the pH is acid. In poikilotherms, such as the frog, the contractile elements of the stomach of adults are inactive during winter and the gastric contents are alkaline (pH 9).

(2) *Gastric.* Intragastric (Frouin, 1907) or intravenous administration of saliva (Guimarais and Tavarnes, 1942) stimulates gastric secretion (histamine?); thus, in addition to its lubricating action, saliva possesses hormonal rather than enzymic activity. Polypeptides (of meat and eggs) and ethanol (C.E. Elwin, 1962) are powerful stimulants of gastric secretion and they act by similar pathways to hormonal stimuli. In addition, distension of the gastric chamber stimulates gastric secretion (Duval and Price, 1960).

Gastrin (Edkins, 1905) produced by the gastric glands, is transported by the blood circulation back to the stomach where it stimulates glands in the fundus. This shows that an *endocrine "shunt" does not exist;* a hormone is transported by normal pathways even if it is produced by the organ on which it acts. This is a true reflex and we shall see also (Volume II) that in neuromuscular proprioceptive reflexes, for example, there are no shunt mechanisms. *Histamine* (produced by the decarboxylation of histidine) also stimulates gastric secretion; gastrin has been identified with histamine but Erspamer has isolated serotonin from the gastric mucosa and Blair, Brahma Dutt and Harper (1962) were unable to detect histamine in the venous blood from the stomach either in the presence or absence of stimulation by gastrin. Gregory and Tracy (1964–66) have described the

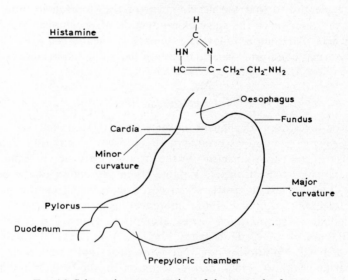

FIG. 16. Schematic representation of the stomach of man.

isolation from antral mucosa (hog and human) of two heptadecapeptide amides: gastrins I and II which were synthesized.

(3) *Intestinal. Enterogastrone*, secreted by the walls of the duodenum under the influence of fatty acids, hormonally inhibits secretion. However, the anterior region of the duodenum participates in stimulation of gastric secretion under the influence of fatty acids and soaps. Anderson (1960) showed that hydrochloric acid, placed in the duodenum, inhibited gastric secretion in dog.

The gastric juice contains 97–99 per cent water. Man secretes 1–1·5 l. gastric juice in 24 h which is equivalent to the volume of saliva. W. Beaumont (1833) suggested and Pavlov demonstrated that there is a relationship between the volume of gastric juice produced and the quantity of food ingested.

Two types of cells participate in the production of gastric secretions.

(1) Cells on the inner surface of the stomach secrete *mucin* lacking in enzyme activity. The function of mucin is believed to be the protection of the mucosa from proteolytic digestion. However, one can observe *directly* the action of mucin cells on solid food. With reference to this I wish to point out that:

The important aspect of the physiology of hollow organs is the nature of the internal physiological processes; however, until recently this aspect of physiology could not be observed directly. Since 1956, I have used techniques which allow one to study these processes for several hours or days depending on the organ. In these techniques the hollow organ is *turned inside out*. This method might well be applied to the heart or uterus. With reference to the stomach, if the organ is removed from the selachian *Scyllium canicula*, it can be everted in a way similar to the finger of a glove. By placing the tissue in a physiological medium developed for isolated tissues of selachians and filling with air the stomach, which is ligatured in the pyloric region, one obtains a kind of floating balloon which *undergoes contractions for many hours* at room temperature (Rybak, 1957). Under a binocular lens one can follow both chemical and mechanical processes, since, if the gastric mucosa is lifted away by cutting tangentially with scissors, the smooth muscle is displayed. Fragments of muscle, for example from the crab, placed in the prepyloric region *rapidly become coated with mucus* but are not digested; generally, it would appear that the mucus covering around food material facilitates their way inside the stomach and thus favours their digestion by pepsin in the presence of hydrochloric acid.

(2) The other cells which produce gastric juice are arranged in simple, branched tubes. *Only those cells in the fundic region of the stomach secrete the essential factors for chemical digestion*, hydrochloric acid and pepsinogen.

7*

Cells which line the lumen of the glandular tubules produce *pepsinogen* (precursor of pepsin which is stored, therefore, in an inactive form in the same way as most factors involved in blood coagulation).

Cells which occupy a peripheral position between the cells producing pepsinogen and the basement membrane, produce HCl.

In order to obtain samples of gastric juice one can: (1) induce the swallowing of sponges which are retrieved when soaked in gastric secretions (Spallanzani); (2) construct gastric *fistulae*; (3) suture the stomach provided with internal fistulae (Heidenhain), preserving the nervous connections ("little stomach" of Pavlov). In the prepyloric chamber, glandular cells produce an alkaline secretion which has no enzymic activity and thus functions in preparation for duodenal digestion by adjusting the pH of gastric contents.

HCl

HCl is produced from NaCl and carbonic acid:

$$H_2O + CO_2 \xrightleftharpoons[\text{anhydrase}]{\text{carbonic}} H_2CO_3 \rightleftharpoons H^+ + HCO_3^- \qquad (48)$$

$$H^+ + HCO_3^- + Na + Cl^- \longrightarrow HCl \quad + \quad NaHCO_3 \qquad (49)$$

secreted in gastric juice (pH = 1·5)	reabsorbed in the blood and able to give alkaline urine

As the mucosal surface of the stomach is negative with respect to the seral coat, a potential difference develops in the gastric epithelium (of the order of 50 mV); this potential is derived from the active transport of Cl^- (Ussing and Zerahn, 1951; Hogben, 1955). Histamine increases the production of HCl by the gastric mucosa and increases oxygen consumption such that the ratio H^+ secreted/O_2 consumed = 11 (Davies, 1957).

Carbonic anhydrase is localized in the peripheral cells (Davenport and Fischer, 1938).

Pepsin

The principal chemical function of the stomach is in the partial digestion of proteins. The inactive zymogen—*pepsinogen* of molecular weight 42,500—is converted to pepsin by the action of HCl and also by an *autocatalytic* process in which a small amount of pepsin activates the conversion of pepsinogen to pepsin. In this conversion inhibitory polypeptides are removed, unmasking the active enzyme. Pepsin which can be activated also by small amounts of ethanol, transforms proteins to *proteoses* and *peptones*. Pepsin is inactivated at pH $\geqslant$ 5 by recombination with inhibitory polypeptides. Two "para-pepsins" can be extracted from pepsin (Ryle and Porter, 1959), one of which—para-pepsin I—may be the gelatinase of Northrop (1932).

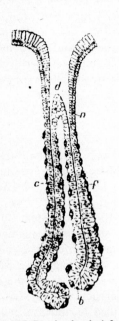

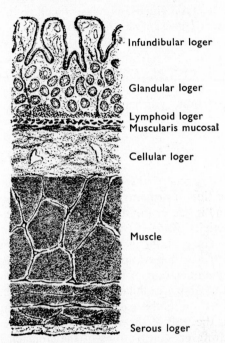

Infundibular loger

Glandular loger

Lymphoid loger
Muscularis mucosal

Cellular loger

Muscle

Serous loger

FIG. 17. Peptic glands (after KLEIN, taken from HÉDON, 1950). b, cul-de-sac; d, excretory duct; n, neck; c, chief cells; p, parietal cells.

FIG. 18. Stomach (pylorus). (DUBREUIL and BAUDRIMONT, 1959.)

Rennin

This enzyme is *secreted by cells in the fundus*; the optimum pH for its action is pH 4 (Fish, 1957).

Rennin coagulates milk such that two fractions separate: one consists of non-coagulated proteins and the other of casein which is precipitated in the form of calcium caseinate. This caseinate is attacked by pepsin. Rennin is very important in digestion in young infants.

Lipase

Gastric lipase generally has little activity and is probably inactive in adult man. The optimum pH is about 5. However, lipase may act in the infant and human milk contains a compound which activates this enzyme.

Other enzymes

The stomach of insectivores is able to produce enzymes which split chitin: *chitinase* which degrades chitin to chitobiose and *chitobiase* or N-acetyl-β-glucosaminidase which splits chitobiose into two molecules of acetylglucosamine. Jeuniaux (1961) has demonstrated the presence of chitinase principally, in a lizard *(Lacerta viridis)*, in a fish *(Carassius auratus)*, in a frog, salamander, sparrow, blackbird, cockerel and in mammals (bat, pig and two insectivores: mole and hedgehog). In contrast, the sloth which is phytophagus and without teeth and other herbivorous vertebrates (such as the tortoise, *Testudo hofmanii*) are unable to synthesize chitinase.

II. Mechanical digestion

Essentially, the food is mixed with mucin and pepsin allowing the digestion of proteins. At rest, there is a potential difference between outside and inside of the stomach which was described by Donné in 1834. During activity of the stomach this potential difference changes.

The motility of the stomach does not appear to be dependent on the extrinsic innervation; although movements of the stomach are augmented by stimulation of the peripheral end of the pneumogastric nerve and inhibited by stimulation of the splanchnic nerve, sectioning these nerves does not inhibit gastric contractions. This is due probably to the presence in the wall of the stomach of nerve plexi of Auerbach and of Meissner. However, contractions of the stomach are modified by intrinsic reflexes initiated mainly in the pyloric region.

When the stomach is empty *the fundus* undergoes rhythmic movements which consist of slow, tonic contractions of low amplitude. Hunger contractions originate in the region of the cardia and participate in the sensation of hunger (cf. Volume II).

Mechanical stimulation of the stomach of cat by increasing the pressure within the lumen modifies the electrical activity of the ventral thalamic nuclei (Varbanov and Papasova, 1962), indicating the existence of a nervous reflex which regulates gastric activity.

When the stomach is full, the hunger contractions cease and the tone of the fundus decreases such that there is little change in the intragastric pressure. Fluids flow along the length of the lesser curvature.

It should be noted that moderate distension of the stomach stimulates contractions while excessive distension inhibits contraction; this appears to be a characteristic of all organic systems showing intrinsic contractility (cf. p. 243), with the exception of hunger contractions (Rybak and Thépaut).

There is some confusion as to the general pharmacology and psychophysiology of contractions of the stomach. Coffee and sodium bicarbonate are stimulants. The role of fear and anxiety has not been defined; there can be either an increase or decrease in the contractions. Physical exercise augments contractions of the stomach. Pain, fever and avitaminosis B_1 inhibit. Morphine is *generally* a stimulant.

It is generally considered that there is a balance between *the stimulatory action of the parasympathetic system and the inhibitory action of the sympathetic system* on the musculature of the stomach and intestine. But atropine, a parasympatholytic and therefore *theoretically inhibitory*, generally stimulates gastric movements. However, tobacco and drugs which prevent sleep ("maxiton", etc.) inhibit contractions of the stomach.

In fact, in the case of the stomach, the important factor would appear to be the tone of the musculature, because, under certain conditions, the effects of nervous stimulation can be reversed and this would tend to explain the paradoxical action of atropine.

Classically, it was believed that gastric antiperistalsis played a fundamental role in vomiting. In fact, in this case the stomach is relaxed and the stomach contents are evacuated by contractions of the abdominal walls; catechol amines can induce vomiting (R. Cahen, 1962).

Normal gastric evacuation

A meal passes through the stomach of man in 2–5 h, evacuation commencing 5–20 mn after the end of the meal.

The mechanisms of this process are not understood fully:

(1) It is generally considered that the *chemical sensitivity* of the pylorus may be a predominant factor. Acidity may lead to opening of the pyloric sphincter which then conditions contraction of the sphincter such that the contents of the stomach pass into the duodenum in fractions.

(2) More recently, it has been postulated that the pylorus is open to a certain extent throughout evacuation; however, this appears to me to be an inaccurate conception as the pylorus shows continuous contractions even in isolated stomach preparations. It would follow nevertheless from this theory that there must be a balance between gastric and duodenal pressures.

The duodenum would function in two ways:

(1) *By physical means:* when the duodenum is empty, the motility of the stomach is optimal and the stomach is emptied, filling the duodenum such that the internal pressure increases and preventing gastric contractions; when the duodenal contents are emptied the contractile processes in the stomach are resumed.

(2) *By chemical means:* enterogastrone, secreted by the duodenal walls under the influence of lipids or high concentration of sugars, inhibits gastric contractions such that, when the duodenum is empty and secretion of enterogastrone has ceased, the contractions of the stomach are resumed.

Taking into account the intrinsic contractions of the pylorus, it is easy to imagine that a compromise of the two theories may obtain in which emptying movements are dependent on neuroendocrine reflexes.

Gastric digestion in ruminants

The capacity of the typical *compound stomach* of ruminants can attain 300 l in Bovidae and the length of the small intestine can be 40 m (it should be remembered that a carnivorous diet is associated with a short intestine,

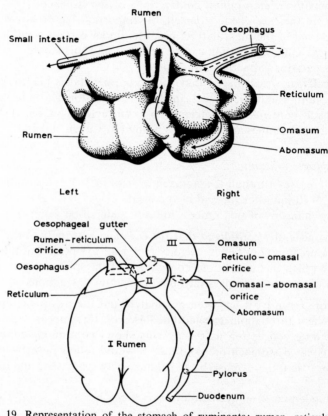

Fig. 19. Representation of the stomach of ruminants: rumen, reticulum, omasum and abomasum. (According to *Scientific American*, p. 198, 1958 and CRAPLET, 1955; Vigot).

cf. p. 78). Ruminants swallow their food without initial mastication. Food is mixed with water in the storage paunch and softened. Bacteria in the rumen attack the cellulose walls of the fodder; this intense activity of symbiotic bacteria allows ruminants to utilize hay and straw which are rich in cellulose and which cannot be attacked by the digestive secretions of carnivorous and omnivorous mammals; in this case, enzymes are borrowed from other organisms. The principal products of fermentation of sugars in the rumen are volatile fatty acids, half of which consist of acetic acid (Tappeiner, 1883). Another result of this bacterial symbiosis is that the daily quantity of milk secreted contains 10 times more riboflavin and twice as much pantothenic acid than does the diet (MacElroy and Goss). Therefore, there is intensive synthesis of vitamins by the symbiotic organisms in the rumen which is simply a fermentation chamber.

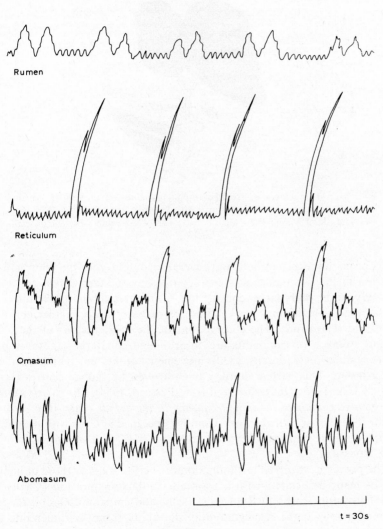

COW

Rumen

Reticulum

Omasum

Abomasum

t = 30 s

Fig. 20

A ruminant eats, or more exactly ingests, food for 6 h of each day and ruminates in a quiet place for 8–10 h; therefore, feeding activity continues for two thirds of the day.

The mechanographical studies of Le Bars, Nitescu and Simonnet, carried out with the aid of a permanent fistula in the rumen, have shown that the stomach undergoes characteristic periodic contractions which consist of an initial contraction of high amplitude followed by rapid movements of low amplitude (Fig. 20). The reticulum shows biphasic contrac-

FIG. 21. Arrangement of the hepato-pancreatic region of man. *I*, Liver; *II*, Gall bladder; *III*, Duodenum; *VI*, Pancreas; *V*, Duct of Wirsung; *VI*, Ampulla of Vater; *VII*, Hepatic duct; *VIII*, Cystic duct; *IX*, Ductus choledochus.

tions occurring every minute and several seconds *before* the contractions of the rumen. The omasum is characterized by the fact that it contracts strongly after each large contraction of the reticulum (Brunaud and Dussardier). The abomasum undergoes a series of fairly regular contractions and relaxations. The contractions of the four sections of the compound stomach of ruminants are co-ordinated. Bilateral vagotomy prevents rumination (Flourens, 1830): movements of the rumen and reticulum are arrested. This gastric motility is controlled by bulbar centres (Iggo, 1951; Clark, 1953; Dussardier and Albe-Fessard, 1954). *The gastric rhythm represents a rhythm of the same frequency in the bulbar centre* (Dussardier, 1958). The bulbar reticular formation is concerned notably with the motor origin of these contractions (Bell and Lawn, 1955; Dussardier, Flinois and Rousseau, 1960).

The processes involved in rumination are very similar to those of vomiting in man; associated with a transient thoracic aspiration (Chauveau and Toussaint; Bost and Ruckebusch), the abdominal muscles play a predominant role in the regurgitation of the bolus from the rumen into the oesophagus; peristaltic movements allow the passage of the bolus to the buccal cavity where it is dehydrated and the water removed is swallowed. The dehydrated bolus is masticated for several minutes and reintroduced to the rumen from where it passes to the reticulum, the omasum and finally into the abomasum; the food can be ruminated several times. Food which has been homogenized does not undergo rumination (Balch; Phillipson, 1961).

Food which has been masticated for the last time is conducted passively from the reticulum to the omasum along the oesophageal groove. In young calves fed on milk, the oesophageal groove is modified in such a way as to allow the passage of milk directly into the omasum and then into the abomasum.

The role of the reticulum is uncertain; it may deal with liquid foods. However, any foreign bodies eaten by ruminants, which sometimes could be injurious, remain in the reticulum which, therefore, may have a protective function. Distension of the abomasum inhibits contractions of the reticulum (Phillipson, 1939); in decerebrated preparations (cf. Volume II) distension of the reticulum stimulates contractions of the rumen and also reflex secretion of saliva by the parotid glands (Comline and Titchen, 1961).

The omasum crushes the ruminated food by compressing it between the laminae such that the water content of the food is reduced.

The abomasum, which is the part of the stomach of ruminants where chemical digestion occurs, contains gastric juice (pH 1·5–3·0) with a *low pepsin content*. However, this is not unfavourable as the diet of ruminants contains very little protein. The abomasum has a predominant role in young ruminants. Note that during the intra-uterine and suckling stages of life the ruminant receives only food of animal origin and later passes to a diet which is exclusively of plant origin.

Digestion in the small intestine

Acidic *chyme*, the product of gastric digestion, is introduced into the duodenum in stages, by the action of the pylorus. In man the small intestine measures 6–7 m and starting from the pyloric region, consists of the duodenum, the jejunum and the ileum.

Chemical processes

Pancreas

(A) *Stimulation of pancreatic secretion.* The secretion of digestive juices by the exocrine pancreas is controlled almost entirely by humoral agents produced in the *duodenum* and *jejunum* in the presence of residual acidity, lipids, proteins and sugars of the chyme. These hormones are carried in the blood stream *throughout the organism* but stimulate only the pancreas, liver and gall bladder.

The following point which concerns endocrinology in general, should be emphasized: the organ on which a hormone acts shows particular sensitivity to that hormone but this is not to say that the hormone acts exclusively on one organ; we shall see that hormones showing a specific action

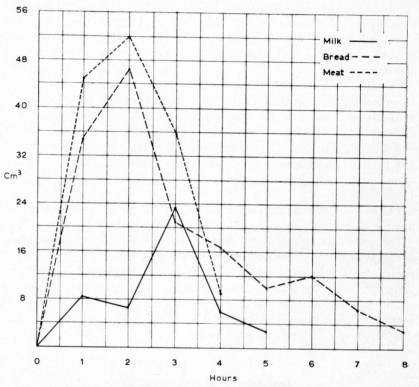

FIG. 22. Pattern of pancreatic secretion following a meal of meat, bread and milk. (Abscissae represent time in h; ordinates, the volume of secretion in cm³). (TERROINE, 1913.)

on one organ can act at different concentrations on other very different organs. These facts can be explained by considering that the hormone molecule can act on the same metabolic system in different organs.

The hormones of the duodenum which were collectively named *secretin* (Bayliss and Starling) are five in number:

(1) *secretin* which stimulates the production of dilute pancreatic juice with low activity;

(2) *pancreozymin* which stimulates the secretion of viscous and enzymically very active pancreatic juice;

(3) *hepatocrinin* which stimulates the *liver* but not the pancreas, leading to the production of dilute bile;

(4) *cholecystokinin* which induces contractions and emptying of the gall bladder;

(5) *enterocrinin* which stimulates the secretion of intestinal juice.

(B) *Properties and constituents of pancreatic secretion.* The pH of pancreatic secretion is on average 7·5–8; it is therefore an *alkaline* secretion.

Enzymes present:

trypsin   cholesterolesterase
chymotrypsin   ribonuclease
peptidase   desoxyribonuclease
amylase   collagenase
lipase

Considering these in sequence:

Trypsin and chymotrypsin are secreted as the inactive precursors which are activated in contact with the intestinal mucosa. *Enterokinase* from the intestinal gland catalyses the formation of trypsin from trypsinogen and a small amount of trypsin activates more trypsinogen (autocatalysis) and also chymotrypsinogen. Thus, *enterokinase does not activate chymotrypsinogen;* this is a function of trypsin. The trypsinogen and chymotrypsinogen (crystallized by Kunitz and Northrop (1935–1936) from beef pancreas) are not the only forms of these enzyme precursors (Laskowski, 1946); Desnuelle and his co-workers (1960) have shown that the structure of chymotrypsinogen can vary and, in trypsinogen from pig, the *N*-terminal residue is phenylalanine while that of beef trypsinogen is valine.

Desnuelle and his co-workers (1957) have studied in detail the activation of chymotrypsinogen. Their results show that depending on the quantity of trypsin there is slow or fast activation in the manner shown below.

"rapidly" activated chymotrypsin
↗
+large amounts of trypsin
↗
chymotrypsinogen
↘
+chymotrypsin
↘
neochymotrypsinogen $\xrightarrow[\text{of trypsin}]{\text{+small quantites}}$ "slowly" activated chymotrypsin

Neochymotrypsinogen is produced from chymotrypsinogen by the hydrolysis of three specific peptide bonds (tyrosyl-threonine, leucyl-serine, asparaginyl-alanine).

The submaxillary glands of canids and felids contain an inhibitor of trypsin and chymotrypsin (Trautschold, Werle and Sebening, 1963).

These enzymes exhibit proteolytic activity and hydrolyse peptones and proteoses to polypeptides. These polypeptides are hydrolysed finally by the action of three enzymes: *carboxypeptidase* (a *pancreatic* enzyme containing zinc); *aminopeptidase* and *dipeptidase* from the *intestine*. Collectively these enzymes were called *erepsin*.

As their name indicates, the carboxy- and aminopeptidases attack polypeptides at the carboxylic and amino groups. The proteins and peptones are reduced finally to free amino acids which can be absorbed by the intestinal mucosa.

Amylase is similar to ptyalin (which converts starch to maltose); the optimum pH for its action is approximately at neutrality (Terroine and Bierry).

Lipase is an important enzyme in digestion; fats are hydrolysed by pancreatic lipase to fatty acids and glycerol. It is probable that this enzyme is activated by bile salts (Terroine). Very pure preparations have been shown to contain lipoproteins (Sarda and Desnuelle, 1957) which leads to the assumption that lipids have an affinity for lipids. Lipase acts only on *emulsions* (it is fixed at the interface of the water–substrate emulsion); in this respect it is different from hepatic esterase. At 4°C pancreatic lipase is stable between pH 3·6 and 9·8 (Borgström, 1955).

Cholesterolesterase (permeation enzyme) is thought to be involved in the processes of absorption of cholesterol because esterification of cholesterol with fatty acids would facilitate the intestinal absorption of cholesterol; this hypothesis should be treated with some reservation.

Ribonuclease and *desoxyribonuclease* hydrolyse the nucleic acid fraction of nucleoproteins. For example, with ribonuclease one obtains:

(1) pyrimidine mononucleotides;
(2) di- and tri-nucleotides;
(3) a nucleic residue or "limit polynucleotide", the middle part of RNA, which resists hydrolysis by ribonuclease and can no longer form a detectable complex with basic dyes.

Collagenase from the *pancreas*, described in 1955 by Ziffrin and Hosie, hydrolyses collagen, the principal constituent of cartilage and fibrous tissue (as for example, in the Achille's tendon). Collagenase acts on the substrate after preliminary hydrolysis by pepsin and trypsin has taken place.

The intestinal contents of snail show chitinolytic activity (Karrer and Hofman, 1929); however, this chitinase could be derived from the associated microbial flora (Jeuniaux, 1950) and, therefore, could not be considered as a constituent enzyme of this animal. This leads one to consider that *the enzymic apparatus for digestion is specific to a trophic type*, involving constitutive, adaptative or symbiotic enzymes and of which, the specificity is greater than that designated by the classification into omnivorous, carnivorous (including insectivorous), herbivorous, graminivorous, xylophages, or frugivorous types.

We shall consider now the chemical processes of intestinal digestion.

In the intestine

The duodenum contains Brünners glands which produce an alkaline secretion with *low enzymic activity*. However, the duodenum, as well as other regions of small intestine, also contains Lieberkühn glands in which are localized the cells of Paneth which are thought to secrete enzymes; but these cells containing granular protoplasm are absent in many mammals including the dog.

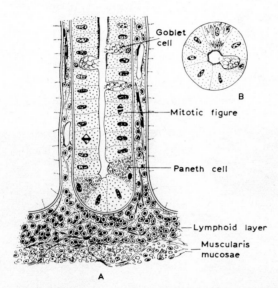

Goblet cell

B

Mitotic figure

Paneth cell

Lymphoid layer

Muscularis mucosae

A

FIG. 23. Gland of Lieberkühn. *A*, longitudinal section; *B*, transverse section. (DUBREUIL and BAUDRIMONT, 1959).

Secretion of intestinal juice is controlled by the hormone, *enterocrinin* (Nasset), produced under the influence of mechanical stimulation of the intestinal contents; thus, there exists here a *typical mechano-humoral relationship*, the mechanical aspect of which is coupled to electrical processes and the humoral aspect is an integral part of enzymic processes. This is an example of functional coupling between physical and chemical processes and such relationships must always be taken into account in an attempt to understand what is meant by *regulation*.

In addition to the pancreatic enzymes, there exists:

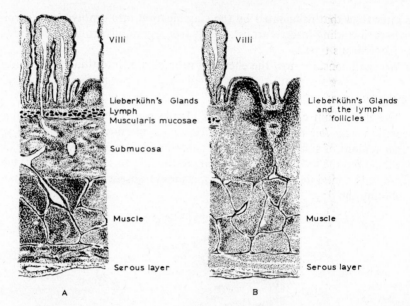

FIG. 24. (A) Jejunum. (B) Ileum. (DUBREUIL and BAUDRIMONT, 1959.)

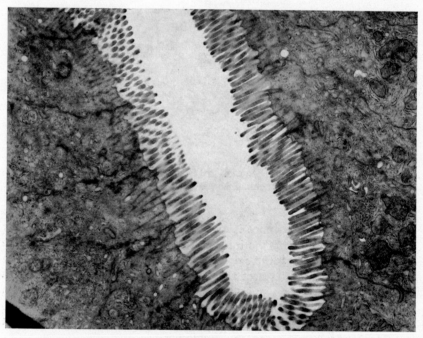

FIG. 25. Crypt of jejunum of mouse (×20,000). (Plate by courtesy of
H. RUSKA, 1963.)

(1) *disaccharidases:*

$$-1 \text{ sucrose} \xrightarrow{\substack{\text{saccharase} \\ \text{or sucrase}}} 1 \text{ glucose} + 1 \text{ fructose}$$

$$-1 \text{ maltose} \xrightarrow{\text{maltase}} 2 \text{ glucose}$$

$$-1 \text{ lactose} \xrightarrow{\text{lactase}} 1 \text{ glucose} + 1 \text{ galactose}$$

(2) *polynucleotidases* (or phosphodiesterases) which hydrolyse polynucleotides to nucleotides;

(3) *phosphatases*, found in an intestinal extract, remove phosphate from certain organic phosphates, probably as they pass across the villi: hexosephosphates, glycerophosphates and nucleotides from the food;

(4) *nucleosidases* carry out the final stage of degradation of nucleic acids by converting mainly purine nucleosides to adenine or guanine and pentose;

(5) *lecithinases* which degrade phosphatides to give principally fatty acids;

(6) *proteinases*, secreted by the mucosa, whose pH for activity varies according to the intestinal region being considered (pylorus, duodenum, etc.).

Bile

Bile plays a very important part in digestion—500 to 1000 cm^3 are produced each day by man. The *liver* secretes bile continuously and this accumulates and is concentrated (absorption of water and salts) in the gall bladder; the sphincter of the gall bladder contracts and relaxes in the presence of *cholecystokinin* and *hepatocrinin*. Meat and fats which facilitate equally the production and evacuation of bile are said to be *cholagogic* or *choleretic* (sugars are inhibitors). The dromedary does not contain a gall bladder.

(a) PIGMENTS. The two principal pigments are:

bilirubin, principal pigment of the bile of carnivores and man;

biliverdin (dehydrobilirubin), principal pigment of the bile of amphibians and birds and exists only in very small quantities in the bile of man.

They are porphyrin derivatives produced by the catabolism of hemin (from hemoglobin) during the destruction of erythrocytes in the cells of the reticulo-endothelial system and particularly in the hepatic cells of Küpffer (cf. p. 168).

These pigments are metabolized in the following way:

In the liver, bilirubin from the catabolism of hemin is *conjugated with glucuronic acid* (cf. p. 307); this glucuronide is excreted in the intestine with the bile. Bilirubin, hydrogenated to *mesobilirubin*, is reduced finally in the intestine to *urobilinogen* which is partially absorbed and distributed in the blood and of which a part (about 2 mg per day in man) is eliminated

8

in the urine, the remainder returning to the bile (*entero-hepatic circulation*; cf. p. 118). That which is not absorbed is rejected and constitutes *faecal urobilinogen* (an average of 160 mg per day in man). Urobilinogen is oxidized in air to *urobilin* which accounts for the darkening of faeces in air. Part of the mesobilirubin can be reduced to *stercobilinogen* in the intestine and gives *stercobilin* in the faeces by bacterial oxidation.

If the level of bilirubin attached to serum albumin in the plasma of man exceeds 10 mg per l, the pigment passes into the tissues which become yellow. This constitutes *icterus* or *jaundice* which occurs notably in cirrhosis and many cases of poisoning, for example by chloroform. Zetterström and Ernster (1956) have shown that *bilirubin* uncouples phosphorylation from respiratory processes in isolated mitochondria and as a result, has a harmful effect on tissues.

(b) BILE ACIDS. These are important agents in the degradation of fats; they act as detergents but, in this way, are inhibited by blood serum (Joannides, 1924).

Four bile acids are known to be present in man:

(1) *cholic acid:* most widely distributed

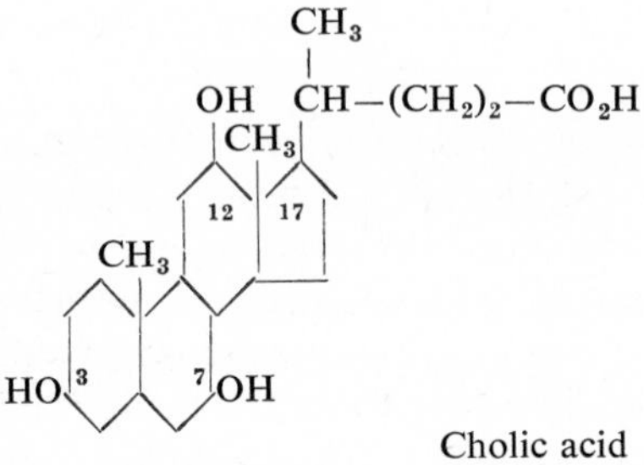

Cholic acid

(2) *desoxycholic acid:* OH absent from position 7; the most widely distributed in faeces;

(3) *chinodesoxycholic acid:* OH absent from position 12;

(4) *lithocholic acid:* one OH group, in position 3.

They are produced in the liver from *cholesterol* from the blood—the liver is the main organ of detoxication (cf. p. 306)—and excreted in the bile as bile acids in *conjugation* with glycine or taurine. They exist in alkaline media in the form of *glycocholates* and *taurocholates* (of sodium or potassium).

This conjugation occurs in the liver microsomes and requires coenzyme A, ATP and Mg^{++}; cholyl-CoA—or active cholic acid—is formed and in the presence of another enzyme conjugation occurs.

The digestive functions of bile

(1) Bile contributes to the neutralization of acid chyme;

(2) bile lowers the surface tension of aqueous solutions and thus ensures the emulsification of fats in the intestine.

If fats are digested poorly, food materials in the intestine cannot be digested completely as they become enveloped with fats, which prevents their degradation by enzymes such that bacterial activity leads to putrefaction and flatulence.

The absorption of vitamins A, D, E and K requires the presence of bile.

Mechanical phenomena

(a) *Progression*. The movement of the intestinal contents towards the large intestine is produced by *peristaltic waves*. The duodenum shows the *most intense peristaltic movements*. The duodenum can exhibit also *antiperistaltic waves* directed towards the stomach which are produced mainly by a chyme containing too much fat.

(b) *Mixing*. This process allows the intestinal contents to come into contact with the surfaces and absorption. The intestinal contents are mixed by the following movements:

(1) *segmentary movements* appear as contractions of equally spaced parts of the intestine, occurring regularly in man approximately twenty times per minute;

(2) *oscillatory movements* of part of an intestinal loop.

Law of the intestine (Bayliss and Starling): *Excitation* (mechanical, chemical, electrical, etc.) *leads to a contraction behind and relaxation in front of the point of excitation*.

This allows the progression of intestinal contents in a physiological (orthodromic) direction.

Mechanisms of contractions

The problem involved is as follows: initiation of automatic contractions can be exclusively muscular in origin—*myogenic origin*—or nervous—*neurogenic origin*. In the case of the intestine, the presence of an intraparietal *nerve plexus*—the plexi of Auerbach between the two muscle layers and of Meissner in the submucosal region—makes it difficult to favour a myogenic system alone. The problem is complicated by the fact that the peristaltic contractions which are themselves coordinated occur only when the integrity of the plexus of Auerbach is maintained (Morin); *initiation* and *coordination* are the central terms in this discussion.

The intrinsic innervation is thus one point to be considered; the extrinsic innervation is another. Excitation of the vagus stimulates intestinal con-

8*

tractions while stimulation of the sympathetic system (splanchnic nerves) inhibits these contractions. However, in the elasmobranchs, adrenaline (or noradrenaline) *stimulates* intestinal contractions (Euler and Östlund, 1957). Serotonin (5-hydroxytryptamine) also stimulates contraction of the intestine of fish. Moreover, U. S. von Euler and Gaddum (1931) and later, Euler and Östlund isolated, from intestine (notably of fish) and brain, a polypeptide called "substance P" which increases intestinal movements and decreases the arterial blood pressure of rabbit. Youmans *et al.* (1943) stated that acetylcholine, produced by intrinsic mechanisms, is the factor which maintains the normal tone of the intestine in mammals.

It can be assumed that the movements of the living intestine will be associated with alterations in the electrical behaviour of the resting state. Bülbring considers that the slow waves of depolarization in the small intestine, which are associated with periodic changes in tension of the muscle, would condition these changes in tension. *Grosso modo* the train of potentials parallels the mechanical recording (cf. Fig. 26). It is possible,

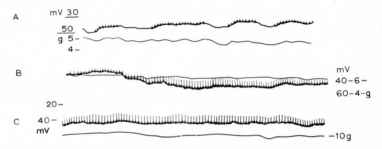

FIG. 26. Recording of the membrane potential and of mechanical tension in a muscle of 7 mm in length (A) which has been stretched to 8 mm (B) and to 10 mm (C). The frequency of spike potentials is 40/mn at 7 mm, 60/mn at 8 mm and 74/mn at 10 mm. (E. BÜLBRING, in *Gastroenterologia*, 1956.)

using micromanometry, to measure the oxygen consumption of fragments of small intestine; an active fragment from rat consumed during 1 h approximately 21 μl O_2/mg total tissue nitrogen while an inactive fragment (which produces no electrical potentials) consumed 2·5 μl O_2/mg total tissue nitrogen during the same period. Therefore, contraction of the small intestine and the associated electrical changes require about 18·5 μl O_2/h/mg total tissue nitrogen (Rybak, 1956).

The small intestine is terminated by the ileocolic sphincter which opens into the caecum or first part of the colon (large intestine). The diameter of this area is about twice that of the small intestine and measures about 6 cm in man.

This sphincter restricts emptying of the small intestine and thus favours the digestive processes by increasing the length of time that the contents remain in the ileum; in addition, it prevents regurgitation of the contents of the colon. Excitation of sympathetic nerves closes the sphincter while parasympathetic excitation is without effect.

Faeces

In man, the faeces are formed in the following way: the 400 ml of fluid passed daily from the small intestine are reduced in the colon to 150 g. *The absorption of water, mineral salts and small quantities of sugars* occurs mainly in the right colon but also in the left. It must be emphasized that *reabsorption of water does not only occur in kidney*, a point which is often forgotten. Moreover, in birds, water is reabsorbed from the intestinal lumen in the region of the cloaca where the urethra and rectum join.

The contents of the caecum of human are composed of:

"*indigestible carbohydrates*" (mainly cellulose);
"*indigestible proteins*" (keratin, elastin).

In addition, a small amount of starch, *digestible* proteins and lipids enter the caecum as a result of incomplete mixing or enzymic degradation in higher regions of the digestive tract. The inorganic compounds found in the faeces are principally of *endogenous* origin with the exception of dietary iron.

About 1 l of *gas* is present in the human colon; the gases are: CO_2 (about 22 per cent), H_2, N_2 (about 20 per cent), CH_4 and H_2S produced from methylmercaptan in the following way:

$$CH_3SH \xrightarrow{+2H} CH_4 + H_2S$$

The faecal odour is due to *mercaptans* and also to *indole* and *skatole* which are derived from deamination and decarboxylation of tryptophan.

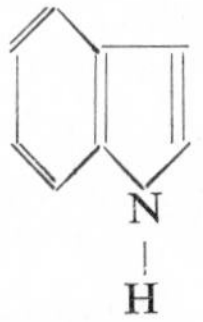

Indole 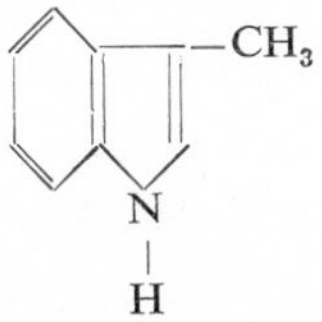   Skatole

These compounds, as well as amines (from the bacterial decarboxylation of amino acids) and phenol (produced in small amounts from phenylalanine and tyrosine) are toxic and may contribute to the nervous symptoms of constipation. However, this has not been shown conclusively as these compounds are not absorbed extensively and the *detoxication systems in liver* which produce non-toxic compounds eliminated in urine do not become saturated (cf. p. 306). It appears that the general disorders associated with constipation arise from mechanical phenomena through the stimulation of sensitive nerves.

Digestion in the large intestine

The large intestine secretes only lubricating mucus which has no enzymic activity and the microbial flora within the intestine is responsible for enzymic processes. The *bacterial flora* which exists *under normal conditions* in the large intestine (and sometimes in the *ileum*) carries out the final stages of biochemical digestion; the large intestine is predominantly an organ containing micro-organisms and thus resembles the rumen.

Generally, in man, the flora of the colon can be considered to *acidify* by fermentation of carbohydrate residues and to *alkalize* by decomposition of nitrogen residues.

In the latter case, *ammonia* is the principal product. Ammonia is absorbed by the portal circulation and removed normally from the blood by the liver. When this hepatic function is disturbed the ammonia content of the blood reaches a toxic level. It is thought that ammonia is responsible for the symptoms of poisoning which characterize *hepatic coma*. A diet rich in proteins may lead to disorders related to *ammonia poisoning* in certain subjects with liver disease.

A balance between the two types of bacterial flora ensures that the faeces are finally neutral or weakly acid or alkaline.

The bacteria can occupy from about a quarter to a third of the volume of the faeces. Approximately 1·5 g faecal nitrogen, principally bacterial, is produced each day, although this is obviously a relative value since the total depends on the amount of "indigestible carbohydrates" (also called *roughage*).

As has been described earlier (cf. p. 94), mechanical tension can affect the motility of contractile systems. Roughage, derived principally from vegetables, stimulates intestinal movements. However, it must be stressed that bran, for example, is far from being inert bulk since it restricts the digestion of proteins.

Movement in the colon

The colon is regulated by identical nervous processes to those involved in the motility of the small intestine (intrinsic plexus of Auerbach, etc.). Stimulation of the sympathetic nerves results in inhibition and of the vagus, in excitation. However, the *vagus* only affects the *right* colon while the *left colon is innervated by the sacral nerves.*

The caecum is practically inactive but the transverse and descending colon exhibit slow peristaltic movements. The contents of the large intestine are moved onwards by massive peristaltic waves occurring two or three times per day and pushed into the pelvic colon. These peristaltic waves are reflex reactions regulated principally by distension of the digestive tract, through the action of baroreceptors (cf. Volume II).

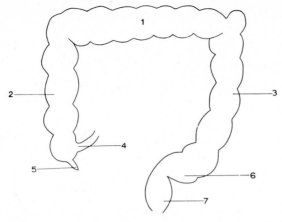

Fig. 27. *1*, transverse colon; *2*, ascending colon; *3*, descending colon; *4*, ileo-coecal valve; *5*, appendix (lymphatic organ); *6*, sigmoid; *7*, rectum.

Defecation

Between the spasms of the colon the rectum is empty. If the contents of the pelvic colon are pushed into the rectum, the reflexes associated with defecation are initiated when the intra-rectal pressure reaches 50 mm Hg (stimulation of baroreceptors).

The defecation reflex consists of contraction of the pelvic region of the intestine produced by the longitudinal muscles. Following this the anal sphincters function: the *internal sphincter opens automatically* and the subject experiences a desire to defecate. Defecation is accomplished voluntarily by contraction of the muscles of the abdominal walls which compresses the colon and increases the intra-rectal pressure, followed by relaxation of the external anal sphincter. Simultaneously, the *levator ani* prevent prolapse (evagination) of the anus.

Evacuation of faeces can be repressed by voluntary control of the external sphincter; the rectal tone decreases, the internal sphincter opens and the sensation disappears until the rectal pressure again increases and stimulates the baroreceptors. The levator ani muscles reinforce the action of the sphincters.

Passage through the gastro-intestinal tract can be followed *radiographically* after the administration of a *barium meal*.

To summarize, the enzymic phase of digestion in mammals (stomach, small intestine) occurs between two principally mechanical phases (buccal, colonic); however, in the colon, two important processes must also be considered: enzymic activity of bacteria and absorption of water.

An examination of digestive processes shows that digestion expresses the ability of the digestive tract to discriminate between food molecules and this is dependent on enzymic composition. This *vegetal intelligence* is rudimentary and therefore constrasts with mental intelligence. The distribution of various types and numbers of enzymes can be thought to be embryological in origin and an expression of "animal–vegetal" polarity (cf. Volume II, Perspectives).

Absorption in the digestive tract

Absorption can occur in the mouth, notably via the sub-lingual route. This is significant in that the absorbed materials pass directly into the superior vascular system (via the jugular) without passing through the liver (which retains materials).

Absorption within the stomach is limited to such compounds as ethanol. Therefore, *digestive absorption is a phenomenon limited almost exclusively*

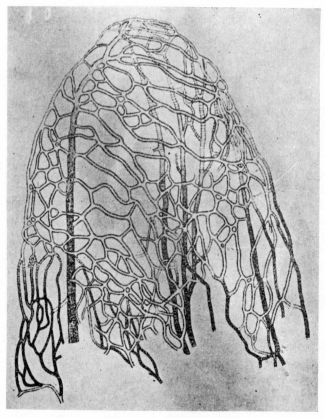

Fɪɢ. 28. Villus of the small intestine of rabbit showing the capillary network of one side, ×68. (Kʀᴏɢʜ, 1959.)

in mammals to the intestine and above all, to the small intestine (cf. p. 99). Absorption is carried out by the villi of which the active surface, in man, measures about 10 m².

Removal of half of the intestine does not lead to extensive disorders, which suggests that, under normal conditions, the area of the surface for absorption is more than adequate. Absorption can take place in the rectum; a phenomenon which is utilized in therapeutics (*suppositories*).

The *epithelial cells* are the structures responsible for *absorption*. Contractions of the villi are controlled by the submucosal nerve plexi of Meissner and can be stimulated by various compounds; for example, bile and peptones. Kokas and Ludany reported that these contractions could also be induced by a hormone, *villikinin*, which may be secreted by the duodenal mucosa.

Sugars and amino acids pass into the portal vein via the capillaries. The portal vein, which also collects blood from the spleen, joins the capillary network of the liver. The hepatic veins lead from the liver and the blood is carried by the inferior vena cava into the right auricle.

Most *fats* are carried by the lymphatics (Asellius, 1627) called *lacteals* into the thoracic duct which opens into the left subclavian vein. The abdominal lymphatic system lies close to the sympathetic ganglia (Kiss, 1930) and the perineural spaces of the plexus of Auerbach are connected with the lymphatic capillaries (Zhdanov, 1959).

Mechanisms of absorption

Formerly, intestinal absorption was thought to be a purely *physical* phenomenon related to numerous factors:

1) osmotic pressure of the intestinal contents, cells and plasma;

2) absorption of substances depending on the presence in the molecule of hydrophilic or lipophilic groups (Devaux and later Langmuir).

The *partition coefficient* describes the distribution, at a given temperature, of a compound placed between two immiscible solvents (for example, linseed oil and water which would determine the lipid solubility and water solubility of a compound). The partition coefficient is expressed by:

$$\frac{C_1}{C_2} = K;$$

3) hydrostatic pressure acting on the intestinal wall;

4) electrocapillarity and interfacial tension.

However, it appears that only the *walls of the capillaries* participate passively in the passage of dissolved compounds of low molecular weight (diffusion). More complex processes must exist although the participation of physical processes cannot be excluded; *active transport* is a necessary process.

In considering osmotic pressure, mention should be made of the phenomenon called the Donnan *equilibrium* which is of particular importance in the understanding of ionic equilibria in biological media.

Let us consider a saline solution (for example, sodium chloride), at room temperature, divided into two parts *a* and *b* by an inert membrane which is impermeable to macromolecules. The salt is dissociated into Na^+ and Cl^- ions which are distributed equally on either side of the membrane. If a certain amount of a colloid NaR is added to one of the compartments, for example *a*, the concentration will be greater than in *b* which contains only small ions. The osmotic equilibrium will be disturbed and a *movement of ions will occur* such that a new equilibrium is reached in which *the products of the concentrations of the two small ions on either side of the membrane are equal:*

$$[Na^+]_a \times [Cl^-]_a = (Na^+)_b \times [Cl^-]_b$$

This is the Donnan equilibrium and implies that

$$[Na^+]_a > [Na^+]_b$$

and

$$[Cl^-]_a < [Cl^-]_b$$

The equilibrium is characterized by a certain imbalance, by an unequal distribution of diffusible ions in the compartments; this unequal distribution creates a potential difference associated with a double electrical layer.

(a)	Na^+	R^-	Na^+	Cl^-	(b)
	Na^+	Cl^-			

The Donnan effect is not *selective* and does not explain *the absence of absorption in the intestine of certain low molecular weight compounds*; for example, sodium sulphate which acts as a laxative by retaining water in the intestinal cavity.

But, the absorption of salts such as NaCl seems to be regulated entirely by osmotic phenomena; the amount of NaCl absorbed per hour increases with concentration up to 15 per cent and then decreases. The absorption of saline can be controlled, therefore, by the establishment of an osmotic equilibrium with the medium. However, *hypotonic* solutions of NaCl can be absorbed and, therefore *one must be careful not to compare a cell membrane with an inert membrane*, to confuse the passive phenomenon of *permeability* with *active transport (permeation)*. Active transport occurs more generally in biological systems.

Man absorbs 5 to 6 l of water per day through the intestine and the absorption of saline is *twice as rapid* as that of pure water (hydration of

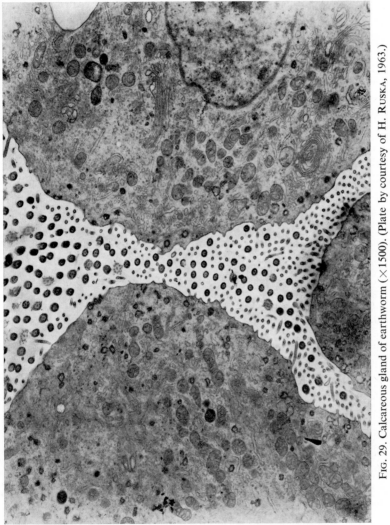

FIG. 29. Calcareous gland of earthworm (×1500). (Plate by courtesy of H. RUSKA, 1963.)

Na ions?). If the kidneys are functioning unsatisfactorily, excess water cannot be reabsorbed and the concentration of the internal medium undergoes pathological changes. The absorption of other salts, notably *phosphates*, does not obey a simple law as is the case with NaCl and, in general, *the selectivity of absorptive processes indicates active absorption by the cells of the intestinal epithelium* (with the exception of pentoses and possibly cholesterol). However, A.M. and A. Monnier have shown that phosphorylated fatty alcohols, for example oleic alcohol, form a thin layer at the surface of water. This layer shows selective permeability to cations and thus, according to the Donnan equilibrium, a potential difference is established (1960, 1961). It is possible that physical phenomena represent the basic processes which are maintained when, for some reason or another, the active, enzymic processes cease.

In *Lumbricus terrestris*, Ca^{++} enters the organism through the integument and not through the intestine and, under conditions of low oxygen tension such as occur in soil, carbon dioxide is not eliminated but is fixed by lime in the calcareous gland (Pora *et al.*, 1961).

Absorption of carbohydrates

The rate of absorption of different hexoses is not constant. If the rate of absorption of glucose is taken as 100, the following values have been found for rat by Cori:

galactose	110	mannose	19
fructose	43	arabinose	9

Hexoses are absorbed more rapidly than pentoses. This selectivity results from the existence of different pathways of absorption:

Pentoses are not phosphorylated before crossing the intestinal wall and appear to penetrate by simple diffusion, a relatively slow phenomenon.

Hexoses are phosphorylated prior to absorption and form hexose phosphates. This phosphorylation is catalysed by the enzyme *hexokinase*; the energy necessary for this process is supplied by ATP.

Intestinal hexokinase differs from *hepatic hexokinase* (which functions in glycogenesis) but is identical with *renal hexokinase* (which catalyses the phosphorylation of glucose in the process of reabsorption from the tubular filtrate). There is a temptation to classify hexokinases as *hexokinases of absorption* (intestine and kidney) and *hexokinases of synthesis* (liver).

When phosphorylation is blocked by monoiodoacetate or phlorizin, the total absorption of hexoses decreases to the level of pentoses and this level represents residual absorption by slow *diffusion*.

L. Garnier (1956, 1961) has shown that the intestinal absorption of glucose is proportional to the concentration of sugar within the lumen. For mixed solutions of glucose and NaCl, isotonic with plasma, the total absorption of the solution is equal to half the sum of the absorption of pure solutions of glucose and NaCl.

Absorption of lipids

Hydrolysis of fats (triglycerides) by lipase yields glycerol and fatty acids. Although the *elimination of the primary fatty acids by lipase occurs relatively easily*, that of the *secondary* and of the *tertiary* acid is carried out *less readily*. Mattson and Beck (1956) have shown that pancreatic lipase specifically attacks *primary* ester linkages $CH_2O-\overset{\overset{O}{\|}}{C}-R$ (cf. Table 27) such that hydrolysis of triglyceride gives a 1,2-*diglyceride* and then a 2-*monoglyceride*. The latter possesses a secondary ester bond and must undergo *isomerization* before the monoglyceride can be hydrolysed; however, this process is very slow. *Therefore, the end products of hydrolysis of fats are 2-monoglycerides and only 30 per cent of dietary lipids are reduced completely to fatty acids and water soluble glycerol.*

TABLE 27

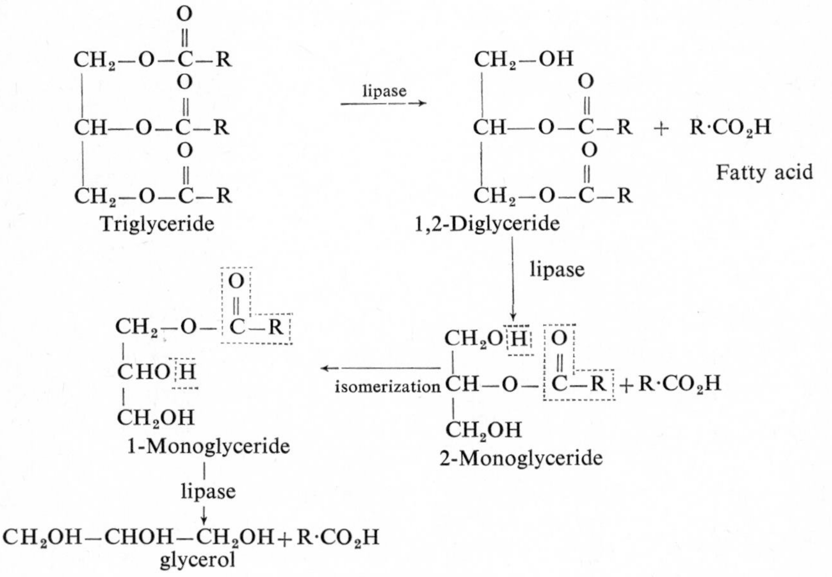

Fatty acids and *monoglycerides* are almost completely insoluble in water and this prevents direct absorption. The combination of fatty acids with bile salts gives water soluble products which can be absorbed. After absorption of the fatty acid–bile salt complex, the fatty acids *are separated* from the bile salts and recombine with *glycerol* to give fats. The bile salts pass in the portal vein to the liver where they are directed to the bile; this process constitutes the *entero-hepatic circulation of bile salts*.

What is the derivation of the glycerol which recombines with fatty acids?

It is probable that at least 95 per cent is not derived from absorbed glycerol formed by lipase hydrolysis (Favarger *et al.*, 1951). Reiser and Williams (1953) did not find ^{14}C-labelled glycerol from hydrolysis in the triglycerides of lymph. Intracellular *dihydroxyacetone* serves as a source of glycerol for esterification of absorbed fatty acids to fats.

$$CH_2OH-\overset{\overset{\textstyle O}{\|}}{C}-CH_2OH \qquad \text{Dihydroxyacetone}$$

The triglycerides, resynthesized after absorption, are not transported in the portal blood. They are found in the lacteals in the form of lipid–peptide particles, the *chylomicrons*, with a diameter of about 1 μ. However, short-chain fatty acids such as lauric acid (C_{12}) pass into the portal blood.

Thus, the fate of fatty acids in the intestine is determined by *molecular weight*: very little *stearic acid* (C_{18}) is absorbed; 90 per cent of *palmitic acid* (C_{16}) is absorbed; *myristic acid* (C_{14}) is completely absorbed and they are found in the lacteals.

Therefore, low molecular weight fatty acids pass into the blood, those of intermediate molecular weight pass into the lacteals and high molecular weight fatty acids are not absorbed or are absorbed in small amounts.

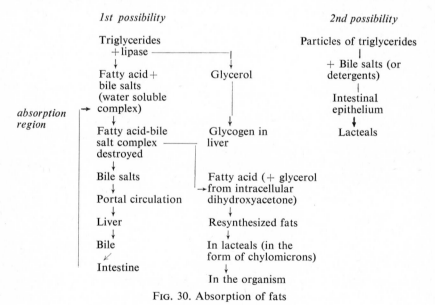

Fig. 30. Absorption of fats

Mention must be made of the fact that very small *unhydrolysed* fat particles can be absorbed in the presence of bile salts or other agents affecting interface tension.

The absorption of fats can be summarized in the schemes shown in Fig. 30.

Orthophosphate favours the absorption of triglycerides and inhibition of phosphorylations by monoiodoacetate or phlorizin inhibits absorption of fats. Verzar has suggested that *phospholipids* are formed as intermediates; however, as yet, only a very slight increase in the phospholipid content of blood and lymph has been demonstrated after fat absorption.

Finally, the absorption of cholesterol appears to be facilitated by esterification with fatty acids (cf. p. 131). This reaction is catalysed by cholesterolesterase.

Absorption of proteins

Small peptides and amino acids from the digestion of dietary proteins are rapidly absorbed by the intestinal mucosa and pass into the portal system (however cf. antibodies in newborn donkey, p. 198).

After separating the intestinal mucosa from adjacent muscle layers using fine forceps, Baillien and Schoffeniels, (1961), mounted the mucosa between two chambers in which different solutions could be placed and studied the passage of chemicals across the membrane. There is a potential difference of 4 mV in the small intestine of the Greek tortoise, the lumen being negative with respect to the serous layer. This potential difference can be annulled by passing a continuous current of given intensity (short circuit current) through the preparation. Using radioactive amino acids, for example glycine, passage from the mucosal to serosal layer in a fragment of colon was the same as in the reverse direction. In the small intestine glycine passes across the membrane more rapidly from the mucosal to serosal surface than from the serosal to mucosal layer. This indicates active transport of glycine in the intestinal epithelium of this region. Glycine, L-alanine, L-histidine are transported actively in the small intestine while L-glutamate is displaced passively; however, all these amino acids move passively through the wall of the colon. L-Alanine partially inhibits passage of glycine from the mucosal to serosal surface; this confirms the work of Hagihira *et al.* (1960).

Amino acids of the natural series (L) are more easily absorbed than D series isomers. This specificity tends to show that intestinal absorption does not depend on purely physical processes and that a certain number of still unknown mechanisms must be involved to explain the ability of intestinal cells to "choose" between stereoisomers.

CHAPTER 7

TRANSPORT OF ALIMENTARY CONSTITUENTS AND INTERMEDIATE METABOLISM

MOLECULES which have been absorbed by the intestine form complexes with other compounds—carriers—and are transported to an organized pellular system where they undergo further degradation (catabolism) o carticipate in synthetic processes (anabolism).

I. Metabolism of carbohydrates

Dietary carbohydrates provide, on average, *more than half* the energy requirements of mammals.

Only a part of the ingested carbohydrates is *stored* in the liver, muscles and extracellular spaces; the remainder is converted to fats (*lipogenesis*) and stored in this form in the liver and *adipose tissue*. The glycogenic function of the liver was discovered by Claude Bernard (1849–1857); hepatic glycogen represents about 6 per cent of the weight of the organ and the total is ten times higher than that found in muscles.

The principal sugars utilized in nutrition are glucose, fructose and galactose. Glucose is the sugar present in blood. In man, the blood glucose level is 1 g/l but after fasting the level decreases to 0·7 g/l. Glucose is derived from three sources:

(1) *digestive:* starch, the dietary sugars constitute the main source of glucose;

(2) *hepatic:* hydrolysis of glycogen (glycogenolysis);

(3) *"molecular"*, from non-carbohydrate precursors:

(a) *glycerol* or *dihydroxyacetone;*

(b) *products of catabolism of glucose* which are also glycogenic (succinic and fumaric acid), the reactions being reversible;

(c) *glycogenic amino acids* via the keto acids produced by deamination of amino acids (for example, pyruvic acid).

This reaction is augmented by adrenal cortical steroids such as cortisone, called "hormone S" ("sugar forming" hormone).

The liver is almost exclusively the only source of glucose during fasting; although the kidney can also provide glucose, *muscle glycogen is not mobilized to form glucose* except *indirectly* through lactic acid. The kidney and liver contain a glucose-6-phosphatase which forms glucose from glucose-6-phosphate but *this phosphatase is not present in muscle*.

The glucose cycle in mammals—the cycle of Cori—can be represented schematically as follows (N.O. Kaplan *et al.*, 1956):

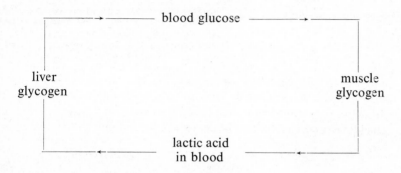

The complete transformation of glycogen to CO_2 and H_2O requires:

the degradation of glycogen;
oxidation of the products of glycolysis.

TABLE 28

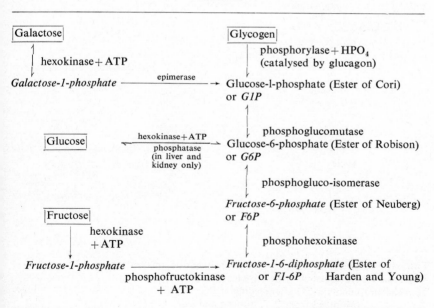

TABLE 29

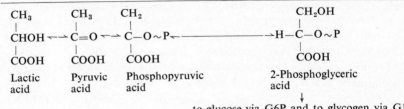

| Lactic acid | Pyruvic acid | Phosphopyruvic acid | 2-Phosphoglyceric acid |

to glucose via G6P and to glycogen via G1P or, according to Leloir and Cardini (1959), via uridine-diphosphate-glucose (UDPG). Glycogen synthesis occurs more readily in a medium rich in K^+ than in a medium rich in Na^+ (Ashmore et al., 1957).

Glucose metabolism which is important in arthropods since they are rich in chitin (containing glucosamine) follows pathways which, although not elucidated completely, can be represented schematically as in Fig 31.

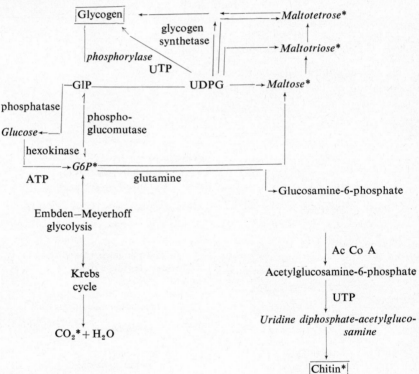

FIG. 31. Probable pathways of glucose metabolism in crustaceans. Compounds in italics have been isolated from crustaceans. Those marked with an asterisk* have been shown to contain ^{14}C after feeding or injecting glucose labelled with this isotope. (Modified from B.T. SCHEER and V.R. MEENASKI, 1961.)

Degradation of glycogen

Glycogen is catabolized by an anaerobic process. All sugars metabolized by the organism are first phosphorylated.

The initial step in the metabolism of glucose is the formation of *glucose-6-phosphate* (ester of Robison) which is the central compound in carbohydrate metabolism (with the exception of fructose).

Table 28 shows the pathways of metabolism of glucose, galactose and fructose in the liver.

In the final stages the reactions are the same as those of anaerobic glycolysis in muscle (first enzyme involved is aldolase). This will be examined later (cf. Volume II); for the moment it is sufficient to say that, following numerous enzymic reactions, lactic acid and pyruvic acid are produced.

Glycolysis is, for the most part, a reversible phenomenon. Table 29 shows simplified pathways of resynthesis of glucose and glycogen from lactic acid. Crabtree (1929) demonstrated that glycolysis inhibits respiration; the Crabtree effect is the reverse of the Pasteur effect (cf. Volume II)·

Oxidative metabolism of sugars

I. The Krebs cycle

The chain reactions of the Krebs *cycle are concerned not only with oxidation of the products of glycolysis but also with the oxidation of fatty acids and amino acids.* This pathway is important as it constitutes *the common terminal pathway of metabolism in mammals*; the pathway involves notably the *interconversion of sugars and lipids.* In general, interconversions are based on the fact that most metabolic reactions are *reversible.*

The critical stage of this cycle is the conversion of pyruvic acid CH_3——CO—$COOH$ to "active acetate" in the form of *acetyl-CoA*, an oxidative decarboxylation catalysed by a *decarboxylase* which requires thiamine and coenzyme A as cofactors. Coenzyme A plays a principal role in many aspects of metabolism.

This important reaction can be written:

$$CH_3-CO-COOH+NAD+CoA\ SH \xrightarrow{\ Mg^{++}\ } CoA\ S-CO-CH_3+NADH_2+H^++CO_2$$

$$(50)$$

The Krebs cycle follows the formation of malate. However, the cycle can be initiated by a carboxylation: phosphorylated enolpyruvic acid

9*

combines with CO_2 to give oxaloacetic acid (Utter). This reaction requires biotin.

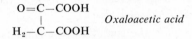

$$O=C-COOH$$
$$\mid \qquad \qquad Oxaloacetic\ acid$$
$$H_2-C-COOH$$

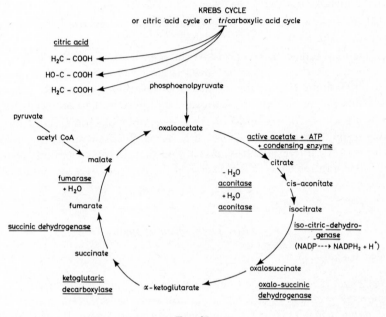

FIG. 32

The tricarboxylic acid cycle produces a *high level of energy*, far more than the anaerobic phase of glucose metabolism. Energy is transferred by a chain of reactions which utilize the system NAD–*flavoprotein–cytochrome*. Therefore, in mammals *the metabolism of different types of compounds* (proteins, lipids, carbohydrates) *must not be separated from respiratory processes*.

II. Oxidation via phosphogluconate

Although oxidation of the products of anaerobic catabolism of sugars occurs principally through the Krebs cycle, another pathway exists originating with G6P. This pathway is called generally the "*hexose monophosphate shunt*".

In rat liver, the major part of glucose metabolism occurs through the

Embden-Meyerhoff and Krebs cycle, the remainder occurring via the hexose monophosphate shunt (Ashmore et al., 1957).

It is worth noting that during fasting, glucose is catabolized via the Meyerhoff pathway while in the conversion of glucose to fatty acids, insulin favours the shunt pathway. The NADP—phosphogluconic acid pathway does not exist in the auricle of the heart of *Myxine glutinosa* and glucose is metabolized via the glycolytic pathway (Rybak and Boivinet, 1959).

The first stage of the shunt is the *oxidation of G6P to phosphogluconic acid* with the intermediate formation of a lactone (Lipmann et al., 1952–1955).

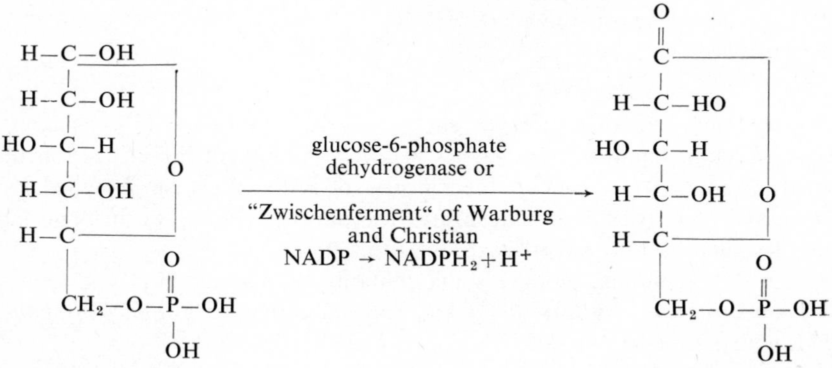

$$
\begin{array}{ccc}
\text{H}-\text{C}-\text{OH} & & \text{C} \\
\text{H}-\text{C}-\text{OH} & & \text{H}-\text{C}-\text{HO} \\
\text{HO}-\text{C}-\text{H} & \xrightarrow[\substack{\text{glucose-6-phosphate} \\ \text{dehydrogenase or} \\ \text{"Zwischenferment" of Warburg} \\ \text{and Christian} \\ \text{NADP} \rightarrow \text{NADPH}_2+\text{H}^+}]{} & \text{HO}-\text{C}-\text{H} \\
\text{H}-\text{C}-\text{OH} & & \text{H}-\text{C}-\text{OH} \\
\text{H}-\text{C} & & \text{H}-\text{C} \\
\text{CH}_2-\text{O}-\text{P}-\text{OH} & & \text{CH}_2-\text{O}-\text{P}-\text{OH} \\
\text{OH} & & \text{OH}
\end{array}
$$

Glucose-6-phosphate   Phosphogluconic acid

(51)

The pathway leads to the formation of ribose-phosphate with the release of CO_2 and is of importance in that it leads to the biogenesis of constituents of nucleic acids.

Metabolism of propionic acid

Propionic acid is glycogenic. Lardy and Adler (1956) demonstrated, in liver of mammals, the pathway by which propionic acid is converted to sugars.

Propionic acid combines with CO_2 to form succinic acid which enters the Krebs cycle (52):

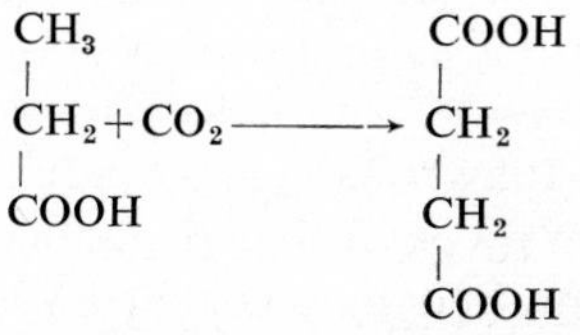

$$
\begin{array}{cc}
\text{CH}_3 & \text{COOH} \\
\text{CH}_2+\text{CO}_2 \longrightarrow & \text{CH}_2 \\
\text{COOH} & \text{CH}_2 \\
& \text{COOH}
\end{array}
$$

Propionic acid   Succinic acid

(52)

"Active propionate" (propionyl-CoA) and "active succinate" (succinyl-CoA) are intermediates in this reaction which requires the following cofactors: CoA, ATP, Mg^{++}, Mn^{++} and probably thiamine and biotin.

The reaction of Lardy and Adler is very important as it ensures the fixation of a carbon atom during metabolism.

II. Metabolism of lipids

Lipids comprise:

neutral fats (triglycerides);
phospholipids (or phosphatides);
cerebrosides;
sterols.

They represent an important energy source although it is considered that a certain quantity of sugars is necessary to ensure the utilization of lipids. About 20 per cent of the calories of a diet must be provided by carbohydrates. Although inaccurate, the phrase "les lipides brûlent au feu des sucres" conveniently expresses this point.

One litre of human blood plasma contains, on average, 6 g total lipids, 1·3 g neutral fats, 5 g fatty acids and approximately 2·5 g total phospholipids which contain:

lecithins (esters of fatty acids with glycerol, phosphorus and a nitrogenous base, in this case choline) 1·25 g/l on average;
cephalins (with same structure as lecithins but the base is colamine or ethanolamine) 900 mg/l on average;
sphingomyelins (the alcohol is sphingosine, the base is choline) 250 mg/l.

The total blood cholesterol is 2 g/l of which 500 mg/l is non-esterified cholesterol.

The clearing factor

Post-alimentary lipemia induces a turbidity of the plasma which disappears after the administration of heparin (Hahn, 1943). This clearing does not occur when heparin is added to lipemic plasma in vitro. In vitro, heparin forms an insoluble complex with β-lipoproteins in the presence of Ca^{++} (Burstein and Samaille). In my opinion, this reaction is between the PO_4^{---} group of phosphatides in β-lipoproteins, calcium ion (bond) and the SO_4^{--} group of the sulphated mucopolysaccharide. Flocculation would result from the increased molecular weight of β-lipoproteins. Moreover, Burstein and Prawerman (1958) induced flocculation of chylomicrons in lipemic serum by polyvinylpyrrolidone.

Nevertheless, Anderson and Fawcet (1950) consider that a substance is present in plasma which is converted to "clearing factor" in the presence of heparin and could be *lipoprotein lipase* (Korn, 1955). This enzyme destroys the chylomicrons rich in β-lipoprotein and gives rise to glycerol and fatty acids which form a complex with plasma albumins. Robinson (1960) suggests that the clearing factor originates in the vascular endothelium. Lipoprotein lipase may be destroyed in the liver (Jeffries, 1954; Lemarchand and Paramelle, 1961).

The importance of lipemia will be understood when it is considered that it has a fundamental role in the aetiology of arteriosclerosis which is one of the more significant causes of ageing in modern man. Lipoprotein lipase can be *inhibited in vitro* by various agents: *protamines* (salification), *sodium cholate, di-isopropylfluorophosphate* (DFP, inhibitor of acetylcholinesterase). *In vivo* these agents lead to hyperlipemia. The plasma of rats receiving cortisone contains an inhibitor (originating in the pituitary?). In adrenalectomized and totally hypophysectomized rats this factor induces hyperlipemia.

Deposition of fats

The so-called "free" fatty acids are transported by albumins (lipoprotein "cénapses" of Macheboeuf) and deposited in the storage areas. (In general, serum proteins play a key role of carrier but this is not analogous of course, to that of coenzyme A). It should be noted that the liver is not concerned primarily with the storage of fats except in pathological conditions.

Fats are stored in *three regions:*

in the subcutaneous connective tissue *(panniculus adiposus)*, about 50 per cent;

in the abdominal cavity (mainly in genital tissue);

in intermuscular connective tissue, about 5 per cent.

Fats are deposited ("lipopexy") *mainly in the form of triglycerides* of saturated fatty acids. By applying radioisotopic techniques it has been found that, in rat, about 10 per cent of fatty acids in fats are renewed each day; therefore, fats are in a dynamic state.

"In adipose tissue fat is transported in two phases. In the first, glycerides are fixed on an active site, by esterification of free fatty acids or by direct assimilation of glycerides. The second phase involves translocation of the glycerides, which may be facilitated by some sort of transesterification in which the original glycerol is discarded and the fatty acids are transferred to "glycerol" from endogenous sources. The chemical form of this "glycerol" has not yet been established, although it has been shown that α-glycerophosphate can fulfil this task." (B. Shapiro, 1960).

Several factors can influence deposition of fats:

(1) *Biotope.* The fats of animals in cold zones generally contain more unsaturated components than those of animals from warmer regions.

(2) *Endocrine glands.* Hyperthyroidism leads to a rapid loss of fats while hypothyroidism increases fat deposition which results in obesity. In Cushing's syndrome (cf. p. 444) there is obesity with an increased excretion of 17-corticosteroids.

(3) *Nervous system.* Neuro-endocrine pathways also participate. G. Clément (1947) sectioned unilaterally the nerves of guinea pig which lead to the peri-renal fat; in starved animals which lost weight, only the peri-renal cells of the denervated side retained their content of lipids.

(4) *Nutrition.* Lipogenesis and oxidation of glucose in adipose tissue are generally *decreased* by fasting or by a diet rich in fats; but a carbohydrate diet favours lipogenesis and oxidation of glucose while during prolonged starvation hepatic glycogen is replaced by fats. There are individual variations in susceptibility to obesity; for example, certain types of humans can eat excessively and never develop obesity while others eat little and rapidly increase in weight. It appears probable that endocrine makeup (in particular of the thyroid) is an important factor in this somatic phenomenon; moreover, this is a phenogenetic question related to lipogenesis in the *panniculus adiposus.* The importance of pulmonary ventilation, which obviously will influence oxidation processes, should be emphasized; Savier, Morel and Lemarchand (1959) stated that, after ablation of one lung of rat, there was a significant increase in the fat content of the liver and carcass.

Although mammals are capable of utilizing dietary sugars as a source of fatty acids (Lawes and Gilbert, 1950) they are unable to carry out directly the reverse process (cf. p. 123). There is little doubt that the main reason for this is that the formation of acetyl—CoA from pyruvic acid is irreversible. Thus, acetyl-CoA is the link in the biogenesis of fatty acids from carbohydrates. According to Kornberg and Pricer (53) the first fatty derivative to be formed is phosphatidic acid:

2 Acetyl-CoA + Glycerophosphoric acid → Phosphatidic acid + 2 CoA

Phosphatidic acid
$$\begin{array}{c} H_2COPO_3H_2 \\ | \\ R'COOCH \\ | \\ H_2COOCR \end{array} \qquad (53)$$

is a precursor of triglycerides and phosphatides.

Utilization of stored fats

The release of free fatty acids requires the presence of serum albumin; that of glycerides is realized to only a limited extent with serum lipoproteins as acceptors (Reshef, Shafrir and Shapiro, 1957).

There are three points to consider:

(1) The mobilization of fats is dependent on endocrine mechanisms involving the pituitary and adrenal cortex;

(2) Mobilization of *unsaturated fats* in the liver is more easily accomplished than of saturated derivatives. This is an advantage for animals living in colder regions.

(3) *In vitro*, peripheral tissues can oxidize fatty acids to CO_2. *This oxidation is an important source of energy.* The oxidation of fats appears to be an alternative source of energy for many living systems (for example frog heart, spermatozoa) and is also a source of water.

Role of the liver in lipid metabolism

When normal carbohydrate metabolism of the liver is inhibited, the animal requires a supplementary source of energy and large quantities of fats are transported to the liver. This occurs in sugar diabetes of depancreatized animals, in poisoned animals (by chloroform, arsenic, phlorizin), in alcoholism. This abnormal accumulation of fats in the liver is called *hepatic steatosis.*

Fatty livers may appear as a consequence of *abnormal nutrition:* for example, in geese and ducks in the Midi de la France receiving a diet rich in cholesterol and deficient in *proteins*—particularly those containing sulphur (cf. p. 130)—and also deficient in vitamin E. The fatty infiltration of the liver in cirrhosis of the liver can be reduced or prevented by a diet rich in sugars and proteins.

Lipotropic factors (Best, 1935)

A phosphatide, *lecithin* (Hershey, 1930), cures fatty liver in depancreatized dogs. The active constituent in lecithin is *choline.* Choline is synthesized in mammals from *amino acid precursors of ethanolamine* and from *methyl group donors* (54).

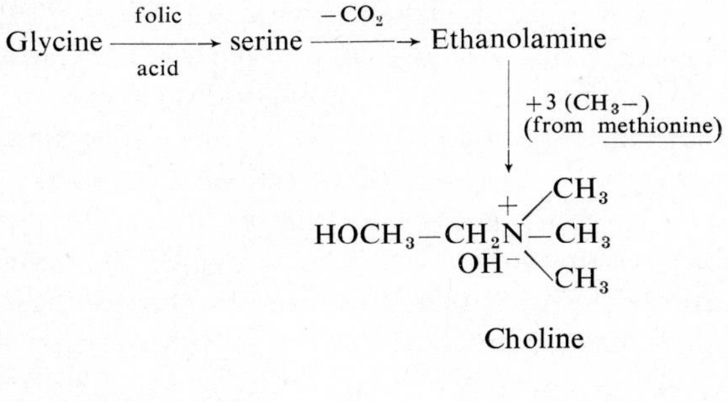

$$\text{Glycine} \xrightarrow[\text{acid}]{\text{folic}} \text{serine} \xrightarrow{-CO_2} \text{Ethanolamine}$$

$$\Big\downarrow \begin{array}{l} +3\,(CH_3-) \\ (\text{from }\underline{\text{methionine}}) \end{array}$$

$$HOCH_3-CH_2\overset{+}{N}{\underset{OH^-}{}}\!\!\nwarrow^{CH_3}_{\searrow CH_3}{\!-}CH_3 \tag{54}$$

Choline

Thus, all the compounds reacting in this metabolic chain are *potentially lipotropic* in that they are precursors of choline.

Birds are unable to effect the initial methylation of ethanolamine.

A source of methyl groups is necessary for transmethylation which is an important reaction in the metabolism of fatty acids. *Methionine* $CH_3 \cdot S \cdot CH_2 \cdot CH_2 \cdot CHNH_2 \cdot COOH$ and *betaines* are *methyl donors*. Betaines are obtained by complete methylation of amino acids; they belong to the *quaternary ammonium* derivatives—of great importance in physiology (cf. acetylcholine)—with the following generalized structure:

$$R-CH-COOH$$
$$(CH_3)_3 \equiv \overset{|}{N^+}-OH^-$$

The most simple betaine is derived from glycine and has the following structure:

$$CH_2-COOH$$
$$(CH_3)_3 \equiv \overset{|}{N^+}-OH$$

(1) *Methionine.* Methionine participates in transmethylations in an "active" form. In this case, however, the "active" derivative is not formed by combination with CoA but is a combination of *methionine with adenosine* (the nucleoside of adenine).

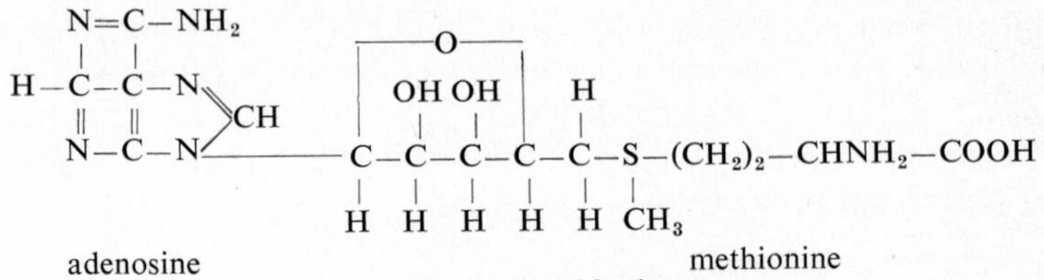

adenosine   methionine

Adenosylmethionine

(2) *Choline* can act as a *methyl donor* after oxidation to betaine. This reaction requires *folic acid, vitamin B_{12}, flavin adenine dinucleotide* and NAD and explains the importance of these compounds in fat metabolism.

It is possible that this reaction may have a role in nervous conduction (starting with acetylcholine). However, the question to be considered is how choline influences fat metabolism.

Using long chain fatty acids, labelled with ^{14}C, Artom (1953) showed that *hepatic tissues* of rats *deficient in choline* oxidize fatty acids less efficiently than normal hepatic tissue. Administration of choline to deficient rats results in normal oxidation of fatty acids. Since this action does not take place *in vitro*, the *lipotropic effect of choline is related to its indirect participation in the oxidation of fatty acids in the liver.*

However, other compounds have been described as being lipotropic and include inositol. The mode of action of inositol has not been elucidated although lipositols may be involved.

The existence of a "lipocaic hormone" with lipotropic activity in pancreas has been suggested (Dragstedt, 1936); however, Best and also Chaikoff have shown that the lipotropic activity of pancreatic extracts is dependent partly on the *presence of choline* and partly on the *presence of trypsin* which liberates methionine, a precursor of choline, from dietary proteins.

Oxidation of lipids

Following the breakdown of triglyceride by lipase to give glycerol and fatty acids, *glycerol* is destined to form glycogen in the liver while the *fatty acids* are oxidized in the *mitochondria* to give CO_2 and H_2O.

A) Oxidation of saturated fatty acids

(1) β-oxidation (Knoop)

Consider a fatty acid:

$$\underset{\alpha}{COOH-CH_2}-\underset{\beta}{CH_2}-CH_2 \ldots CH_2-CH_2-CH_2-\underset{\omega}{CH_3}$$

Oxidative cleavage would occur at the level of the β carbon such that *at each oxidation a two-carbon fragment is lost:* stearic acid with an 18-carbon chain gives palmitic acid with a 16-carbon chain which in turn gives a C_{14} acid, etc. The first stage of β-oxidation, which occurs in the liver, involves activation of the fatty acid by combination with coenzyme A (which again indicates the widespread importance of CoA in metabolism). A series of reactions involving notably flavoproteins leads to the formation of acetyl-CoA which can enter the Krebs cycle. Alternatively, two molecules of acetyl-CoA can condense to give acetoacetyl-CoA which is converted in the liver to free acetoacetic acid. Acetoacetic acid passes into the blood and is transported to different tissues where it can be oxidized (cf. ketogenesis).

(2) ω-oxidation (Verkade)

Fatty acids with an *even* number of carbon atoms are *ketogenic*. Fatty acids with an *uneven* number of carbon atoms are supposed *not to be ketogenic* and could be included in the diet of diabetics.

When *undecanoic acid* (C_{11}), obtained by esterification of three alcohol groups of glycerol, is injected to dog or man a dicarboxylic acid *still containing* C_{11} *(undecanedioic acid)* is found in the urine *(diaciduria)*.

$$CH_3-(CH_2)_9-COOH \rightarrow HOOC-(CH_2)_9-COOH \qquad (55)$$
$$\text{Undecanoic acid} \qquad \text{Undecanedioic acid}$$

This is an example of ω-oxidation, i.e. oxidation of the terminal CH_3 group.

This dicarboxylic acid itself can undergo β-oxidation with the removal of two-carbon fragments to give finally *diacetyl* $CH_3 \cdot CO \cdot CO \cdot CH_3$.

ω-Oxidation also exists for fatty acids with an even number of carbon atoms. However, the length of the chain of fatty acids determines the susceptibility to β-oxidation or ω-oxidation; compounds containing C_{10}, C_{11} or C_{12} are more susceptible to ω-oxidation while C_{18} acids first undergo β-oxidation and then ω-oxidation. The mechanisms of ω-oxidation have not been fully elucidated.

B) Oxidation of unsaturated fatty acids

Unsaturated fatty acids (ethenoid) can represent the oxidized forms (by dehydrogenation) of *saturated fatty acids*.

Ingested *palmitic acid* leads to the synthesis of *stearic acid* which on dehydrogenation will give *oleic acid* while some palmitic acid will be dehydrogenated directly to *palmitoleic acid*.

An *unsaturated* fatty acid can be *saturated* in the organism (for example, oleic acid forms stearic acid). These compounds can undergo β- and ω-oxidation.

Ketogenesis

Under normal conditions, the blood of man contains 20 mg ketone derivatives per l and about 1 mg passes daily into the urine. Thus, ketonemia and ketonuria occur under physiological conditions. *The ketones are derived from the degradation of lipids.* They include:

acetoacetic acid
acetone
β-hydroxybutyric acid:

$$CH_3-\overset{\overset{\textstyle O}{\|}}{C}-CH_2-\overset{.\ .\ .\ .}{\vdots COO\vdots}H \xrightarrow{\ -CO_2\ } CH_3-\overset{\overset{\textstyle O}{\|}}{C}-CH_3 \qquad (56)$$

$$\text{Acetoacetic acid} \qquad\qquad\qquad \text{Acetone}$$

In fact, β-hydroxybutyric acid *is not a ketone derivative* but as it accompanies acetone and acetoacetic acid from which it originates (equation 57) β-hydroxybutyric acid is classed *incorrectly* as a ketone body.

$$\text{Acetoacetic acid} \underset{NADH_2}{\overset{-2H}{\rightleftharpoons}} CH_3-CHOH-CH_2-COOH \qquad (57)$$

Ketone bodies accumulate in the blood in large quantities *(ketonemia)* and are eliminated as well in the urine *(ketonuria)* when *carbohydrate utilization is disturbed* and particularly in sugar diabetes or excessive fasting when the carbohydrate reserves are depleted and *the organism derives energy from lipids. When the production of acetoacetic acid by condensation of acetyl CoA* (cf. p. 131) *exceeds the capacity of the tissues to oxidize ketone bodies* they are distributed in the blood and pass into the urine. *Ketogenesis* occurs principally in the *liver*.

The result of excessive ketonemia is that a potentially lethal *acidosis* (hyperacidemia) develops.

Glucose is considered to be antiketogenic since excessive ketogenesis can be reduced by the administration of carbohydrates (in the case of prolonged fasting) or carbohydrates and insulin (in sugar diabetes), glucose administration leading to economy in the utilization of lipids.

Metabolism of ketone bodies

Acetoacetic acid can be metabolized by various pathways:

(1) by the Krebs cycle to give CO_2;
(2) it can form hepatic *cholesterol* in rat. This was demonstrated by Brady and Gurin (1951) using ^{14}C-acetoacetic acid;
(3) it can function in *lipogenesis;*
(4) decarboxylation of acetoacetic acid gives *acetone*.

Acetone can be metabolized in two ways:
(a) *oxidation to pyruvic acid*

$$\begin{array}{ccc} CH_3 & & CH_3 \\ | & & | \\ C=O & \longrightarrow & C=O \\ | & & | \\ CH_3 & & COOH \end{array}$$

(b) *transformation to propanediol:*
Evidence for the following reaction was obtained by Rudney (1954) using ^{14}C-labelled acetone:

$$\left(\underset{\text{Acetone}}{CH_3-\overset{\overset{\textstyle O}{\|}}{C}-{}^{14}CH_3} \rightleftharpoons \underset{\text{Enol acetone}}{CH_3-\overset{\overset{\textstyle OH}{|}}{C}={}^{14}CH_2} \right) \xrightarrow{+H_2O} \underset{\text{Propanediol}}{CH_3-\overset{\overset{\textstyle OH}{|}}{C}H-{}^{14}\overset{\overset{\textstyle OH}{|}}{C}H_2} \quad (58)$$

Administration of ^{14}C-propanediol to rat results in the appearance of radioactive carbon in the methyl group of *choline* (cf. p. 130) and also in the amino acid, *serine*. Thus, the metabolic pathway of propanediol is important and in addition:

(1) propanediol can enter into the synthesis of carbohydrates, oxidation giving pyruvic acid; as a result, the formation of propanediol constitutes a route for the interconversion of lipids and carbohydrates;

(2) serine is an amino acid and thus, propanediol formation constitutes a pathway for the interconversion of lipids and proteins.

In considering the mechanisms of interconversions, mention should be made of the fact that many *amino acids are glycogenic*. By this pathway amino acids and proteins can contribute to the formation of lipids. Moreover, as we shall see, certain amino acids are *ketogenic* (cf. p. 142) and by this route they can also form fats. We have just seen that the three fundamental types of chemicals in living organism can be transformed, one into the other, during the course of metabolism. The multiple pathways for interconversion represent the most efficient form of metabolic regulation.

Cholesterol

Egg yolk, animal fats, liver oils are the main sources of dietary cholesterol which is absorbed in the intestine in the presence of bile salts and after esterification by cholesterol esterase (cf. p. 119). However, plant sterols *(sitosterols)*, although they can be esterified in the intestine, are not absorbed. It is now recognized (Favarger *et al.*) that esterification is not an indispensable condition for the absorption of sterols.

Blood transport of cholesterol

Cholesterol passes from the *lacteals* into the *blood* which becomes turbid. Cholesterol is carried by high molecular weight β-lipoproteins (1,300,000).

The deposition of cholesterol and other lipids in the arterial wall is a condition called *arteriosclerosis* (hardening of arteries). In my opinion, the β-lipoproteins, in the presence of Ca^{++}, precipitate with sulphated mucopolysaccharides (metachromatic) of the inner wall of the vascular system. This disorder is more severe in individuals receiving a diet rich in animal fats and it should be noted that herbivores, as for example rabbits, are particularly sensitive to a diet rich in cholesterol.

It had been claimed that polyethenoid fatty acids are efficient anticholesterolemic agents.

Metabolism of cholesterol

Biogenesis of cholesterol occurs principally *in the liver* but also in skin, adrenal cortex, aorta, testes etc.; a total of 1–2 g/day. *Acetate* is the *direct precursor* of cholesterol; this synthesis is *extremely rapid* as shown by experiments using ^{14}C-acetate.

A mechanism by which cholesterolemia is regulated must exist in the liver; the post-absorptive cholesterol content of the liver increases considerably after a meal rich in cholesterol but the blood level of cholesterol does not increase until later. However, the mechanism for this has not been elucidated.

Cholesterol is excreted in the bile in the form of bile salts and also by cutaneous and intestinal routes but not to any extent in the urine.

Cholesterol levels are modified by certain hormones. Hyperactivity of the thyroid lowers the total cholesterol in blood. Parallel administration of cholesterol and oestrogens to chicken results in increased susceptibility of aortic regions to arteriosclerosis compared with the peripheral arteries. This demonstrates the existence of differential sensitivity in related tissues to chemical impregnation.

Phosphatides

These important compounds are constituents of cells and nerves (the myelin sheath is rich in sphingomyelins), release choline, etc.; they are synthesized in the organism from phosphatidic acid and phosphorylcholine acting in the presence of cytidine triphosphate on a 1,2-diglyceride. Phosphatides are hydrolysed by *lecithinases* localized principally in kidney, liver and pancreas.

Lecithinase A, which is found in the venom of certain snakes and in pancreas, removes a fatty acid from a molecule of lecithin to give *lysolecithin*. Lysolecithin is strongly "surface" active—and therefore, hemolytic—from which it can be thought that it would favour the intestinal absorption of various compounds such as cholesterol and triglycerides, (cf. p. 118). Lysolecithin is rendered non-hemolytic by lecithinase B which removes a fatty acid residue from lysolecithin.

III. Metabolism of proteins and amino acids

After absorption, amino acids pass almost exclusively into the portal circulation to give a plasma concentration of about 80 mg/l in man. This value is higher than the normal content of 30 mg/l and excess amino acids are resorbed within several hours principally by the liver.

1. Biosynthesis of proteins

One of the important functions of amino acids in the organism is the the reconstruction of proteins. The hypothesis formulated independently by Casperson (1941) and J. Brachet (1942) in which ribonucleic acids are associated with the biosynthesis of proteins has been shown to be correct. The animal cell contains *ribosomes*, associated with the endoplasmic reticulum (Borsook; Hoagland and Zamecnik; T. Hultin), which represent the sites of protein formation. The mechanisms of protein biosynthesis can be summarized as follows:

(1) in the presence of an energy source, ATP, and activating enzymes, amino acids are converted to acyl-adenylates;

(2) each acyl-adenylate is fixed by the activating enzyme to a molecule of ribonucleic acid (discovered by Hoagland *et al.*). This has been called *soluble ribonucleic acid* (RNA$_s$) or *carrier RNA* (it can represent up to 15 per cent of total cellular RNA). Each amino acid has its own carrier ribonucleic and therefore, at least 20 forms of carrier RNA must exist. (The activation of an amino acid by attachment to RNA resembles the activation of acetate by coenzyme A);

(3) each RNA$_s$–amino acid complex is transferred to a ribosomal particle (more exactly a complex called *polysome*); this transfer requires guanosine triphosphate.

The biosynthesis of proteins must be considered in relation to the specificity of proteins. Let us consider that protein specificity depends on gene specificity ("one gene → one enzyme"); desoxyribonucleic acid communicates its code in such a way that, at the ribosomal level, the amino acids are arranged in the required sequence at the moment of the formation of the peptide bonds. The information is carried by a ribonucleic acid image of desoxyribonucleic acid; this RNA is called "messenger RNA" (F. Gros *et al.*; Brenner *et al.*). The code uses at least three nucleotides to determine the fixation of one amino-acid (triplet). Assuming the symbol U for uridylic acid residues, poly-U (synthetic) gives the indefinite repetition of the same sequence *phenylalanine* (Nirenberg and Matthaei, 1961); the system U-U-U- is then called the *codon* ("code word") for phenylalanine. The theory on control of protein biosynthesis at the genetic level was given by Jacob and Monod (1961) but it is out of the scope of this book to develop that point.

In 1953, Snoke, Yanari and Bloch synthesized glutathione *in vitro*, this is to say that they realized, for the first time, *the synthesis of a peptide bond*, R·CO·NH·R′. This tripeptide (glutamyl-cysteyl-glycine) can be synthesized in the presence of acellular extracts of pigeon liver and with ATP as an energy source (59):

$$
\text{Glutamic acid} + \text{Cysteine} \xrightarrow{\hspace{3cm}} \text{Glutamyl-cysteine}
$$

$$
\text{peptide synthetase} + \left(\begin{array}{c} \text{ATP} \\ \text{Mg}^{++} \\ \text{K}^+ \end{array} \right) \xrightarrow{\hspace{1cm}} + \text{glycine} \;\;\downarrow \;\; \text{Glutathione} \qquad (59)
$$

(2) Transpeptidation

Pancreatic extracts contain enzymes, *transpeptidases*, which can transfer the glutamic acid from glutathione to another amino acid with the formation of a new peptide bond; thus, with leucine, glutamyl-leucine is formed.

Tripeptides and even polypeptides can be obtained *in vitro* by similar mechanisms. The reaction requires, in addition to glutathione, one molecule of ATP for each peptide bond formed.

Hormones influence protein synthesis. The incorporation of leucine and valine into hepatic proteins is activated by thyroxine (Sokoloff and Kaufmann). The nature of the processes by which *pituitary growth hormone* and testosterone influence anabolic mechanisms has not been fully elucidated but the number of polysomes varied with the amount of growth hormone available to the rat and growth hormone stimulated the labelling of liver nucleic acid (Talwar *et al.*, 1962) including that of messenger RNA (Korner, 1964). However, it is thought that insulin acts by facilitating glycolysis which provides the energy (ATP) for the synthesis of peptide bonds. *Adrenaline* and *thyrotropic hormone* act *indirectly* in that they decrease the total circulating amino acids. Administration of aldosterone to rat increases the total serum proteins to a greater extent in the intact animal that in adrenalectomized animals (Pérès and Zwingelstein, 1961).

"Dynamic equilibrium" (Whipple, 1938) describes the *continuous exchange between plasma proteins and tissue proteins.*

A certain fraction of plasma proteins penetrates the walls of the capillaries to be incorporated into the interstitial fluid and some proteins then pass into the blood stream by way of the lymphatic system. Using proteins labelled with ^{15}N, the duration of this circulation, blood–lymph–blood, can be estimated and is of the order of 30 minutes in mammals.

General metabolism of α-amino acids

Transamination

In 1937, Braunstein and Kritzman showed the presence in liver, heart and brain of birds and mammals of a reaction of considerable biological importance which regulates the biogenesis of amino acids: certain enzymes, the *transaminases*, catalyse the transfer of an amino group from *glutamic* acid and *aspartic* acid to α-keto acids.

The general reaction is written (60):

$$\underset{\substack{|\\R-CH-CO_2H}}{\overset{NH_2}{|}} + \underset{\substack{\|\\R'-C-CO_2H}}{\overset{O}{\|}} \;\rightleftharpoons\; \underset{\substack{\|\\R-C-CO_2H}}{\overset{O}{\|}} + \underset{\substack{|\\R'-CH-CO_2H}}{\overset{NH_2}{|}} \qquad (60)$$

Thus, not only is there formation of an amino acid but also of a new keto acid. The details of a reaction are as follows:

Glutamic acid + Pyruvic acid $\underset{\text{transaminase}}{\rightleftharpoons}$ α-Keto glutaric acid + Alanine (component of Krebs cycle)

Pyridoxal phosphate (cf. vitamin B_6, p. 69) is the *cotransaminase*. Amino acids such as *lysine* and *threonine* never undergo transamination while *glutamine* and *asparagine*—in the place of aspartic and glutamic acids—can serve as donors of amino groups: a special *amide* derivative is formed as an intermediate and *ammonia* is the final product (Meister, 1954.) The se-

10

quence of reactions from glutamine is shown in equation (61). This equation gives an indication of the value of radioactive elements (^{15}N in this case) which allows one to follow the fate of molecular constituents.

$$
\begin{array}{ccc}
CO^{15}NH_2 & CO^{15}NH_2 & COOH \\
| & | & | \\
CH_2 \qquad O & CH_2 & CH_2 \\
| \qquad \parallel & | & | \\
CH_2 + R-C-CO_2H \rightleftharpoons & CH_2 + R-CHNH_2-CO_2H \rightarrow & CH_2 + {}^{15}NH_3 \\
| & | & | \\
CHNH_2 & C=O & C=O \\
| & | & | \\
CO_2H & CO_2H \quad \text{Amino acid} & CO_2H \\
\text{Glutamine} & \alpha\text{-Keto} & \alpha\text{-Keto} \\
& \text{glutaric amide} & \text{glutari cacid}
\end{array} \tag{61}
$$

Analogues of vitamin B_6, such as deoxypyridoxine, which are antivitamins specifically inhibit transaminases.

Decarboxylation

This is a reaction characteristic of bacteria (putrefaction)

$$
\begin{array}{cc}
NH_2 & NH_2 \\
| & | \\
R\,CH\cdot COOH \rightarrow & R\,CH_2 + CO_2 \\
& \text{Amine}
\end{array} \tag{62}
$$

In this way, cadaverine is formed from lysine, tyramine from tyrosine, taurine from cysteic acid, putrescine from ornithine, 5-hydroxytryptamine from 5-hydroxytryptophan and histamine from histidine.

Deamination (Neubauer, 1909)

In this process an α-amino group is removed from amino acids with the liberation of ammonia (this reaction differs from transamination and shows the analogy between amino-acid transaminases and peroxidases on the one hand and deaminases and catalases on the other). This oxidative reaction occurs principally in the *liver* but also in *kidney*. The first step is a dehydrogenation in the presence of amino acid oxidase (a flavoprotein, cf. p. 67) producing water and an imino acid $[R—C=NH—(CO_2H)]$ which finally is hydrolysed spontaneously with the liberation of ammonia and the formation of an α-keto acid. The overall reaction is given in equation (63)

$$
\begin{array}{cc}
NH_2 & O \\
| & \parallel \\
R-CH-CO_2H + \tfrac{1}{2}O_2 \xrightarrow{\text{amino acid oxidase}} & R-C-CO_2H + NH_3
\end{array} \tag{63}
$$

Let us now consider the fate of the products of deamination.

(1) *Keto acids* can react in many ways to give CO_2 and H_2O, carbohydrates, lipids or even amino acids by reamination.

Administration of large amounts of leucine, tyrosine or phenylalanine to animals leads to the formation of increased amounts of acetoacetic acid and other ketone bodies. Isoleucine and hydroxyproline are *both* ketogenic and glycogenic although both properties are weak.

(2) *Ammonia* can take part in aminations or can be transformed to ammonium salts in mammals (ammonemia is high in crustaceans and molluscs). However, the *largest fraction of ammonia is utilized in the formation of urea*.

Ureogenesis

Urea (discovered in 1773 by Rouelle) is the main product of *protein* catabolism in man (cf. p. 148) and represents the highest percentage of nitrogen eliminated in the urine. [The synthesis of urea was accomplished by Woehler (1828) and was the first organic synthesis from inorganic materials; it had great importance in chemical history and philosophy].

The liver is the only organ capable of synthesizing urea; in hepatectomized mammals there is a rapid decrease in total urea in blood and urine and a parallel increase in total ammonia and amino acids. The formation of urea is a detoxication process (with respect to ammonia). Krebs and Henseleit (1932), while studying the formation of urea in liver slices in the presence of ammonia at 37°C and pH 7, found three amino acids to be involved: *ornithine, citrulline* and *arginine*. *Arginase* (Kossel and Dakin, 1904) which is localized exclusively in liver in mammals acts on arginine to give urea and ornithine. Arginase is not present in the liver of birds and no urea is formed; *uric acid* is the principal end product of protein catabolism in all animals which do not contain arginase; the mechanism by which uric acid is formed from ammonia is not known. Mammals, batrachians and elasmobranchs produce urea and are said to be *ureotelic* (cf. p. 148).

Arginase is present in bony fish but in these animals the principal end product of protein catabolism is ammonia. The ammonia is derived from glutamine. Baldwin states that teleosts and marine elasmobranchs also excrete trimethylamine.

The Krebs and Henseleit cycle (which must not be confused with the tricarboxylic acid cycle of Krebs) can be represented diagrammatically in the following way:

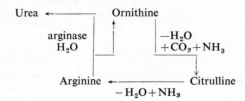

The introduction of CO_2 and NH_3 is achieved through the intermediary of "carriers" and ATP. The processes are as follows:

(1) *In the formation of citrulline from ornithine* the first reaction is a *transcarbamylation* in which *carbamylphosphate* is the intermediary (Lipmann *et al.*, 1955) (64):

$$CO_2 + NH_3 + ATP \rightarrow H_2N - \overset{\overset{O}{\|}}{C} - O - \overset{\overset{O}{\|}}{P} = (OH)_2 \qquad (64)$$

The carbamyl group is transferred to ornithine in the form of carbamyl-aspartate (carrier) and citrulline is formed.

(2) *There are two stages in the transformation of citrulline to arginine:*

(a) citrulline condenses with *aspartic acid* (in the presence of ATP and Mg^{++}) to give *argininosuccinic acid* (Ratner, 1953);

(b) *fumaric acid* is eliminated from argininosuccinic acid by the action of an enzyme related to aspartase. Fumaric acid enters the citric acid cycle and gives malic acid and finally oxaloacetic acid which, on transamination, forms aspartic acid which can re-enter the cycle.

$$CO_2H - CH = CH - CO_2H$$
Fumaric acid

Ammoniogenesis

Mention has already been made of the fact that metabolic activity in mammals leads to the production of large quantities of ammonia but this does not accumulate in the blood as the liver has a high capacity for ammonia fixation. However, the intestinal bacteria acting on nitrogenous compounds produce a large amount of ammonia such that the *ammonia content of portal blood is high*. Antibiotic treatment (for example, with aureomycin) blocks intestinal ammoniogenesis (J. Stahl *et al.*, 1961). In cirrhosis of the liver a coma can develop which is related to accumulation of ammonia (Pavlov *et al.*, 1893–1896; Burchi, 1927); hyperammoniemia leads to electroencephalographic disturbances (J. Stahl *et al.*). In addition, the kidney produces ammonia which participates in the regulation of acid-base balance; production increases during acidosis and decreases during alkalosis.

Contrary to the concept held for many years, *renal ammonia is not produced from urea*; if radioactive urea is injected no labelled ammonia can be detected in the urine. The source of renal ammonia is *glutamine* which is widely distributed in blood plasma and is deaminated in the presence of renal *glutaminase* (65):

$$\underset{\text{Glutamine}}{\overset{\overset{\displaystyle NH_2}{|}}{CO} - (CH_2)_2 \overset{\overset{\displaystyle NH_2}{|}}{CH} - CO_2H + H_2O} \xrightarrow[\text{ATP}]{\text{glutaminase}} \overset{\overset{\displaystyle OH}{|}}{CO} - (CH_2)_2 - \overset{\overset{\displaystyle NH_2}{|}}{CH} - CO_2H + NH_3 \qquad (65)$$

We shall consider *ammoniotelics* later (p. 149).

We shall examine now certain aspects of the metabolism of amino acids. As this study has not here to be exhaustive but conceptual, I have chosen, for reasons which will become apparent, the following amino acids: *glycine, cysteine, phenylalanine, tyrosine* and *tryptophan.*

Metabolism of glycine

(1) Glycine is *glycogenic* and *antiketogenic.*

(2) Glycine is synthesized by the oxidation of *choline* to give betaine which, after demethylation, gives glycine.

(3) Glycine participates in numerous syntheses and notably in the biogenesis of purines, glutathione and creatine.

(4) Glycine can be oxidised in *two ways:*
   (a) by conversion to serine $HOCH_2 \cdot CHNH_2 \cdot COOH$;
   (b) by the formation of *glyoxylic acid* $CHO \cdot CO_2H$ in the presence of an amino acid oxidase, *glycine oxidase*, a flavoprotein.

Metabolism of cysteine

Cysteine metabolism has been studied extensively by Claude Fromageot.

(1) *Synthesis by transulphuration:* the stages of this reaction which occurs in the presence of *transulphurase* are shown in equation (66):

$$\text{Methionine} \xrightarrow{\text{demethylation}} \text{Homocysteine} + \text{Serine} \xrightarrow{-H_2O} \text{Cystathioneine}$$

pyridoxal phosphate $\uparrow$ $\rightarrow$ $+H_2O$

Cysteine + Homoserine

(66)

(2) *Oxidation of cysteine is spontaneous* and occurs rapidly in aqueous solutions; this is an example of a reaction which, by virtue of its energetics, does not require an enzymic system. Two molecules of cysteine combine to give one molecule of cystine with the transformation of the thiol groups to a disulphide bridge (67).

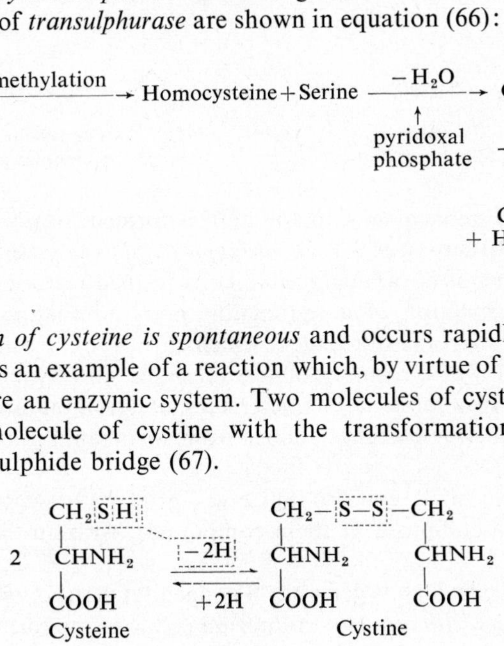

$$2 \begin{array}{c} CH_2SH \\ | \\ CHNH_2 \\ | \\ COOH \\ \text{Cysteine} \end{array} \xrightarrow[+2H]{-2H} \begin{array}{c} CH_2-S-S-CH_2 \\ | \qquad\qquad | \\ CHNH_2 \qquad CHNH_2 \\ | \qquad\qquad | \\ COOH \qquad COOH \\ \text{Cystine} \end{array} \qquad (67)$$

(3) The *glycogenic activity* of cysteine arises from its transformation to pyruvic acid, one of the mainstays of carbohydrate metabolism (68):

$$
\begin{array}{ccc}
CH_2SH & & CH_3 \\
| & & | \\
CHNH_2 & \rightarrow & C=O \quad +H_2S+H_3N \\
| & & | \\
COOH & & COOH \\
\text{Cysteine} & & \text{to glucose}
\end{array} \tag{68}
$$

(4) Cysteine activates numerous enzymes by the presence of SH^- groups (as does glutathione, discovered by Hopkins).

(5) It participates in the formation of skin and hairs.

Metabolism of phenylalanine and tyrosine

Classically, these two amino acids are studied together although phenylalanine is an essential amino acid in man and tyrosine is non-essential. In fact, these two compounds are glycogenic and ketogenic via fumaric acid and acetoacetic acid; moreover, phenylalanine can be converted to tyrosine (in the presence of a hydroxylase, O_2, $NADPH_2$ and Fe^{++}). This reaction is not reversible which explains the inability of tyrosine to replace phenylalanine in the diet.

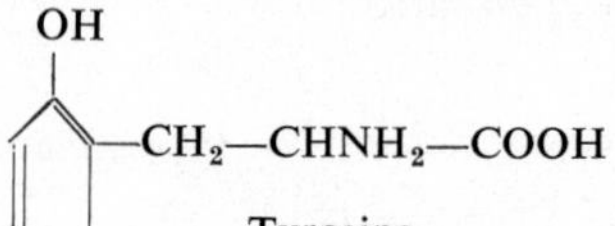

OH

—CH_2—$CHNH_2$—COOH
Tyrosine
(phenolic group)

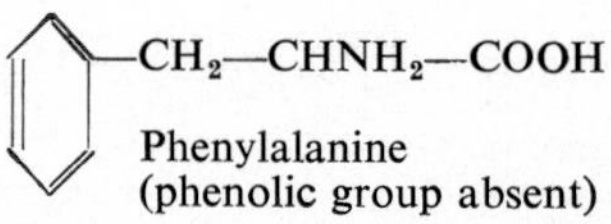

—CH_2—$CHNH_2$—COOH
Phenylalanine
(phenolic group absent)

Melanin, the pigment of skin and hair, is formed from tyrosine with the intermediate formation of *3,4-dihydroxyphenylalanine* or dopa. Tyrosine is also a precursor of noradrenaline which is formed via dopa and hydroxytyramine. Methylation of noradrenaline gives adrenaline. Tyrosine is a precursor of thyroid hormones which are iodotyrosine derivatives. Decarboxylation of tyrosine produces *tyramine* with hypertensive properties; the sympathomimetic characteristics of tyramine are derived from its ability to liberate catechol amines from chromaffin granules of certain cells.

However, we shall concern ourselves principally with an abnormal aspect of the metabolism of these compounds, with an "inborn error of metabolism".

In *alcaptonuria* which is generally a congenital disease, *the urine becomes brown on contact with air*. This coloration is due to the formation of *homo-*

gentisic acid which derives from metabolism of tyrosine and phenylalanine and which oxidizes to give acetoacetic acid and fumaric acid.

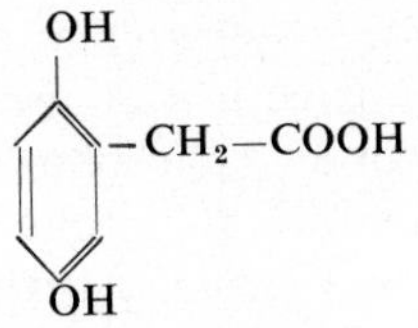

OH
$-CH_2-COOH$
ÓH   Homogentisic acid

Alcaptonuria also occurs in scorbutic guinea pigs; this induced alcaptonuria disappears on administration of *vitamin C*. Vitamin C is ineffective in the treatment of congenital alcaptonuria.

Metabolism of tryptophan

(1) Tryptophan, the only amino acid to contain an *indole nucleus*, is an essential amino acid for man but since it can be transaminated the corresponding keto acid—*indolyl-3-pyruvic acid*—can provide an exogenous source.

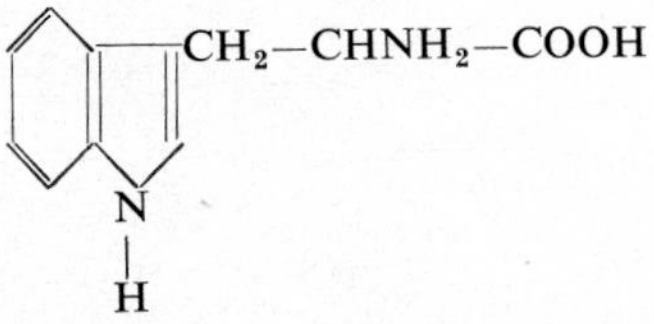

CH_2-CHNH_2-COOH
N
H

(2) The urine of dogs and rabbits contains *kynurenic acid* (isolated by Liebig in 1853) and *kynurenine* which are derived from tryptophan. Both these compounds are of interest in physiological genetics. Objective and, where possible, chemical criteria are required for genetic interpretation and in addition to the labelling of genes by enzymes, other compounds can be utilized. It is essential that these compounds be recovered as easily as possible. Kynurenine is the precursor of a pigment in the eye of certain insects. Butenandt (1954) has demonstrated, using compounds labelled with ^{14}C, that one of these pigments, *xanthommatin*, in particular from *Calliphora erythrocephala*, is derived from kynurenine.

(3) In man, dog, rat, rabbit and pig, *tryptophan can be converted to nicotinic acid* and, therefore, functions in the synthesis of NAD and NADP. In pellagra there is generally a deficiency in both tryptophan and niacin; this can be remedied not only by the administration of tryptophan but also of indolpyruvic acid.

(4) Oxidation of tryptophan gives *5-hydroxytryptophan*. This compound, on decarboxylation, produces *5-hydroxytryptamine* or *serotonin* (Page, 1948; Erspamer; Rapport, 1954), a powerful *vasocontrictor*.

Creatine and arginine

Finally, in this brief examination of the metabolism of certain amino acids, mention must be made of the synthesis of a compound of considerable biological significance—*creatine*. Creatine is present mainly in muscle and brain, either in a free state or as a phosphorylated derivative (phosphocreatine = *phosphagen* of Eggleton). It is found also in blood and traces of creatine are present in urine. *Creatinine* is the anhydride of creatine; it is also found in blood and urine. In man, there is a constant relationship between the total excreted and size; thus, the law of sizes (cf. p. 25) could also be expressed as a function of creatinuria.

The biosynthesis of creatine and creatinine is accomplished by the following sequence of reactions (69)

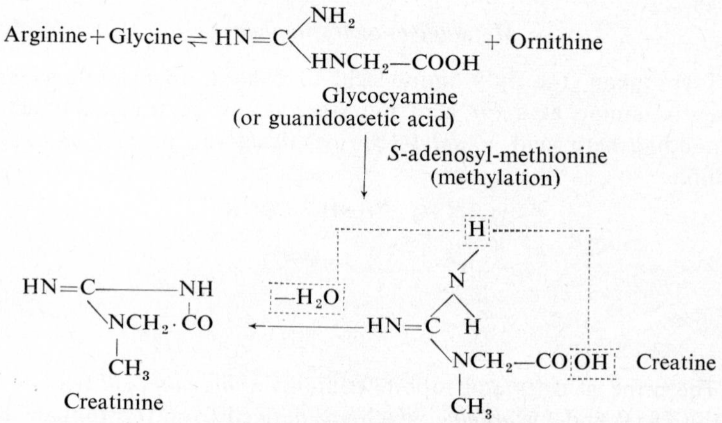

Arginine participates in this synthesis but in the muscles of invertebrates *arginine-phosphoric acid* replaces *creatine-phosphoric acid*. It would appear that invertebrates possess a more simple pathway for the synthesis of a compound whose function is equivalent to a compound which, in the vertebrates, is synthesized by a more complex pathway. J. Roche, Nguyen-Van Thoai and Robin (1957) have shown that creatine is more widely distributed than believed by certain English workers (Needham, Baldwin). Creatine is found in sea urchins, polychaete annelids and sponges but is not found in arthropods, molluscs, gephyrians and nematodes.

The catabolism of arginine in invertebrates can be presented diagrammatically as in Fig. 33 (after J. Roche and Y. Robin, 1962):

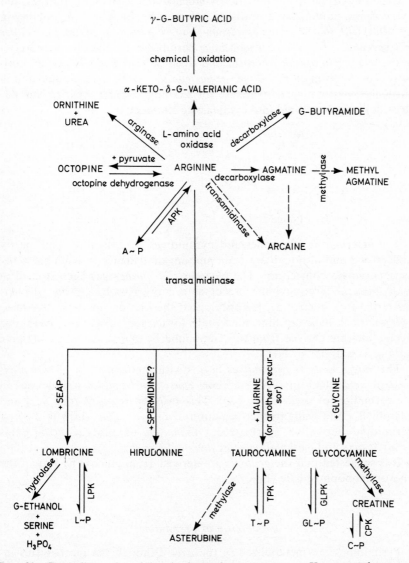

FIG. 33. Catabolism of arginine in invertebrates. ———— Known pathways. −−−−−− Hypothetical pathways. G. = guanido. A ~ P, C ~ P, GL ~ P, L ~ P and T ~ P = arginine-, creatine-, glycocyamine-, lombricine- and taurocyamine-phosphates. APK, CPK, GLPK, LPK and TPK = arginine-, creatine-, glycocyamine-, lombricine-, taurocyamine phosphokinases.

IV. Metabolism of nucleic derivatives

Experiments performed with ^{15}N-labelled derivatives of purines and pyrimidines, administered orally or parenterally, have shown (notably in rat) that, *with the exception of adenine, no free purines or pyrimidines either of exogenous* (dietary) *or endogenous* (produced by nucleic metabolism in tissues) *origin are incorporated into tissue nucleic acids.* However, derivatives in the form of *nucleotides* or *nucleosides* can be incorporated and in particular, pyrimidine derivatives; for example, *cytidine* (pyrimidine nucleoside) is incorporated into cytoplasmic ribonucleic acid.

Purine   Pyrimidine

The *biosynthesis* of purines and pyrimidines by animals is only partly understood and although a certain amount of evidence is available many aspects remain conjectural. The *biogenesis of purines* has been studied in man, rat and birds and the precursor is known to be *glycine* which is converted to a nucleotide. It appears that the *first* purine derivative *to be synthesized* is hypoxanthine nucleotide (or *inosinic acid)* and guanidine nucleotides are derived from this compound by *amination* with glutamine serving as a donor of NH_2.

The *biosynthesis of pyrimidines* begins with the formation of *carbamylaspartic acid* which involves *carbamyl phosphate* which also functions in the formation of urea (cf. p. 139). This reaction requires folic acid and vitamin B_{12} and leads to the formation of *uridylic acid* which is the first pyrimidine derivative synthesized. *Amination* of uridylic acid yields *cytidylic acid* and *methylation* gives *thymidylic acid.*

It is suggested that the *pentoses* are derived from glucose by way of the "hexosemonophosphate shunt".

Catabolism of pyrimidines

Pyrimidines are metabolized in the *liver* although the mechanisms involved are *hypothetical. Uracil* (one of the three pyrimidine bases, the others being cytosine and thymine) forms respiratory CO_2 from carbon 2 of the pyrimidine nucleus. β-Alanine and carbamyl-β-alanine are excreted after administration of uracil to rats; NH_3 is formed also.

Catabolism of purines

Degradation of purines also occurs principally in the liver.
The process can be presented diagrammatically as follows:

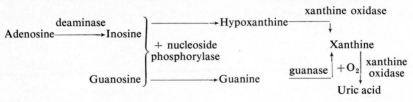

TABLE 30

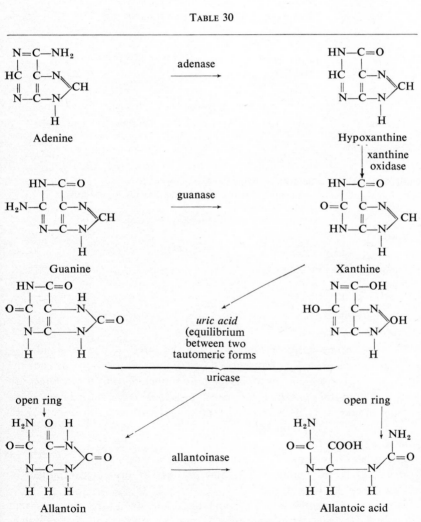

Xanthine oxidase (or the enzyme of Schardinger) is found in liver and in milk (Polonovsky); this enzyme requires riboflavin, iron and molybdenum.

Table 30 shows the reactions of the purine bases, adenine and guanine.

Table 31, from Florkin and Duchateau (1943), shows the main stages of purine catabolism in different groups of animals.

TABLE 31

↓ uric acid |	Man and primates, birds, terrestrial reptiles. Cyclostomes. Insects (with the exception of Diptera).
(uricase) ↓ allantoin |	Mammals (with exception of man and primates). Diptera. Gasteropods.
(allantoinase) ↓ allantoic acid |	A group of teleosts (salmonids, pleuronectids, anguillids).
(allantoicase) ↓ urea |	Selachians, Dipnoi, crossopterygians. A group of teleosts (cyprinids, esocids, scombrids). Batrachians. Freshwater lamellibranchs.
(urease) ↓ ammonia	Sipunculids, marine lamellibranchs, crustaceans.

Ureotelism and uricotelism

These phenomena are dependent: (1) on the *origin* of the terminal product; (2) on the presence or absence of one or another catabolizing enzyme. More precisely, the end products of *protein metabolism in mammals* are NH_3 and *urea* (these animals are called *ureotelics*) while *purine catabolism in man*, monkeys, birds, terrestrial reptiles, cyclostomes and insects (with the exception of Diptera) leads to uric acid (from which the name *uricotelic* is derived).

The solid excrement of snakes consists of a mixture of urine, rich in urates, and faecal material. The end product of *protein* catabolism in *birds* and reptiles is uric acid.

Man can be considered to be a ureo-uricotelic since urea and uric acid are found in urine. However, urea is derived from protein catabolism and uric acid from purine catabolism and it seems to me that it is preferable, therefore, to use the name *ureo-pururicotelic*.

Most mammals, Diptera and gasteropods contain a *uricase* (Batelli and Stern, 1909) which opens the purine nucleus with the formation of *allantoin*, the end product of purine metabolism in these animals (from which the name *allantoinotelic* is derived).

Certain teleosts (notably the migratory teleosts: *Salmonidae, Anguillidae*) possess an allantoinase which opens the purine ring further to give *allantoic acid*, the end product of purine catabolism (these animals are termed *allantoicotelics*).

In other teleosts *(Cyprinidae, Scombridae)*, selachians, batrachians and freshwater lamellibranchs the end product of *purine* metabolism (but not of protein metabolism) is urea produced by the action of an *allantoicase*. These animals are *special ureotelics* and could be called *purureotelics*.

Finally, uricolysis proceeds another stage in sipunculids, marine lamellibranchs and crustaceans to give ammonia by the action of *urease (ammoniotelic* animals). "It would appear, therefore, that the evolution of purine catabolism is expressed by a shortening of the enzymic chain of uricolysis, a process which could be called *enzymapheresis* (or loss of an enzyme), the primitive chain being shortened progressively by the loss of the last link" (Florkin, 1944; cf. p. 12). In short, it would appear that higher animals do not require the energy contained in uric acid. *I would suggest that there would be some balance between genes giving rise to energy processes (predominant in lower animals) and genes giving rise to morphological processes (which become progressively more predominant in higher animals).*

Finally, Needham believes that there is a *formal relationship between the type of nitrogenous excretion and the method of reproduction*: ureotelic terrestrial vertebrates would be viviparous and uricotelic vertebrates oviparous. The case of ovo-viviparity has yet to be explained.

Gout

The normal end product of *purine* metabolism in man is *uric acid* which is found mainly in urine (Scheele, 1776) but also in bile and faeces.

Gout which was mentioned by Hippocrates (460–370 B.C.) results from the *deposition of urates* in articular (digits) and non-articular cartilage (ears). It is surprising that, although gout is associated with an increase in the uric acid content of blood, hyperuricemia (such as occurs in leukemia) is not a systemic cause of gout. One can suppose, therefore, that a local factor in cartilage is involved in the deposition of urates and leads to gout. I would suggest that this selective reactivity of cartilage may be associated with some modification of the distribution of the three types of chondroitin sulphate, A, B and C, described by K. Meyer. The affinity of urates for cartilaginous tissue again indicates the existence of selectivity in the reactions of living systems. J. D. Benedict *et al.* (1952) have demonstrated, using ^{15}N-glycine, that the excess uric acid produced in subjects with gout is derived, not from purine derivatives but from amino acids and therefore, from *protein* catabolism as in the case of birds. In addition, gout symptoms can be produced in chicken by feeding with a diet rich in proteins (Oppenheimer, 1941); one wonders, therefore, to what extent the gallinaceans of Houssay (cf. p. 78) developed gout, at least in those birds in which a diet of meat proved fatal.

CHAPTER 8

THE INTERNAL ENVIRONMENT ("MILIEU INTÉRIEUR")

IN HIS *inaugural lecture* to the Cours de Physiologie générale at the Muséum d'Histoire naturelle (1876), Claude Bernard said:

"There are very simple creatures, organisms reduced to a single or small number of anatomical elements, unicellular creatures, infusoria and animals placed higher in the animal scale which are profoundly influenced by their immediate surroundings. For these, the essential environment which presents the conditions for life is the external medium which surrounds them.

"Creatures higher in organization, formed from an assemblage of elementary organisms, of histological elements, do not enter into a direct relationship with the external environment. The living elements which are more deeply situated within the organism, separated from the cosmic medium, must find, in the depths where they are situated, the same indispensable conditions surrounding them. There is a true *milieu intérieur* which is intermediate between the cosmic medium and the living substance."

Later one reads the definition of *vie constante* which Claude Bernard distinguishes from *vie latente* and *vie oscillante*.

"In lower animals or plants there is no *milieu intérieur;* in others this medium is not independent; in these two cases the creature is immediately influenced by the external medium: when this provides favourable conditions, life follows its regular course; when it ceases to provide these conditions, life is suspended in a temporary or definitive manner and the creature falls into a state of latent life or it dies. Seeds, spores of cryptogamic plants . . . rotifers, tardigrades exhibit this particular condition of life which is expressed by the name *vie latente*.

". . . all plants . . . all invertebrates and cold blooded vertebrates fall into the category of creatures with *vie oscillante*.

"Finally, *vie constante* or free life . . . belongs to higher mammals".

In fact, the concept of Claude Bernard is rightly limited since even poikilotherms exhibit a certain constancy of internal environment, as we shall see later, and the fact that they are *influenced* by external conditions does not prevent them from being *adapted* to this medium, without which life would not be possible.

151

The liquid medium which bathes the organs is the *milieu intérieur*, the internal environment. The study of the *internal environment* amounts, therefore, to a study of the internal ecology of the organism being considered. For an animal to survive and multiply it is adapted to its external environment and the internal environment is adapted to the external conditions. For *each genetic type, the adaptive conditions can be more or less strict* and the extent of the *geographical distribution* of species is related entirely to this autonomy which is developed to a greater or lesser extent in each species.

The liquid medium which surrounds the cells is derived from the *transudation* of blood from the capillaries. Starling considers that there is an osmotic equilibrium between the blood plasma and interstitial spaces which are separated by the capillary walls. The interstitial space contains an *ultrafiltrate* of blood plasma, containing no proteins and in which the concentration of electrolytes follows the Donnan equilibrium. This concept must be modified to some extent following the results obtained with radioactive elements. L. Sapirstein (1953) reached the conclusion that "the capillary wall does not act as a barrier to any compounds", macromolecules, dyes, etc. Electron micrographs of the walls of the capillaries showed the presence of internal vesicles (Moore and Ruska) formed by the engulfment of external medium and notably of proteins: this phenomenon which is exhibited widely by cellular members of different organs in mammals is called *pinocytosis* (Lewis, 1936). According to Starling, there would be ultrafiltration towards the external medium in the arteriolar extremities and ultrafiltration towards the interior in the venous extremities.

In man, the internal medium which constitutes the *extracellular fluid* corresponds to about 15–20 per cent of body weight. In birds and mammals the internal environment is regulated principally by two pairs of organs:

the lungs, which control the levels of O_2, CO_2 and H_2O (water vapour);

the kidneys which regulate the quantitative and qualitative composition with respect to non-gaseous compounds. These organs are capable of absorption and excretion.

The *glands producing internal secretions* (endocrines) also regulate the composition of the internal environment. The specific functions of these organs will be considered later. We shall concentrate now on a study of the *osmotic relationships which link the internal and external media*.

Measurement of the molecular concentration of the internal environment

(1) The *total molecular concentration* is measured using a *cryoscope* which gives the *freezing point* of the medium (an impure aqueous solution freezes at a temperature below 0°C and the higher the concentration of dissolved molecules the larger the depression of freezing point).

If $|\Delta|$ is the depression of freezing point

$$\pi = \text{osmotic pressure} = 12\cdot03 \, |\Delta|$$

Example: For seawater from the coast of France $|\Delta| = -2°C$, such that $\pi = 12\cdot03 \times 2 = 24\cdot06$ atm or $240\cdot6$ m water.

(2) The *electrolyte* concentration is determined by *measuring resistance* (with a Kohlrausch bridge and the results are related to standard curves for saline solutions of known concentration);

(3) The values from (1) and (2) are subtracted to give the concentration of *non-ionized organic molecules*.

Thus, the molecular concentrations of a medium can be determined. Table 32 contains several values of $|\Delta|$ for the internal environment (blood) of vertebrates (average values) provided mainly by P. Portier.

TABLE 32

Blood of terrestrial mammals	0·60°C
Blood of marine mammals	0·70°C
Blood of marine palmipedes	0·66°C
Blood of carp (freshwater teleost)	0·50°C
Blood of cod (marine teleost)	0·6–0·8°C (depending on biotope)
Blood of dogfish (selachian)	2·2°C

COMMENTS: (1) Marine or terrestrial mammals (with an aerial ventilatory system) can live in water but since there are no *surface membranes* where *exchange with this medium* can occur, the molecular concentration of the blood is different from that of the aqueous medium. However, a small amount of water may be absorbed when they feed (Baleinoptera);

(2) Fish possess *surface membranes where exchange* with the external aqueous medium can occur: the *gills* and *digestive tract. Selachians are normally uraemic*; marine elasmobranchs contain up to 25 g urea per l of blood but there is less NaCl; freshwater elasmobranchs contain 6 g urea/l blood and more mineral salts than marine species.

Mention should be made of the role of the digestive tract in the maintenance of physico-chemical equilibria; this tract is not only concerned with digestion and we have seen already that in the colon of mammals and birds there is reabsorption of water and mineral salts (p. 109) (the colon thus exhibits some renal type of function). The so-called digestive tract also provides the organism with water and water-soluble molecules. *The correlations between functions in the organism as a whole are apparent not only in the interdependence of metabolic processes at the molecular level but also in the interdependence of different organs whose evolution must have been related.*

(3) Problems concerning osmoregulation also arise in palaeontology, in particular with reference to vertebrates, in so far as one is concerned to know whether the first fish were freshwater or marine. The question is debatable: Romer and Grove (1935), MacFarlane (1923) consider that the Ordovician vertebrates were freshwater and that the Agnatha and fish migrated secondarily to the sea; on the other hand, Gross (1950), Denison (1956) and J.D. Robertson (1957) consider that vertebrates passed from salt water to freshwater in the upper Silurian.

Osmoregulation

If $| \Delta_i |$ = depression of freezing point of the internal medium
and
$| \Delta_e |$ = depression of freezing point of the external medium,
it has been shown that:

(1) $| \Delta_i | = | \Delta_e |$ in marine invertebrates
(2) $| \Delta_i | < | \Delta_e |$ in marine teleosts
(3) $| \Delta_i | > | \Delta_e |$ in freshwater invertebrates, freshwater teleosts, batrachians, reptiles, birds and mammals,
(4) $| \Delta_i | \geq | \Delta_e |$ in marine elasmobranchs.

These considerations are important in defining a so-called physiological solution: that which is "physiological" for one type of organism is not so for another (cf. p. 47).

Mechanisms of osmoregulation

Osmoregulation in mammals will be examined during a study of the kidney and hypothalamo-hypophysopituitary system. We shall consider at this stage osmoregulation in freshwater teleosts and marine teleosts.

(1) Freshwater teleosts

In these animals $| \Delta_i | > | \Delta_e |$ such that, according to the law of osmosis, *water will enter the organism* through the gills (Krogh) and digestive tract; there is a tendency for the internal medium to be diluted but *water is eliminated continuously* by the kidney in the form of urine which is *hypotonic* to the blood.

A *continuous current of water* passes through these organisms, decribed by Dastre as a "lavage du sang". The important function of the kidney is maintained by a *large number of arterial glomeruli* (approximately 100,000 Malpighian corpuscles), which form a large volume of renal ultrafiltrate. This is modified by reabsorption and secretion.

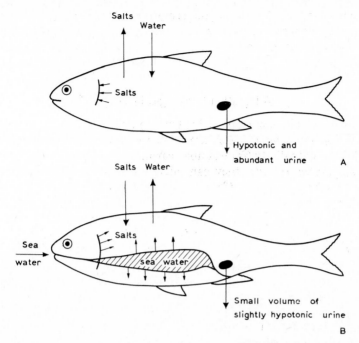

FIG. 34. Scheme of the mechanisms of osmoregulation in teleosts. (A) Freshwater teleost. (B) Marine teleost. (FLORKIN, modified from BALDWIN).

(2) Marine teleosts

In these organisms $|\varDelta_i| < |\varDelta_e|$ and there is a *tendency for the animal
to become dehydrated.* To counteract the loss of water (W. Smith) *the animal absorbs seawater through the digestive tract* (salts are eliminated by the
gills), the water is retained and the internal medium is diluted. The kidney
contains a *small number of arterial glomeruli* and in the sea horse there are
not even any glomeruli.

To summarize, *the freshwater teleost eliminates water and the marine
teleost retains water.*

In *marine elasmobranchs*, the concentration of urea in urine is lower than
in blood (Baglioni), *there being reabsorption of urea by the kidney* (Smith).
In addition, the membranes of the gills and mouth are only slightly permeable to urea (Duval and Portier). It is of interest that this uraemia is
compatible with the normal functioning of different cells and organs and
particularly that the hemoglobin of selachian erythrocytes is not denatured by a high concentration of urea in the blood. Is the fact that elasmobranchs are uraemic and cartilaginous a simple coincidence or an expression of a more profound phenomenon in which urea inhibits ossification?
Urea and urates are elements in the catabolism of purines and proteins

and we have seen (p. 149) that gout is a process which involves both urates and cartilage; consequently urea–urates–cartilage might involve some significant physiological correlation.

Adaptation to changes in salinity

A number of terms must be defined before we proceed further:

(1) *stenohaline animals:* they are unable to withstand any great variations in external salinity (for example, carp).

(2) *euryhaline animals:* they can migrate from freshwater to seawater and in the reverse direction (for example, eel and salmon).

With reference to breeding migrations, the eel is said to be a *catadrome* (migrates from rivers to the sea) and the salmon is said to be an *anadrome* (migrates from the sea to rivers).

What are the mechanisms which explain, for example, *the euryhalinity of eels* allowing their catadromic migration towards the Sargasso Sea?

There are two immediate mechanisms and one delayed mechanism.
(a) 1. In the passage from freshwater to seawater the *mucus of the skin* of eel plays a physico-chemical role in protection (combination of salts with mucoprotein?);

2. in addition, chlorides are eliminated by the *gills* (as in marine teleosts).
(b) When the eel passes from fresh into salt water, *25–40 per cent of blood proteins disappear within 8 days.* The reverse phenomenon occurs in the passage from seawater to freshwater (Boucher–Firly, 1935).

The eel in freshwater has a lower blood salinity than the eel in seawater but also contains more proteins such that the resultant osmotic pressure is the same.

It is possible that the production of corticosteroids (favouring protein anabolism) via stimulation of the pituitary (corticotropin) is responsible for this hyperproteinemia. As a result of these adaptations the organism is not damaged by the change in external environment.

Variations in salinity and metabolism

Working with *Gammarus locusta* (a brackish water form), my student G. Chéoux (1951) has shown this species to be euryhaline; however, when the *ambient salinity decreases respiratory exchange increases.* There is no doubt that this increase in oxygen consumption can be explained by the *metabolic and osmotic work* required during the active absorption *of ions from the medium: the animals strive to prevent hypotonicity* which acts as a stress (cf. p. 226).

Marine and estuarine crustaceans are capable of withstanding dilution of the medium only in cases where the blood concentration (osmotic pressure) is maintained at a higher level than the concentration (osmotic pressure) of the external medium.

In general terms, these adjustments are in a sense unique and express the adaptative possibilities of each type of organism. It is only when the species being considered acts in a positive way—mainly by the individuals as a group but also by the effects of an incoercible realization: nest building by fish and birds, human organization—that the environment is *passively* adapted to the living creature. If the external medium is modified spontaneously, for example, by geological changes, the living creature which is found there is unable to face successfully the new conditions; either the creature would die or it would migrate and could be replaced by other creatures or the internal environment could be adapted by functional changes and the species or individual resume activity in a new form which would condition anew the environment of which the organism would remain more or less dependent.

The Darwinian concept of the survival of the fittest would appear to me to mean the survival of *compatible* types such that, in a given biotope, animals with different physiology—but not animals which are physiologically incompatible with each other—can exist; the fact of natural selection is expressed in the distribution of animals (past and present zoogeography); the living types (individuals or populations) are such because of their genetic makeup and ecological distribution.

The constancy of the internal environment is not a static state but one of *dynamic equilibrium*. The external ecological factors shape living material. Time must also be taken into account as an ecological factor. It should be regarded as the principal factor in biological changes, in the understanding of living potentialities.

CHAPTER 9

FLUID COMPARTMENTS OF MAMMALS

A COMPARTMENT is an homogenous space, i.e. where each constituent has the same probability of distribution.

This chapter will not include *secretory* fluids, for example *milk*, or *excretory* fluids, for example *urine*, or a fluid system such as *sperm or seminal fluid*. The aqueous and vitreous humours of the eye and the endolymph and perilymph of the inner ear will be examined when we study these organs.

The *extracellular fluids (humoral pathway)* are principally:

(1) *cerebrospinal fluid,*
(2) *lymph* which corresponds to interstitial fluid or *interstitial plasma,*
(3) *blood* or more exactly *blood plasma.*

I. Cerebrospinal fluid

In man, this is a clear alkaline fluid (pH = 7·9, more alkaline than blood plasma) with a density of 1·008; the volume is approximately 125–150 cm^3 in adults.

Cerebrospinal fluid is an ultrafiltrate of blood plasma from the *choroid plexus*, which consists of *blood vessels* lying beneath the pia mater and lined with secretory epithelium, the cells being rich in *mitochondria*. Certain differences in the chemical composition of blood plasma and cerebrospinal fluid suggest that *active* processes (possibly secretory but selective in every case, cf. barrier, p. 161) may occur in the region of the choroid plexus. Principally, it appears to be the lateral cerebral ventricles which are active in the formation of cerebrospinal fluid.

Blood in the sinus of the dura mater undergoes pulsations which are synchronous with the heart beat and constitute the *venous pulse*. This pulse is derived from compression of the large cerebral veins during dilation of the cerebral arteries, the cranium being rigid.

The circulation of cerebrospinal fluid begins in the *lateral ventricles*. Reabsorption of fluid occurs in the granulations of Paccioni or *arachnoid villi* which evaginate into the venous sinuses.

There are two main fluid spaces: the *central compartment* and the *peripheral compartment*.

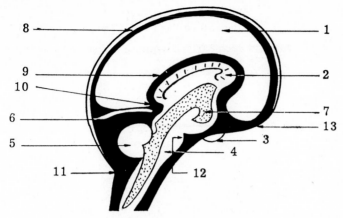

FIG. 35. Schematic representation on a sagittal section of the topography of cephalic reservoirs. *1*, cerebral hemisphere; *2*, *corpus callosum*; *3*, turcic zone; *4*, peduncle of cerebellum; *5*, cerebellum; *6*, awning of cerebellum; *7*, IIIrd' ventricle; *8*, periviscefal layer; *9*, pericallosal cistern; *10*, ambient cistern; *11*, *cisterna magna*; *12*, prepeduncular cistern; *13*, opto-chiasmatic cistern.

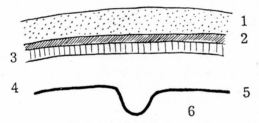

FIG. 36. *1*, Bone. Meninges: *2*, Dura mater; *3*, Arachnoid; *4*, Sub-arachnoid *5*, Pia mater. *6*, Brain.

FIG. 37. *1*, Dura mater; *2*, granules of Paccioni; *3*, superior longitudinal sinus; *4*, arachnoid; *5*, sub-arachnoid space; *6*, pia mater.

(1) *The central compartment consists of:*

(a) *the ventricular cavities*, or:

   — two lateral ventricles corresponding to each hemisphere,

   — the third ventricle, single and medial,

— aqueduct of Sylvius which links the third and fourth ventricles,
— the fourth ventricle, in front of the cerebellum.

The labyrinthine and cochlear perilymph is composed of cerebrospinal fluid.

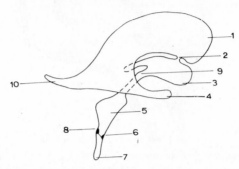

FIG. 38. Central compartment. Diagram of lateral view. *1*, frontal horn; *2*,
foramina of Monro; *3*, IIIrd ventricle; *4*, temporal horn; *5*, IVth ventricle;
6, ependymal canal; *7*, ventricle of Krauss; *8*, foramina of Magendie; *9*,
aqueduct of Sylvius; *10*, occipital horn.

(b) *the medullary cavity of the ependymal canal* which is little developed except at the base where it opens out to form the ventricle of Krauss. The medulla spinalis does not extend below the second lumbar vertebra and therefore, spinal puncture is possible below L_2.

(2) *The peripheral compartment.*

It consists essentially of the *subarachnoid spaces* (between the pia mater and the arachnoid) which are narrow in the cerebral curvature but are generally greatly enlarged to form the cisterns. The principal *cisterns* are:

(a) the *cerebello-bulbo-medullary cistern* or *great cistern;*

(b) *the basal cisterns:*
  prepontine cistern *in front* of the protuberance,
  prepeduncular cistern *in front* of the cerebral peduncles,
  ambient cistern, *behind* the cerebral peduncles,
  opto-chiasmal cistern *around* the optic chiasma;

(c) the *pericallosal cistern* between the hemispheres.

The subarachnoid spaces are continuous one with another. The central compartment is linked with the peripheral compartment by the *foramen of* Magendie.

There are also *neural sheaths*, associated with nerves, and perivascular extensions of the subarachnoid spaces or *prolongations of* Virchow-Robin.

Pressure of cerebrospinal fluid

The pressure is determined using a water manometer (Strauss tube). The pressure varies:

(1) *with the position of the body*; therefore:

(a) in a *sedentary* man the pressure in the lumbar region measures 35 cm and about 1 cm in the suboccipital region;

(b) the pressure is the same in both regions and measures about 10–20 cm in a man *lying on one side*;

(2) with *ventilation, coughing* and *movements* in general;

(3) *with venous pressure*, there being a parallel increase in the pressure of cerebrospinal fluid which can reach 20 cm water during abdominal and in particular, jugular compression (Queckenstedt's test).

Increased pressure in the great cistern, for example in rats, favours reabsorption of injected ^{35}S-methionine (Wasterlain, 1963).

Composition in man

(1) *Cytological:* Under normal conditions it is rare to find lymphocytes in the fluid but during inflammatory processes polynuclear appear.

(2) *Chemical:* The urea and bicarbonate content of cerebrospinal fluid is the same as that of blood plasma but the total calcium is independent of plasma calcium levels and whereas the plasma contains 6 per cent NaCl the cerebrospinal fluid contains 7·3 per cent NaCl; in addition, 1 per cent glucose is found in plasma and only 0·5 per cent in cerebrospinal fluid; siderophilin is also present (Galli, Jeamaire and Gérard, 1960). The protein content of cerebrospinal fluid is 250 mg/l compared with about 75 g/l in blood plasma. The low protein content can be attributed principally to the absence of fibrinogen. Cerebrospinal fluid does not coagulate spontaneously.

There is a barrier between blood plasma and cerebrospinal fluid. For example, *antibiotics cannot pass this barrier*—the *meningeal barrier*. It protects the central nervous system from the effects of toxic chemicals.

Role of cerebrospinal fluid

(1) *Essential.* This is a mechanical function (hydrostatic). The cerebrospinal fluid represents a *protective cushion* preventing the compression of the nerves against bone and provides a *fluid suspension* which is of fundamental importance in the functioning of the brain.

(2) *Accessory.* The fluid functions in the removal of toxic products from nervous tissue and in *nutrition* of the tissue. It also provides a humoral pathway (encephalocriny).

Hydrocephalus

This condition demonstrates the results of a pathological disturbance in a dynamic system; in this case, a superabundance of fluid leads to a morphological abnormality. Hydrocephalus can result from both physiological and anatomical conditions:

(a) from blockage of the connecting foramina, the foramen of Monro and the foramen of Magendie; this is *internal hydrocephalus.*

(b) from excessive production or inadequate reabsorption of cerebrospinal fluid. This constitutes *external hydrocephalus* which is equivalent in physiological terms to economic crises arising from overproduction or underconsumption.

II. Lymph

Lymph *irrigates* and *drains* the tissues and its functional importance is related to its large quantity. Ludwig considered the lymph to constitute a *quarter of the total weight* of mammals; lymph can account for half the body weight.

Secretion of lymph can be stimulated by *lymphagogues:*

(1) with *an action on the liver,* for example *peptones,* which increase the pressure in the portal region, disturb the permeability of hepatic capillaries and are toxic, causing a decrease in arterial pressure (cf. peptone shock, p. 190).

(2) with an action on other systems, for example, *hypertonic solutions* (NaCl or glucose) which cause a loss of water from extracellular spaces equivalent to the "lavage du sang" of freshwater teleosts (cf. p. 154). The osmotic disequilibrium produced between blood plasma and extracellular fluid leads to dilution of blood and the production of lymph is weak; finally, there is a massive capillary transudation; the blood is concentrated once again.

Increase in pressure within capillaries leads to an increased production of lymph. This occurs when one passes from a lying to standing position: venous stasis occurs in the lower limbs through the action of plasma transudation and the blood is concentrated.

Canalization

Interstitial fluid passes through the lymphatic capillaries in which it becomes canalized; the lymphatic vessels which are provided with valves in birds and mammals (with the exception of the lymphatic capillaries) are generally in close association with blood vessels as is seen for example, in the thyroid gland (Rusznyák, Földi and Szabó, 1960) and the vascular beds of the skin of digits (D. A. Zhdanov, 1952).

In lower vertebrates there are well-innervated lymphatic hearts which beat independently (Johannes Müller) and in which the movements, dependent on medullary innervation, are paralysed by curare (Kolliker, 1856). The effect of curare results from the fact that the nerves to the lymphatic hearts terminate at *motor end plates* (observed in grass-snake by Ranvier, 1878). Ranvier also found that direct excitation of the lymphatics hearts by repeated electrical stimulation leads to arrest in systole and strong stimulation of the spinal cord induces arrest in diastole.

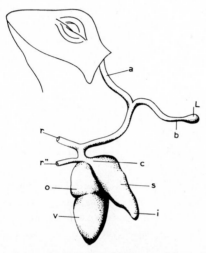

Fig. 39. Relation of the anterior lympathic heart of frog with the blood system. *L*, anterior lymphatic heart; *a*, jugular vein; *b*, vein from the anterior lymphatic heart; *s*, sinus venosus; *i*, inferior vena cava; *c*, superior vena cava; *v*, ventricle; *o*, auricle; *r* and *r''*, brachial veins. (RANVIER.)

In mammals, lymph is carried by two large canals, the *great lymphatic vein* and the *thoracic duct* which join the blood system via the subclavian veins.

Composition

Chemical

Lymph is a *yellow* liquid formed by *transudation* of blood through the capillaries. The composition depends on the region from which the sample is taken. The composition is similar to that of blood plasma although lymph contains *less* protein (in particular, there is very little fibrinogen and coagulation of lymph is slow). In the lacteals the lymph is rich in lipids (cf. p. 118). It generally contains more Cl^- than blood plasma (Donnan effect).

Cellular

Since this is a study of extracellular media, strictly speaking cellular elements should not be included either in the case of lymph or blood. However, such an extreme position would rapidly become untenable both from a didactic point of view and that of understanding functional mechanisms. However, this reservation appears to me to be useful in that it emphasizes the limitations of classification.

The origin of cells in lymph will be examined when we consider hemopoiesis. For the moment it is sufficient to say that human lymph contains approximately *8000 white corpuscles/mm³* which is more than in blood. These leucocytes are *lymphocytes* and mononuclear cells from the *lymphatic ganglia* (for example, Peyer's patches distributed along the lacteals) and lymphatic organs (such as the thymus which is resorbed at puberty).

The leucocytes in the lymphatic canals are not mobile since the partial pressure of O_2 is low and that of CO_2 (anaesthetic) is relatively high. Muscular exercise activates the lymphatic circulation and consequently activity such as gymnastics favours detoxication of the organism.

An example of the role of lymphocytes in invertebrates

The circulation in *Cancer pagurus* (a decapod crustacean) is open. Madeleine Marrec and I (1947) have found evidence, in certain cases, of a lymphocyte network having the composition of a *wound scab*. The conditions of the experiments were as follows: when a section of carapace together with the inner membraneous layer is removed experimentally using an

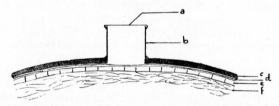

FIG. 40. *a*, collodion membrane covered with paraffin; *b*, cell of Van Tieghem; *c*, carapace; *d*, membranous layer, *e*, epithelium; *f*, connective.

electric cutter or naturally when the animal is damaged, the *epithelium is exposed.* Lymphocytes (eosinophils) pass into the injured region and a scab is formed by progressive stratification. Deposition of melanin carried in the blood plasma causes the scab to darken. When the injured area is completely occluded a layer of calcified chitin forms beneath the scab. At the time of moulting the scab is lost leaving this layer which now forms part of the new carapace. If the membraneous layer remains intact when the carapace is removed a lymphocyte scab *is not formed*; under these conditions only a chitinous layer is formed.

If the carapace is removed *leaving the membraneous layer intact* and the opening is covered by a van Tieghem cell (closed at the lower edge by the membraneous layer and at the upper edge by a collodion membrane covered in paraffin) in which has been placed, for example, peptone, bacteria or glycogen (a constituent of epithelial cells), after 3 days at 17°C, or 15 days at 7°C, a *submembraneous scab* appears formed by a massive infiltration of lymphocytes. Seawater was placed in control cells and in these cases only a submembraneous chitino-calcareous layer had formed.

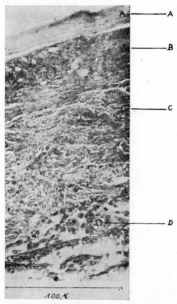

FIG. 41. *A*, membranous layer; *B*, clot formed of eosinophilic lymphocytes; *C*, connective tissue infiltrated by lymphocytes; *D*, eosinophilic lymphocytes.

Madeleine Marrec had shown previously that the lymphocyte organ of decapod Crustacea (discovered by Cuénot in 1893) is situated (in *Carcinus maenas*) on the dorsal and lateral walls of the masticatory stomach at the end of the pyloric region. She had also provided evidence of *variations in the activity of this organ during intermolt;* this cycle has been studied extensively by P. Drach. The activity of the lymphocyte organ was particularly marked during stage D_2 (secretion of new chitinous layer) and D_3 (resorption of skeleton).

Thus, in Crustacea, lymphocytes participate in the reconstruction of integument, mainly during the first stage of this process, cicatrization.

III. Blood

As in previous cases, only the acellular fraction of blood, the plasma, should be considered. However, the blood is a *fluid tissue*. It could even be called a *fluid organ* since it consists of many types of cells which *differ in function and morphology* in spite of the fact that many of the cells have an apparently common origin.

The following subjects will be considered:

    (A) Hemopoiesis,
    (B) Blood volume,
    (C) Plasma proteins,
    (D) Blood coagulation,
    (E) Blood groups.

Oxygen and carbon dioxide transport in blood will be considered with respiration and *acid–base balance* in a study of renal excretion.

A. Hemopoiesis

The cellular elements of blood of mammals originate in three regions: *bone marrow, spleen* and *lymphatic organs.*

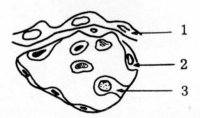

FIG. 42. *1*, external cells; *2*, internal cells; *3*, bud. (BESSIS, 1957.)

In the early stages of development, *"blood islands"* appear in the *mammalian embryo* in the extra-embryonic space of the *umbilical vesicle* (equivalent to the yolk sac of birds and reptiles).

In human embryos the umbilical vesicle atrophies rapidly and the vitelline circulation disappears at an early stage. In the embryos of other mammals, for example rabbit, the vitelline circulation persists almost until parturition.

It should be mentioned that the existence of species differences can complicate the task of the student but they help the research worker in that they allow him to choose the most suitable material for each study.

The *peripheral cells* of the *"blood islands"* form the walls of blood vessels while the *internal cells* represent the mother cells of erythrocytes. With the appearance of blood circulation in the embryo *erythrocytes are formed in the liver and spleen.*

The *white corpuscles—leucocytes—*only appear about one month after the beginning of embryogenesis, which indicates the priority of respiratory detoxication.

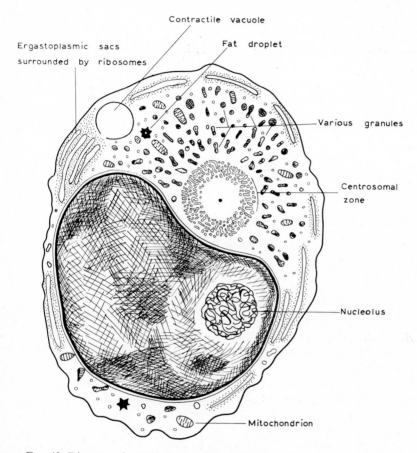

Contractile vacuole

Ergastoplasmic sacs
surrounded by ribosomes

Fat droplet

Various granules

Centrosomal zone

Nucleolus

Mitochondrion

Fig. 43. Diagram of a section of a neutrophilic promyelocyte as seen with the electron microscope. (Bessis, 1956.)

Leucocytes are formed in liver, spleen and in connective tissue cells which constitute the *reticulo-endothelial system* (Aschoff, 1913) or *reticulo-histocytary system* or simply the *reticular system* (a recent name). These names derive from the fact that the cells are often arranged in a *network*. It is preferable to use one of the former names as this avoids confusion with the neural reticular system.

In man, the *embryonic bone marrow* exhibits hemopoietic activity during the fifth month of gestation and finally almost completely takes over the hemopoietic function of liver and spleen (zones of erythrocyte retention).

Reticulo-endothelium is found in man in:

bone marrow,
Küpffer cells of liver (intra-lobular capillaries),
splenic sinuses and lymph nodes,
as free histiocytes (macrophages),
adrenal.

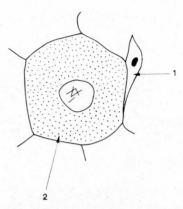

FIG. 44. *1*, Küpffer cell; *2*, polyhedral cell of hepatic parenchyma.

Granulopexic power (macrophagia)

All the cells of the reticulo-endothelial system are capable of absorbing *vital stains* (such as lithium carmine and trypan blue) and particles, including bacteria (phagocytosis can be considered as a macroscopic form of pinocytosis).

All blood cells originate in the reticulo-endothelium and return there, when old or diseased, to be destroyed (I would like to call this the Cronos cycle).

Hemopoietic marrow

Only the *red marrow is hemopoietic*. Red marrow is found in the flat bones (such as ribs or sternum) and in the *extremities of long bones*. The central yellow marrow of long bones contains fat and has no hemopoietic ability. *In the embryo*, the whole of the long bones contains red marrow which is cytogenic. However, after excessive hemorrhage in adults the yellow marrow becomes red and the whole of the bone marrow contributes to hemopoiesis.

In red marrow, it is not possible to differentiate between the cells which are precursors of red cells and those which are precursors of white cells; this has led to the controversy of the "unitarian" and "pluralistic" theories of hemopoiesis.

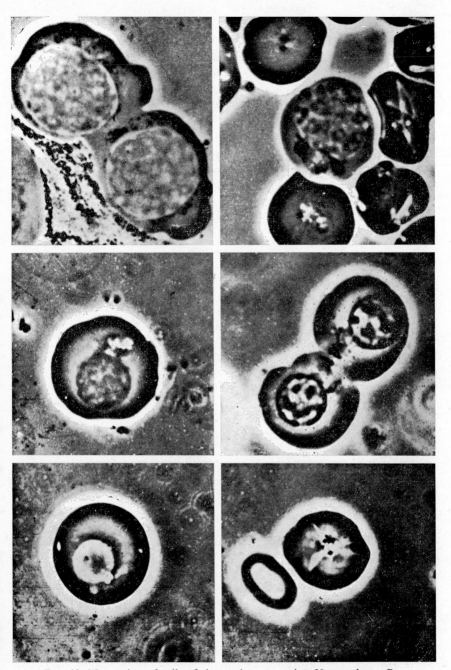

FIG. 45. Maturation of cells of the erythrocyte series. Upper plates: Pro-erythroblasts and basophilic erythroblast. Centre plates: Polychromatophilic erythroblasts; the presence, on the left, of vacuoles formed *in vitro* should be noted. Lower plates: Acidophilic erythroblast and reticulocyte. (Normal sternal puncture. Phase contrast.) (BESSIS, 1954.)

Cellular types

The reticular mother cell of red marrow gives rise to two cells which become spherical and free within the network of the reticulum. These cells are the non-circulating, undifferentiated *hemocytoblasts*, the precursor cells (the suffix *blast* describes a precursor cell).

(1) Development of red corpuscles

The hemocytoblast becomes *basophilic;* this property is related to the presence of *ribonucleic acid* (erythroblast stage). Later this cell acquires a red coloration following the *active synthesis of hemoglobin* (cf. p. 330).

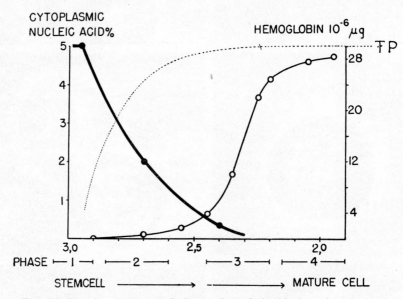

Fig. 46. Cytoplasmic changes during erythropoiesis. Abscissa: logarithm of cell volume. ← per cent of cytoplasmic ribopolynucleotides; –o–o– total quantity of hemoglobin in the cell; – – – – total cellular protein obtained from the total volume of cells in each development phase. (Bo Thorell, 1947.)

At the same time, the size of the cell decreases and, in mammals, the nucleus disappears. The cell at this stage is called a *reticulocyte*. The reticulocyte rapidly develops into a mature erythrocyte from which it differs only in the presence of *granules*. Therefore, maturation is achieved through a loss of structural components with the transition from a nucleated to a *non-nucleated cell*. The resorption of the nucleus has been studied by Bessis using accelerated microcinematography. The nucleus is not destroyed within the reticulocyte but is *ejected* and engulfed by macrophagic histiocytes. The erythrocytes of fish, amphibians and birds are nucleated. Maturation of the reticulocyte to give an erythrocyte occurs in man within a few hours. The urine and plasma of anaemic humans or animals contains a factor which activates erythropoiesis, *erythropoietin* (Erslev, 1953). It has been suggested that this compound is a mucopolysaccharide derived from the kidney (Gordon, 1960). It can easily be concentrated from urine of anaemic subjects by ultra-filtration on a collodion membrane (D. C. van Dyke, 1960).

In man, 10^{11} reticulocytes are produced each day. The importance of this cytogenesis can be appreciated by comparing this figure with that for spermatogenesis in rabbit: only 80,000 spermatozoa are produced per minute.

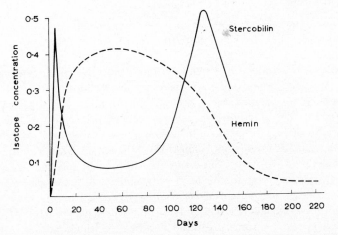

FIG. 47. Duration of life of human red corpuscle measured by labelled hemin after ingestion of glycine labelled with heavy nitrogen (broken line). Elimination of labelled stercobilin (continuous line). (After SHEMIN and RITTENBERG.)

The average life of an erythrocyte is 100 days (Shemin and Rittenberg, cf. Fig. 47) and, as has already been stated, the erythrocyte is destroyed in the reticulo-endothelium. Each erythrocyte is orange and biconcave. The diameter of erythrocytes in various species is as follows:

7·5 μ in man,
4 μ in goat,
9 μ in elephant
55 μ in Proteus (a cavernicolous amphibian).

Since the major part of oxygen in blood is carried by erythrocytes, it could be assumed that the larger the erythrocytes the greater the amount of oxygen carried but they may be less numerous such that the total surface for oxygenation is less. In fact, in Proteus there are only 45,000 erythro-

12*

cytes /mm³ and 250,000/mm³ in frog. These quantitative relationships apply to poilikilothermic animals such as amphibians in which venous and arterial blood become more or less mixed in the heart and in which respiratory metabolism is relatively weak and directly related with external temperature.

(2) Development of white corpuscles

There are three types of white corpuscles:

(a) granulocytes (70 per cent).
(b) lymphocytes (25 per cent).
(c) monocytes   (5 per cent).

(a) Granulocytes are formed in bone marrow which also gives rise to erythrocytes. The cause of differentiation into erythrocytes or leucocytes is not known.

In the development of a hemocytoblast into a granulocyte, the cytoplasm *becomes granulated* and the granules, according to their affinity for dyes, are acidophilic, basophilic or neutrophilic.

A remarkable phenomenon can be seen in the region of the nucleus during this maturation: the nucleus divides into several lobes (2 to 5) *which are joined by very thin connectives*. This led early microscopists, who were unable to see these connectives, to think that the nucleus had fragmented and gave rise to the term *polynuclear. In mammals, this change in the structure of the nucleus in white cells occurs at the same stage as ejection of the nucleus in red cells.* I consider this synchronization of nuclear phenomena to have some profound significance which has yet to be elucidated.

The granulocytes exhibit *amoeboid movements* with extension of *pseudopodia* and it is by these movements that the cells leave the bone marrow.

Splenectomy produces a transitory hypergranulocytosis in blood and lungs which leads one to consider that the spleen contains a system which inhibits differentiation of leucocytes (cf. p. 426, adrenal steroid hormones).

(b) Lymphocytes are formed in lymph nodes; in addition, the structure of the nucleus is different from that of granulocytes. The thymus produces lymphoid cells (thymocytes) which are carried by the blood to the spleen and lymph nodes. Sainte-Marie and Leblond have established that thymocytes come from cells of the reticulo-endothelial endoderm and initially form large and medium-sized lymphocytes, later dividing to give small lymphocytes. *The pouch of Fabricius* of birds, situated dorsally near the cloaca, is equivalent to the thymus and coexists with the thymus in these animals. In a dog weighing 10 kg, about 5000 million lymphocytes pass from the thoracic duct into the blood each day (Yoffey, 1936). The parasympathomimetic alkaloid, pilocarpine (muscarinic), induces lymphocytosis which is characterized by an increase in the number of lymphocytes in the thoracic duct (Harvey, 1906; Rous, 1908), an effect which may be due to contraction of smooth musculature of spleen and lymph nodes leading to extrusion of lymphocytes.

(c) Monocytes are formed in the reticulo-endothelial system and exist as free cells in the blood. These are *histiocytes*.

There is morphologically little difference between the cells of the last two categories.

The average life of a leucocyte is two days. Each minute approximately 7200,000 white corpuscles are formed in the blood of man. The diameter of these cells is between 10 and 20 μ.

(3) Development of platelets (thrombocytes or globulins)

The origin of these generally very small blood elements (2–5 μ) was discovered by Wright (1906). The red bone marrow (hemopoietic) contains larger basophilic cells with trapezoid nuclei, called *megakaryoblasts*. The *megakaryoblasts* give rise to *megakaryocytes* with a diameter of 150–300 μ and which are *densely granulated*; these structures being the future platelets; (this stage is synchronous with the extrusion of nuclei in cells of the red series and lobulation of the nucleus in cells of the white series). The megakaryocytes migrate towards the capillaries which they partially enter and the membrane disappears allowing the platelets to enter the circulation: "The giants of bone marrow give rise to the dwarfs of blood" (Bard).

Man normally contains about 300,000 platelets/mm^3 of blood; this is the number formed per minute.

Action of ionizing radiation

Local irradiation destroys the red marrow and lymph nodes but vicarious hemopoiesis can maintain the required level of blood cells.

Irradiation of the whole animal with 200 R* decreases the number of blood cells. With a dose of 500 R a rabbit dies. Cell divisions cease, particularly those of erythroblasts.

With high doses, the bone marrow exhibits a form of fatty degeneration.

After sublethal doses regeneration can appear after 5 days with the formation of hemocytoblasts. The anaemia can be treated by giving repeated blood transfusion.

Hemopoiesis is gravely affected in persons subjected to atomic radiation and this condition can be treated by injections of bone marrow suspensions.

---

* The Röntgen (R) is the quantity of X-rays which liberate, by ionization of a cubic centimetre of air, one unit of electrostatic charge (derived from the formula of Coulomb).

Massive doses of radiation lead to leukaemia which is an explosive proliferation of leucocytes; the subject contains 300,000 leucocytes per mm³ compared with 6000–8000 under normal conditions. Many of the inhabitants of Hiroshima developed leukaemia after the first atomic bomb explosion.

The normal, average number of blood cells in 5 l of blood in man is as follows:

> 25,000 milliard red corpuscles,
> 35 milliard white corpuscles,
> 1500 milliard platelets.

Daily it forms about:

> 220 milliard erythrocytes,
> 15 milliard leucocytes,
> 500 milliard platelets.

Thus, there is not equal production of red cells and white cells and the predominance of red cells implies that under normal conditions leucocytopoiesis is restrained (role of the spleen or of the steroid hormones? cf. p. 431).

These figures leads to a consideration of blood volume.

B. Blood volume

The first techiques used in measurement of blood volume were very inaccurate and included such methods as bleeding and colorimetric estimation of hemoglobin in homogenized tissues.

Today, three techniques are used principally, all of which are based on the principle of dilution:

(1) colorimetric techniques;

(2) techniques involving the use of non-toxic radioactive tracers;

(3) techniques utilizing impedance measurements.

(1) A known quantity of dye is injected into the circulation. After a certain time interval a known quantity of blood is removed and centrifuged to separate the cellular elements. Using the plasma fraction, the concentration of dye diluted by blood is measured colorimetrically and from this the total blood volume can be calculated. *Evans blue* is used mainly because it combines with serum albumins and, therefore, remains in the circulation for a long period of time.

(2) In radioisotope techniques a tracer is used which has a duration of life compatible with the duration of the experiment. Either the plasma proteins or the red corpuscles can be labelled.

In the first case, radioactive ^{131}I is added to a known volume of plasma *in vitro*. This is administered by intravenous injection and after an interval

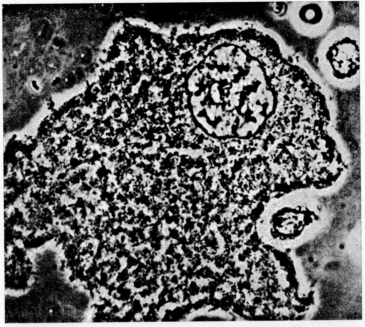

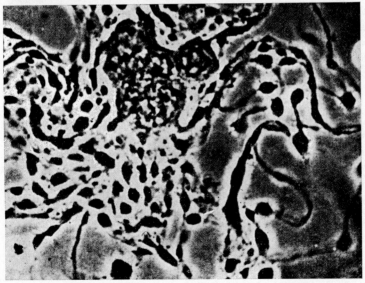

FIG. 48. Formation of thrombocytes. (BESSIS, 1957.)

of time, a sample of blood is removed and radioactivity measured. Knowing the dilution of the tracer the blood volume can easily be calculated as in the case of dyes.

In the second method, the erythrocytes are labelled with various radioactive elements: ^{55}Fe, ^{32}P, ^{42}K. Hevesy who developed this technique (1951) has stated that the loss of ^{42}K from erythrocytes is negligible and considerably less than the loss of ^{32}P. The method using ^{42}K would be the method of choice. Hevesy (1951) has also shown that the erythrocytes can be labelled with thoron (a derivative of thorium); in rabbit, activity is retained for several hours.

Modification of Ashby: Ashby used, as an indication of plasma dilution, a known quantity of red corpuscles from a *different* blood group from that of the recipient (different but compatible) and finally, specific and quantitative agglutination of erythrocytes; from a measurement of dilution the total volume can be calculated.

(3) In the technique involving impedance measurements (electrical resistance) of blood, a hypertonic or hypotonic solution (either with or without electrolytes) is injected and the impedance of blood samples is estimated before and after injection. This technique appears to be difficult —even risky—since the reactions of the organism to hypotonic or hypertonic solutions can interfere with the results (cf. p. 162, with reference to lymphagogues).

Comment. Similar techniques can be used to measure *the volume of extracellular water and intracellular water.*

(1) *Extracellular water.* A measurement is made of the degree of dilution of a non-toxic compound which satisfies the following conditions:

(a) it must pass easily across the endothelium of capillaries;
(b) it must not penetrate the cellular membrane;
(c) it must not be metabolized or excreted.

Both ^{24}Na and ^{38}Cl are used and the volumes obtained are defined as the "sodium space" or the "chloride space". Some error will be introduced to the measurements by the fact that *both these ions will penetrate cells to a certain extent.*

(2) *Intracellular water.* Heavy water D_2O is administered. This penetrates *all* biological membranes (cells and capillaries) and gives a measure of *total water.* By subtracting value (1) from value (2), a value for the volume of intracellular water is obtained.

Normal blood volume

In man or woman there is about 90 cm^3 blood per kg body weight and about 50 cm^3 *plasma* per kg body weight.

If these volumes are related to m^2 body surface the value of 3250 cm^3/m^2 is obtained for blood and 2000 cm^3/m^2 for plasma.

This volume depends on the *age* and *physiological state* of the individual: the blood volume related to body weight is lower before puberty than in the adult;
the blood volume increases during gestation.

The plasma volume is also influenced by regulation of water metabolism (cf. p. 457). Blood is unequally distributed within the organs, particularly during such processes as digestion, excercise, etc. For example, the four limbs contain about 20 per cent of the total blood volume at rest.

The ratio of blood volume to plasma volume is the *hematocrit;* this is 46 in man and indicates that the volume of blood cells is equal to 46 per cent of the total blood volume or 54 per cent of the volume of plasma.

C. Plasma proteins

The plasma of man contains approximately 70 g proteins per litre. Certain proteins possess *enzymic* activity, (for example, phosphatases), while other fractions show no activity.

Plasma proteins are classed according to their *molecular weight.* In order of decreasing molecular weight they are:

β-lipoproteins (with a molecular weight of about 1,300,000);
fibrinogen (with a molecular weight of about 350,000);
α-lipoproteins (with a molecular weight of about 200,000);
γ-globulins (with a molecular weight of about 155,000);
β-globulins (with a molecular weight of about 90,000);
serum albumin (with a molecular weight of about 70,000).

The different types of proteins can be separated by fractionation.

Fractionation:

Several methods are available which are carried out at low temperatures to prevent denaturation. Generally, plasma proteins are separated either by precipitation with neutral salts such as $(NH_4)_2SO_4$ (salting out), liquid phase electrophoresis* (Harris, Tiselius), paper electrophoresis or ultra-centrifugation (Svedberg).

The relative distribution of different proteins is modified in pathological conditions. Evidence for this was produced by G. Sandor (1950). *Normally,* in man, there are more albumins than globulins: albumins constitute 55·2 per cent of total plasma proteins and globulins 44·8 per cent of which fibrinogen represents 5–6·5 per cent and globulins 11 per cent.

*Migration in an electrical field of molecules placed in a conducting solution.

Origin:

Plasma proteins are derived from *ingested* amino acids. Almost all plasma proteins are formed in liver but γ-globulins—which include the *antibodies*—appear to be formed in the reticulo-endothelial system. These are present in large quantities in leukaemic serum and in serum of patients with infectious hepatitis.

Functions of plasma proteins

The existence of lipoproteins was demonstrated by Michel Macheboeuf and he called them *"cénapses lipo-protéiques"*.

We have seen that proteins can serve as carriers for metals (for example, siderophilin, the iron-binding globulin). Lipids are also carried in the form of *lipoproteins* in a medium—the plasma—in which the lipids are virtually insoluble. *In some ways lipoproteins are homologous with respiratory pigments, such as hemoglobin in the form of oxyhemoglobin, in that they bind fundamental compounds which are almost or completely insoluble in water or plasma.* In one case oxygen is carried and in the other, lipids. Among the respiratory pigments which will be considered in detail later, one can distinguish circulating carriers (such as hemoglobin) and intracellular carriers (such as cytochromes); one group transporting O_2 and CO_2 and the other electrons.

Lipoproteins also transport lipid-soluble vitamins and steroid hormones. Thus there exists both simple and complex *molecular transport.*

(1) Proteins function, therefore, in *humoral transport* and represent a source of protein.

(2) They act as *buffers:* proteins are amphoteric and combine with acids and bases. At the pH of mammalian blood (pH 7·4) proteins behave as acids and combine with cations such as Na^+; they are, however, weak buffers.

(3) Plasma proteins participate as macromolecules in the *regulation of osmotic pressure* by the Donnan effect and thus play an important role in the aetiology of water retention oedema in which there is an accumulation of interstitial water. At this stage each type of protein should be considered independently; osmotic pressure depends on the number of molecules and since serum albumin is smaller than a globulin then the osmotic pressure will be raised to a greater extent by 1 g of serum albumin in solution than 1 g of globulin.

Starling's hypothesis provides an explanation of the exchange mechanisms between blood and interstitial spaces. It considers the balance between osmotic pressure and hydrostatic pressure.

Within the capillaries, *the osmotic pressure of plasma proteins* is 25 mm Hg against which the proteins of the interstitial spaces exert an osmotic pressure of 10 mm Hg. The resultant pressure of 15 mm Hg favours the passage of interstitial fluid into blood plasma.

However, *the hydrostatic pressure of blood* must be taken into account; this favours filtration of blood plasma into the interstitial spaces and opposes the osmotic pressure effect. Now:

in the arteriole, the hydrostatic pressure measures 30 mm Hg and the hydrostatic pressure of interstitial fluid is only 8 mm Hg giving an effective hydrostatic pressure of:

$$30 - 8 = 22 \text{ mm Hg};$$

in the venule, the hydrostatic pressure of interstitial fluids is still 8 mm Hg but the hydrostatic pressure of blood is no more than 15 mm Hg giving an effective pressure of:

$$15 - 8 = 7 \text{ mm Hg}.$$

Under these conditions, the difference between absorption and filtration in the arteriole and venule will be determined by the direction in which fluid movements occur:

in the arteriole, the difference between effective osmotic pressure and effective hydrostatic pressure is:

$$22 - 15 = 7 \text{ mm Hg}$$

which is a *filtration pressure*; in this case the hydrostatic pressure dominates and fluids will tend to pass from the circulation into the extracellular spaces;

in the venule, the difference between effective osmotic pressure and effective hydrostatic pressure is:

$$15 - 7 = 8 \text{ mm Hg}$$

which is an *absorption pressure,* that is to say, since osmotic pressure predominates interstitial fluids will tend to pass into the circulation.

The circulation of fluids within the capillaries assures efficient fluid exchange; this is paralleled by the exchange of O_2 and CO_2 in arterioles and venules.

An excess of interstitial fluid results in *oedema.* This occurs when there is a decrease in the level of proteins in the internal medium or when there is an increase in the concentration of NaCl, mainly in the extracellular fluid (cf. p. 114, Donnan equilibrium) or when the venous pressure increases as in certain cardiovascular disorders. There are two types of heart disorder associated with thyroid dysfunction: one associated with hyperthyroidism and a loss in body weight and the other with hypothyroid myxoedema and an increase in weight. It is important to consider the thyroid-adrenal cortex balance in relation to water metabolism and also fat metabolism. As we shall see (cf. p. 431), adrenal corticoids favour reabsorption of sodium and chloride in the renal tubule and administration of cortisone leads to marked oedema. In considering the distribution of water, not only must the role of the kidney in normal elimination of water be taken into account but also the reabsorption of water in the colon (under normal conditions) (cf. p. 114). Reabsorption in the colon would appear to be under hormonal control. This was shown by M.H. Rees in 1920, later by Isawa, Denis, Wood and by L. Garnier (1959) using [post-] hypophyseal extracts, thyroxine, female sex hormones, and in the case of a chloride deficient diet, using mineralo-corticoids.

(4) Finally, the plasma proteins, prothrombin and fibrinogen partici-pate in *blood coagulation*.

D. Blood coagulation

The ability of blood to form a clot was discovered, without doubt, by early man. The red mass of the *clot* contracts after a certain time and a more or less translucent fluid is expressed: the *serum*.

The stages of spontaneous hemostasis are as follows:

(1) *a vascular* or parietal *phase* characterized by capillary constriction;

(2) *a platelet phase* during which platelets agglutinate to form a "he-mostatic plug" ("white thrombus" of Zahn, 1872) *without fibrin*;

(3) *a plasma phase* with development of fibrinous structures;

(4) *a thrombodynamic phase* associated with clot retraction.

General morphological characteristics of coagulation

Under a microscope, needles appear centred around agglutinated plate-lets. They multiply and form a network which after a few minutes contracts and serum is exuded.

The degree of retraction of the clot is proportional to the number of platelets and below 50,000 platelets per mm^3 retraction does not occur. The role of the platelets in blood coagulation was demonstrated by the Frenchman Hayem and the Italian Bizzozero about 1860. Aggregation of platelets ends with changes in viscosity ("métamorphose visqueuse") dependent on the presence of thrombin (Lüscher and Bettex-Galland, 1961).

General chemical characteristics of coagulation

The fibrils which form the network are composed of fibrin. Fibrin is produced from a plasma protein, fibrinogen, which represents about 5 per cent of total plasma proteins.

Fibrinogen is converted to fibrin by the action of thrombin which exists in plasma in the form of *prothrombin*.

Prothrombin is converted to thrombin by the action of thromboplastin and calcium ions.

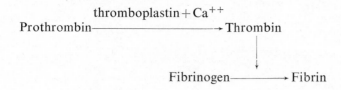

This relatively simple scheme is, in fact, complicated by the involvement at each stage of numerous other factors.

Origin of thromboplastin

Compounds with thromboplastic activity are found in blood plasma, platelets, and tissues.

A. BLOOD PLASMA.

(1) Thrombokinase: this is relatively thermostable; deficiency leads to *hemophilia* characterized by hemorrhages which can last for many days and lead to death.

Hemophilia is transmitted by females and appears exclusively in males. The hereditary characteristics of hemophilia were described by the American Otto (eighteenth century).

In the human species, from the heterochromosomal aspect, ♀ is XX and ♂ is XY.

The gene for hemophilia is carried by X and in the female its action is neutralized by the presence of the other heterochromosome X such that hemophilia does not appear in the female but she can *transmit* it. A female having this gene is called a *carrier*. However, in the male the action of this gene is not counteracted and the individual exhibits hemophilia.

Table 33 compounded by Bessis, shows the distribution of hemophilia in the descendants of Queen Victoria.

Today hemophilia can be treated with blood transfusions (anti-hemophilic globulin or factor VIII, in the case of hemophilia A, which appears to be associated with prothromboplastin). Hemophilia B (classified according to Pavlovsky) is associated with a deficiency in *Christmas factor*. Anti-H_A and anti-H_B activity has been detected in saliva, milk and amniotic fluid (Nour-Elding, 1957). In von Willebrand's disease (combined factor VIII deficiency and capillary defect) the production of blood factor VIII is intact, but its *evolvement* is impaired (Nour-Elding, 1966).

(2) Plasma thromboplastic factor is found in both serum and blood plasma. This factor appears with a fraction of γ-globulins and its concentration in plasma is less than 10 mg/l. The activity of this factor is reduced by vitamin K deficiency.

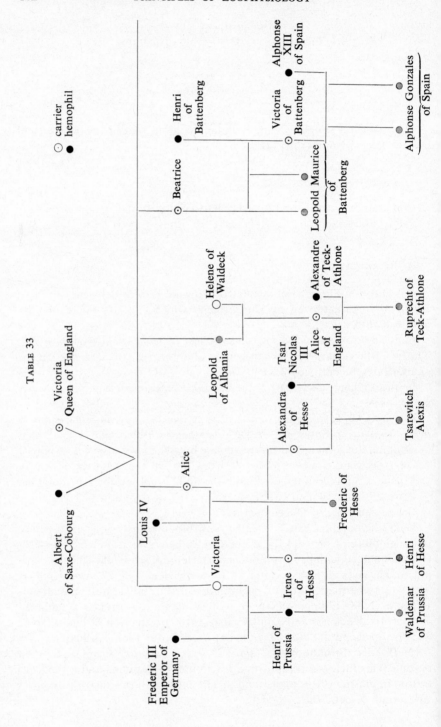

TABLE 33

B. PLATELETS. J. Bordet (1921) found a factor with thrombokinase activity in alcoholic extracts of platelets. (Since all enzymes are proteins and animal proteins are insoluble in ethanol, the term "kinase" is incorrect in this context.)

The work of Chargaff shows this factor to be a *cephalin* (phosphatide). It is also obtained by disintegration of platelets on contact with rough surfaces (glass, metal, rubber, dry cotton lint). For this reason the coagulation of blood samples can be prevented by placing them in containers with *non-wettable surfaces* (which prevent the spreading of platelets which would otherwise occur through interface tension effects and lead to disintegration). Containers made of plastic or lined with *silicone* are suitable.

C. TISSUES. The presence of a thromboplastic factor in tissue is important in that, when a tissue is wounded, this factor is liberated and favours the arrest of hemorrhage by activating blood coagulation.

Working independently, E. Chargaff (1945) and Howell (in the same year) isolated a purified thromboplastin from lungs of ox (the lung is rich in thromboplastin). Such independent discoveries frequently occur in science when a subject is ripe for development. Chargaff showed that this factor was a high molecular weight *lipoprotein*.

Production of prothrombin

Prothrombin is formed exclusively in the liver and requires vitamin K ("Koagulations-Vitamin" of Dam, 1934). In vitamin K deficiency hypoprothrombinemia develops with *uncontrollable hemorrhages*. The intestinal flora of a newborn child produces insufficient amounts of vitamin K and since, during gestation, little vitamin K passes to the embryo from the mother, hypoprothrombinemia exists in the newborn until the intestinal flora become active.

The anti-vitamin K — bihydroxycoumarin (or dicoumarol) — is hypoprothrombinemic and is used in the treatment of *thromboses*.

Total prothrombin is generally determined by the technique of Quick ("prothrombin time") in which the rate of coagulation of the test plasma is compared with that of plasma supplemented with Ca^{++} and thromboplastin, generally from rabbit brain.

Conversion of prothrombin to thrombin

This reaction is catalysed by numerous substances and requires Ca^{++}.

I. *The first stage is the conversion of inactive prothrombin to active prothrombin*

Two catalysts take part in this reaction:

(1) *Factor V (Owren) or accelerator I of prothrombin conversion or proaccelerin* is present in plasma in an almost inactive form. It is converted

to the active form or *accelerin* in the presence of traces of *thrombin*. This factor originates in the liver.

(2) *Thrombin* acts in a cyclic manner and activates the processes of blood coagulation once they have been started.

II. *Finally, active prothrombin is converted to thrombin in the presence of thromboplastin*

This reaction is catalysed by:

Ca^{++},

a platelet accelerator, factor I;

Factor V,

Factor VII, accelerator II of prothrombin conversion or preconvertin which is present in plasma in an inactive form and is converted to the active form or *convertin* by thromboplastins in the presence of Ca^{++}. It is synthesized in the liver in the presence of vitamin K.

An additional factor, factor X ($=$ Stuart–Prower factor) must be mentioned; it is absent when the subject is treated by dicoumarol (vitamin K deficiency).

Conversion of fibrinogen to fibrin

Fibrinogen consists of six polypeptide chains linked to mannose, galactose and sialic acid (Fig. 49); but only the third of the chains is spirally coiled.

In the presence of thrombin which is *proteolytic*, fibrinogen gives a molecule of the monomer fibrin and of two fibrinopeptides A (of molecular weight 2000) and B (of molecular weight 2400) such that the monomer of fibrin is not very different from fibrinogen as regards molecular weight (School of Laki, 1960). The difference is sufficient, however, for the monomers of fibrin to polymerize and form a clot of physiological fibrin (insoluble in urea). This requires an enzyme *(fibrinase* of Loewy, adsorbed by fibrinogen) and Ca^{++}. I would suggest that Ca^{++} combines with SO_4^{--} groups of peptide B (which is rich in tyrosine).

Table 34 shows a simplified scheme for the molecular stages of blood coagulation in mammals.

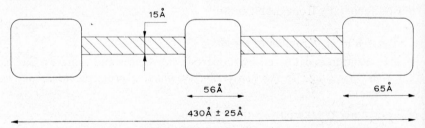

FIG. 49. Diagram of a fibrinogen molecule from electron micrograph. (HALL and SLAYTER, 1959.)

TABLE 34

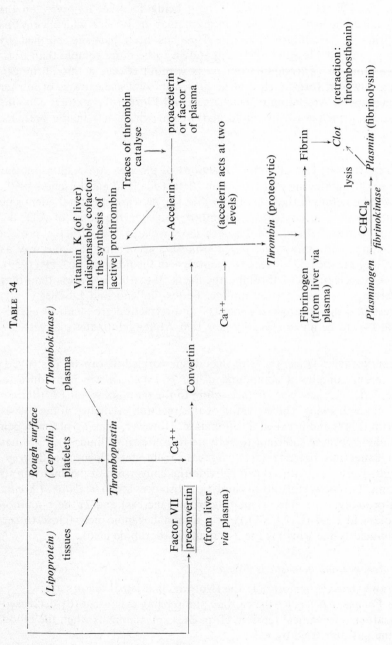

It is notable that all these factors normally exist in an inactive form but are rapidly activated with the onset of bleeding.

Our knowledge of the mechanisms of blood coagulation in invertebrates is less well developed. Interest is now being focused, however, on the mechanism in arthropods and in particular in lobster which contains a fibrinogen (Halliburton, 1885). This has been partially purified by Duchâteau and Florkin (1954) and shown to be more soluble than fibrinogen from ox. Thrombin from ox is without effect on this fibrinogen which however, forms a clot in the presence of muscle extracts of lobster ("coaguline"). According to Duchâteau and Florkin this extract contains an active principle of low molecular weight which acts in the presence of Ca^{++}.

RETRACTION: The glycolytic reactions of platelets are responsible for the change in viscosity and retraction (Bettex-Galland and Lüscher, 1960). Thrombin, acting in the presence of glucose, increases the total adenosine triphosphoric acid content of thrombocytes, later the total of ATP decreases. Bettex-Galland and Lüscher have isolated a contractile protein from human thrombocytes, *thrombosthenin*, which, in the same way as myosin of muscle (cf. Volume II), contracts in the presence of ATP (1959). The ATPase activity of thrombosthenin is 50 to 100 times less than that of actomyosin from striated muscle. Bettex-Galland and Lüscher (1961) suggest that the "change in viscosity" and retraction are identical phenomena. It is to be noted specially here that ADP agglutinates platelets.

FIBRINOLYSIS: Once the clot has been formed it can be destroyed. The serum contains a compound called *plasminogen* which produces a proteolytic enzyme, *plasmin* (according to the terminology of Christensen and McLeod, 1945). The activator of this reaction originates in the tissues (Astrup, 1947) and is called fibrinokinase. However, shaking plasminogen with chloroform is sufficient to activate it (removal of lipids ?). Enzymes from dangerous bacteria (*staphylokinase* from staphylococci and *streptokinase* from streptococci) can activate the conversion of plasminogen to plasmin. Two powerful anti-fibrinolytic substances are 1) *inhibitor of Künitz* (anti-proteolytic enzyme originating from pancreas) and 2) the ε–amino-caproic acid $H_2N–(CH_2)_5 CO_2H$ or 6 amino–hexanoic acid (close to the amino-acid lysine which is the 2–6 diamino–hexanoic acid).

Hemorrhages and hemostasis

In an extensive hemorrhage the two principal lethal factors are:

(a) *Volume deficiency* — the cardiac pump being exhausted (by decreased intracardiac mechanical tension, cf. p. 243), in mammals when the blood volume has decreased by half;

(b) *Oxygen deficiency*—hemoglobin loss leading to *anoxia.*
Severe blood loss is treated by infusion of physiological fluid and oxygenation. Administration of vasoconstrictors is dangerous since, at the

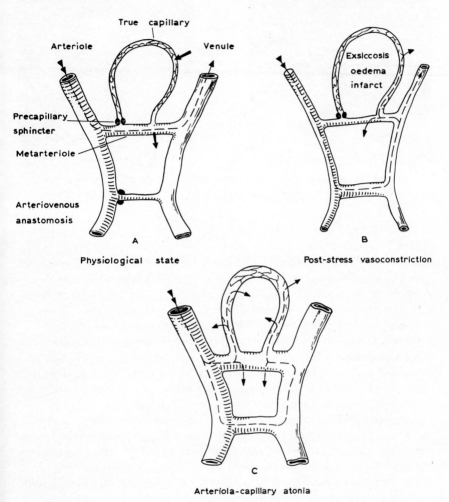

Fig. 50. Arteriola-capillary system, site of exchange between plasma and interstitial compartments. (LABORIT, 1958.)

beginning of hemorrhage shock at least, the agents accentuate tissue anoxia.

In order to define this shock it is necessary to understand certain aspects of the *anatomy and physiology of capillaries.*

With the exception of cardiac capillaries, although this has been con-

13*

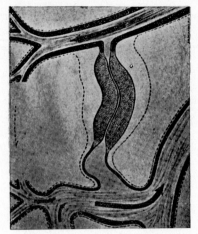 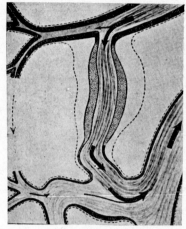

Fig. 51. (a) Diagram of normal glomus with vessel contracted (in systole).
(b) Normal glomus with vessel relaxed (in diastole). (P. Masson, 1937.)

tested (Provenza and Schulis, 1959), the wall of the *arterial* part of capillaries (spleen, kidney, uterus, etc.), is richly innervated by non-myelinated fibres and differentiated to form a muscular precapillary sphincter or *neuro-myoarterial glomus* (P. Masson). This glomus regulates the movement of blood from the precapillary arteriole to the postcapillary veins and thus determines the amount of blood in the capillary bed.

In hemorrhage shock the *capillaries* react in the following way:

Firstly, there is *reflex vasoconstriction* through the action of adrenaline and a compound from the kidney which acts on the glomi (this is called renin and must not be confused with gastric rennin). It is possible that this vasoconstriction results also from the action of a compound released by breakdown of platelets, *serotonin* (5-hydroxytryptamine).

If vasoconstriction is maintained the capillaries become congested such that the glomi are forced open by the hydrostatic pressure of the blood. The capillary bed fills with large quantities of blood which has a low oxygen tension and serious reactions develop.

The reactions which occur immediately after hemorrhages can be summarized as follows:

(a) acceleration of the heart through stimulation of the sympathetic nervous system;

(b) pulmonary hyperventilation leading to increased oxygen exchange;

(c) reduction in urine volume which decreases fluid loss;

(d) marked hemopoietic activity of spleen and bone marrow appears later.

To summarize, the immediate mobilization of adrenaline and renin allows the cardiac, pulmonary and renal systems to adapt to the reduced blood volume and later another reflex reaction leads to increased production of blood cells via the production of *hemopoietin* (Carnot and Deflandre, 1906).

In order to arrest hemorrhages:

(1) ligatures are used where possible;

(2) coagulants are administered, one of the most active being thrombin. At the same time, the wound is compressed and covered with a rough material (cotton lint) to facilitate the release of platelet thromboplastin. In the presence of antihistamines the glomi remain closed.

ANTICOAGULANTS. Anticoagulants are used—mainly during surgical operations—where, to prevent intravascular clotting, it is necessary to prevent blood coagulation for a certain period. There are five classes of anticoagulants:

(1) *decalcifying agents:* oxalates, citrates and to a lesser extent, phosphates;

(2) *antiprothrombins:* notably heparin, earths rare salts, dicoumarol;

(3) *antithromboplastins:* protamines and histones;

(4) *antithrombins:* heparin, hirudin, thiol derivatives (cysteine, glutathione);

(5) *antifibrinogens:* neutral salts, $KMnO_4$.

If blood is stirred the fibrin is removed and it will not clot but this is only of interest in the laboratory.

The most well known, least toxic and most physiological anticoagulant is *heparin*.

Heparin was discovered by Howell in 1918. We now know it to be *polysaccharide esterified by* SO_4 *groups* (Jorpes). It is therefore a strong acid and can form salts, notably with basic proteins, such that, *in vitro*, these two anticoagulants are neutralized.

Heparin is soluble in water; 1 mg of heparin inhibits coagulation of 100 ml of human blood.

Heparin is not absorbed by the intestine and is active only when administered by parenteral or perlingual routes. Heparin rapidly disappears from the circulation.

It is of interest that other sulphated polysaccharides are anticoagulants; notably dextran, but also a compound which is often forgotten, *fertilisin* from *sea urchins* (cf. p. 535) which is a sulphated mucopolysaccharide. The anticoagulant action of fertilisin was demonstrated by Immers and Vasseur in 1949.

Burstein (1956) showed that *in the presence of CaCl₂, heparin or synthetic heparinoids precipitate all β-lipoproteins from blood serum* (cf. p. 177). This fact is of interest as it suggests a mode of action of heparin and it also provides a method for the preparation of β-lipoproteins since the precipitated lipoproteins can be redissolved in sodium citrate solutions and characterized and isolated by such methods as electrophoresis. About the same time, the Czech workers (Hladovec *et al.*, 1956) found that heparinoids possess *antilipemic* activity. This is of importance in considering the treatment of atherosclerosis.

Heparin is found mainly in the liver but large quantities are also present in lung tissue (which contains thromboplastin, cf. p. 181).

Specialized histiocytes exist which contain metachromatic granules*; these histiocytes are called *mastocytes (mast cells)* or *labrocytes*. They are present in areas containing connective tissue and particularly in the liver. L. Arvy (1955) has shown that, in dog, the vessels of relatively inactive tissues contain very few mast cells while in active tissues (thyroid, *post partum* uterus), "les vaisseaux sont engainés par un véritable treillis de labrocytes". At the present time it is difficult to explain this distribution.

Jorpes and his co-workers have shown that *mast cells contain heparin* which accounts for the metachromatic granules. Julen, Snellman and Sylven (1950) suggest that heparin is attached to a protein substrate. In 1952 Riley and West showed the presence in mast cells of *histamine* and it is possible that they also contain *serotonin* (Rowley and Benditt, 1956). Thus, the problem of hemorrhage shock can be considered on a new basis. *Peptone shock*, produced by rapid intravenous injection of peptone, is accompanied by increased lymphatic circulation and lymphatic vasodilation with passage of mast cells into lymph nodes *where they are disrupted* (L. Arvy, 1956); in this way both histamine and heparin can be liberated.

In peptone shock,

(1) *there is liberation of histamine* (Feldberg and O'Connor, 1937);

(2) Schmidt-Mulheim (1880) stated that *in vivo* an anticoagulant factor is released in dog and this was later shown to act *via the liver* and, in fact, to be heparin.

Finally, Wilander (1938) observed the disappearance of hepatic mast cells during peptone shock.

I would suggest that breakdown of mast cells within lymph nodes is responsible for the phenomena associated with peptone shock through the release of histamine, heparin and possibly, the vasocontrictor serotonin.

Hypothesis on atherosclerosis

A certain number of isolated facts are available and these can be brought together to provide a coherent concept.

It seems that dietary factors are important in the aetiology of atherosclerosis. Compounds which may be involved are: ethenoid fatty acids,

---

* Dyes which stain certain histological or cytological elements are termed metachromatic after Ehrlich. For example, thionine stains basophilic elements blue and acidic polysaccaharides red (Lison).

vitamins such as pyridoxine which lowers cholesterolemia to a small extent, plant sterols—sitosterols—including β-sitosterol which inhibits intestinal absorption of cholesterol and thus reduces cholesterolemia (Beher *et al.*; Heptinstall and Porter, 1957), etc. However, atherosclerosis does not develop only in well-nourished individuals receiving a high fat content diet. F. Blaha, an internee working in the infirmary at Dachau concentration camp, found from more than 10,000 autopsies that both young and old prisoners developed atherosclerosis. These prisoners were receiving a diet containing less than 1000 cal per day and no alcohol or animal fats. Therefore, in man at least, diet does not appear to be the systematic factor in atherosclerosis; the situation may be different in rabbit which is highly susceptible to development of aortic atheroma (Ignatowski; Saltykow, 1918). Moreover:

(1) adrenaline released by stimulation of splanchnic nerves or *emotional stimuli* reduces the time for coagulation (school of Cannon, 1914, 1931) and there is an increase in total glucose (the role of glucose in platelet metabolism has already been mentioned). Within a few minutes after sympathetic stimulation there is a distinct increase in total cholesterol. Ischlondsky (1960) believes this endogenous hypercholesterolemia of nervous origin to be the general factor in atherosclerosis;

(2) β-lipoproteins contain 17 per cent proteins (this is low compared with α_2- and in particular α_1-lipoproteins), 25 per cent triglycerides, 21 per cent phosphatides and 37 per cent cholesterol of which 29 per cent is esterified and 8 per cent free. Thus, heparin, which inhibits blood coagulation, combines with β-lipoproteins (through Ca^{++}) while adrenaline, which increases cholesterolemia, shortens the bleeding time. There appears to be a correlation between blood coagulation and cholesterol and between cholesterol and atherosclerosis. I consider that the damage to the endothelium caused by release of catechol amines (Ischlonsky) would allow deposition of blood β-lipoproteins, through the intermediary of Ca^{++}, on the vascular sulphated mucopolysaccharides (Kirk and Dyrbye, 1957; Murata, Kirk and Asawa, 1964) exposed by vascular changes. During stress, the fate of thrombocytes which attach themselves, according to Roskam and Bounameaux (1961) to aortic connective fibres (the aorta being particularly susceptible to atheroma development) should be considered and, in addition, the fate of mast cells containing heparin. In addition I have drawn attention (1966) to the fact that pressive phenomena in atherome have to be related to the localization of chemo- and baro-receptors in the aortic arch (cf. p. 293).

E. Blood groups

In dealing with this subject we shall be considering the problem of biological specificity and in particular, the specificity of the internal environment.

There is a tendency to view blood groups in terms of *erythrocyte groups* and to exclude them from a study of the internal medium. However, as we shall see, erythrocyte groups are directly associated with *serum groups*.

With the discovery of the circulation of blood by Harvey, physicians attempted to transfuse blood from one human to another weakened by hemorrhage. However, these transfusions proved fatal. Did this result from blood coagulation with occlusion of vessels? To some extent this is correct but the coagulation is not of the type that has just been described.

In 1900, Karl Landsteiner provided the explanation when he discovered the existence of *blood groups*.

What are blood groups? To answer this question it is necessary first to define *antigens* and *antibodies*. An *antigen* is a macromolecular compound, generally a protein, which when administered to mammals causes the appearance of a compound which will combine with and is *specific* for the antigen; this compound is called an *antibody*. The antibodies are protein in nature or more precisely, *γ-globulins*. Wurmser and his co-workers (1950) have shown that the *energy changes* during combination of agglutinogens and agglutinins parallel those deduced from observation of agglutinants. Filitti-Wurmser and Jacquot-Armand (1947) established that isoagglutination of human erythrocytes is a *reversible* phenomenon dependent on temperature and dilution.

The four blood groups of Landsteiner

There are two erythrocyte "antigens" called *agglutinogens* (A and B) and two plasma antibodies called *agglutinins* (α or anti-A and β or anti-B). Neutralization of the antigen by the antibody occurs with *agglutination* of erythrocytes (J. Bordet, 1895; Landsteiner, 1900). When this agglutination occurs *in vivo* there is embolism (the cause of accidents during blood transfusion). Consequently, *agglutinins can only exist in blood which lacks the corresponding agglutinogens; in the case where the agglutinogen is absent, the agglutinin is always present*. The mechanisms of this phenomenon are unknown. This exclusion makes the assimilation of the agglutinogen with an antigen more difficult. The discussion of haptens (cf. p. 196) will throw some light on this subject (cf. p. 201).

The four blood groups are:

group A, *anti*-B (or *β*),
group B, *anti*-A (or *α*),
group AB, without agglutinin,
group O, *anti*-A and *anti*-B (or *αβ*).

Table 35 summarizes the erythrocyte–serum relationships.

Under these conditions, blood groups can be determined in two ways:

(1) *directly* by identifying the *agglutinogen* of erythrocytes by serum tests;

TABLE 35

Group	Agglutinogen (in erythrocytes)	Agglutinin (in serum)	Erythrocytes react with serum	Serum reacts with erythrocytes
A	A	anti-B (or β)	anti-A (or α) or the serum of B α and O $\alpha\beta$ subjects	A AB
B	B	anti-A (or α)	anti-B (or β) or the serum of Aβ and O α β subjects	A AB
AB	A and B	none	anti A (or α) anti-B (or β) or all sera: B α, Aβ, and O$\alpha\beta$.	no reaction
O	neither A or B (O) exception (cf. p. 195)	anti-A (or α) anti-B (or β)	no reaction	A B AB

(2) *indirectly* by identifying the agglutinin of plasma or serum using washed A and B erythrocytes.

Although blood groups are inherited (we shall examine this later, p. 199) the *total* of isohemagglutinins can vary under certain conditions; for example, an individual with lymphoid leukaemia produces less anti-B antibodies than normal subjects (André and Salmon, 1961).

DISTRIBUTION OF BLOOD GROUPS ABO: In France the approximate distribution in man is as follows:

40–45 per cent are O,
42–47 per cent are A,
5–10 per cent are B,
3–4 per cent are AB.

The brahman caste at Pondichery shows the following distribution according to Kherumian and Nioguy (1960):

41·18 per cent O
25·34 per cent A
26·24 per cent B
7·24 per cent AB.

The agglutinogens and agglutinins of *anthropoid apes* are the same as in man. Gibbons and chimpanzees are generally group A, orangoutangs are either group A or group B. Human serum contains agglutinins active against erythrocytes of chimpanzees.

In relation to cross circulation experiments (cf. p. 294, 357) one will note that only 25 per cent of dogs react unfavourably to blood from another dog (Melnick, Burack and Cowgill, 1936).

The incidence of agglutinogen B increases statistically from western Europe to the East (Table 36a).

TABLE 36a. *Percentage of group B in human populations*

In western Europe	12–14%
In the Balkans	20–23%
Among Turks and Arabs	25%
Among Indochinese, Chinese and Indians	34–49%

It is frequently stated that agglutinogens A and B arose *separately* and in different regions of the world. Hirzfeld has inferred that there are two fundamental human races, A and B, with the stable hybrid AB and the mutant stable recessive O. The hypothesis of a double origin of A and B raises the important problem of the mono- or polyphyletic origin of human species. From the distribution of blood groups among oceanic populations, it is possible to explain the dispersion of man in the northern Pacific and western Atlantic by sea currents and prevailing winds (Avias, 1949). Dispersion from east to west—from Peru to Polynesia—under the influence of the equatorial Humboldt current and trade winds is possible, as was shown by the famous Norwegian expedition aboard the Kon-Tiki raft (Thor Heyerdahl, 1947). Thus blood groups provide a means of labelling populations in the same way as they serve as experimental labels in the technique of Ashby (cf. p. 176).

It is remarkable that specific agglutinogens are not localized only in blood. They are found in milk, saliva, gastric juice, bile, sperm, and cell homogenates.

Among 75–84 per cent of white individuals agglutinogens A and B are secreted in the *saliva* (Yamaki); these subjects are called *secretors*. Among the remainder secretion is negligible *(nonsecretors)*. 60 per cent of black Americans are secretors. Horse saliva contains human group A and B substances (Friedenreich, 1937).

The demogeographical distribution is of interest. Table 36b shows the frequency of ABO in different human populations.

TABLE 36b

	% O	% A	% B	% AB
Irish	56	32	9·5	2·5
English	46	42	9	3
American Negroes	48	27	14	6
Portuguese	41·6	47·2	8·3	2·9
Hindus	30	24	38	8
Indians of Peru	100	0	0	0

SUBGROUPS OF A: Within group A von Dungern and Hirzfeld (1911) have identified the subgroups A_1 and A_2. These are characterized *by agglutination after adsorption*. The following example will clarify this reaction:

When certain anti-A sera are adsorbed with erythrocytes of subgroup A_2 these sera contain an agglutinin which does not act on A_2 cells but can still agglutinate other blood cells called A_1.

The frequency of A_1 is about 80 per cent and that of A_2 about 20 per cent.

Landsteiner and Levine demonstrated that certain normal human sera contain agglutinins which can distinguish between the two subgroups. One, designated by α_1, rarely exists in A_2 sera but is found most often in A_2B sera; it agglutinates A_1 and A_1B erythrocytes.

The other, designated by α_2, is rarer than α_1 and exists in certain carriers of agglutinogen A_1 (in particular A_1B); it agglutinates A_2 erythrocytes.

Agglutinins α_1 and α_2 *are rare*. Moreover, since they react at temperatures below 37°C they have no importance in transfusions. The incompatibility of A_1 and A_2 has little *practical* significance although this would have to be considered in transfusions carried out at low temperatures.

Agglutinogen O

The red corpuscle is a complex structure containing a large number of proteins and therefore, a large number of antigens, forming according to Charles Nicolle a "mosaïque d'antigènes". This applies to all cellular antigens and is related to their structural complexity. The poly-antigenic nature of a cellular structure implies that different antibodies can react with one type of cell.

Human erythrocytes can be agglutinated by numerous animal sera and this is *independent of group* ("anti-man agglutinins").

At low temperatures (below 37°C), *human* agglutinin α_2 can, to a certain extent, agglutinate not only erythrocytes A_2 but also erythrocytes of group O. The name anti-O agglutinin is given to α_2 and it follows that even group O corpuscles contain a reactive substance, agglutinogen O.

Chemistry of groups ABO specificity:

The chemistry of agglutinins is that of γ-globulins.

The chemistry of *agglutinogens* will be considered here. From a study of erythrocytes and commercial pepsin obtained from the gastric mucosa of pig and horse (Tomarelli, 1954; W.T.J. Morgan, 1956, 1958) evidence was obtained for the presence of polypeptides (20 per cent), mucopolysaccharides (80 per cent) containing acetylglucosamine, acetylgalactosamine, L-fucose, D-galactose (Landsteiner; Kabat; Schiff; Freudenberg).

$$
\begin{array}{l}
\text{HC}{=}\text{O} \qquad \text{L-Fucose} \\
\;\;\;| \\
\text{HOCH} \\
\;\;\;| \\
\text{HCOH} \\
\;\;\;| \\
\text{HCOH} \\
\;\;\;| \\
\text{HOCH} \\
\;\;\;| \\
\text{CH}_3
\end{array}
$$

The polysaccharides are responsible for *specificity*. Although they have no antigenic properties—or are only weakly antigenic—they can react with the corresponding antibodies; these polysaccharides are given the name *haptens* (cf. p. 219).

Plant inhibitors:

There are certain macromolecular substances of plant origin which inhibit group agglutination by the homologous antisera. The work of G.F. Springler (1958) has shown that such substances are found in organisms as different as *Fucus vesiculosus* and *Escherichia coli.* They are *non-nitrogenous polysaccharides*, the most active being a polysaccharide from *E. coli* (strain 0 86) which acts on group B. It is interesting to consider these polysaccharides as homologues of heparin-like anticoagulants or as competitive inhibitors of group polysaccharides.

MN Groups.

These were discovered by Landsteiner and Levine in 1928. Anti-M serum is obtained by injecting 0M erythrocytes of man to *rabbit* and anti-N serum by injecting 0N erythrocytes.

Apart from specific antibodies, because of the antigenic diversity of erythrocytes, one can also obtain agglutinins called "species agglutinins" which agglutinate *all* human blood. These are removed by *adsorption* on erythrocytes which cannot interfere with M and N and in this way anti-M sera and anti-N sera are obtained.

To demonstrate groups M and N it is necessary to use serological immunization techniques (cf. p. 216); hetero-immuno-sera are produced. *Natural isoagglutinins only exist for the ABO system.* In addition, *agglutinogens M and N cannot act as antigens in man as agglutinins for these are absent from human sera* (this distinguishes MN from Rh).

The absence of natural M and N agglutinins (except for rare cases) as immune agglutinins in man means that the MN system has no bearing on problems of incompatibility between donor and receiver in blood transfusions. M and N agglutination can function in the labelling of erythrocytes (cf. p. 176, Ashby's technique for measurement of blood volume).

Agglutinogens M and N form three groups: M, N and MN. The distribution of these groups in Europe is:

M   30 per cent,

N   20 per cent,

MN 50 per cent.

Another group exists which was discovered by Landsteiner and Levine at the same time as M and N. This is *group P;* horse, pig and rabbit sometimes contain an anti-P agglutinin.

Rh groups:

This system was also discovered by the great immunologist Landsteiner while working in the U.S.A. with his pupil Wiener (1940).

Experiments carried out by another famous immunologist, J. Bordet (1895) of Belgium, *suggested that the serum of a given animal species contains agglutinins which can agglutinate the red corpuscles of all other animal species.* Thus:

Rabbit serum in high concentrations will agglutinate erythrocytes from the monkey *Macaccus rhesus.* If rabbits are injected with erythrocytes from this monkey in the hope of obtaining large quantities of specific antibodies, an *anti-Rhesus rabbit serum* is obtained (or anti-Rh) which strongly agglutinates erythrocytes of monkey but which is *also capable of agglutinating blood from 80 per cent of white individuals.* These subjects are called *Rh positive.* Unreactive subjects are described as Rh negative. In France, 30 per cent of Basques are Rh — while among Chinese, Australian aborigines and American Indians the frequency of Rh + is 95 to 100 per cent. These figures suggest a great variety of anthropological problems related to blood groups.

Hemolytic disease of newborn.

This has been studied in man by Levine (1940) and in donkey by Caroli and Bessis (1947).

The *principal* observation of Levine resulted from an accident during the transfusion of group O blood from a husband to his pregnant wife. Contrary to the expected result the blood from the wife agglutinated red cells from the husband. Since Landsteiner and Wiener have recently produced evidence for the Rh factor, Levine supposed that the patient must have been Rh negative and possessed anti-Rh agglutinins, the husband being Rh positive. However, it was established that the patient had not been transfused previously and the problem arose of how she developed this immunity. Levine formulated the hypothesis that the pregnant woman had developed anti-Rh agglutinins through *contact with the foetus* which was Rh positive like the father. To substantiate this idea Levine searched the scientific literature for all transfusion accidents during gestation and he found that these transfusions were given because of hemorrhage related notably to *icterus gravis* of the newborn. From these facts Levine developed the idea that this disease of newborn infants resulted from the passage of anti-Rh agglutinin of the mother—produced in contact with the Rh+ foetus—*across the placenta* where they agglutinated the Rh positive erythrocytes of the infant leading to a hemolytic disease. This idea has been confirmed.

Nowadays this hemolytic condition in newborn can be treated by inducing premature parturition at the eight month and replacing the blood of the infant by transfusion. Therefore, marriages between $Rh_\female^-$ and $Rh_\male^+$ have nothing to fear. In the case of $Rh_\female^-$ $Rh_\male^+$ there are 5 chances in 100 of producing an erythroblastic infant since this is dependent on the aptitude of the subject for forming antibodies. In the hemolytic disease of newborn donkeys, transference of agglutinins from the mother to the offspring does not occur across the placenta but through colostrum (first secretion of mammary glands); suckling must be prevented at first (the absorption of proteins by the intestine at this early stage should be noted).

There are numerous hemolytic diseases of newborn: Tovey, Lockyer Blades and Flavell (1962) have shown that one such disease is associated with the ABO system and appears in about the same proportion of subjects as that associated with Rh system (1 case in 200 pregnancies).

Other Rh agglutinogens:

The Rh factor of Landsteiner and Wiener is now called "standard Rh" or "Rho". Anti-Rh antibodies are formed most often during heterospecific pregnancy but other antibodies can be produced having different specificity, the immunizing erythrocytes possessing different agglutinogens from Rho. In addition to Rho serum which acts in 85 per cent of blood, two other sera have been discovered, one acting on 70 per cent of blood and the other on 30 per cent of blood. According to Fisher and Race the 85 per cent serum is anti-Rho or D, the 70 per cent serum is anti-Rh′ or C and the 30 per cent serum is anti-Rh″ or E.

Inheritance of blood groups:

I must mention first of all that this subject provides a particularly clear example of the association between *gene → enzyme → specific compound* or, in other words, of genetic, physiological and biochemical interrelationships. This subject provides a new example of that which is called *physiological genetics* which also includes the study of pigments and metabolic requirements.

Inherited characteristics are transmitted according to the laws of Mendel.

1. INHERITANCE OF GROUPS ABO. From the work of von Dungern and Hirzfeld (1910) and particularly of Bernstein (1925) *the properties of the groups are considered to be transmitted by three allelomorphic genes:* A (transmitting A), B (transmitting B) and R (transmitting O); A and B are dominant, R is recessive.

The facts can be summarized in the *two following rules* (formulation of Moullec):

(1) The *A and B properties* are inherited according to Mendelian rules as two dominant characters while their absence is a recessive character. Therefore, they cannot appear in infants if they did not exist in the parents. However, they can be absent from the children although existing in the parents.

(2) Parents belonging to group O cannot produce group AB children and parents of group AB cannot give rise to group O children.

There are three catagories of gametes in this respect:

— gametes carrying gene A,
— gametes carrying gene B,
— gametes carrying gene R.

Table 37 shows the six categories of genotypes which can be obtained from matings:

TABLE 37

♂ \ ♀	A	B	R
A	AA	AB	AR
B	BA	BB	BR
R	RA	RB	RR

There are thus six genotypes: AA, AR, BB, BR, AB and RR.

A practical application of these genetic considerations is to be found in testing cases of disputed paternity.

In the *inheritance of subgroups of A* the theory of Bernstein has been extended by considering there to be four allelomorphs in place of three: A_1, A_2, B and R; A_1 dominates A_2 and A_2 dominates R.

2. INHERITANCE OF THE SECRETORY CHARACTER. The "secretor" character, S, is dominant and the "nonsecretor" character, s, is recessive. As a result, secretors are either homozygotes SS or heterozygotes Ss, while nonsecretors are all homozygotes ss.

It should be remembered however that the difference between secretors and nonsecretors is due to the fact that group substances exist in two forms: *liposoluble* and *hydrosoluble* (Schiff, 1931). The liposoluble form is localized in the erythrocytes and tissues of secretors and nonsecretors. The hydrosoluble form is almost completely absent from aqueous extracts of nonsecretors and is absent from body fluids. The hydrosoluble substances are abundant in glands of secretors. Thus the genes transmit not only morphological characters but also physicochemical characters. This extends the concept of taxonomy.

3. INHERITANCE OF M AND N. *There is no recessive character among these.* The existence of the allelomorphs M and N is accepted and these are transmitted independently of group ABO. Genetically, a phenotype M is always MM and a phenotype N is always NN.

An infant can only show factor M or factor N if it existed in the parents which explains the medicolegal application of the identification of M and N.

P is inherited in a simple manner similar to S.

4. INHERITANCE OF RH FACTORS. The inheritance of *Rho* (standard Rh) was described by Landsteiner and Wiener in the following way:

Rh^+ is transmitted by a dominant gene R (this must not be confused with R, the recessive of group ABO, which should be written R_{ABO}, cf. p. 199). The recessive allelomorph r would transmit the absence of Rh. *Inheritance of Rh is independent of ABO or MN.*

There are two ideas concerning the inheritance of Rh' and Rh", that of Wiener and that of Fisher and Race. To describe this controversy fully would take too long and would be more appropriate in a book on genetics. I can only give some indication as to the problem.

I should like to emphasize that although the problem will be dealt with only briefly it has, in my opinion, considerable importance.

Generally, one only considers the inheritance of agglutinogens and inheritance of agglutinins is neglected. The mechanism is little understood.

How can one explain the fact that when agglutinogen A is present agglutinin A is absent?

My hypothesis is as follows:

Only agglutinogens are transmitted by hereditary pathways and the agglutinins would be induced by the following mechanism:

(a) For example, A erythrocytes during their formation (hemocytoblast or reticulocyte stage) would contain a large quantity of agglutinogen B which is serologically very active. Agglutinogen A which only exists in the hapten form would be masked by the antigen agglutinogen B.

(b) The young erythrocytes would form anti-B antibodies which would be fixed by agglutinogen B and inactivated, while hapten A would remain free in the circulating erythrocytes. Thus, A characteristics would be apparent; and similarly for B.

The existence of group AB may be explained by the unique hapten-like character of group substances in this case. Group O would result from the inactivation of antigenic group AB substances. This would explain the presence in serum of the opposing agglutinin and of the agglutinogen.

Under these conditions transmission in the ABO system would occur in the following way: the individual carrying A would be recessive for A (presence of hapten only) and dominant for B (presence of antigen).

A method of verifying this hypothesis would be to remove B antibodies from A erythrocytes. This hypothesis does not contradict the results of Filitti-Wurmser, Jacquot-Armand and Wurmser in which natural antierythrocyte agglutinins and immune agglutinins are distinct; the former being homogenous from the point of view of their affinity for the respective agglutinogen and the latter being heterogeneous. These authors conclude that "the genetic label imposed by organisms does not exclude the possibility that natural isoagglutinins also depend, at a certain stage, on an immunization mechanism".

Serum groups

In addition to the agglutinins of serum which correspond with the erythrocyte groups—no leucocytes groups being known—O. Smithies (1955) has shown the presence of a class of serum proteins of which the structural variations are determined genetically and, as a result, it allows a study of human populations (Moullec *et al.*, 1961). These groups are defined by three protein fractions: α_2-globulins with haptoglobins, β-globulins with transferrins and α_1-globulins [the haptoglobins are serum proteins which have the ability to combine with hemoglobin (M. Polonovski and M. F. Jayle, 1938)]; the serum groups can be distinguished by starch gel electrophoresis. There are two groups: one called Hp_1 (consisting of a low molecular weight monomer) and the other containing two subdivisions Hp_2 (polymer) and ($Hp_1 + Hp_2$). The frequency of these haptoglobin types in the population of France is 14·7 per cent for Hp_1,

37·4 per cent for Hp$_2$ and 47·9 per cent for (Hp$_1$+Hp$_2$) (Moullec *et al.*, 1960) and among Zulus 31 per cent for Hp$_1$, 25 per cent for Hp$_2$, 38·8 per cent for (Hp$_1$+Hp$_2$) and 2·6 per cent contain no Hp (Barnicot *et al.*, 1959).

Blood transfusion

Blood of groups A, B, AB and O can be injected into individuals of the same group without any danger.

Blood from a different group can also be injected on condition that it is *compatible* with the blood of the recipient, in other words, blood containing erythrocytes which are not agglutinated by the plasma of the recipient.

Group O individuals possess anti-A and anti-B plasma agglutinins and can only receive group O blood. Group O erythrocytes are not agglutinated by anti-A or anti-B and therefore group O individuals can donate blood to *all* other groups and are termed *universal donors*.

O can be donated to all groups.
A and B can be donated only to
universal recipients (after Ottenberg).

These considerations are of value when *relatively small quantities are transfused and injected slowly*. Under these conditions, the agglutinins of the injected blood are immediately inactivated by the large volume of red cells of the recipient and their effect is negligible.

In the case of massive transfusions the agglutinins of transfused blood can combine with the erythrocytes of the recipient and cause transfusion accidents. Moreover, there are group O donors which possess abundant agglutinins. These donors are clearly dangerous even if they give small quantities of blood. It is necessary to estimate the agglutinin content of all group O blood.

In the technique of blood transfusion stored blood of a compatible group is injected by means of a needle attached to a plastic tube. Transfusions are given particularly during anemia and after hemorrhage.

Blood conservation

In order to store blood, coagulation must first be prevented by using sodium citrate or heparin or by placing previously sterilized blood in containers coated with silicone. These containers are kept at 0°C to prevent bacterial growth. Paul Becquerel (1932) introduced the technique of conserving living elements by freezing and drying. In order to conserve blood for long periods at low temperatures disruption caused by the formation of ice crystals must be prevented. Disruption of cellular structures is related to influx of water at the moment of freezing and the appearance of ice microcrystals. If thawing is carried out slowly movement of molecules occurs leading to further cell damage. One would assume from these facts that freezing should be carried out slowly to prevent the appearance of large crystals within cells and thawing carried out rapidly. However, this is not the case.

The technique of conservation of blood particularly by cooling has been described by Luyet (since 1948).

The formation of ice crystals is prevented and the ice formed is described as *amorphous* or *vitreous*. Two methods are available:

ultra-rapid freezing (there is not sufficient time for molecules to become arranged in crystals);

or the viscosity of the medium is increased such that movement of molecules is delayed and cellular dehydration improved (using glycol; J. Rostand has used glycerol).

The rates of freezing necessary to prevent crystallization have been estimated by Luyet:

1000°C per s for a cell containing 80 per cent water

100°C per s for a cell containing 60 per cent water.

Below 30 per cent water the rate of freezing is not significant. The temperature at which devitrification occurs increases with the molecular weight of the solution: glucose, sucrose, gelatin, albumin and starch increase the temperature.

Luyet and Menz (1951) stated that practically no hemolysis occurs when the temperature of blood is maintained around the freezing point (0·6°C).

Sloviter (1951) demonstrated that erythrocytes frozen at —79°C with glycerol, thawed and reinjected into rabbit or man, survived normally in the circulation of the recipient. The necessity of carrying out repeated dialysis to remove the glycerol reduces the practical application of these results.

At higher temperatures cold reveals a curious property of blood: Hayem (1888) found that a suspension of red cells of certain subjects when in contact with serum of the same subject induced agglutination of erythrocytes after several hours at 0–7°C (however, 8°C is the lowest reversible temperature for deep hypothermia during cardiac surgery (Dubost, 1960). This phenomenon is reversible at about 25°C. The reaction is dependent on "cold agglutinins".

Using the integrity of glycolysis as a measure of metabolic activity in erythrocytes, Paysant (1952) found that blood was best conserved at 4°C in solutions containing 0·5 per cent glucose and ATP at a pH of 6·7. The presence of a reducing compound such as glutathione inhibits hemolysis and improves blood conservation (Fegler, 1952).

14*

Fleish and Frei (1961) delayed hemolysis in stored blood by continuous slow rotation (from 4 revolutions per day to 45 revolutions per hour for blood stored in bottles three quarters full); this was observed at 5°C, 20°C and 37°C.

Substitution of blood or exsanguino-transfusion

Essentially, *the total blood volume of a subject is replaced by a mixture of blood from other subjects.* Replacement can be *total* or *partial.*

Initially this method was introduced by Bessis as an experimental technique (in dogs, 1939) but was extended to newborn and adult man (1946–1947).

Blood is perfused to the *newborn* via the vein in the umbilical cord. This technique has mainly been used in cases of hemolytic disease; (the transfused blood must be Rh−, such that injected cells are not destroyed by anti-Rh agglutination).

In the *adult* 8 litres of blood are required, a large volume necessitating 16 to 24 donors. Substitution is carried out mainly for purposes of detoxication, in particular, to eliminate urea in acute nephrites.

From the experimental aspect, this technique to some way extends the *crossed circulation* technique which will be considered with reference to the experiments of Heymans (cf. p. 294).

CHAPTER 10

SPECIFICITY OF THE INTERNAL
ENVIRONMENT AND IMMUNITY

The problem of specificity is one of the most important aspects of biology.
The individual characteristics of each group (and within each group, those
of each creature) express the ability of each living creature to make use
of elements from the external environment and to incorporate them into
specific material. C. Richet described this as the "personnalité humorale"
but it is principally a question of *cellular individuality.*

Biochemical and physiological individuality

Biological individuality would appear to be related to *morphology* (an
example of this being finger prints) but it is also related to function, corresponding to the correlation between gene, enzyme and structure. This
individuality is apparent in the chemical composition of different animal
species. Needham and later Florkin, in laying down the bases of comparative biochemistry, did no more than establish this survey of chemical
composition and its physiological implications. The value of comparative
biochemistry is that it is a *molecular expression of genetic development*
from which *similarities* and *differences* in structure and function of different lines, which are related morphologically, can be established. The concept of physiological specificity can be included in that of functional
taxonomy. Numerous examples of biological specificity can be cited.
We have already seen that the food requirements (vitamins, amino acids
etc.) of one group of animals differ from those of another. In addition to
apparent experimental variations, individuals of the same species show
differences in, for example, the amount of a compound secreted and such
a compound expresses an aspect of biological individuality.

For example, the concentration of the various constituents of human
blood is variable such that a given concentration cannot be described as
pathological since it may relate to some anomaly within the organism.

Thus, the level of glucose in blood can vary between 84 and 125 mg
per cent, these extreme levels being an expression of biochemical individuality and not of any abnormality.

In addition, plasma can contain 5 to 12 mg per cent glutamine, 0·4 to 2 mg per cent pyruvic acid, etc.

In the natural history of parasites there are numerous examples of adaptation of a parasite to a host and thus, *specific sensitivity* of the host to a certain parasite. This is another example of biological specificity.

In *strict parasite specificity* or *stenoxenia* there is some degree of affinity between the host and parasite; for example, in the case of cestodes of birds, intestinal Protozoa of amphibians.

In other cases the parasite shows a certain degree of ubiquity or *euryxenia*. Although parasitism can be dangerous for the host it can also be benign as in the case of malarial parasites during the stage within *Anopheles* (host-vector). The pathogenic property of parasites depends on numerous factors, particularly on the "environment" which when favourable constitutes a *preferendum* for the parasite.

Another example of biological specificity (a particular aspect of parasitism) is provided by *grafts:*

It could be thought that this specificity is exclusively *cellular*. There is no doubt that cellular specificity plays a part in determining whether a graft will implant successfully but humoral specificity is also important. Medawar (1945) emphasized that *interactions of the antigen–antibody type between the graft and recipient tissue are involved.* In adult mammals grafts are only possible between animals of *the same race* having a genetic relationship (homografts). In the case of skin grafts for example, only grafts from the same individual (autografts) can be successful and homografts are eliminated. This expresses a degree of intolerance in nature which is closely linked to genetic dissimilarities. However, *homografts between identical twins are particularly successful.*

As a general rule, the higher the animal (for example, mammals compared with amphibians) and the greater the differentiation (as in adults or senescence) the more difficult it becomes to obtain positive grafts (in addition, the possibilities of regeneration are less). These facts led to the very successful technique of grafting young elements to adults. For comparison with terms such as autoplasia, heteroplasia and homoplasia R. M. May has introduced the term *brephoplasia* to describe "*functional and durable* transplantation of tissues from embryos or newborn to young or adult mammals" (1934).

May and Arpiarian (1954) have shown that brephoplastic grafts of spleen and liver from newborn mice are successful in the peritoneal cavity of adult mice and brephoplastic grafts of bone marrow not only developed but protected the adult against irradiation with large doses of X-rays (1200 R) by augmenting hemopoiesis.

The specificity of the internal environment in the case of grafts applies also to *grafts in the anterior chamber of the eye.* These grafts are often used for several reasons: experimental access is easy; the transparent nature of the cornea allows the progress of implanted tissue to be followed; the graft can be vascularized from the iris and in particular, the ionic composition of the aqueous humour is similar to that of blood serum and it contains *less antibodies* (and less proteins in general).

The importance of the internal environment in grafting will be further understood after examining the excellent work of Filatov on the use of *corneal grafts* in the case of corneal thickening. Not only has this technique of homografting (from the eyes of corpses preserved at low temperatures) prevented blindness or restored the sight of many individuals but it has increased our knowledge of conditions for graft implantation. Filatov observed that grafts were more successful when they have been taken from conserved tissues and in particular, when aqueous extracts of conserved tissues were injected. These extracts contain compounds, not yet identified, which prepare the environment. This is the basis of *tissue therapeutics*.

Even fertilization is an expression of biological specificity (successful or sterile mating). The concept of environment can be understood most clearly by examining the case of susceptibility—and its corollary resistance—to infectious diseases. Firstly, we shall examine the concept of environment in relation to that which I call *edaphic immunity*, that is to say, in relation to the nature of the environment.

1. Numerous *non-specific factors* are involved:

The penetration of airborne bacteria is inhibited and their virulence is destroyed or modified by the presence of *lysozyme* in tears (discovered by Laschtchenko). Lysozyme is also found in human milk (although not in cow's milk). In addition, conjunctive tissue constitutes a barrier against microbial attack. Hyaluronic acid contributes to the structure of intercellular cement (Meyer and Palmer, 1936) (it should be mentioned that synovial fluid is considered to be a dilute form of basic intercellular substance, W. Pigman, 1962). Duran-Reynals in 1933 demonstrated that the invading forms of staphylococci and streptococci contain a soluble factor which facilitates the diffusion of various substances (such as dyes) and organites in tissues and which facilitates infection by non-invading forms of staphylococci and streptococci or by other bacteria or virus. This factor is *hyaluronidase* which hydrolyses hyaluronic acid decreasing its viscosity (Robertson, Ropes and Bauer, 1940).

Thermal conditions: most people have heard of the experiments of Pasteur (1878) in which cooling from 42°C to 37°C induced sensitivity to cholera bacteria in chickens.

Climatic factors are of some importance. Perkins found that the diptheria antitoxin content of man is lowest in March, April and May, the period in which diptheria occurs with greatest frequency. This introduces the problem *of seasonal factors*. In the northern hemisphere the erythrocyte content (and therefore hemoglobin content) of man is maximum in July and the number of eosinophilic leucocytes increases in spring. All these factors define the environment.

Induced sleep (by sodium amytal) decreases resistance to tuberculosis (in rabbit, guinea pig) and phagocyte function is diminished (Drabkina and Koutchak, 1952).

Age of the subject: generally, young individuals are more susceptible to infection than older individuals of the same species, which can perhaps be explained by the lower level of antibodies;

Diet influences development of infectious diseases as has been shown by Bretey in the case of guinea pigs inoculated with tuberculous bacilli: experimental tuberculosis was aggravated in animals deficient in *ascorbic acid*. Lycopene (red saturated hydrocarbon from tomatoes) increases the resistance of mice to bacteria and X-rays (Lingen, 1958). Moreover, many bacterial lipopolysaccharides augment resistance to bacterial infection without exhibiting antibiotic properties.

Erythrocytes of chicken blood are agglutinated by influenza virus (Hirst, 1942). This hemagglutination is inhibited by a carbohydrate from serum sialoproteins. However, this inhibitory ability disappears on incubation of sialoproteins with virus; *N*-acetylneuraminic acid or *sialic acid* (Blix, 1936, 1957) or nonulosaminic acid is liberated. These sialoproteins are found in mucin of submaxillary gland, vitreous humour, cerebrospinal fluid, colostrum and lastly brain where they appear in the form of *gangliosides*, sphingolipids.

$$H{-}ROCNH$$

OH—|—H   H—|—CH—CH—CH$_2$OH

H—|—H   O OH OH

OH—CO$_2$R′

General formula of sialic acids

The liver contains large amounts of sialoproteins. This raises the question of the possible importance of high concentrations of hepatic sialoproteins in natural protection against certain aspects of influenza infection.

II. Finally, *more specific factors* are involved. The presence of lysozyme in milk is a *specific* factor since it is dependent on species. However, the existence of such factors is not exclusively related to species and they can be considered to be non-specific or not directly specific factors.

An example of such a factor is *properdin* (Pillemer, 1954), a serum euglobulin of mammals, which, in the presence of Mg^{++}, destroys certain bacteria. The level of properdin in blood is directly proportional to the resistance of the individual to Gram-negative bacteria; this level decreases after irradiation. Properdin was first isolated with the aid of "zymosan" (crude yeast cell wall preparations consisting chiefly of protein–carbohydrate complexes, Pillemer and Ecker, 1941) following the observation of von Dungern (1900) that the addition of yeast to blood serum inactivated complement (cf. p. 220). Let us now consider the legendary resistance of the mongoose to snake venom. The mongoose is just harmed by an injection of cobra venom eight times greater than the lethal dose for rabbit. Similarly, the hedgehog resists the toxic effects of viper venom. Moreover, the blood of viper is toxic in mammals but is without effect in viper which produces antitoxic autoantibodies. It should be mentioned in this respect that intravenous injection of bile to man results in toxic phenomena; it is advisable, therefore, only to compare phenomena which have properties in common. Another example of possible specific resistance is that of chickens for wich the lethal dose of tetanus toxin is 100,000 times greater than the lethal dose for guinea pig.

The pigeon is insensitive to pneumococci; the bacteria are rapidly eliminated from the general circulation and become localized in liver and spleen where they are destroyed by phagocytosis (P. Kyes, 1916).

In the intestinal cavity, in contact with damaged cells of a species of *Drosophila*, certain spirochetes are *agglutinated* and dissolved (E. Chatton, 1912).

These facts concern *natural immunity*. Facts which relate to specific immunity have general application to *acquired immunity*.

Immune reactions are a means of protecting, if not the constancy, then at least the functional integrity of the internal environment.

Immunity is conditioned by *humoral factors* and by *cellular factors*.

General considerations

Specific humoral factors are related to the production of *antibodies*. Cellular factors are associated essentially with *phagocytosis*.

Let us now examine certain points relating to comparative physiology which will allow us to form a more complete picture when we compare the facts.

The production of antibodies is not a property of all living matter. Only in homoiotherms has the production of antibodies been clearly demonstrated. In poikilotherms, including poikilothermic vertebrates, injection of antigen does not result in the production of antibodies at temperatures below 25°C (Metchnikoff). E. Wollman (1938) demonstrated that under normal conditions (without previous injection) a one-tenth dilution of serum of the giant frog from Chile *(Calyptocephalus gayi)*, a useful animal because of the volume of blood, agglutinated typhous bacilli; this result was obtained at 16°C. A similar result was obtained after repeated injection of typhous bacteria at 16°C. However, repeated parenteral injection of bacteria at 30°C led to the production of antibodies such that the agglutinating ability of serum increased and a 1/200–1/400 dilution gave agglutination of bacilli (compared to 1/10 under normal conditions). Similar experiments carried out in dogfish and caiman have confirmed the importance of an ecological thermal factor in the appearance of antibodies. This is important both from the point of view of the significance of the production—or liberation?—of antibodies and from an *evolutionary* aspect. As indicated by Wollman, under normal conditions of ambient temperature the production of antibodies cannot contribute significantly to immunity in poikilotherms. "A different situation must have existed in the second era when the giant saurians reached the height of their development and the temperature of the earth was higher. It can be thought that the drastic disappearance of giant reptiles was partly the result of epidemics associated with the decrease in ambient temperature and loss of ability to produce antibodies". The present day localization of giant reptiles in warm regions substantiates this view. Thus geographical distribution of animals to some extent also expresses biological specificity.

If one examines organisms such as higher plants which are situated lower on the phylogenic scale, it is found that resistance to infection is not linked with immunity through the production of antibodies. This can be demonstrated by considering the case of *crown-gall*, a plant tumour.

This tumour is produced by inoculation of young regions of a dicotyledon with a bacterium *(Agrobacterium tumefaciens)*. Within two or three months an extensive, lobed tumour is produced. Inoculation of old regions of the plant gives negative results.

FIG. 52

If after three months apparently healthy regions of the stem situated on either side of the tumour are inoculated with highly pathogenic *Agrobacterium tumefaciens*—as shown by tests on a similar plant—no new tumours appear. Instead, at the point of inoculation, scars appear with vulnerary cork. This phenomenon, discovered by the Italian phytopathologist Arnaudi (1925), is shown clearly in Geraniaceae. No agglutinating or precipitating antibodies of the type described for animals are formed as I have shown (1948). The agglutinations obtained, or the precipitations obtained with extracts of the bacterium, show no specificity and are due to plant tannins. In 1949 I was able to show that resistance to reinoculation is related to metabolic factors which are summarized in Table 38 for *Pelargonium zonale*.

TABLE 38

Regions analysed	% total N	Ratio % $\dfrac{\text{total tannins}}{\text{total nitrogen}}$
young	1·58	$\dfrac{10}{1\cdot58} = 6\cdot3\%$
old (near root neck)	0·93	$\dfrac{8\cdot8}{0\cdot93} = 9\cdot4\%$
healthy 5 (cm above and below tumour)	0·74	$\dfrac{18\cdot0}{0\cdot74} = 24\cdot3\%$
tumours	1·52	$\dfrac{20}{1\cdot52} = 13\cdot1\%$

This table can be explained as follows:

Tissues resistant to reinoculation with pathogenic A. tumefaciens possess the highest ratio value. This has led me to speak of *"immunity through deficiency".* Total nitrogen represents both nutritional and structural nitrogen and thus, nitrogen deficiency in resistant regions prevents the development of tumours. Moreover, the high level of tannins favours cicatrization at the points of reinoculation.

In summary, tumours develop on the stem and metabolic requirements for intensive growth are such that related tissues become *deficient* in metabolites and enriched with factors for cicatrization.

Crown-gall is itself a parasite and the immunity which develops is anti-crown-gall immunity (anti-tumour) and not immunity against A. tumefaciens, the primary parasite. This immunity is based essentially on nutritional factors although it can be associated secondarily with the development of anti-*tumefaciens bacteriophages.*

The multiplicity and diversity of phenomena associated with immunity show that this subject produces evidence for the unity of science which would reflect the higher unity of nature.

Phagocytosis

Among poikilothermic vertebrates and invertebrates (Metalnikov) phagocytosis is the predominant means of assuring resistance to microbial infection.

In general terms, *phagocytosis constitutes the most rapid and most efficient method of defence found in animals and also in plants*—the excellent work of Noël Bernard on fertilization in orchids established the presence of phagocytosis—in *defence against micro-organisms*. Defence by phago-cytosis is immediate while that by antibodies requires several hours or days to become apparent.

Phagocytosis was discovered by Elie Metchnikoff (1892).

Phagocytosis is a particular example of the behaviour of amoeboid type cells (for example, mycophagous amoeba) but some phagocytes are fixed. The method of nutrition of amoeba becomes the method of defence of leucocytes or more exactly, of the organism containing the phagocytes. This is also a particular case of the ability of white blood cells to engulf colloidal and granular particles. The German biologist Ribbert (1911) was the first to observe the retention of lithium carmine by the reticulo-endo-thelial system after intravenous, intraperitoneal and subcutaneous injec-tion of dye. The importance of these experiments lies in the fact that for the first time cytoplasm, which was reputably unstainable *in vivo*, was shown to take up dyes; such dyes were given the name *vital stains*. Ribbert gave the name "chromophilie" to cells showing this property and found that such cells were present in hepatic Küpffer cells, endothelial cells of bone marrow, vascular endothelial cells of adrenal medulla, lymph nodes, endothelial cells of ganglia and cells of splenic pulp. In 1913 Aschoff demonstrated the functional unity of these apparently distinct cells which, in fact, constitute the *reticulo-endothelial system* (R.E.S.). It is of note that *synovial cells* can engulf a large variety of particles: haemosiderin, gold, erythrocytes (Collins, 1951; Ghadially and Roy, 1967).

Among vertebrates the most active phagocytes are *polynuclear* cells or *microphages* and the *large mononuclear cells* or *macrophages*, representing the free elements of the R.E.S. These three cell types make up the mobile phagocytes while histiocytes are the fixed phagocytes of the R.E.S. Lym-phocytes are less efficient as phagocytes. In both types of phagocytes microbes are engulfed by movements of *pseudopodia* and digested within the cell by numerous enzymes, notably proteases. A. Delaunay and his co-workers have shown that the glycogen content of leucocytes decreases during phagocytosis of microbes and the uptake of glycogen increases with more intense phagocytic activity. This indicates that the energy for phagocytosis is derived from glycogen.

The precise nature of the microbial agents which release the chemotactic response of phagocytes is not yet known. However, when a solution of sterilized peptone is injected into the peritoneum of guinea pigs, after several hours, an exudate from the peritoneal cavity contains a large num-ber of polynuclear neutrophiles. Starch grains induce a positive chemotac-tic response in polynuclear cells. It is possible that fixation of granules (granulopexy) is partly as a result of attraction when phagocytes are in contact with solid structures *(thigmo-reaction)*.

Certain microbes can destroy phagocytes: for example, staphylococci secrete *leucocidin* with bacteriolytic properties, other bacteria induce a *negative* tactile response: anthrax and cholera bacteria, etc.

Diapedesis

This property of white corpuscles was discovered by Cohnheim (1867) and is most remarkable. The process can be followed by observing mesentery from frog under a microscope:

Within capillaries the erythrocytes circulate continuously while various leucocytes (polynuclear cells, lymphocytes, mononuclear macrocells) aggregate along the internal face of the vascular wall (this phenomenon, the *margination of leucocytes*, would appear to be a characteristic *thigmo-*reaction). Through movements of the pseudopodia the leucocytes become displaced from the walls of the vessel and pass through the endothelium by gliding between the endothelial cells. This movement is termed *diapedesis* and takes several minutes, sometimes 30 mn.

In *septic inflammatory processes diapedesis is active* and the first combatants to die—the polynuclear cells—are destroyed by phagocytic activity of macrophages. This accumulation of bacteria, phagocytes and phagocytic debris forms *pus*.

Peptones are chemotactic; in addition, oil of turpentine induces aseptic abscesses which are of therapeutic value (abscess filled with polynuclear cells). Bacterial polysaccharides, constituents of glyco-lipo-polypeptide antigens of Gram-negative bacteria (Boivin) produce chemotropic responses in leucocytes. The *endotoxins* of Gram-negative bacteria (Coli-typhic- dysenteric), that is the glyco-lipo polypeptides themselves, *inhibit diapedesis of leucocytes* (Boivin and Delaunay) which explains the action of endotoxins in enhancing infection. Moreover, Delaunay and Lebrun (1954) have shown that in *experimental hypothermia* in rat, in which the rectal temperature falls to 15°C, leucocyte diapedesis is inhibited; thus, hypothermia favours bacteria invasion in the homoiotherm.

Opsonization

Phagocytosis can easily be followed *in vitro* (for example, in a hanging droplet at 37°C as described by J. Bordet; Fig. 53). In blood serum of homoiotherms, there are substances which *enhance phagocytosis* and these are found particularly in *sera from immunized animals*. Metalnikov and Toumanoff (1925) reported that immunization in guinea pig modifies the proportion of different leucocytes; *immunization also stimulates phagocytosis and intracellular digestion*.

Therefore, the cells of an immunized animal become more active, that is, more sensitive to a given micro-organism. The greater the number of cells sensitized the greater the reaction, providing better defence of the animal. For example, in the presence of antistreptococci serum, the engulfment of streptococci is facilitated. If leucocytes are washed after being dipped in antistreptococci serum, the phagocytic ability is that of normal phagocytes. However, if the streptococci are washed after being dipped in antistreptococci serum, they are more easily engulfed by leucocytes.

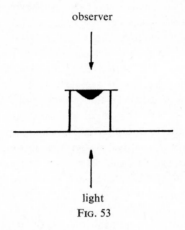

observer

light

FIG. 53

Thus, immunoserum contains:

(1) *substances active against micro-organisms: antibodies* tend to lower the vitality of micro-organisms and thus favour phagocytosis;

(2) *substances which enhance the action of phagocytes*, termed *opsonins* by Wright and Douglas (1903). In the presence of opsonins contact between microbe and leucocyte is favoured.

The evolution of types of immunity in living organisms can be summarized as follows:

(1) resistance to reinfection in *plants* is linked primarily with *metabolism* and a series of non-specific chain reactions;

(2) specialization is first seen in *poikilotherms* in which resistance to microbes consists essentially of *phagocytosis* (a kind of trophic reaction —modified amoebae);

(3) in *homoiotherms* the defence system is more specialized; it is more complex and involves both *phagocytosis* and the *production of antibodies* and local or centrally integrated metabolic changes. It would appear that the relatively high temperature of homoiotherms was associated with development of the ability to produce antibodies.

Site of antibody formation

Antibodies are formed by elements of the *reticuloendothelial system* (R.E.S.), a fact which was suggested by Metchnikoff and first demonstrated by Cary and Kye. Antibodies are produced in abundance by organs rich in phagocytes, for example the spleen (after splenectomy the antibody content decreases for a period but later the liver and lymph nodes take over the production of antibodies). This indicates the importance of the R.E.S. in immune reactions since it assures both the production of anti-bodies and phagocytic ability.

Production of antibodies can be inhibited by blocking the R.E.S.; injection of trypan blue decreases and even completely suppresses the formation of anti-Eberth bacilli antibodies (Gay and Clark); after injection of colloidal iron oxide rabbits are unable to produce hemolysins (Bordet). However, because of the functional diversity (vicariance) in the R.E.S. this block can only be partial.

Burnet (1962) considers the thymus to be "the principal centre of im-munological homeostasis". The thymus forms lymphocytes which will produce antibodies in spleen but the thymocytes *in situ* are not able to produce antibodies. It is as if the thymocytes possessed potential immuno-logical information—ablation of the thymus after birth induces immunity deficiency—but they cannot themselves express this information inside the thymus.

In the *instructive theory* the antigen plays a role of "template" at the site of antibody production.

In the *clonal selection theory* it is postulated that when certain cells with a special genetic endowment to synthesize specific antibody meet with antigen they multiply to form colonies of cells (clones) actively producing antibodies.

It is noticeable indeed that, after a second dose of antigen, antibody is produced at a faster rate; this kind of "memory" phenomenon is known as the "anamnestic response". The selection theory explains such reactions on the grounds that when antigen is injected into an animal *some* competent cells are stimulated to multiply and develop into antibody producers, other cells being only activated or "potentiated" and when antigen is re-injected these "potentiated" cells rapidly become antibody-producing units. The instructive theory could, I suppose, explain the anamnestic response by the "recruitment" of RNA inside the antibody-producing units. In any way γ-globulin is synthesized at the ribosomal level in the rat lymph node (Manner, Gould and Slayter, 1965), and the formation of antibodies can be transferred with the mesenchymal tissues in which they are being formed to a non-immunized recipient (Štuzl, 1955). Since lymphocyte does not undergo mitotic division (Shelton, 1962) the selection theory has to give more arguments to be accepted. In fact lymphocytes are of multiple types—

and functions (even trophic in the case of the rapidly proliferating villi of the intestine)—and great care has to be introduced in the study of this "heterogeneous collection of cells masquerading under the same morphological guise" (M.W. Elves, 1966). But what is new is that Ada (1966) considers that there is a sequential relationship of the type, antigen–macrophage–lymphocyte, where either the altered antigen (or another factor derived from antigen) *induces* the lymphocyte to synthesize immune bodies and that macrophage may "process" the antigen to form the factor working as the actual stimulus of the lymphocyte.

Furthermore, *the production of antibodies is not limited to whole organisms.* The spleen of rabbit immunized against human γ-globulins produces antibodies *in vitro*. This has been demonstrated by incubating spleen with ^{14}C-labelled glycine which is incorporated into the antibodies (Keston and Dreyfus, 1951).

The formation of antibodies begins *immediately* after injection of antigen and small amounts can be detected in the circulation after 8 h (Ramon, 1928). Antipneumococci antibodies can be found in man 8 years after injection of antigen (Heidelberger *et al.*, 1950). This prolonged production of antibodies explains the durability of vaccinations. However, a problem is posed in that protein turnover makes it necessary to explain the prolonged action of antigens on the cells forming antibodies.

Antigenicity

Not all proteins are antigenic, an exception being gelatin; in addition, some polysaccharides are antigenic such as those from pneumococci.

The problem of which factor or factors determine antigenicity has not been solved. It is complicated by the fact that certain substances can excite immune responses in the organism by a non-specific action ("depot-action", S. H. Stone, 1963). There is also specific inhibition of immunogenicity which is called "immunological tolerance". This has been defined by Bussard (1963) as the state of "an organism which in the presence of normally immunogenic substances is specifically incapable of an immune reaction specific for this substance".

Different types of antibodies

A study of the antigen–antibody reaction *in vitro* seems to indicate that there are two types of antibodies: *complete* or *bivalent* antibodies which induce the visible phenomena of agglutination, lysis, etc., and *incomplete* or *univalent* antibodies whose presence can only be shown after previous "impregnation"; for example, by ox serum albumin or dextran (agglutinoids) or even by indirect techniques (in the case of cryptagglutinoids) such as the antiglobin test of Moreschi (1908) and Coombs, Mourant and Race (1945). Complete antibodies are thought to be able to form the linkages between antigen-receptor and antigen-receptor while incomplete antibodies can only be fixed to the antigen by a side chain, the other side of the antigen remains free and only a substance, such as ox serum albumin, allows the visible reactions by forming a linkage between two free side chains of incomplete antibody.

The chemical nature of antibody sites is not known. D. Pressman (1962) considers tyrosine to be important. The test known as *Coombs test* (which is applicable in the preparation of blood for transfusion and in the study of hemolytic diseases) is an "anti-antibody–antibody" reaction (Dunsford and Grant). This test is carried out as follows: human serum is injected into rabbit and antibodies are produced which will agglutinate all human erythrocytes; the agglutinates removed, the rabbit serum contains anti-globulin antibodies which have been produced by antigenic γ-globulins of human serum. If for example, after blood transfusions, incomplete antibodies (such as Rh) are fixed to erythrocytes without causing agglutination, after washing the erythrocytes of the subject, agglutination will occur in the presence of anti-human globulin rabbit serum.

This test can be associated with the concept of *allotypes* (Oudin, 1956; Dray, Dubiski, Kelus, Lennox and Oudin, 1962). The *isotypic specificity* of a protein is the antigenic specificity which is the same for all normal individuals of the same animal species. (It is the classical antigenic specificity). *Allotypic specificity* is not the same for all normal individuals of the same animal species. In the principal example of allotypic specificity observed in rabbit by Oudin "the first rabbit is immunized against a certain antigen, for example, ovalbumin. The immunoserum obtained in this way serves in the preparation of a specific precipitate which is injected into other rabbits together with certain adjuvants. In serum of animals treated thus, precipitating antibodies against the immunizing serum and against other rabbit sera often appear" (Oudin, 1963). Allotype specificity is genetically determined.

Antibodies are classified according to the type of reaction they produce. Principally, they include:

antitoxins,
agglutinins,
precipitins,
lysins : bacteriolysins and hemolysins,
opsonins,
cytotoxins or more exactly, cytotoxic properties of antibodies.

To this list could be added:

antiblastins which inhibit proliferation of bacterial cells or the production of certain bacterial elements, such as pigments,
coglutining ability, etc.

15

These various properties provide an explanation for *vaccination* (Jenner) —induced mild disease—and *serotherapy* (passive immunization, introduced by Behring and Kitasato, 1890) in which the antiserum corresponding to the infectious germ or toxin is injected.

(1) Antitoxins

In the administration of tetanus or diphtheria toxin to laboratory animals, vaccination cannot be obtained because these toxins are very potent and generally the animals do not survive these injections. There are some exceptions; the chicken is insensitive to tetanus toxin and the rat to diphtheria toxin. By heating the toxic substances at 70°C (C. Fraenkel) the potency is decreased and a practically non-toxic product is obtained which stimulates the production of immunizing antitoxic antibodies. These antibodies are capable of inhibiting the toxicity of natural toxins untreated by heat. Treatment of protein toxins with formol results in the production of factors which preserve their specific immunizing ability (*anatoxins* of Ramon).

Anti-venom sera (of great practical significance) can be produced and also *anti-enzyme sera* (for example, anti-emulsin produced in 1893 by Hildebrandt). The lecithinase in venoms (which converts lecithin to hemolytic lysolecithin) can give rise to anti-lecithinase antibodies. A traditional application of mithridatism is found among snake charmers who allow themselves to be bitten from time to time by young cobras having insufficient venom to be dangerous. This is an example of the practical arcana of the fakirs.

(2) Agglutinins

In 1889, Charrin and Roger found that *Bacillus pyocyaneus*, which multiplies in normal rabbit sera giving uniform infection, develops by forming particles which are precipitated when placed in culture with rabbit antiserum formed with this bacillus.

In 1895, Bordet recognized that a homogeneous suspension of cholera Vibrio in physiological saline forms clumps in the presence of serum from animals immunized against Vibrio.

Agglutination proceeds rapidly at 37°C but is inhibited by a temperature of 55°C. Agglutination is a general phenomenon which develops during immunization of homoiotherms with various pathogenic and non-pathogenic micro-organisms. The specificity of this process allows it to be used in diagnosis of infectious diseases—*serodiagnosis*; a certain type of bacteria being agglutinated by one antiserum and not by another. Widal and Sicard (1896) were the first to use this of type diagnosis (for typhoid fever). The determination of blood groups should be remembered at this point.

Bacterial agglutination by specific antibodies requires the presence of electrolytes (NaCl, $CaCl_2$, etc.) but high concentrations of electrolytes and non-electrolytes (such as sucrose) inhibit agglutination by antibodies.

(3) Precipitins

Agglutinins form clumps with *cellular* antigens while precipitins form precipitates with *soluble molecular* antigens. In the case of micro-organisms these antigens [macromolecules: proteins, complex 0 (a glyco-lipo-polypeptide) of Boivin and specific toxins] are precipitated. Inhibition of a *toxin* (such as snake venom) by the corresponding immunoserum if often characterized by *flocculence* related to *loss of hydrophilic properties* by the antigen–antibody complex. In this case, antitoxic ability is an integral part of precipitating ability.

Since antigenicity is a systematic property of proteins (with a few exceptions, such as gelatin) and appears to be related principally to the presence in the molecule of cyclic amino acids, anti-protein antibodies can be obtained with numerous body fluids: for example, anti-milk precipitins with milk. Thus, the antigen–antibody reaction provides a useful method for analysis of biological specificity.

However, non-protein compounds without or with low antigenic properties, such as polysaccharides from pneumococci, can react *specifically* with anti-pneumococci antibodies to form precipitates. Such compounds are termed *haptens*—or *incomplete antigens*; they include the specific compounds of ABO blood groups, (those of the MN system being true erythrocyte antigens associated with protein molecules).

In 1940, Heidelberger developed a *quantitative serological test* in which total nitrogen and sugars in each known constituent were measured. By this method it was estimated that one molecule of antigenic ovalbumin combines with 10 to 12 molecules of antibody and that one molecule of antigenic thyroglobulin combines with 10 to 50 molecules of antibody. Therefore, each antigen molecule appears to have several antigenic sites.

(4) Lysins (bacteriolysins and hemolysins):

The memorable experiments of J. Bordet (1895) indicated that immune bacteriolysis results from the participation of two distinct and independent factors. Methodologically, the search for these factors seems to arise from an attempt to introduce anti-thesis of the theories of Metchnikoff: *can serum totally deprived of phagocytes destroy bacteria?*

Buchner (1890) provided a positive answer and called the active substance which produced lysis of bacteria, and also erythrocytes (Landois, 1875), *alexin* (or *complement*, Ehrlich).

It appears that normal serum exhibits bactericidal properties but these properties are more prominent in serum from immunized animals which thus acquires *antiseptic* ability.

15*

PROPERTIES OF ALEXIN: In the presence of fresh serum from mammals immunized against cholera Vibrio, these bacteria form granules. The lytic property of serum is destroyed by heating at 55°C but the agglutinating activity is unchanged. If normal fresh serum is added to heated immune serum lytic properties reappear.

J. Bordet considered that "the bacteriolytic effect of fresh cholera serum is due to the interaction of two substances; one specific compound which resists heating at 55°C and even at 60–65°C and is characteristic for the immunoserum, the other destroyed at 55°C and also by prolonged storage of serum and which exists in both normal and vaccinated animals". The latter compound is *alexin* of Buchner (*complement*) while the first, said to be a *sensitizing factor* (Bordet) or *amboceptor* (Ehrlich), represents non-bactericidal antibody. The amboceptor becomes fixed to cellular antigen which can then adsorb complement resulting in lytic activity. These two substances act together. The amboceptor appears to act as a *mordant* (as in dyeing).

These facts can be applied to the case of *paroxysmal hemoglobinuria*. Landsteiner and Donath demonstrated that during this disease the serum contains a compound related to a hemolytic amboceptor but which combines with erythrocytes only at *low temperatures* resulting in hemolysis, in a similar way to "cold agglutinins" (cf. p. 203). This point must be considered in the use of deep hypothermia.

(5) Opsonins

One can say that in the same way that amboceptor favours *lysis* of cellular antigens by complement, opsonins favour *phagocytosis* of cellular antigens by leucocytes. Once again we find the complementary notions of humoral immunity and cellular immunity. Opsonins were discovered by Wright and Douglas (1902).

Amboceptors and opsonins can be considered to be "sensitizing factors", except for the fact that complement also plays a part in opsonization. It has been shown that opsonic properties express the combined action of classical amboceptor and complement.

Cytotoxic ability

Hemolytic ability is a particular example of this. Injection of mammalian cells results in the production of antisera such as *leucotoxic* sera, *spermatoxic* sera (which immobilize but do not cause lysis of spermatozoa).

Great hopes were placed on cytotoxic properties in the treatment of cancer but unfortunately the results have been negative. It should be mentioned, however, that U. S. von Euler and L. Heller (1949) found that sera from normal and sarcomatous rats possess *caryolytic properties* (on rat liver cells and tumour cells) which were absent from rabbit and human sera. Perhaps the problem should be approached from another direction; a virus consists of ribonucleoprotein and if cancer results from a virus, by fractionating the ribonucleoprotein of cancer cells, it should be possible to obtain an antigen which would prevent multiplication of the virus and thus cellular multiplication; penetration of this antibody is to be considered.

Application of serology

The specificity of immune reactions can be used experimentally, not only to induce resistance to infection but also as a technique for the *analysis of biological structure* as an extension of serological diagnosis.

Ebert (1955) followed the morphogenesis of certain organs such as heart, using dilute specific antisera. Perlmann (1951) traced the changes in antigenic structure of sea urchin egg during embryonic development up to the pluteus stage. This analysis was quantitative and Perlmann stated that the level of certain antigens decreased regularly from stage 16 to the pluteus stage. Another excellent experimental technique is that of *fluorescent antibodies* (Coons *et al.*, 1942). Fluorescin for example, as isocyanate, becomes fixed to free amino groups of lysine in proteins and particularly in antibodies. Thus, using a pure compound as the antigen, the fate of antibodies in tissues can be traced (with a fluorescence microscope with a light source of 4100–4750 Å). Using this method detailed histochemical analysis can be realized and it is even possible to trace the fate of certain bacteria or virus, etc.

However, *there are limitations to immunological specificity*; it is not so well defined as was originally thought. Two examples illustrate this point:

(1) Serodiagnosis of syphilis (Wassermann–Bordet reaction)

The adsorption of complement in the presence of amboceptor can be the basis of a diagnostic test, fixed complement disappearing from the serum (a reaction called *complement deviation*). In 1906, Wassermann developed a method which he believed to be of value in detection of syphilis. Cultures of pathogenic spirochetes were not available and he used as the antigen extracts of tissues infected with spirochete—mainly liver tissue—and stated that complement is fixed *in vitro* in the presence of these extracts. This technique was used with apparent success until it was discovered (notably by Marie and Levaditi) that products from the syphilis micro-organism did not enter into the reaction which occurred just as well in the presence of extracts of *healthy* human or animal organs (liver, heart, etc.) The active agents were identified as *lipids*.

(2) The heterophilic group of Forssman

Table 39 summarizes the fundamental experiments of Forssman.

The name Forssman *antigen* is given to all antigens in tissues, cells or body fluids, which, when injected into rabbit, induce the formation of *hemolysins for sheep erythrocytes*. The name Forssman *hemolysin* is given to all *hemolysins for sheep erythrocytes*, either natural or acquired, which are capable of being adsorbed by kidneys of horse and guinea pig or by residues from alcoholic extracts of these organs.

TABLE 39

Injection of sheep ery- throcytes	into rabbit	results in production of	anti-sheep hemolysins	*these hemolysins are* (a) *toxic* for guinea pig
				(b) *totally* adsorbed by sheep erythrocytes
				(c) *partially* adsorbed by kidney of pig or horse.
Injection of kidney of guinea pig	into rabbit	also results in produc- tion of	anti-sheep hemolysins	*these hemolysins are* (a) *toxic* for guinea pig
				(b) *totally* adsorbed by sheep erythrocytes
				(c) *totally* adsorbed by kidney of guinea pig or horse
moreover: injection of horse kid- ney	into rabbit	results in production of anti- sheep he- molysins		
but: injection of horse kid- ney	into guinea pig	*does not re- sult* in for- mation of anti-sheep hemoly- sins		

Lipids are also involved in this reaction.

The distribution of Forssman *antigen is extensive*; it is found notably in human erythrocytes of groups A and AB, in carp eggs, horse and guinea pig heart, muscle of camel and white mice and in many bacteria including tuberculous bacilli. The antigen is not present in group O and B human erythrocytes, *Macacus rhesus*, rabbit, rat, pig, ox, frog and many bacteria including *E. coli*, etc.

Moreover, phenomena associated with the response of an organism to microbial invasion or toxins are complicated by "excessive" reactions which exceed their protective capacity and can lead to death of the organism. These are the associated reactions of *anaphylaxis* and *allergy*.

Anaphylaxis

This term which signifies the opposite of protection (phylaxis) is due to C. Richet (1902) who, together with P. Portier, discovered this phenomenon. It must be mentioned that Magendie in 1893 showed that injection of egg white into rabbits sensitized them such that they reacted unfavourably to a second injection given after a period of time. Anaphylaxis must not be confused with tachyphylaxis which is an adaptation phenomenon shown by an organism to repeated administration of a pharmacologically active compound such as serotonin or adrenaline (for example, the former ceases to show hypertensive ability in dog after 4–5 intravenous injections separated by one hour; adrenaline, however, remains hypertensive). The mechanism of tachyphylaxis is unknown.

The anaphylactic reaction is characterized by two conditions:

(1) *increased sensitivity* to a compound induced by earlier injection of the same compound (sensitizing and "shocking" injections).

(2) *a period of incubation* which is necessary for the development of hypersensitivity.

As a whole, this is the opposite of *mithridatization*.

The first observation was reported by C. Richet (1923). "During a cruise on the yacht belonging to Prince Albert of Monaco, the prince and G. Richard advised P. Portier and me to study the toxic properties of *Physalia* found in southern waters. Aboard the yacht we carried out several experiments which proved that aqueous or glycerine extracts of *Physalia* filaments are extremely toxic (for ducks and rabbits). On returning to France and being unable to obtain *Physalia*, I decided to carry out a comparative study on tentacles of actinians which we could obtain in abundance (coelenterates which in certain respects resemble *Physalia*). Through the kindness of Y. Delage, I obtained large quantities of actinians from Roscoff. The isolated tentacles were placed in glycerine and thus we had in Paris several litres of an extremely toxic liquid, the glycerine having dissolved the active principles from the actinian tentacles.

"In determining the toxic dose of this liquid it was apparent at once that it was necessary to wait several days to obtain a result as many dogs did not die until the fourth or fifth day or even later. Dogs which had been injected with a sublethal dose were kept as we wished to use them for a second experiment when they had completely recovered.

"It was then that an unforseen result was obtained. These recovered dogs were extraordinarily sensitive and succumbed to very low doses within several minutes.

"The experiment which showed this phenomenon very clearly was carried out on the dog Neptune. This was an exceptionally vigorous and healthy dog. Initially it had received 0·1 (ml?) of the glycerine extract without becoming ill. Twenty-eight days later since it was in excellent health I gave it an injection of the same dose. A few seconds after the injection the dog became extremely ill; respiration was spasmodic and laboured. It could hardly move and lay on its side; it had diarrhea and vomited sanguinolent matters. It lost consciousness and died within 25 mn".

The Arthus phenomenon (1903) is an anaphylactic *skin* reaction produced not by a toxic compound bu tby horse serum injected intravenously to the skin of rabbit; it produces generalized shock and can lead to death.

Anaphylaxis is the cause of so-called serum shock in serotherapy and is produced even in decerebrate animals showing that this is not a phenomenon related to emotion.

Anaphylactic shock is characterized principally by:

(1) *marked decrease in arterial pressure,*
(2) *appearance of scattered hemorrhage zones,*
(3) *congestion of liver and pulmonary oedema,*
(4) *sensitization of smooth muscle.*

A fragment of isolated intestine sensitized to a foreign serum will contract in the presence of that serum (Schultz, 1910). A similar result is obtained with isolated uterus of guinea pig (Dale, 1912).

Shock is thought to be due to the combined action of:

(1) *antigen–antibody complex* producing anaphylatoxin,
(2) *histamine.*

The hemorrhage reaction—Sanarelli-Shwartzman phenomenon—which can be clearly demonstrated in rabbits is related to anaphylaxis. It is characterized by local or intestinal hemorrhage in animals receiving, for example, a filtrate of *E. coli* culture. The reaction appears several hours after an injection of the filtrate (in anaphylaxis there is a delay of several days). Acute hypothermia develops and can lead to death.

Pautrizel (1950) has shown that substances with vitamin P activity (epicatechol, coumarin, rutin, cf. p. 64) *inhibit* anaphylactic shock although they have no antihistamine activity and no action on the antibody–antigen reaction. Therefore, the "terrain" must be modified and this inhibition can be explained to some extent in terms of increased capillary resistance.

Allergy

The term *allergy* signifies modified reaction and was proposed by von Pirquet (1906). The animals show intolerance not only to antigens as in anaphylaxis but also to substances which are not necessarily antigenic and which are called *allergens.*

Allergies are generally acquired but they may result from the appearance of a congenital or hereditary predisposition. The terms congenital and hereditary must not be confused. Hereditary relates to a genetic manifestation while congenital relates to conditions during gestation (for example, passage of antibodies across the placenta).

Allergens can be inhaled, ingested or received by injection.

The most common allergies are associated with *dietary factors: urticaria* resulting from ingestion of milk, eggs, fish, crustaceans, strawberries, etc. Allergies also include *hay fever* induced by various pollens and *asthma* induced by mineral dusts or fine particles from animals (cat, horse, etc.) or even mushroom spores. Asthmatic dyspnea is due to spasm of smooth muscle surrounding the bronchioles (muscles of Reisseissen), which can be compared with the muscular sheath of the neuro-myo-arterial glomus (cf. p. 188).

Many individuals shown an allergic reaction after injection of antibiotics such as penicillin. Certain allergies, such as those described by Pautrizel (1950), are induced by bedding fluff and acyclic terpene compounds utilized extensively in certain industries. I would add to this list allergies of psychic origin (dyspnea induced by crowds, claustrophobia, etc.).

Chemical allergy is diagnosed by *skin reactions* to a range of allergens. A persistent red coloration signifies a positive reaction. Skin reactions are applied to the diagnosis of tuberculosis. In the tuberculin test the epidermis of the arm is pricked and a drop of *tuberculin* (sterilized filtrate of culture media of tubercular bacilli) placed in the wound. The result is analysed after 48 to 72 h and is positive if hardening of the tissue exceeds that of the control (> 2 mm ϕ). An intradermic reaction can also be used.

The exact composition of tuberculin (containing a protein and polysaccharides) and the nature of the reaction are little understood. Tuberculosis leads to an allergic condition but the pathways remain unknown. Canetti who has studied this problem extensively, stated that *anaphylaxis* and *tuberculin hypersensitivity* are opposed for the following reasons:

(1) *Anaphylaxis to protein can most easily be induced in rabbit.* Anaphylaxis can be induced in guinea pig less easily; however, the guinea pig is more susceptible to tuberculin hypersensitivity than the rabbit.

(2) The tuberculin hypersensitivity reaction is *slower* in development than anaphylactic reactions.

(3) While anaphylaxis is characterised by *hypothermia* and *smooth muscle spasm* (bronchial spasm), tuberculin hypersensitivity is characterized by *hyperthermia* and absence of smooth muscle spasm.

(4) Tuberculin hypersensitivity *cannot be passively transferred* while this is possible in anaphylaxis (injection to normal rabbit of serum from a sensitized rabbit).

(5) *It is impossible to produce tuberculin hypersensitivity with the same bacterial substances* that release the reaction while in anaphylaxis the sensitizing and releasing substances must be the same.

It would appear that it is the environment ("terrain") that is modified. But what is meant by "terrain"? We have already examined several aspects of this problem but the concept may be clarified by considering that allergy is passively transferable by *lymphocytes* to normal subjects (Chase). The work of Frey and Wenk has shown that lymphatic elements (ducts and nodes) are necessary for sensitization and that the cutaneous lymphatic system is important when the allergen is applied to the skin. In addition, the development of an allergic reaction requires a normal blood flow to the zone receiving the allergen. Innervation does not appear to be necessary.

From these facts the importance of lymphatic elements in various immune reactions to micro-organisms or grafts becomes apparent.

The concept of "terrain" will be clarified further by considering the phenomenon called "*stress*" (H. Selye). I would translate "stress" as *physiological tension*. P. Decourt (1952) has compared stress to the Reilly phenomenon (1934). I consider the founder of this concept to be A. Guieysse-Pelissier ("L'état réactionnel").

The state of stress or "*general adaptation syndrome*" (Selye) is a state of crisis caused by the summation of non-specific organic reactions, induced in the organisms by attack by some aggressive agent.

Reilly and his co-workers demonstrated the appearance of non-specific reactions after aggression. During studies on the pathogenesis of typhoid fever, the *peripheral nervous system was found to be of prime importance* in the development of intestinal lesions analogous to typhic lesions occurring after injection of *various agents*—microbial toxins, nicotine, lead salts, arsenic, etc. or even after physical stimulation such as faradic stimulation, mechanical irritation.

Non-specific agents can induce common organic reactions through the intermediary of the peripheral nervous system.

Selye distinguished three phases in the adaptation syndrome following aggression which lead to loss of homeostasis and hypotension:

(1) alarm reaction,
(2) resistance reaction,
(3) exhaustion.

(1) *The alarm reaction.* The "call to arms" is characterized by *defence reactions* involving general and rapid mobilization of energy with the development of hypoglycemia, hemoconcentration, lymphatic and reticulo-endothelial reactions, degranulation of cells in the adrenal cortex (hormone mobilization).

(2) *Resistance phase.* If the aggression continues there is hyperglycemia due to breakdown of hepatic glycogen, hypertension (mediated by the sympathetic nervous system) and liberation of pituitary adrenocortico-tropic hormone.

(3) *Exhaustion phase.* With continued aggression, the organism enters a phase of generalized catabolism leading to death.

Both hormonal and nervous mechanisms are involved in *adaptation to stress* and these constitute a kind of *generalized reflex:*

(1) the nervous mechanisms are rapid and mainly *hypothalamic* in origin. They lead to peripheral vasoconstriction and hypertension (via splanchnic nerves).

(2) *The endocrine mechanisms* are characterized by:

(a) hypersecretion of *renal hypertensive hormone*, renin;

(b) hypersecretion of *adrenocorticotropic hormone*;

(c) hypersecretion of adrenal *mineralocorticoids* and *glucocorticoids*:

mineralocorticoids such as *desoxycortisone* favour the elimination of K^+ and fixation of NaCl and, then, water retention;

glucocorticoids such as *cortisone* activate the reticulo-endothelial system, augment conversion of proteins and lipids to sugars and augment protein catabolism. They also have antiallergic properties.

Not only is there hyperactivity of the adrenal cortex, the adrenal medulla is activated (adrenaline release during the hypertensive phase) and probably the thyroid but this needs further study.

Finally, *histamine* is important in anaphylaxis and certain allergies such as asthma. Is this compound formed easily or is it stored in mammalian tissues? Histamine is toxic and therefore must be synthesized *ex tempore*. Histamine can be found in nearly all tissues with the exception of bone. J. L. Parrot (1958) has shown that the mammal is protected against exogenous histamine by acetylation which occurs in the intestinal lumen and by histamine binding in intestinal secretions and liver (histaminopexy). Normal plasma contains a γ-globulin—"plasmapexin I"—with histamine binding properties while serum of allergic subjects does not contain this factor (Parrot and Laborde). Binding requires calcium or magnesium and is inhibited by potassium and copper (1960–1964). One has to emphasize that serotonin can be bound by normal human serum (Parrot and Flavian); heparin can also bind histamine and this fact is of significance in considering the physiology of mast cells (cf. p. 190) and also in considering that an injection or aerosol of serotonin induces bronchospasm in the guinea pig (Herxheimer, 1953). Digitoxin (cardiotonic glycoside) is bound by serum albumin (Gengoux) as are several other therapeutic compounds (Bastide *et al.*, 1961). These compounds are transported in the form of the complex

$$HC\!=\!\!=\!C\!-\!CH_2\!-\!CH_2\!-\!NH_2$$

HN   N

\HC/

Histamine

Histamine produces *vasodilation* and *hypotension*. The French pharmacologists Halpern, Bovet and Staub discovered the histamine antagonists, the *antihistamines*, which are used extensively to improve states of

shock ("Phenergan", "Neo-antergan" etc.). The most important antihistamines are derivatives of phenothiazine.

$$S \underset{}{\overset{}{\bigg\langle}} N - CH_2 - \overset{\overset{\displaystyle H}{|}}{\underset{\underset{\displaystyle CH_3}{|}}{C}} - N(CH_3)_2$$

Phenergan
3277 RP

With special reference to bacterial infection, A. Delaunay and his co-
workers (1948) showed, using typhoid endotoxin (the glyco-lipo-polypeptide antigen of Boivin), that injection of toxin into rabbits, guinea pigs or
mice resulted in vasomotor disturbances in the mesenteric and cutaneous
circulation. These disturbances were similar to those of traumatic shock.
The authors describe the phenomenon in the following way: "by decreasing
blood output, vasoconstriction deprives the internal tissues of nutritive
materials and oxygen and leads to anoxia and anemia of tissues". Inhibition
of diapedesis also occurred. It should be mentioned that in other cases of
shock (carbon monoxide, burns) it has been demonstrated by Romani
(1952) that at the time of the alarm reaction *intratissular migration of
polynuclear* cells occurs in connective tissues, lungs and lymph nodes. This
migration is non-specific and is probably related to the formation of
histamine-like substances and secondarily to an increase in mineralocorticoids and a decrease in glucocorticoids.

Thus the organism reacts to aggression by the appearance of specific
and non-specific phenomena. There is a transitory change in the composition of the internal environment; the organism adapts but sometimes these
regulatory mechanisms are insufficient and shock follows. Artificial hibernation presents a method by which the reactions of the organism to
aggression can be modified, particularly during surgery (role of Q_{10}).

CHAPTER 11

BLOOD CIRCULATION

INTRACELLULAR movements of compounds represent the most primitive form of circulation in living organisms. Circulation is vital to life since it allows the transport of metabolites and detoxicating compounds and the removal of catabolites. Blood circulation is related to the processes of nutrition and detoxication.

This chapter will be divided into two sections:

(I) Physiology of the heart (the motor),

(II) Physiology of vessels (the vascular tree).

It should be mentioned that in animals employing undulating movements for locomotion (cyclostomes, ophidians, etc.) somatic muscular activity improves blood circulation.

Historical introduction

After observing the emptiness of arteries *in a corpse* Hippocrates deduced that these vessels were part of the trachea. In the second century Galien noted the presence of blood in arteries. It was not until 1553 that the existence of the pulmonary *circulation* was established by Michel Servet—some claim that this was discovered in the thirteenth century by the Arabian doctor Ibn An Nafis. Servet described the *"little"* circulation (right auricle → right ventricle → pulmonary capillaries → left auricle). In 1628, an historical date, William Harvey discovered the *great circulation* (left auricle − left ventricle → aorta → veins → right auricle— a scheme which, according to the eminent writer L. Chauvois, was formulated by Harvey to explain the path of blood which entered the heart from the vena cava and reappeared in the vena cava). Though the microscope dates back to 1610, it was not until 1661 that it was used by Malpighi to show the passage of blood from arteries to veins and it was not until 1885 that the existence of vasomotor nerves was demonstrated by Claude Bernard (the chorda tympani acting on the submaxillary salivary gland).

There was much argument before the existence of blood circulation was accepted and Harvey had to defeat chiefly the objections of the Parisian doctor, Jean Riolan. I have extracted the following passage from one of two letters from Harvey to Riolan, collected by L. Chauvois, to provide an example of the simplicity of his expression: "On examining a fully opened vein and artery, no matter who has carried out this operation, it can be seen that the portion of the vein nearest the heart does not emit any blood while the most distant part pours forth nothing but blood. On the other hand, the cut artery produces very little blood from its distal portion while the proximal portion violently pours forth bright red blood".

Why had the evidence for circulation of blood not been recognized earlier? In fact, Plato has already written in the "Timaeus": "While the heart... the source of blood which circulates impetuously in all limbs"; the truth of the statement was forgotten. But why did Servet go no further than the pulmonary circulation? In any case, Harvey's method of deduction should be remembered because it reappears in many aspects of physiology. It will be seen that in a study of nerves by degeneration, in principle, the method differs very little. The concept of cyclic events can also imply a to-and-fro movement (before–after, up-stream–down-stream, distal–proximal, afferent–efferent) and is thus equally valuable when applied to tunicates for example in which blood flows alternately in one direction and then in the other.

I. Physiology of the heart

Evolution of cardiac anatomy

Among the vertebrates one finds:

(a) *in fish:* one auricle and one ventricle (the heart propels venous blood and differs from the heart of molluscs and arthropods which contains only oxygenated blood);

(b) *in amphibians:* two auricles and one ventricle;
(c) *in reptiles:*

    (1) *chelonians:* two auricles and two ventricles with an incomplete septum (Fig. 54);

    (2) *crocodiles:* two auricles and two ventricles.

(d) *birds and mammals:* two auricles and two ventricles with *variations related to the mode of life.* The heart of birds is characterized by considerable development of the anterio-posterior diameter of the right ventricle. This heart is adapted for aerial life with a pronounced pulmonary circulation.

In addition, the sinus venosus which is anatomically more distinct in lower vertebrates collapses into the right auricle in higher vertebrates. This expresses my rule of *concentration of somatic structures* with increasing evolutionary position of vertebrates; a rule which will be illustrated by other examples in this book.

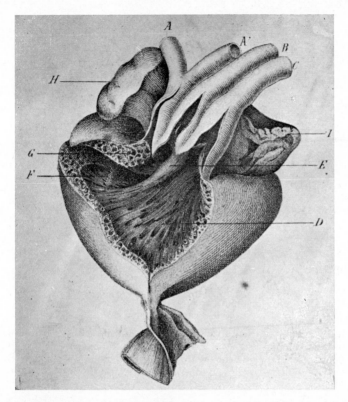

FIG. 54. Heart of tortoise, preserved for a long period in alcohol and hardened and shrunken (natural size). *A*, right aortic arch; *B*, left aorta; *C*, left pulmonary artery; *D*, false septum and its lip, pulmonary cavity; *E*, cartilaginous core of the lip of the false septum; *F*, slit situated between the posterior face of the lip and the posterior fleshy mass, allowing communication between the pulmonary cavity and the aortic vestibules; *G*, cut surface of posterior fleshy mass and particularly the common right bundle; *H, I*, auricles. (SABATIER, 1873.)

The heart of mammals

As an example we shall consider the heart of man. Figure 55 is a diagram of a transverse section of the heart. In fact, there are *two hearts* to consider, the right and the left.

The right auriculo-ventricular orifice is closed by the *tricuspid valve* (with three flaps). The left auriculo-ventricular orifice has a similar system, the *bicuspid* or *mitral valve* (with two flaps). The base of the aorta and pulmonary artery are closed by the *semilunar valves*. These valves are part of the endocardium, formed by an endothelial and a fibrous layer. The valves arise from the wall of the auriculo-ventricular ring and are attached by their free margins to the *papillary muscles* of the ventricles by tendinous

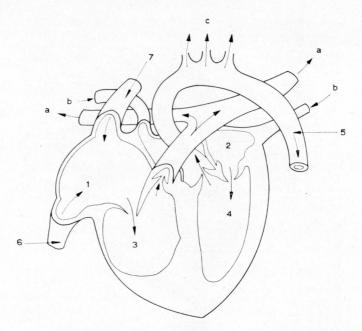

FIG. 55. *1*, right auricle; *2*, left auricle; *3*, right ventricle; *4*, left ventricle; *5*, aorta; *6*, inferior vena cava; *7*, superior vena cava; *a*, venous blood going to lung; *b*, arterial blood coming from lung; *c*, arterial blood going to head and superior limbs.

cords. I have demonstrated with my student H. Cortot (1958) that in the open empty heart of mammals movements of the mitral and tricuspid valve are largely independent of intracardiac hemodynamics. More recently (1959), my work with the sheep's heart has shown that movements of tricuspid valves are governed by three mechanisms:

(1) passive movements due to intracardiac blood currents between auricular and ventricular lumens;

(2) at the moment of ventricular intersystole the valves are held in position by muscles situated at the base of these valves; tension exerted by the papillary muscle–tendinous cord system apparently prevents extroversion of the valves by pressure of ventricular blood;

(3) facing with warping of valves.

The latter two types of movement allow complete closure of the valves.

The individual valves are not inherently contractile but in fish, particularly in elasmobranchs, the sinoauricular valves exhibit spontaneous contractions (Rybak and Cortot, 1956).

The heart consists of a hollow muscle—the myocardium–lined on the inner surface by the *endocardium* which extends throughout the auricular and ventricular cavities and to the intima of blood vessels.

The endocardium consists of:

(1) an *endothelium* with polygonal cells;
(2) a *fibrous layer* of elastic connective tissue containing a few smooth
muscle fibres;
(3) a *thin deep layer*.

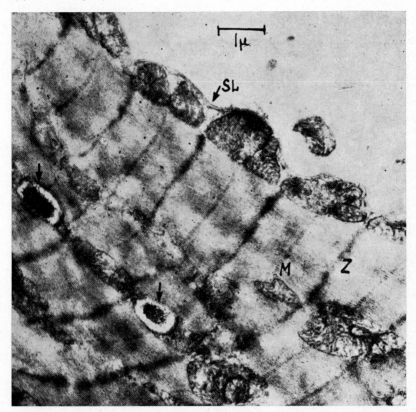

FIG. 56a. Myofibrils with typical striations. The *Z* and *M* zones are clearly
visible; sarcosomes with an internal lamellate structure fill the space between
the sarcolemma (*SL*) and the myofibrils as well as the space between different
myofibrils (× 12,000). (KISCH, 1960.)

The structural complexity of the endocardium shows that to use the
term "myocardium" to describe the heart is a simplification. The heart is
surrounded by the *pericardium* which is a serous membrane consisting
essentially of elastic connective tissue.

The ventricular mass is irrigated by the *coronary vascular system*. The
openings to the coronary arteries are situated above the aortic semilunar
valves, the coronary veins opening into the right auricle *(coronary sinus)*.
The coronary veins also extend into the ventricular cavities, forming the

veins of Thebesius (1708), such that catheterization of the coronary sinus only yields about $^3/_4$ of the coronary output (Markwalder and Starling, 1913–1914). The coronary arteries are of extreme importance to the life of the individual and to understand their physiology it must be remembered that the longitudinal vasomotor muscles fibres are *smooth muscle fibres* (which explains the vasodilator action of adrenaline on these vessels). The pCO_2/pO_2 ratio affects the coronary resistance perhaps because CO_2 dilates coronary vessels.

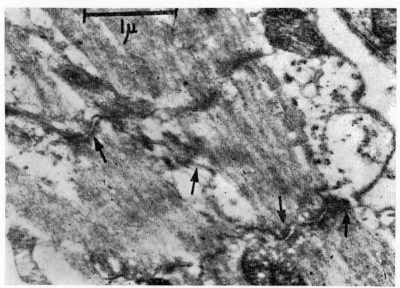

FIG. 56b. Auricle of mouse. The arrows indicate a part of a probable intercalated disk; ×25,000. (KISCH, 1960.)

Until recently, the *myocardium was considered to be a syncytium of striated muscle* consisting of *nucleated myofibrils* in an abundant *sarcoplasm*. The fibres anastomose to give Y-shaped structures forming a network which is crossed here and there by transverse striations: the *intercalated disks* of Eberth which thicken with age. However, the *electron microscope has shown cardiac muscle to be a cellular structure and not syncytial* (B. Kisch, 1951–1960; Sjöstrand and co-workers, 1954; Poche and Lindner, 1955; Couteaux and Laurent, 1957). These polynucleated cells are joined in the *regions of the intercalated disks*. This important new finding has been confirmed (Sperelakis *et al.*, 1960) and must lead to a revision of many aspects of myocardial physiology.

An endoplasmic reticulum has been observed in myocardial cells of man and rabbit (D.A. Nelson and E.S. Benson, 1963). This consists of a system of narrow, longitudinal tubules anastomosing around each sarcomere and

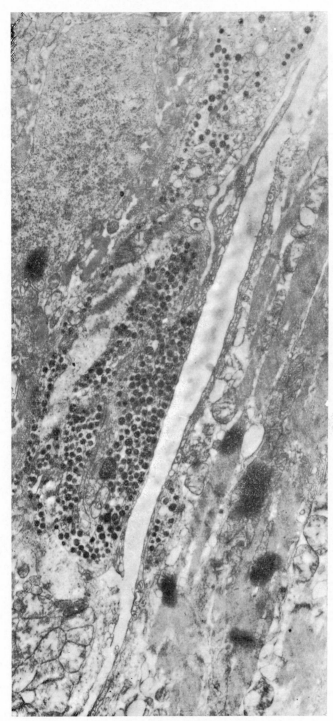

FIG. 57. Secretory granules bordering a capillary in the branchial myocardium
of *Myxine glutinosa*; ×15,000. (HOFFMEISTER, LICKFELD, RUSKA and RYBAK,
Ztschf. f. Zellf., 1961.)

FIG. 58a. Common trunk of the bundle of His. Specific conduction tissue. Predominance of dense bodies in the perinuclear sarcoplasm. They are less predominant in the region of the myofibrils where they are situated along the border of band I. ($\times 18,000$). (BOMPIANI, ROUILLER and HATT, *Excerpta medica*, 1960.)

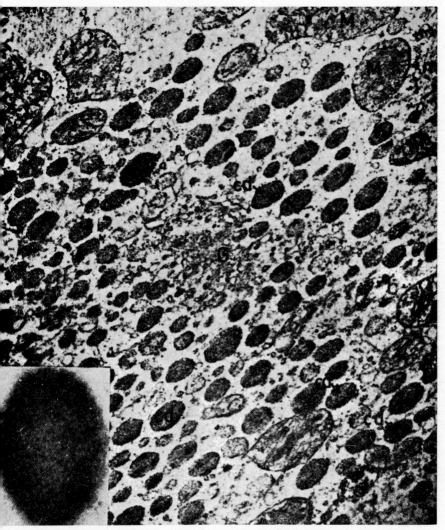

FIG. 58b. Common trunk of the bundle of His. Numerous "dense bodies" and mitochondria (M) (× 24,000). Lower left: "dense body" (×110,000), note the absence of a membrane in this structure. (BOMPIANI, *Excerpta medica* 1960.)

a system of thick tubules which cross the myofibrils transversely in the regions of the Z striations and which appear to be direct extensions of the cell membrane. These structures seem to play an important role in chemical and electrical activation of heart (cf. p. 256).

The myocardium is characterized by the presence of elements of fundamental importance in cardiac physiology: the *nodal tissue* formed by anas-

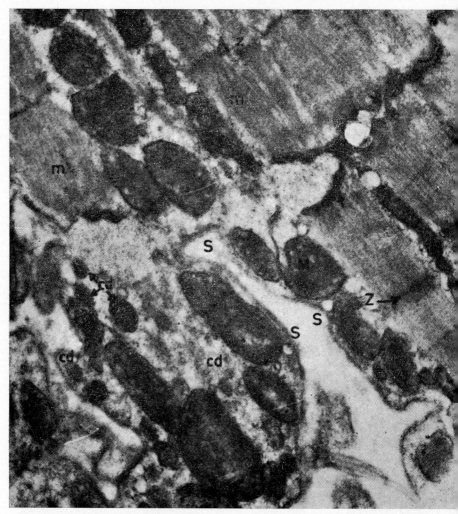

FIG. 58c. Right auricle. External muscular layers. Intercalated disk. Its structure is simpler than in the region of the ventricular myocardium (×48,000). (BOMPIANI, ROUILLER and HATT, *Excerpta medica*, 1960.)

tomosed muscle elements showing an apparently embryonic, undifferentiated structure, of which the sarcoplasm is particularly rich in glycogen.

The heart of man contains:

the node of Keith and Flack or *sinus node* situated in the region of the *sulcus terminalis* at the point where the superior vena cava opens into the right auricle;

the node of Ashoff and Tawara or *auriculo-ventricular node*, situated above the point of insertion of the tricuspid valve near the opening of the coronary sinus into the right venous heart. The bundle of His extends from this node, passes along the interventricular septum and ramifies through the two ventricles as the Purkinje *network* (Purkinje "fibres" are not fibres but chains of cells).

Certain histologists have described bundles passing from the sinus node to the left auricle (bundle of Bachmann) and to the right auricle (bundle of Thorel). However, *the existence of these bundles is open to debate.*

In lower vertebrates, such as frog, the equivalent of the sinus node is situated in the sinus venosus, the antechamber of the right auricle. In fish the node is situated at the base of the sinoauricular valves (Ripplinger, 1951; Jullien and Ripplinger, 1952). The bundle of His in frog is called the *bundle of* Gaskell. These bundles are the principal, but not the only route for conduction of motor impulses: if these bundles are prevented from functioning by injury or the action of drugs, contractions still occur showing that motor impulses pass by non-specific muscle pathways.

The ultrastructure of the specific conduction tissue in the bundle of His of rat (Fig. 58a, b, c) has been shown to consist of small fusiform cells, poor in myofibrils and containing cytoplasmic granules of $0\cdot2$–$0\cdot6$ μ diameter. The significance of these dense bodies is conjecturable (Bompiani, Rouiller, and Hatt, 1959) but they could be secretory granules similar to those found in the auricle and ventricle of the branchial heart of Myxine (Jensen, 1960; Östlund *et al.*, 1960; Hoffmeister *et al.*, 1961).

Intracardiac innervation:

The heart of mammals is richly innervated and contains:

the subpericardial plexus,
the myocardial plexus,
the subendocardial plexus.

The plexi consist of *sensitive nerve* basis (endocardial network of Smirnov) and terminations of inhibitory (parasympathetic) and excitatory (sympathetic) nerves.

Nerves cells are abundant in the sinus node and the sinoauricular node as well as in the interventricular septum.

In higher vertebrates the nerves cells are generally *dispersed* throughout the myocardium but some are localized at the base of the heart to form the parasympathetic ganglion called the ganglion of Wrisberg.

In frog the nerves cells are *grouped in numerous distinct ganglionic masses:*

ganglia of Remak in the sinus venous;
ganglia of Ludwig in the auricles;
ganglia of Bidder at the base of the ventricle.

The molecular structure of cardiac muscle is similar to that of striated skeletal muscle (presence of myosin, etc.) and will be examined when we study voluntary muscle (cf. Volume II).

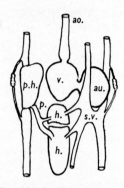

FIG. 59. Diagram of the heart and adjacent structures in *Myxine glutinosa:* *ao*, ventral aorta; *au*, auricle; *h*, liver; *p*, portal vein; *ph*, portal heart; *sv*, sinus venosus; *v*, ventricle. (AUGUSTINSON, FÄNGE, JOHNELS and ÖSTLUND, *J. Physiol.*, **131**, 2, 1955.)

Analysis of the cardiac cycle

The heart is a muscle exhibiting spontaneous contractility. Under certain conditions, the isolated organ undergoes regular depletive (systolic) and repletive (diastolic) contractions. There are two conditions for this activity:

1. Perfusion by normal vascular pathways: (1) using an isotonic fluid (principally that of Ringer) for most poikilotherms, but the elasmobranch heart requires a solution containing a high level of urea (Baglioni; Fühner; Rybak); (2) using a solution containing not only electrolytes but also a source of energy such as glucose (for example, Tyrode's solution).

The heart of poikilotherms can be perfused at room temperatures through the cavity via the veins (vena cava). However, in the case of homoiotherms the physiological saline must be at the same temperature as that of the animal and requires oxygenation, the heart being perfused via the coronary vessels (intraparietal).

2. Extension of an opened heart using techniques which I have realized during the past few years. These techniques have the advantage of excluding the need for perfusion in the case of hearts from poikilotherms and requiring little apparatus. Above all they allow *direct visual observation* —including cinematographic observation—of the inside of the heart during *spontaneous and induced activity* and allow an analysis of electrical, mechanical, biochemical and pharmacological properties of the inner sur-

Fig. a)

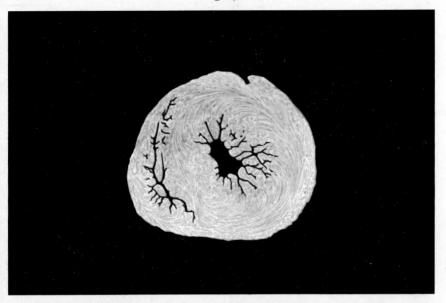

Fig. b)

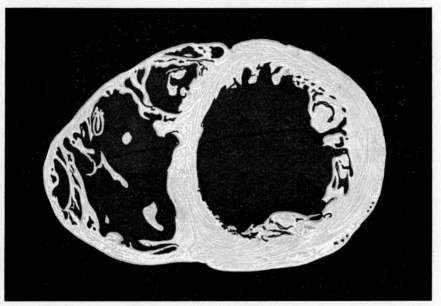

FIG. 60. Transverse section of the human heart. (a) systole; (b) diastole. The lumen of the right ventricle disappears completely in maximum systole, that of the left ventricle diminishes (after KREHL). (HÜRTHLE. 1920.)

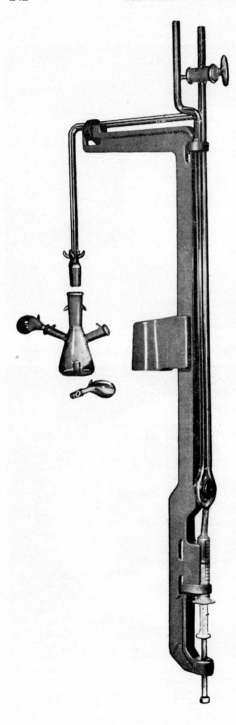

Fig. 61. Manometer and flask with ground glass side-arms, for Warburg apparatus (s.e.c.a.s.i., France).

face of the heart. In particular, it was possible, for the first time, to study and record directly the valve mechanism.

I will not go into details as to the method of opening and displaying the heart which can be used for mammalian heart both *in vitro* and *in vivo* and which, in the latter case, allows the extrinsic nerve supply to be maintained (Rybak, 1960). In general terms, the heart of poikilotherms is opened from the *ventral surface* (such that the sinoauricular valves are not injured) and the heart of homoiotherms is opened from the *dorsal surface* to prevent excessive damage to nodal tissue and coronary arteries. In the latter case, for the heart to survive for long periods, the coronary vessels must be perfused. The perfusion is incomplete because the heart has been opened but it is sufficient in that spontaneous contractions can be recorded for 6 hours without auriculo-ventricular dissociation. In both these techniques it is of fundamental importance that *a certain amount of mechanical tension has to be applied to the heart in order to maintain regular contractions* (pressure of fluid if perfusion or tension applied around the edge of the opened walls). According to the "law of the heart" (Starling, 1918) *the energy of contraction is a function of the mechanical tension on the cardiac muscle fibres.* This progressive phenomenon is different from the phenomenon of *impulsion* which occurs when a certain *threshold* is reached and which proceeds according to the "all or nothing" law. Moreover, the spontaneous contractions of the frog heart or right ventricle can be initiated or maintained by *rhythmic agitation*. I was able to demonstrate this in 1955 using the Warburg apparatus which is a microspirometer working at constant temperature and volume. The manometers are attached to flasks containing the biological system which respires in the chosen medium. The apparatus is agitated to and fro to obtain adequate mixing of the atmosphere within the flask. To ensure contraction of the heart the agitation does not need to be continuous. Regular contractions of dogfish auricle can be obtained for 30 mn at 10°C by shaking for 20 s. Andjus and Rajevski (1960, 1961) revived opened frog hearts after freezing at -196°C; continuous mechanical stimulation was essential and absence of calcium during treatment favoured the return of contractility (cf. p. 256).

The law of Starling is a *particular case* of a more general law to which I have given the name *mechanical catalysis: for the initiation, maintenance or amplification of the spontaneous contractions of heart a mechanical impulse is required*. The cause of cardiac contractility is not simply a biochemical problem; it is a mechanical and biochemical problem. Mechanical tension determination is necessary for all quantitative studies on the isolated heart. The apparatus which I have constructed *(extensometer)* allows the tension to be controlled continuously (Fig. 62).

Hence the circulatory system is an assisted system. The heart pumps blood through the organism and the returning blood fills the heart causing distension and nourishes the heart favouring contraction. In this way, the heart is adapted to its function, this adaptation depending on the volume of blood returning to the heart. Diastole is a mechanically *active phase* and determines systole.

FIG. 62. Extensometer for the heart.

But what is the origin of this contractility?

From the biochemical point of view this problem has not yet been solved. On the basis of results obtained I have formulated a working hypothesis (Table 40). From the morphophysiological aspect the areas which initiate the contractions have been localized. They are:

the sinus venous in frog;

the Keith—Flack node in mammals (the pacemaker).

The oxygen consumption of the pacemaker is less than that of auricular muscle (estimated from polarographic recordings of isolated auricle of sheep by Kunze, Lübbers and Rybak, 1961).

The pacemaker constitutes the driving system of the heart.

(1) The position of the pacemaker in frog heart can be located either by systematic sectioning of the isolated heart or by placing *ligatures* around different regions of the heart according to the method of Stannius (1852).

The seventh experiment of Stannius consisted of isolating the sinus venosus from the remainder of the heart by *placing a ligature around the sinoauricular junction:* the sinus continues to beat while the rest of the heart remains in diastole.

The ninth experiment of Stannius was *to ligature the heart around the auriculo-ventricular junction:* the sinus and auricles continue to beat at a certain rhythm while the ventricular beat is slow and independent of the sinoauricular beat (*allorhythmia*).

In his tenth experiment Stannius prevented contraction of the auriculo-ventricular system by placing *one ligature around the sinoauricular junction and the other around the auriculo-ventricular junction:* the ventricle begins to contract with a slow rhythm (the idioventricular rhythm) while the auricles remain quiescent.

When the ventricle of the heart of *Myxine* is surgically separated from the auricle, it can continue to beat for 20 h *in vitro* at 18°C at a rhythm of 18 beats per mn or half the auricular frequency (Rybak, 1960). This is the only case of prolonged ventricular contractility known among adult vertebrates.

(2) In general, the sinus venosus of mammals has disappeared in so far as the morphological entity has been condensed into the right auricle. (In rats the sinus venosus persists, being situated above the right ventricle; Prakash, 1954). By elimination procedures such as sectioning and cauterization, the Keith–Flack node has been shown to initiate rhythmic contractility. This can be confirmed by electrographic recordings since as each part of the heart becomes active it is electronegative with respect to inactive regions.

The normal rhythm of cardiac contraction initiated by the sinus venosus in frog and the sinus node in mammals is called the *sinus rhythm*. If the sinus node is suppressed the heart can continue to beat but at a slower rhythm called the *septal rhythm* (initiated by the auriculo-ventricular node). When the auriculo-ventricular node is suppressed the heart beats with a slow irregular rhythm called the *idioventricular rhythm* (cf. the tenth experiment of Stannius). This rhythm results from the automatogenic ability of the bundle of His.

Acceleration of cardiac rhythm is termed *tachycardia* and slowing is termed *bradycardia*; the absence of contraction is called *asystole*.

Normally contraction is initiated in the sinus region but how is this information conducted, resulting in activation of systole?

There do not appear to be any definite pathways in the auricles and the impulse proceeds in a diffuse manner through the whole auricle. In the human auricle the impulse is transmitted at a rate of 900 mm/s such that the auricles are completely activated within 0·08–0·12 s. However, the *rate of conduction is much faster when the auricles are dilated:* 1500 mm/s or more. There can be no doubt that this fact is related to the law of mechanical catalysis by tension but the form of this relationship has yet to be defined. The rate of conduction in the right auricle of dog varies between 487 and 1300 mm/s.

Conduction in the ventricles occurs along the bundle of His via the interventricular septum to the ventricular apex passing up through the ventricle walls to the base of the ventricle. When the impulse reaches the auriculo-ventricular node there is a time lag such that the ventricular mass beats about 1/10 s after the auricles. Normally the human ventricles are completely activated in 0·08–0·1 s. This is made up in the following way:

0·02 s in the interventicular septum;

0·04 s in the ventricle walls;

0·02 s from endocardium to epicardium (transmural conduction).

The nodal tissue *initiates the systolic impulse* and *constitutes the main pathway of conduction.*

But what is the significance of intracardiac innervation?

This leads us to a discussion of two theories of genesis of contraction: *the myogenic theory* and *the neurogenic theory.*

The *myogenic theory* suggests that the contractility originates within the cardiac muscle fibres or at least within the nodal tissue.

The *neurogenic theory* postulates that the muscle fibres are incapable of initiating a contraction and that isolated nerve cells or groups of cells constitute the excitomotor centres.

The neurogenic theory could only be applied to arthropods. The presence of a dorsal nervous ganglion in the heart of *Homarus, Astacus, Ligia,* etc. has been demonstrated by Alexandrowicz (1932). Carlson (1907) has shown that rhythmic contractions of the heart of *Limulus* (the king crab) are initiated by the cardiac ganglion. It should be noted that low concentrations of *acetylcholine* which inhibit vertebrate hearts *accelerate* the heart of arthropods.

It seems probable that a unitarian theory of automatism cannot be formulated. Here, as in other fields, the truth lies between extreme positions which help, by promoting discussion, to establish the true situation. The following arguments favour the myogenic theory: *the embryonic heart of vertebrates* (in particular of chicken) *beats before the appearance of innervation* and is insensitive to acetylcholine; by treatment with trypsin it is possible to obtain a suspension of isolated *cardiac cells* of the embryonic chick heart and these cells will beat independently (Moscona, 1952; Agrell, 1956). In addition, the heart of certain adult cyclostomes such as *Myxine*

glutinosa has no innervation but shows perfect rhythmic contractility (Östlund and Fänge, 1955); this heart is insensitive to both acetylcholine and
adrenaline but can be inhibited by enzymic inhibitors such as monoiodo-
acetic acid and 2-deoxyglucose (Rybak), a competitive inhibitor of glucose-
6-phosphate.

The problem of myogenic and neurogenic contraction in vertebrates has
been developed considerably in the light of work on the branchial heart of
Myxine. This heart consists of muscle cells containing few granules and
non-muscle cells rich in granules (U.S. von Euler *et al.*, 1960; Jensen, 1961;
Hoffmeister *et al.*, 1961) (cf. p. 235); these granules contain catechol amines
(School of U.S. von Euler). We have just seen that the normal branchial
heart is insensitive to exogenous adrenaline. However, after depletion of the
catechol amine content by reserpine the heart becomes sensitive to adrenaline (Bloom, 1962); this indicates that for normal functioning the heart must
be "saturated" with catechol amines. Moreover, granules of catechol
amines four times smaller than those within the cells can be detected in the
intercellular spaces (Hoffmeister *et al.*, 1961). I consider the non-muscular
cells rich in catechol amines to be *paraganglionic cells* (cf. p. 423) and thus
embryologically of nervous origin. The granules within the adrenal medulla
—containing large quantities of adenosine triphosphate which according
to Hillarp (1955, 1959) acts an as anionic cement—resemble those of the
branchial heart of *Myxine*. As a *working hypothesis*, I have formulated an
integrated concept, as shown in Table 40, which takes into account the periodicity of contractions and the contraction–relaxation cycle of the branchial heart of Myxine. This is based on the facts and ideas currently presented and notably on the concept of *catechol amine liberators* (tyramine, etc.).
To these I would tentatively add "eptatretin" isolated from the branchial
heart of a Californian myxinoid by Jensen (1963) and which activates the
frog heart.

Thus, the old question of a "heart hormone" (Demoor, 1924) is again
posed but further research must be carried out to determine the general
application of such a hormonal action.

Techniques for the study of cardiac cycle

Mechanographic recording

This is the least costly technique. The heart movements are recorded on
a revolving smoked drum by means of a lever attached to the heart usually by a clip on the apex of the heart. This apparatus possesses a certain
amount of inertia. Cardiac activity is transmitted either by a thread attached
directly to the lever or by an electro-mechanical transducer (variable
capacitance, resistance, etc.) or, still, by a thread attached to a sensitive
tambour, designed by Marey, which transmits pneumatically. There is

generally more inertia in the latter system and the tambour is only used in cardiology for special cases such as a *study of intracardiac pressure changes*. Chauveau and Marey (1861) described a technique for intracardiac catheterization which has since been used in man for the study of intracardiac electrogenesis (cf. p. 253).

Chauveau and Marey introduced a hollow supple tube to the jugular vein of horse. The catheter terminated in two small bulbs which penetrated the ventricular and auricular cavities and were compressed by increases in intraventricular and auricular pressure. The two bulbs were linked to two Marey tambours and pressure changes in the right heart were

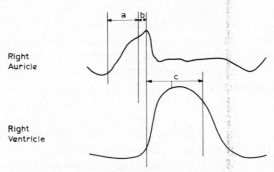

Fig. 63. Intracardiac pressure (schematic representation). *a*, systole; *b*, closure of auriculo-ventricular valves; *c*, systole.

registered graphically. To obtain a recording of pressure changes within the left ventricle a catheter is inserted via the carotid artery. Although the diagrammatic traces shown in Fig. 63 do not indicate all the movements of the auricle and ventricle, they serve chiefly to show the distribution of pressure changes.

Muscle tone is defined as the resistance to stretch and during normal function of the heart there are more or less rhythmic changes in tone. These can be demonstrated by mechanographic recording (as shown by Rybak, 1960, for papillary muscle *in situ*). These spontaneous variations in tone can also be detected by continuous recording of cardiac tissue p_{O_2}: when the tone decreases the electrode (cf. Fig. 78) becomes relatively heavy such that coronary circulation beneath the electrode is reduced and the p_{O_2} decreases, and inversely for an increase in myocardial tone (Kunze, Lübbers and Rybak, 1961).

Phonocardiography

During the function of heart of mammals in particular, sounds are emitted. The determination of anomalies in these sounds has been the basis of *auscultation* invented by Laennec, in which the *stethoscope* is applied to the chest of the subject. There are two principal heart sounds:

the first is the strongest and corresponds to closure of the *auriculoventricular valve*;

the second, separated from the first by a short silence, corresponds to closure of the *semilunar valves*. Between the second sound and the first sound of the following cycle there is a long period of silence called the "great silence".

Auscultation can be improved by using a microphone and an electronic amplification system terminating in a cathode ray oscilloscope or tape recorder. In 1957 Lenègre and his co-workers produced a record called "l'Auscultation cardiaque".

Electrocardiography

Cardiac activity is accompanied by changes in electrical potential (Mateucci, 1843; Kölliker and Müller, 1856; Waller, 1887; Einthoven, 1903). This property is of value in the detection of slight changes in the myocardium during normal and pathological activity. There have been considerable developments in electrocardiography but I can only give an indication of the general principles of the discipline by considering the technical aspect and a few results.

At one time the electrical changes were detected either by a Lippmann *capillary electrometer* or an Einthoven *string galvanometer*. Today a *cathode ray oscilloscope* is used which has little inertia or even a recording microgalvanometer which has low mechanical inertia and is adequate for most electrocardiography with the exception of work on the heart of small birds, such as humming birds, in which the cardiac frequency can be 1000 beats/mn. The heart of rat contracts at a frequency of 300 beats/mn (cf. law of sizes, p. 25). According to the number and position of electrodes different types of electrocardiogram can be obtained: external bipolar or unipolar contact recordings and intracellular recordings *in situ* or on isolated heart or fragments of heart.

The *standard external E.C.G.* of man is recorded in three ways:

Lead I—right arm to left arm (transverse)
Lead II—right arm to left leg (oblique)
Lead III—left arm to left leg (longitudinal).
Precordial leads (in the heart region on the chest) are also used.

The distribution of leads can be explained by the fact that the course of a train of potentials is modified according to the *axis* of recording and consequently, in a serial examination of the electrical state of the heart it is necessary to obtain the maximum number of results.

There are three essential components of the E.C.G.: $P-(Q) R (S)$ and T to which a fourth component, U, is sometimes added.

The terms PQRST were chosen by Einthoven (1895) since they have not previously been included in physiological terminology and thus would not lead to confusion with other systems.

17

In the intact heart:

the P wave corresponds to auricular activation;

the R (fast wave) and T (slow wave) system correspond to ventricular activity;

the significance of the U wave is obscure; it frequently occurs in enlarged hearts of athletes showing bradycardia.

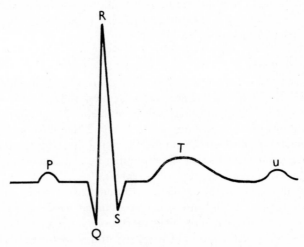

FIG. 64

Taking into account the mechanical inertia of mechanographic recordings and the greater sensitivity of electrical recording, the T wave can be considered to correspond with the moment of *ventricular contraction-decontraction process. The T wave can be inverted.* Coraboeuf (1955) explained the T wave in terms of metabolic processes within myocardial cells.

In man, the beginning of the first sound occurs 0·03 to 0·07 s after the appearance of the R wave (intersystole) while the second sound occurs 0·0 to 0·04 s after the end of the T wave.

The duration of each electrical event is as follows: 0·12 to 0·20 s for the PR interval (time for auricular ventricular conduction), about 0·07 s for QRS and 0·02 s for the duration of T.

In dog, QRS precedes the maximum ventricular contraction by about 0·02 to 0·038 s and the start of sinus activity precedes the start of the P wave by 0·05 to 0·15 s.

The *amplitude* of the electrical changes in normal man are

0·1 mV for P
0·95 mV for R   } using Lead I recording (right arm–left arm)
0·30 mV for T

These values must not be considered to be absolute. There are important variations in cardiac potential recorded at a distance from the heart; these variations depend on Ohm's law (electrical resistance): the greater the external resistance the greater the potentials recorded.

Both the amplitude and the form of the electrical potentials vary with the method of recording. This is illustrated in the tracings in Fig. 65; for example, the P and T waves are of low amplitude in recordings from Lead

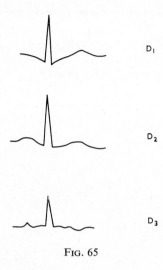

D_1

D_2

D_3

FIG. 65

III (T can be inverted). *These differences depend on the direction of the electrical axis of the heart with respect to a straight line joining the two recording electrodes.*

The electrical axis more or less corresponds with the anatomical axis and during cardiac activity the orientation of the axis changes. The electrical axis is defined as *the line joining the two points of the heart between which there is the maximum potential difference.* Einthoven has provided a geometric model called the Einthoven triangle (Fig. 66). This is an *equilateral* triangle of which each side represents a lead position, Lead I, Lead II and Lead III. On each side is a straight segment, e_1, e_2 and e_3 which is proportional to the potential difference of deflection R for each lead position. The projections e_1, e_2 and e_3 are such that algebrically $e_1 + e_2 + e_3 = 0$.

It follows that: (1) *If the electrical axis is parallel with a certain lead position, the potential difference is maximum in this position* (in this case, D_2); $e_2 = e_1 + e_3$ (Einthoven's law).

(2) *If the electrical axis is perpendicular to a certain lead position there will appear to be no difference in potential between the two recording electrodes.*

Strictly the heart (which has a volume) should be considered in three dimensions and represented by a tetrahedron rather than a triangle.

17*

I have adapted the method of E.C.G. analysis for use with the opened isolated heart of frog by placing a pair of electrodes around the preparation. The results of this *circular analysis* are exemplified in Fig. 67, recorded approximately 2 cm from the heart. The mirror images (Eyster *et al.*, 1934) indicate that the Q, R, S elements are *interconvertible*.

With my student, Madame Trépeau (1956), I have systematically sectioned the heart of frog and can state that heart fragments as small as the longitudinal strips of sinus venosus can produce a complete E.C.G., that is,

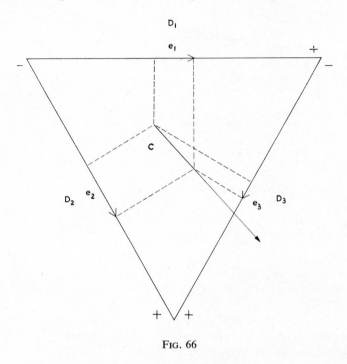

Fig. 66

with P, R and T waves. In conjunction with H. Cortot (1956), I observed that one isolated sinoauricular valve from the dogfish could produce a complete E.C.G. Using isolated auricles of poikilothermic animals and in particular, the tortoise, H. Fredericq (1921) found that these elements could produce a train of RT potentials. From these facts it can be seen that the E.C.G. is not related to the anatomical structure of the heart. An explanation of these results is provided by microscopical observation during the recording of electrical activity:

when a cardiac fragment contracts in one phase only R and T deflections are produced;

when the same fragment contracts in two phases—one before the other—(i.e., when a new *ectopic centre* arises, a new generator centre) a complete train of potentials is recorded which is equivalent to P, R and T. This is the case in the whole heart where there is an auricular and ventricular phase.

$P = RT$ but where T is based on S_a or R_v and an E.C.G. of the *whole* heart results from auricular RT and ventricular RT. This has already been considered by Eiger (1923).

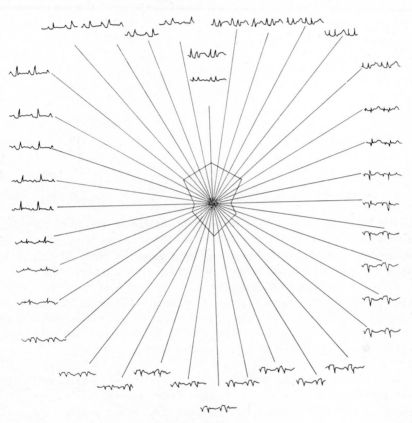

FIG. 67

Summarily, *RT is the electrical expression of one myocardial contraction.* Two alternating coupled contractions give rise to a train of waves equivalent to PRT (cf. Fig. 68).

In the method of *unipolar recording*, one electrode is placed on a part where the potential is constant and a second electrode acts as the recording electrode.

Unipolar recording within the auricle is made possible by catheterization and using this technique P. Puech (1956) has mapped the configuration of

auricular waves. There are three stages in the cycle of auricular activity
in dog:

Initially, the right auricle is active for 0·015 s and this is followed by
activity in both auricles and within the septum which commences after
0·015 s and lasts 0·045 s. Finally, the left auricle becomes active.

This technique involving the use of a *catheter-electrode* (the position
of the catheter within the heart is determined by radioscopy) has been
developed by Forsmann, Cournand, Lenègre, etc. (For an examination of
the right heart the catheter is passed through an arm vein).

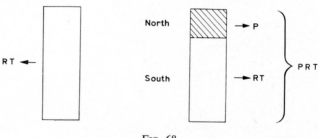

FIG. 68

Electrical depolarization of the heart *in situ* can be studied by recording
within the oesophagus (Cremer, 1908) and also *within the bronchi* (Savjaloff,
1926) although the interpretation of these results is more difficult. With
these techniques electrical fields in the cardiac region are detected.

It is now possible to maintain an *active* opened heart for long periods
(Rybak) and therefore, an analysis of electrical, mechanical, chemical and
pharmacological properties can be carried out for the *whole* endocardium
and transparietal and transeptal regions. In this way, an electro-mechano-
chemical map of the *whole* heart can be established.

MONOPHASIC ACTION POTENTIAL AND AFTER POTENTIAL OF HEART.
(1) We have just seen that the normal electrogram of cardiac muscle is
polyphasic: RT. These two deflections unite to form a simple *monophasic
wave* if the excitation appears to be abolished at a point in the heart
which does not correspond with the natural termination of the fibre. The
active part of the fibre remains negative with respect to the resting part
which is polarized throughout systole and produces a potential which,
according to the position of the electrodes, leads to an increase or decrease
in the RS-T segment of the E.C.G. Thus, this plateau appears to behave as
an "equivalent" of T (Fig. 69). This wave is monophasic. This "injury"
potential can result from lesions or any other form of damage: KCl, digi-
talis, etc. This potential can be obtained by using a *suction electrode* (an
electrode placed inside a branched tube attached to a vacuum pump).

An *injury wave* occurs in cardiopathological conditions such as *angina pectoris* and myocardial infarcts.

The action potential is due to depolarization of the muscle membrane (cf. physiology of striated muscle; Volume II) and damage to the muscle membrane would be expected to produce a diastolic injury potential of the

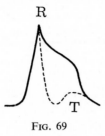

FIG. 69

same amplitude as the monophasic action potential. However, this is not the case. To explain this fact, L.N. Katz considered that around the *completely* depolarized damaged muscle there is a *partially* depolarized region which cannot conduct the excitatory impulse.

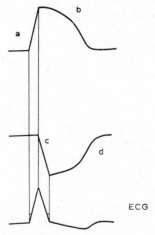

FIG. 70. (a) sub-endocardial depolarization; (b) sub-endocardial repolarization; (c) sub-epicardial depolarization; (d) sub-epicardial repolarization.

It is often stated that the fundamental elements of the E.C.G. (RT without Q and S) represent the difference of *two monophasic action potentials* (Fig. 70).

In the differential concept of the electrocardiogram it is generally considered that the RT complex results either from the opposition of basal and apical monophasic action potentials (De Boer, 1917) or from the combination of an action potential from the right ventricle and an action potential from the left ventricle (T. Lewis, 1916). Thus, the right ventricle, the interventricular septum and the left ventricle can each be compared with a simple cell exhibiting successive polarizations and depolarizations.

Before accepting this theory it is necessary to explain why there is a *delay in subendocardial repolarization*. It has been suggested that this delay results from the pressure of blood within the heart. From the experiments carried out on the isolated opened heart where the pressure of fluid is equally distributed on the endocardium and epicardium, this explanation would not appear to be valid. According to the unitarian theory, *the T wave corresponds only to the wave of repolarization of myocardium*. However, I support the *dualistic theory* which envisages the train *of RT potentials as two functionally associated but causally distinct events:*

(a) *The rapid R wave* must be considered to develop at the same time as a change from the resting metabolic state (diastole) to the metabolic state associated with *activation*. Metabolism of muscle activation (as we shall see for skeletal muscle) is not necessarily associated with activation of the enzymic processes which determine the resting potential. Depolarization can correspond to new energy-producing reactions or even to an energy gap. In any case one has to consider that changes in metabolic state always occur with variations in oxidation–reduction potentials and thus with local variations in electron flux (normally electrons do not exist in a free state) and ion (Na^+, Ca^{++}) flux.

(b) *The slow T wave* is associated with the contraction–relaxation phase and the U wave is known to be influenced by mechanical and hemodynamic factors. In my opinion, the T wave could be related in some way to rearrangement of myosin and actin molecules (cf. Volume II), with changes in the dimensions of these macromolecules and thus, changes in the permanent electrical moments (μ = electrical charge $\times$ length of molecule or particle); nuclear magnetic resonance experiments with turtle heart show proton concentration changes contemporaneous with contraction–decontraction and development of T wave (Rybak *et al.*, 1966); in addition ion transfer phenomena (in particular, of K^+ and Na^+) are involved principally [in part through changes in the Donnan equilibrium due to local changes in protein concentration? (cf. Vol. II)]. However, Danielson, Öbrink and Sjöstrand (1962), using radioactive sodium and potassium, have not yet been able to correlate movements of ions across membranes with distinct stages in the electrocardiographic cycle and during T phase specially. *The existence of a T wave in heart and ist absence from striated skeletal muscle could be associated with the presence of intercalaled disks in cardiac muscle.* Bourne (1960) considers that the intercalated disks, which are rich in enzymes, would act as boosters for the wave of contraction and it could be suggested that these synapses of muscle–muscle conduction would contribute to the development of the prolonged electrical events constituting the T wave (cf. p. 266).

The injury wave would correspond to delayed repolarization, coinciding in the terminal phase with macromolecular changes of the type discussed above.

An argument for the dualistic theory is that when the heart is stopped by moniodoacetic acid (A. S. Dale, 1935), the T wave disappears but the R wave persists for some time and the decrease in the T wave parallels the decrease in mechanographic recording.

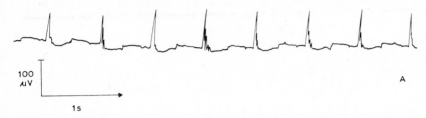

100 μV

1 s

A

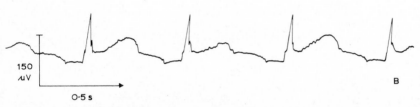

150 μV

0·5 s

B

FIG. 71a (A) Female rabbit, 2·3 kg. Weighted right hemi-heart. External bipolar recording along the auriculo-ventricular groove. (B) Same preparation. External bipolar recording with one electrode placed in the region of the auriculo-ventricular groove, the other on the basal third of the ventricle. Note the accentuation of the T wave.

In 1962, I put forward a series of arguments which criticize the differential theories on bipolar external electrocardiograms:

Surgery is not only a therapeutic technique but can be used as a method of analysis. Having used this method in a study of the E.C.G. of poikilothermic animals I extended its use to more delicate mammalian hearts, perfused *in situ* with oxygenated, warm Tyrode solution. Parts of the heart can be systematically eliminated without affecting contractility such that, for example, half a right heart is obtained—a preparation consisting of the interventricular septum and right auricle—and also a preparation in which the ventricular apex has been eliminated. Sectioning is carried out slowly and the time for restoration of the membrane is observed: in all cases trains of potentials constituting the PRT wave have been recorded (Fig. 71). Although "monophasic action potentials" can appear they always disappear within 45 mn which coincides with restoration of the membrane.

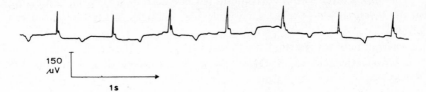

150
μV

1 s

FIG. 71b. Male rabbit, 2·5 kg. Auriculo-septal preparation (right auricle-inter-ventricu . . 'um). External bipolar recording from the auriculo-septal groove. (RYBAK, 1962; *Life Sc.*)

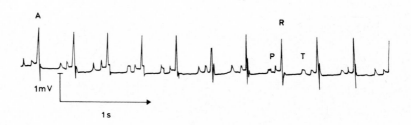

A

R

P   T

1 mV

1 s

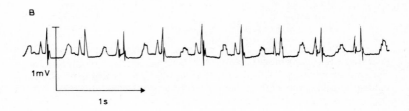

B

1 mV

1 s

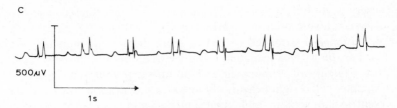

C

500 μV

1 s

FIG. 71c. (A) Normal electrocardiogram (intact heart at t_0) obtained by ex-ternal bipolar recording from the auriculo-ventricular groove; (B) same recording 15 minutes after section eliminating the ventricular apex; (C) re-cording (B), 55 minutes after section. Note the decrease in amplitude of R and T, the augmentation of P and the maintenance of the ternary pattern (RYBAK, 1962; *Life Sc.*).

(2) The monophasic action potential of frog heart does not rapidly return to the resting level but returns gradually to form a *negative after potential* (the term negative signifies negativity of the intact muscle; in galvanometric recordings the deflection passes above the isopotential line). The negative after potential which corresponds with the descending part of the mechanogram does not appear in fresh preparations but is marked in fatigued or anoxic tissues.

A *positive after potential* can be recorded from frog heart perfused with Ringer solution and fatigued by intensive activity.

An electron micrograph of the fatigued frog ventricle shows the presence of swollen mitochondria and changes in the configuration of mitochondrial membranes. In addition, hyaline degeneration of fibres and disappearance of glycogen have been reported (Rybak, H. Ruska and C. Ruska, 1964) (cf. Hajdu and Leonard, p. 285).

INTRACELLULAR MICROELECTRODES (UNIPOLAR RECORDING). In 1949 Coraboeuf and Weidmann applied the technique of intracellular microelectrodes to cardiac tissue. These electrodes have been used in a study of *Valonia* cells (Osterhout, 1928) and of the axon (Hodgkin and Huxley, 1939). Fine glass capillaries with a diameter of less than 0·5 μ are filled with 3M KCl (Ling and Gerard, 1949). The microelectrode descends vertically onto the preparation and is inserted.

Figure 72 shows the different types of traces which can be obtained with dog heart according to the type of fibre impaled (after Coraboeuf, Distel and Boistel, 1955).

The potentials obtained in this way are of greater significance than those obtained by external bipolar recording as in the standard E.C.G.

Coraboeuf, Kayser and Gargouil (1956) simultaneously recorded the external E.C.G. using electrodes on the tongue and anus and the intracellular electrogram of the ventricle. Traces in Fig. 73 represent the results obtained: *the apex of the T wave corresponds with the end of repolarization.* However, in certain conditions of ionic imbalance the plateau of the monophasic action potential elongates or is amplified while the contractions diminish (Hoffman and Suckling, 1956; Coraboeuf and Garnier, 1960–1961). With the present state of the problem it is difficult to correlate exactly electrical and mechanical events in mammalian hearts. One may suggest that temporal displacement of the plateau is an expression of movement of microions across the membrane depending on *booster effect* of intercalated disks. In *normal conditions* electrical events—which belong to the membrane—are closely and necessarily correlated with mechanical events, which belong to myofibrils, and chiefly in considering the T part of the PRT cycle. The T wave must have something in connection with the changes in *size* or shape of the heart. A point to remember is that the action potential in the pacemaker (for example, sinus venosus of frog) exhibits rapid diastolic repolarization while in non-pacemaker regions the diastolic membrane potential remains constant (Brady, 1952; Hutter and Trautwein). Electron micrographs of frog sinus venosus show it to be composed of narrow cells containing few myofibrils and having, therefore, a greater membrane surface which would favour depolarization. The sarcoplasm of these cells is rich in granules and vesicles and thickened structures, called *desmosomes*, situated at the ends of Z striations (cf. Volume II). These may facilitate conduction; however,

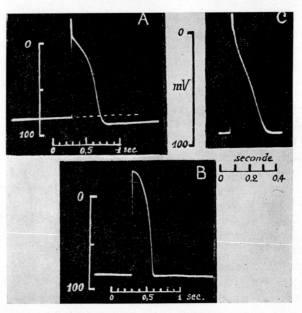

FIG. 72. Electrical activity of cardiac tissue recorded using an intracellular microelectrode. (A) ventricular myocardial tissue in spontaneous activity; (B) ventricular myocardial tissue stimulated electrically; (C) auricular myocardial tissue stimulated electrically. Vertical scale: 100 mV. Zero level corresponds to the potential of the microelectrode when outside the tissue.
(CORABOEUF, DISTEL and BOISTEL, *C. R. Acad. Sci.*, 1955.)

intercalated disks are rare (Rybak, H. Ruska and C. Ruska, 1964), remarkably the plateau is poor.

The membrane potential (measured with intracellular electrodes) decreases slightly with a decrease in temperature from 37°C to 20°C (Trautwein, 1953; Coraboeuf and Weidmann, 1954) and in deep hypothermia the potential is considerably reduced (Délèze, 1960).

Electrical phenomena appear to be related principally to ionic processes. There is an unequal distribution of potassium ions between the inner and outer surface of the membrane (cf. p. 45) and the membrane potential

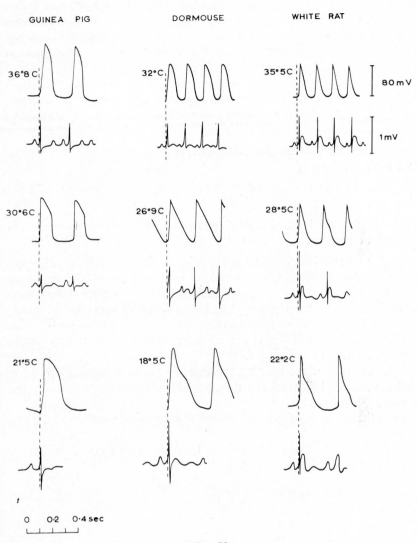

GUINEA PIG DORMOUSE WHITE RAT

36°8 C 32°C 35°5C 80 mV

1mV

30°6C 26°9C 28°5C

21°5C 18°5C 22°2C

0 0·2 0·4 sec

FIG. 73

difference can be explained in terms of ionic assymetry according to the Nernst equation. This equation is important for all excitable biological systems:

Nernst (1889) was concerned with the mechanism of production of electrical potential in a conducting system where a difference of potential will occur at the interface of two phases (for example, metal-solution of metal ions). The metal will tend to donate positive ions to the solution due to the influence of the solution pressure (P) of the metal which is then negative. This is a particular example of *the general case of partial pressure* and it is not surprising that the Nernst equation is comparable with the Peter equation (cf. Chapter 1) or with that from which the osmotic pressure laws of Van't Hoff are derived.

Liberation of ions from the metal is regulated by an equilibrium which is established when a small quantity of ions has passed into solution; Nernst considers that the osmotic pressure (π) of ions in solution opposes the ionization of the metal. Therefore:

if $P > \pi$ (as in the case of alkali metals) the metals are negatively charged in their salt solution;

if $P < \pi$ (as in the case of gold) the metals are positively charged in their salt solution;

if $P = \pi$, there is no double electrostatic layer and consequently, no potential difference.

Such a scheme is also valuable for compounds producing negative ions when the inverse of the above relationships applies. Considering the gas law $(PV = RT$ or $V = RT/P)$ applicable to solutions and if V is the volume of the solution containing one gram ion of metal with valency m, n faradays are produced when this reversible electrode is coupled with a more positive electrode and the electrical energy expended by the reversible electrode will be: $En\,\mathcal{F}$ joules.

With dilution of the solution the osmotic pressure decreases to $\pi - d\pi$ and the potential difference of the reversible electrode potential becomes $E - dE$;

thus, the energy required to dissolve one gram ion of metal is $(E - dE)n\,\mathcal{F}$ joules.

At equilibrium and according to the first principle of Carnot, the osmotic pressure is equal to $V d\pi$ such that:

$$En\,\mathcal{F} - (E - dE)n\,\mathcal{F} = V d\pi$$

and
$$n\,\mathcal{F}\,dE = V d\pi$$

Now
$$n\,\mathcal{F}\,dE = RT\frac{d\pi}{\pi}$$

thus, after integration
$$n\,\mathcal{F}E = RT \ln \pi + K$$

We have seen that when $P = \pi$, $E = 0$, and therefore,

$$RT \ln P + K = 0$$

thus
$$K = -RT \ln P$$

such that
$$n\mathcal{F}E = RT \ln \frac{\pi}{P}$$

or
$$E = \frac{RT}{n\mathcal{F}} \ln \frac{\pi}{P}$$

which is the Nernst equation for the electrode potential.

In the case of different concentrations the electromotive force results from the differences of potentials E_1 and E_2 at the interface of electrode and electrolyte and E at the junction of the two solutions. This latter potential is called the *liquid junction potential* or *diffusion potential* and when two electrolyte solutions of different concentrations are in contact, the ions of the more concentrated solution will tend to diffuse into the more dilute solution. However, if the cation diffuses towards the dilute solution more rapidly than the anion (i.e. if it is more *mobile*) the dilute solution will become *positively* charged with respect to the concentrated solution. The reverse process occurs when the anion is more mobile. In these two cases a double electrical layer will be produced at the junction giving rise to the potential difference. (It is remarkable that it is the more concentrated solution which possesses the charge of the less mobile ion). A general equation applicable to physiology is obtained:

$$E = \frac{RT}{n\mathcal{F}} \ln \frac{\text{(Ionic concentration}^{+\text{or}-}\text{) in the most concentrated region}}{\text{(Ionic concentration}^{+\text{or}-}\text{) in the most dilute region}}$$

According to the theory of Hodgkin and Katz, the ascending phase of the intracellular monophasic action potential is considered to be related to increased intracellular penetration of sodium ions. Because of the ratio of concentrations of sodium inside and outside the muscle fibre, this hyper-permeability must cause a change in potential greater than the resting potential and of opposite sign; then when the external medium has a low sodium content this reversal of the membrane potential (overshoot) should no longer appear. However, Coraboeuf and Otsuka (1956) and Wallon, Coraboeuf and Gargouïl (1959) did not observe a decrease in the initial size of the overshoot when the isolated heart of rabbit was perfused with sodium-free solution. The overshoot persisted in spite of the appearance of other effects of low sodium on the heart: increase in contraction, shortening of plateau. Efflux of potassium during the descending phase of the intracellular action potential of myocardium (the shoulder of which is depressed during anoxia of Purkinje fibres in cat; Trautwein, Gottstein and Dudel, 1954) has yet to be demonstrated clearly. The "voltage clamp" technique, used by Hodgkin, Huxley and Katz for giant axons (cf. Volume II), has been applied to cardiac Purkinje cells by Trautwein (1964). In this technique, changes in the intracellular potential of a fibre are prevented. A microelectrode is placed in the cell and this records the membrane potential. A second electrode is introduced through which a square wave stimulus is passed and of which the potential is fixed at a certain level by a negative feedback system. Trautwein was able to fix the intracellular potential of a Purkinje fibre at between -150 mV and $+50$ mV. He obtained no evidence for a significant increase in potassium conductance which would account for repolarization of the cell. He concluded that potassium conductance did not behave in an analogous way to that described for giant axon of squid.

During the course of an intracellular action potential in *conducting tissue*, the electrical resistance of the membrane is 175 times lower at the apex of the ascending phase than at the base (Weidmann, 1951). Resistance then increases to become 335 times greater than at the apex at the point at which a shoulder appears in the terminal phase (corresponding to the apex of the T wave of bipolar recordings). At the end of the potential the membrane resistance is about two thirds of that found at the start. However, Coraboeuf and his co-workers (1958) measured the electrical resistance of the membrane of contracting *ventricular tissue* during the plateau. They state that at the moment of depolarization the resistance is low but immediately following this, while the potential is unchanged, the resistance increases. They suggest that permeability to potassium—as to sodium—decreases during the plateau phase. This hypothesis was repeated by Hutter and Noble (1960).

These facts do not oppose the hypothesis which I proposed in which the subterminal and terminal phase of the intracellular action potential of the heart correspond to rapid changes in macromolecular state (systolic concentration–relaxation dilution) associated to a booster effect of intercalated disks. This change in concentration of macroions leads, as a result of the Donnan effect, to a change in concentration of microions (probably K^+). However, the facts suggest that in nodal tissue instability of the membrane (diastolic depolarization) would be preceded by hyperpolarization.

There is no doubt that the *ratios of concentration* of Na^+, K^+ and Ca^{++} are important and are a function of the type of fibre, the animal species, physical and chemical conditions. Moreover, since we are considering ion transfer, one must refer to movement of ions which implies that during the different contractile phases there are variations in ionic flux ($d\phi_{Na^+}, d\phi_{K^+}, d\phi_{Ca^{++}}$) and rates of movement. The question is still in basic progress.

I shall end this discussion of electrocardiography by examining two points:

(a) The *embryonic* heart exhibits a special type of electrogenesis. The chick heart has mainly been studied, principally by Patten and his co-workers (1939). During the 33rd–36th hour of incubation cardiac movements are *peristaltic* in nature and at this stage, only *slow electrical waves* of the type found in smooth muscle, can be recorded by external electrodes. A complete electrocardiogram (PRT) appears towards the 42nd hour of incubation (about 20 somites). Each slow contraction is associated with a slow wave similar to the T wave. In the *foetal heart* (studied by Sachs since 1923) the plateau of the ventriculogram of rat, obtained by intracellular recording, is not as well defined as that from the mother; this may result from weaker contractility (Gargouïl, 1958). However, from the 19th day the pattern of the potential during ventricular activity approached that observed in the newborn (Bernard, Monnereau-Soustre and Gargouïl, 1963). Between the 18th and 20th day of foetal life pituitary–adrenal interrelationships assume importance (Jacquot, 1959) and there is a change in ionic concentrations (Maniey, 1961). For the rat foetus this period represents a point of inflexion in morphogenetic development and the E.C.G. provides a clear indication of this change.

(b) The exact form of the E.C.G. of crustaceans is not clear. Hoffman (1911) found that the E.C.G. of *Limulus* was oscillatory in nature with rapid volleys of potential fluctuation. Since then this question has been studied by numerous authors, some confirming the results of Hoffman such as Garrey (1932), Armstrong (1939), others finding no evidence of such activity, for example, Eiger (1913), Dubuisson and Monnier (1931). Kizo Matsui (1955) working with the lobster *Palinurus japonicus* found, on some occasions, an E.C.G. of the vertebrate type and on other occasions, an oscillatory E.C.G. The latter type of recording appeared to be abnormal and could be induced by an increase in potassium concentration or mechanical tension or by damage to the cardiac wall. Laplaud and his co-workers (1961) used intracellular electrodes to study the electrical activity of the heart of *Carcinus maenas* and showed the existence of miniature potentials during the phase of repolarization.

Responses of the heart to direct excitation

(1) General points

The heart of tunicates contracts periodically and in alternate directions. Ebara (1954) found that after removal of the dorsal ganglion this type of contraction persisted but prolonged faradic stimulation of the generator centre induced repetitive contractions in *one direction*. This is undoubtably the result of fatigue processes which have yet to be defined.

Rhythmic stimulation of the isolated heart of rat at a frequency of 12 shocks per s produces tachycardia followed by ventricular fibrillation (L. Grumbach, 1956). In ventricular fibrillation the ventricle appears to tremble and this phenomenon is serious when it occurs spontaneously in man.

(2) All-or-nothing law

An inert ventricular strip responds to a single induction shock by a contraction which is at once maximal or the tissue does not respond at all. In other words, there is no relationship in the vertebrate heart between intensity of stimulus and amplitude of contraction. This was originally thought to represent a property of the syncytial structure, all fibres being excited simultaneously or not at all. However, the recent reports on the structure of cardiac muscle make it necessary to revise this concept (cf. p. 234) although the heart can be considered *functionally* as a syncytium, which is to correlate with the *T* wave (cf. p. 256).

(3) Latent summation

When a non-beating, isolated ventricle of frog is stimulated with frequent subthreshold shocks, a contraction of the ventricle can be obtained. The subliminal shocks can summate to produce an effect such that there appears to be a build-up of energy. It is reported that tetanus cannot be induced in heart tissue. However, imperfect tetanus can be produced in isolated dogfish heart (Coraboeuf and Breton). This property of dogfish heart may be related to the low frequency of contractions (6 beats per mn in the experimental conditions of these authors).

(4) Staircase phenomenon (Bowditch, 1871)

When the ventricle of an isolated frog heart ligatured in the sinoauricular region is stimulated by induction shocks every 10 s, the first five or six contractions show increasing amplitude (Fig. 74, recorded on a stationary drum).

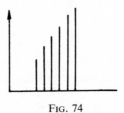

FIG. 74

A. Szent-Györgyi (1953) found that progesterone and desoxycortisone *abolish* the staircase phenomenon; 19 other sterols had no action in lower vertebrates. Season, general metabolic activity and sexual maturation influence cardiac physiology. This result is important in that it indicates that so-called *non-specific actions of hormones are, in fact, an expression of the widespread actions of hormones* which are dependent on the threshold for stimulation of the organ being considered.

(5) Periodic inexcitability

A stimulus applied during systole is without effect, (*refractory period*; Marey, 1876). From the start of diastole the heart will respond to a stimulus by a premature contraction, an extrasystole, which is followed by a *compensatory pause*. This is a period in which the heart is inactive for a longer period than between normal contractions. There are also *relative* refractory periods and the course of restoration of complete excitability exhibits fluctuations ("dips", M.C. Brooks *et al.*).

(6) Excitation by static restraint and mechanical catalysis

An isolated heart stops beating relatively rapidly (in about 30 mn for frog heart at 20°C). However, if a frog heart is opened from the ventral surface and subjected to tension at 6 points by 1 g weights arranged 60 degrees apart using the apparatus described on p. 244, the heart continues to contract for 24 h in certain cases. This is a fiftyfold increase in the duration of spontaneous contractility (Rybak). Traction influences the parameters of conduction and excitability (Penefsky and Hoffman). An extensiometric study of frog heart demonstrated that: (a) the extensibility of auricle is greater than of ventricle; (b) longitudinal extensibility of the ventricle is greater than lateral extensibility. It is remarkable that a frog heart weighing 90 mg wet weight can support up to 6×12 g.

Mechanical catalysis by traction is defined by *critical loads for rhythmic contractility* (Rybak, 1960). For example, the frequency of contraction of an opened heart of frog, weighing 70 mg and supporting a load of 6×10 mg, falls by half within 27 mn at 20°C. However, by increasing the load to 6×60 mg the initial rhythm is stabilized within 28 mn (54 beats per mn).

The pattern of the train of potentials obtained by bipolar external recording does not change markedly as long as the heart is in a reversible state. Thus, for small loads, an increase in load above the critical load for stabilized rhythm and decrease to the initial load can be carried out without marked hysteresis and without changes in the pattern of the electrocardiogram. Above a certain load the amplitude of each train of waves decreases and the intensity of contraction decreases; this is the *critical amplitude load* (Rybak).

Problems of cardiac physiology can be related in quantitative terms to the resistance of structures and it is of interest to note that certain loads produce an electrocardiogram from the frog heart which resembles certain traces during human ventricular hypertrophy (cf. also, fibrillation p. 276).

Aspects of cardiac biochemistry

Two aspects must be distinguished: the *biochemistry of automatism* and the *biochemistry of contraction and relaxation*.

The first aspect which concerns the cause of periodic impulses leading to contraction is still unknown (however, cf. Table 40). The second aspect is similar to that of skeletal muscle although certain differences are apparent. Biochemistry of contraction and relaxation cannot be studied until we have considered certain points:

Usually in biochemical techniques the molecular organization of the organ is studied by separating it into various fractions: thin slices, powders, homogenates. This analysis of the primary functioning of an organ may appear paradoxical in that on the whole one is studying *post-mortem* life. It is necessary, therefore, to develop a *functional* biochemistry.

Secondly, the distinction between the biochemistry of automatism and that of contraction must be clearly recognized in order to prevent confusion. This relates particularly to the heart which is primarily a muscular system and the substances concerned with automatism may be lost in the mass of substances concerned with contractile processes.

Finally, therefore, one must consider a *differential biochemistry* in that:

(1) it is related to function. In other words, it must be studied systematically with reference to numerous blanks, according to zero techniques;

(2) it distinguishes, in the heart for example, between *metabolism of impulsion* (metabolism of the node of Keith and Flack or of the sinus venosus in frog, the ultrastructural basis of which has been discussed on p. 239), and *metabolism of contraction* (cardiac mass). This concept which I put forward in 1955 can be illustrated by the following example:

I modified the flasks used in Warburg manometry by diametrically inserting two platinum electrodes such that respiration of the isolated heart or heart fragment and the E.C.G. could be recorded *simultaneously* (Fig. 75). The curves obtained for tortoise auricle are represented in Fig. 76.

Comments. Using the heart–lung preparation of Starling, it is possible to measure the oxygen consumption of the heart with reference to work. However, the values obtained are less accurate when related to biochemical methodology. In fact:

(1) The upper curve of Fig. 76 shows the oxygen consumption of an isolated *functional* auricle (the traces next to each point on the graph correspond to electrical changes in the sinoauricular system). The quantity of oxygen required for auricular functioning is not given by this curve. A sinoauricular system which has been inactivated by injury to the pacemaker also consumes oxygen (the lower curve takes account of basal metabolism). Thus, it is *the difference between the quantity of oxygen consumed by the functional myocardial system and the quantity consumed by the inactive myocardial system* which gives the amount of oxygen neces-

sary for automatic contractions. This concept is fundamental and has been re-stated more recently, notably by R.J. Bing (1958).

It follows from this that death is a relative notion. There is a hierarchy of death: not all organs die at the same time (this allows a creature to survive functional death such as menopause). Moreover, within an organ not all

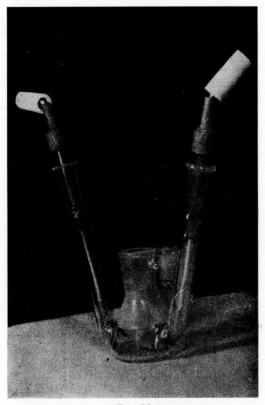

FIG. 75

parts die at the same time (which allows a functional analysis of different parts of an organ). Even within an organite all tissues, cells and enzymatic systems do not die at the same time (which allows classical biochemical analysis).

Thus, how should death be defined? My definition is that *death corresponds to total and irreversible denaturation of proteins within an organism.*

(2) Secondly, for the values for oxygen consumption shown on these curves to be significant *they must be related to some chemical criterion:* total nitrogen, total phosphorus, etc. Although "one auricle" or "one heart" has morphological significance, it is without quantitative meaning from the chemical aspect.

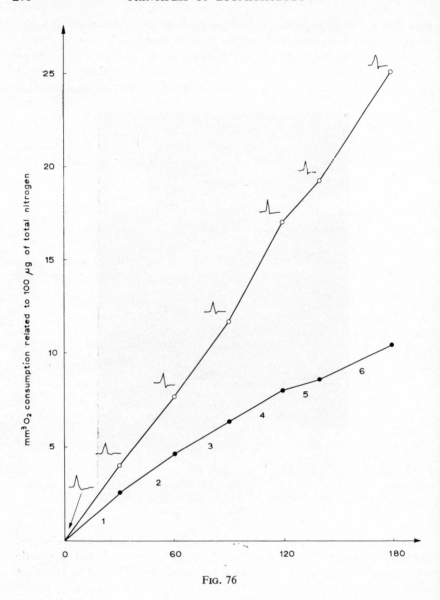

FIG. 76

Polarographic estimation of p_{O_2}

The polarographic estimation of oxygen using platinum electrodes of the following type described by L.C. Clark (1956) is based on the principle: platinum acts as the cathode and oxygen is reduced at the electrode surface to give hydrogen peroxide which with two other electrons forms water according to the equation established by Kolthoff and Lingane (1952):

$$O_2 + 2H_2O + 2e^- \rightarrow H_2O_2 + 2OH^-$$
$$\downarrow$$
$$+ 2e^- \rightarrow 2OH^- + 2H^+ \rightarrow 2H_2O$$

The greater the number of oxygen molecules reduced the stronger is the reduction current. Using a stable double membrane glass electrode (Fig. 78) we have followed the changes in p_{O_2} of the internal surface of a beating opened heart from mammal (rabbit) with the following conditions: decreased oxygen content of the perfusion fluid by displacement with nitrogen

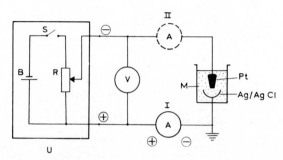

Fig. 77. Circuit for the polarographic measurement of oxygen tension. The voltage source U produces the polarization voltage V which can be modified by resistance R. The polarization voltage is applied to the polarographic cell (M) consisting of a platinum electrode (Pt), the electrolyte and the reference electrode (Ag/AgCl). The instrument A which is situated either at I or II registers each change in current. A, Apparatus for measuring current (sensitivity about 10^{-9}A/s); B, battery; M, polarographic cell; R, resistance (about 1 k-ohm); S, switch; U, voltage source (up to 1 V); V, apparatus for measuring voltage. (LÜBBERS.)

(decrease in p_{O_2}), administration of coronary vasodilators (increased p_{O_2}), stimulation of the vagus or bulbus maintained by external circulation. The following points emerge from a study of autonomic regulation: (1) administration of adrenaline leads to tachycardia and increased coronary output such that the p_{O_2} increases; however the amount of oxygen consumed often exceeds that supplied, as a result, an oxygen debt can develop in mammalian myocardium which is markedly oxygen dependent. Under these conditions, the p_{O_2} decreases and can lead to cardiac collapse; (2) the bradycardia or asystole resulting from administration of parasympathomimetics is associated with a decrease in p_{O_2} followed by a marked increase. This is probably related to the fact that at the time of recovery of cardiac beats there are variations in tone of ventricular muscle and excess oxygen compared with consumption (Fig. 79) (Kunze, Lübbers and Rybak, 1961).

With reference to respiratory phenomena in heart, in 1932 Theorell isolated the hemoglobin of muscle—myoglobin—in crystalline form from horse heart.

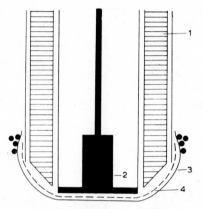

FIG. 78. Sagittal section of electrode used. *1*, reference electrode Ag/AgCl; *2*, platinum electrode surrounded by glass; *3*, Teflon membrane 12 μ thick; *4*, Cuprophan membrane 12 μ thick. (KUNZE, LÜBBERS, RYBAK, 1961, *C.R. Acad. Sci.*).

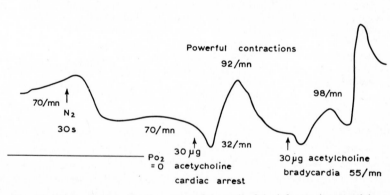

FIG. 79. Electrode placed at the centre of the wall of the left anterior ventricle (KUNZE, LÜBBERS and RYBAK, 1961, *C.R. Acad. Sci.*)

Action of ions and hormones

The effect of calcium ions on heart was reported by Locke, 50 years ago. Low calcium solutions depress cardiac contractions. There are variations between species but generally the amplitude of the ascending phase of the intracellular monophasic action potential does not change in normal or low calcium solutions. It is generally considered that cardiac contraction is associated with the entry of calcium ions to the fibres during the descending phase of the intracellular action potential. For example, the force of contraction of the frog heart would be determined by the ratio of the concentration of calcium and sodium in the surrounding physiological medium. The sodium and calcium ions would alternately combine with an anionic group R on the cellular interface such that RNa^+ would be involved during the depolarization (ascending phase) while RCa^{++} would be involved during contraction (H.C. Lüttgau, 1964). Calcium ions and anaesthetics "stabilize" excitable membranes such that, for example, more intense stimuli are required to excite tissues bathed in Ca^{++} rich solutions (Fleckenstein and Hardt, 1952; Brink, 1954; Weidmann, 1955). Calcium ions are thought to act on the sodium ion transport system which would be equivalent to R (when strontium ions replace calcium ions in physiological solutions there is a positive effect on cardiac tone).

In the presence of glycosides such as ouabain the membrane of Purkinje cells is depolarized but if the concentration of potassium in the external medium is doubled the toxicity of ouabain is decreased. In detail, ouabain $(0.5 \times 10^{-6} \text{ mol/l})$ causes a two-fold decrease in the potassium content of these fibres (P. Müller, 1963). An ouabaine-sensitive membrane adenosine triphosphate should be considered. However, potassium ions and calcium ions can be bound by myosin molecules (cf. Volume II) and P. Wasser (1963) has stated that the cardioactive glycosides (such as strophanthoside) increase potassium fixation while glycosides which are not active on the heart do not have this ability. Within cardiac fibres there would be *many macromolecular acceptors* of ions: ions are cofactors in enzymic reactions; for example, K^+ activates phosphofructokinase, it is necessary for the activity of pyruvate phosphotransferase and the binding of K^+ by mitochondria is 25 times greater than of sodium ions (Gamble, 1957).

On the other hand, Bachrach and Reinberg (1951, 1953) presented evidence for a phenomenon of general importance called the *thermo-ionic equilibrium*. An increase in the concentration of potassium ions antagonizes the effect of increased temperature on the isolated ventricle of snail. An increase in the concentration of calcium and magnesium ions antagonizes the effect of decreased temperature. This represents a *thermo-cationic* equilibrium in which variation in one factor can be compensated by variation in another. A *thermo-anionic* effect can also be demonstrated and is characterized, using the isolated ventricle of snail, by the fact that an *increase* in temperature can be compensated for by an increase in the concentration of iodide ions and inversely, while an increase in the concentration of bromide, sulphate or phosphate ions is antagonized to a certain extent by *a decrease* in temperature.

Stolkowski and Reinberg (1955) have reported that, for snail hearts *(Helix pomatia)* bathed with a solution containing both glucose and ATP, the *cells lose potassium*. (In the absence of glucose there was no potassium loss). The authors attribute this loss to the production of H^+ by the cells which displaces potassium from protein or carbohydrate binding sites. Competition between K^+ and H^+ has earlier been indicated by these authors during a study of the effects of corticoid hormones on the heart of *Helix:*

(1) when the medium contains ATP and glucose there is a decrease in tissue K^+;

(2) when the medium contains ATP, glucose and cortisone the K^+ content of cells increases;

(3) when the medium contains ATP, glucose and desoxycortisone there is no change in the K^+ content;

(4) when the medium contains ATP, glucose and hydrocortisone the loss of K^+ by tissues is augmented.

The authors propose that corticoids which lower the K^+ content would change the intracellular H^+ level.

Unfortunately, these experiments do not help in explaining the mechanism of action of desoxycortisone in abolishing the staircase phenomenon as demonstrated by Szent-Györgyi.

It should be mentioned that adrenalectomy decreases the amplitude of mechanograms and repolarization occurs without the appearance of a plateau (Tricoche *et al.*, 1961). Adrenalectomy in rat results principally in depolarization of myocardial fibres (Monnereau-Soustre *et al.*, 1963). These actions could be direct on the fibre or could result from changes in the water–mineral balance of adrenalectomized animals. Conversely, adrenal hormones (G. Hoffman, 1854; Emele and Bonnycastle, 1956), progestational and androgenic corticoids (R.D. Tanz, 1963) have a positive inotropic action (cf. p. 281) on myocardium. However, high doses of corticosteroids in association with NaH_2PO_4 cause myocardial necrosis—particularly after parathyroidectomy—(Lehr and Krukowski, 1963); these authors state that there is a direct relationship between the decrease in renal function and frequency of myocardial necrosis, calcemia and phosphatemia reaching toxic levels.

Ventricular fibrillation can be induced in the isolated perfused rabbit heart by injection of $CaCl_2$ as well as by perfusion with a potassium-free physiological saline (L. Grumbach *et al.*, 1954). In addition, L. Grumbach (1956) has obtained ventricular fibrillation of the isolated perfused rabbit heart by injection of KCl but only after pretreatment with adrenaline. There is still much to be explained in this field (cf. p. 265).

Adrenaline and thyroxine are necessary for the maintenance of normal metabolism in mammals. Gargouïl and his co-workers (1958) have shown that the intracellular electrogram of rat ventricle exhibits a weakened plateau after ablation of the thyroid and that the plateau returns after administration of thyroxine. By autoradiography of heart slices of rat after intravenous injection of thyroid hormones labelled with ^{131}I, J. Roche and co-workers (1963) observed that L-thyroxine diffuses into the muscle mass while 3,5,3′-tri-iodothyronine is localized principally in the conducting system. This affinity may be related to cardiac contraction. The action of adrenaline is represented in Fig. 80. Raab (1953) considers that the cardiovascular effects of thyroid hormones and adrenaline are interrelated.

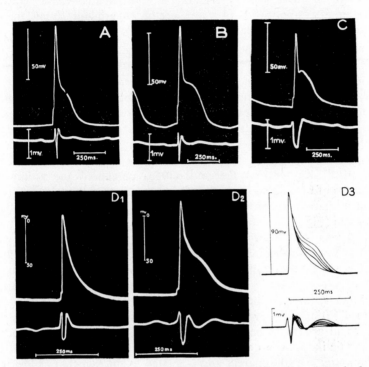

Fig. 80. (A) normal heart at 22°C; (B) injection of adrenaline: 2·5 ml of a 1/200,000 dilution, 1 mn after injection; (C) injection at same concentration as in B, but injected more rapidly (2 ml). Note the appearance and development of the shoulder of the potential which modified the plateau. The E.C.G. is also modified. The decrease in amplitude is probably due to progressive extrusion of electrode; (D₁) and (D₂), another type of electrical activity; (D₁) normal; (D₂) 1 ml of 1/200,000 adrenaline; (D₃) superimposed traces showing, following the injection, the progressive changes in the plateau of the monophasic response of the E.C.G. The phenomenon lasts about 1 mn and is completely reversible. (From GARGOUÏL, TRICOCHE, FROMENTY and CORABOEUF, *C. R. Acad. Sci.*, 1958.)

Pituitary hormones modify the dimensions of the heart and cardiac work (Beznak, 1960); after total hypophysectomy the volume of the heart decreases.

Action of certain physical factors on the heart

(1) Pressure

Gavaudan and his co-workers (1954) have shown that the heart of embryonic chick aged 9–10 days is remarkably resistant to *increased pressure*. Twenty mn after application of a pressure of 900 atm for 10 mn, the cardiac rate is still 100 beats per mn at 39 °C.

(2) Hypothermia

(A) *On the whole animal:* I shall consider briefly the effects of hypothermia of hibernating animals (a slowly developed phenomenon) and in comparison, induced hypothermia (a rapidly developed phenomenon) brought about by ice pads or cold chambers or by immersion of the anaesthetized or immobilized animal in cold water (anaesthesia by curare, ethylurethane, etc).

Coraboeuf, Kayser and Gargouïl (1956) compared the myocardial electrogram of rat and spotted dormouse *(Cricetus citellus)* during hypothermia by parallel recording of external bipolar E.C.G. and intracellular electrogram (Fig. 72). The most clearly defined effect obtained by bipolar recording was *the division of the T wave* to give T and an *earlier* wave called the *"wave of* Osborne" (0 in standard E.C.G.) and by intracellular recording, the extension of the plateau. The Osborne wave seems characteristic of hypothermia in most mammalian hearts. At low temperatures the spotted dormouse does not behave in the same way as rat in that division T+0 of the wave only appears at lower temperatures, 18·5°C. There would appear to be a difference between the thermal *threshold* for hypothermia induced in non-hibernating and hibernating animals.

Bradycardia is also apparent and is related to the fact that at low temperatures reactions are slower (Q_{10}). Hypothermic bradycardia has been recognized by numerous authors, notably Pereira, Tender and Amarante (1957), Osborne (1953), Benoît *et al.* (1958) for dog.

At temperatures below 25°C *hypothermic ventricular fibrillation* occurs, particularly in dog. During surgery on the opened heart, the use of electrical stimulators called defibrillators eliminates the serious inconvenience of ventricular fibrillation (a single "countershock" of 300 V in man).

The technique of cooling to slow or even arrest the heart is of value in facilitating surgery. In 1949 Bigelow suggested the application of this method to operations on the heart *in situ*.

(B) *On the isolated heart* (cf. p. 273, thermo-ionic equilibrium.)

With the rat heart *in vitro*, H. Cortot and I (1958) observed asystole after cooling below 18°C with progressively increasing bradycardia (the heart must be continuously perfused with oxygenated Tyrode's solution). *On no occasion was there ventricular fibrillation. Therefore, ventricular fibrillation would be conditioned by nervous (or neuro-humoral) connections to the heart and initiated by asphyxia.* With progressive warming the opened heart of rat returns to the rhythm recorded *in situ* (300c/mn). There is no *A–V* disso-ciation and the heart can function for many hours.

Cardiac fibrillation appears to be related to asphyxia, to local hypercap-nia, nervous stimuli and for sinus disequilibrium, to changes in ion ex-change. It should be mentioned that Howell and Duke (1908) observed a progressive loss of potassium ions during vagal stimulation (cf. p. 281) and it was found that fibrillation induced by perfusion of acetylcholine can be overcome by slow perfusion of KCl (Burn, Gunning and Walker, 1960); in addition, dinitrophenol causes persistent fibrillation.

Cardiac output:

There are two measurements to be considered:

(1) *the output of a single systole or systolic shower V;*
(2) the minute-output $D = VF$ where $F =$ frequency/mn of heart beats. In mammals the right and left heart have the same output.

Techniques for measuring output.

All the techniques have certain disadvantages. Under laboratory condi-tions a flow meter is placed directly on the aorta.

Those applicable to man are:

(1) the technique of gas exchange;
(2) inhalation technique.

(1) Ventilatory exchange

This method was proposed by Fick (1870) for use with O_2 and modified for CO_2 by Grehant and Quinquaud.

Let N cm³ be the volume of CO_2 exhaled by the lungs in 1 mn;
    n cm³ the CO_2 content of 100 cm³ arterial blood;
    n' cm³ the CO_2 content of 100 cm³ venous blood;

with $n' > n$ and $n - n' =$ loss of CO_2 in the lungs.

Therefore $N/(n' - n)$ expresses the number of times that 100 cm³ blood flow through the lungs in 1 mn.

N is easily measured,
n is determined from an arterial sample,
n' is determined from a sample obtained by catheterization of the right heart (Cournand).

This is the best method.

(2) Inhalation

A gas with a low solubility is introduced into the blood. The gas must be harmless and not normally found in blood, for example, ethyl iodide, acetylene.

If V is the quantity of gas absorbed per minute; and
v is the quantity of gas in 1000 cm^3 arterial blood, it follows that

$$\frac{V}{v} = \text{output/mn}$$

Value for cardiac output.

The *systolic output* is ~ 75 cm^3 in adult man. This value related to kilograms body weight gives the "systolic index" which is of the order of 1–1·5.

The *minute-output* in adult man at rest for $v = 75$ and $F = 80$ is 6 l/mn (D) (Cournand, 1949).

For a man weighing 78 kg the blood volume is equal to 1/13 body weight (or 6 l of blood) and in 1 mn all the blood passes through the heart. The rapidity of the circulation is related to its importance in metabolism of man.

The cardiac output varies with voluntary muscle activity (and can reach 25 l/mn) or visceral muscle activity (digestion). Cardiac output is altered during cardiac disease.

By radiocardiography using measurement of emission from ^{85}Kr on a scintillation counter, J. Cournand and his co-workers (J. Durand *et al.*, 1961) have found that untrained subjects at rest have a lower ventricular diastolic volume (particularly of the left ventricle) than trained subjects. During the first seconds of physical exercise the left ventricle of untrained subjects shows an increase in diastolic and residual volumes with a reduction in systolic volume; this represents the brief heterodynamic phase. During the beginning of exercise in trained subjects the two ventricles behave similarly: maintenance and then augmentation of systolic volume, decrease in residual and diastolic volume.

Cardiac work.

This is principally *work of resistance* and corresponds to the force exerted by the myocardium to expel blood at each systole. Cardiac work is apparent even in opened beating hearts containing no blood where the applied mechanical tension acts in the same way as the pressure of blood (Rybak). It is necessary for the heart to work against a force since in the absence of mechanical tension the heart ceases to beat.

(1) In the organism at rest, neglecting the kinetic energy of blood, $1/2$ MV^2,

$\mathcal{W}/mn$ = intra-aortic pressure (in g) × minute-output (in cm^3) (70)

(2) for the opened heart, isolated or *in situ:*

$\mathcal{W}/beat = \Sigma$ load (in g) × systolic displacement of each point under load (in cm). (71)

This work can be considered as being essentially ventricular (the ventricle of frog, Myxine, etc. contains more glycogen than the auricle). Ventricular work in mammals, in particular, depends on coronary irrigation (provision of oxygen). Since coronary circulation is dependent on intra-aortic pressure for filling the coronary vessels and via nervous mechanisms, it follows that cardiac work requires a coronary circulation regulated by reflex (feedback) action. The location of the coronary vessels in a structure undergoing rapid periodic contractions makes it necessary to distinguish between coronary output and coronary dilation. A substance which increases cardiac rhythm produces an increase in coronary output without necessarily having a vasodilator action on coronary vessels. Although coronary output is generally adapted to cardiac work, it is possible that a substance causing tachycardia, even though it dilates coronary vessels, can lead to asphyxia of myocardium, particularly in anaemic subjects. The relationship between cardiac work and coronary output complicates the interpretation of values for coronary output.

For man *at rest*, with values of 110 g-m/systole and 80 beats/mn then:

$$\mathcal{W}/day = 110 \times 80 \times 60 \times 24 = 12672 \text{ kg-m}$$

During exercise $\mathcal{W}/mn$ can increase to 45 kg-m. This may explain the *cardiac hypertrophy* of athletes who also exhibit bradycardia.

Thyroxine, adrenaline and emotion cause tachycardia and increased ventricular output (for example, Basedow's disease).

Thus, cardiac work normally depends on arterial pressure, volume of the systolic wave and cardiac frequency.

Rhythmic activity of the heart is characterized by variations in maximum elongation with frequency changes. A simple mechano-electronic model of these changes is presented by any spot follower (galvanometer-photocell). Systematic examination of the frequency–force relationship reveals that it follows a triphasic curve, i.e., at very low frequencies amplitude is fairly high and decreases first with the acceleration of the imposed rhythm up to a certain critical rate, which may be called "pessimal" frequency, then increases up to a summit (optimal frequency) and finally decreases again until the limit of refractoriness and reduction of rhythm by half is reached (Kruta, 1937). Concerning the interpretation of this rather complicated pattern of contractility changes, Kruta (1937, 1938) suggested that it might result from the interaction of two opposing effects which accompany each change of frequency. The steady state amplitude at a given frequency thus depends on the mutual relation of these two effects. The second one, augmentation of response following activity, has been compared with a fairly general phenomenon in skeletal muscle and some nervous elements known as *potentiation*, the restraining effect of the shortening of interval on the next contraction being termed *restitution*.

Changes in cardiac work of mammalian hearts can be studied experimentally using the *heart–lung preparation* of Starling. Dog and also rat can be used (Malinow *et al.*, 1953). Defibrinated, circulating blood is oxygenated by the lungs. The experiment is carried out with isolated organs. The pressure of blood which is warmed by circulation through a thermostatic system, is varied by changing the diameter of the lumen of the rubber tubes making up part of the heart–lung circuit. Measurements of oxygen consumption show that the heart *adapts to work* through:

(1) *intrinsic mechanisms* (with increased work the ventricle is emptied incompletely, diastolic work increases, the fibres are stretched and their contractile force increases: Starling's law).

(2) *extrinsic mechanisms:* the vagal parasympathetic nerve is cardioinhibitory and the sympathetic nerves produce cardio-acceleration.

This aspect will now be considered in more detail:

Extrinsic innervation of the heart of mammals

We shall examine:

motor control
{
(1) extracardiac nervous centres (bulbo-medullar) — { cardio-inhibitor centres / cardio-accelerator centres

(2) extracardiac motor nerves (centrifuge)

(3) sensitive nerves (centripetal) — { cardio-inhibitor nerves / cardio-accelerator nerves
}

I. Centres

(1) *Cardio-inhibitor centres*

These are symmetrically situated in the *medulla oblongata* below the floor of the fourth ventricle. Their position can be demonstrated either by *destruction* of the *bulbus* which leads to loss of cardio-inhibitory ability or by *excitation* of the bulbus which causes cardiac arrest (Weber, 1846), or by section of the *spinal cord below the bulbus*. In the latter case, excitation of spinal nerves no longer provokes cardio-inhibitory reflexes while excitation of cranial nerves continues to induce cardio-inhibition.

The Goltz reflex should be mentioned as an example of a cardio-inhibitory reflex. This results in transitory arrest in *diastole* (characteristic of the cardiac action of the vagus) following brief stimulation of splanchnic nerves in the intestinal region. Compression of the eyes of mammals also leads to bradycardia via the ophthalmic branch of the trigeminal nerve (oculo-cardiac reflex).

(2) Cardio-accelerator centres

These are situated in the *medulla oblongata* and give rise to *accelerator* fibres which accompany the *vagus*. The vagus is a complex nerve containing both inhibitor and accelerator fibres.

II. Nerves

The cardio-inhibitor nerves were discovered by the Weber brothers in 1845 (this discovery was important in that previously it was thought that nerves could only be excitatory).

The *cardio-accelerator nerves* were discovered by the de Cyon brothers in 1866.

(1) Cardio-inhibitor nerves

These are the *vagi* or *pneumogastric* nerves. Repeated stimulation leads to:

a decrease in cardiac frequency, arrest in diastole or true asystole (negative *chronotropic* effect, Engelmann);

a decrease in force of beat (negative *inotropic* effect, Engelmann);

a decrease in cardiac conduction (negative *dromotropic* effect, Engelmann);

a decrease in cardiac tone (negative *tonotropic* effect, Engelmann);

an increase in cardiac excitability (positive *bathmotropic* effect).

Bivagotomy leads to immediate tachycardia and thus the vagus normally has a restraining effect on the heart. Continuous excitation of the vagus leads to total cardiac arrest. However, if stimulation is continued the ventricle begins to contract while the auricle remains inactive. This phenomenon is called *ventricular escape* (Sherrington).

Using the opened heart of tortoise *in situ* with the nervous connections intact, I have shown that when a heart is arrested by faradic stimulation of the vagus, the sino-auricular valves generally resume contractions before the rest of the heart and notably before the auricles (1959). I have called this phenomenon *valve escape*.

Sudden emergence of the head region of a teleost fish which has been held immobile in circulating water for several hours, leads to immediate and marked bradycardia. This is dependent on vagal action, associated with a decrease in the oxygen content of blood (Serfaty and Raynaud, 1947; Serfaty and Waitzenegger, 1963).

19

Mode of action of these nerves. (1) Howell (1905) suggested that the inhibitory effect was mediated by potassium ions. Excitation of the vagus would lead to liberation of K^+ from non-diffusible complexes. However, K^+ paralyses the heart (the myocardium becomes inexcitable) and this is not the case in vagal inhibition.

(2) Loewi (1921) made an extremely important discovery using the frog heart: *stimulation of the vago-sympathetic trunk of the isolated perfused frog heart results in the liberation of a substance which, when applied to a second heart, has the same effect as nervous excitation.* The second heart is *slowed* by the perfusate from a heart arrested by stimulation of the vagus. The second heart is *accelerated* when the perfusate comes from a heart accelerated by excitation of the *sympathetic* nerve fibres to the heart.

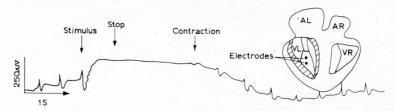

FIG. 81. Effect of vagal stimulation on the contact electrogram of a left papillary muscle of rabbit (RYBAK, 1960, *C. R. Acad. Sci.*). Note that total opening of the heart does not functionally modify vagal innervation

The experiment can also be carried out with two hearts linked by a cannula such that the connection between them can only be humoral. *Excitation of the vagus to one heart results in arrest or slowing of the rhythm of both hearts.*

The active products are called chemical mediators. The compound released by the vagus is *acetylcholine* which is rapidly hydrolysed in the heart, in particular, by *acetylcholinesterase* (the existence of this enzyme was suggested by Dale in 1914). Inactive choline is produced and thus the

Acetylcholine

$$CH_3-N(CH_3)(CH_3)(OH)-CH_2-CH_2-OCOCH_3$$

action of the vagus is transitory. *Eserine* (or physostigmine), an alkaloid from the Calabar bean, prolongs the action of the vagus by preventing enzymic breakdown of acetylcholine. Eserine is described as a *parasympathomimetic*. Atropine which inhibits the action of acetylcholine is termed a *parasympatholytic*.

Acetylcholine hyperpolarizes the myocardial membrane potential and shortens the duration of the action potential.

The acetylation of choline to give acetylcholine is catalysed by *choline acetylase*. Let us consider these enzymes in more detail.

(1) *Esterase*. Cholinesterase from *serum* hydrolyses acetylcholine but is *not specific for this substrate* since other choline esters can be hydrolysed. This enzyme is called a pseudocholinesterase.

Acetylcholinesterase is specific for acetylcholine which is hydrolysed to give choline and acetic acid. This enzyme is found in many tissues including non-nervous tissue such as placenta and spleen. Acetylcholinesterase is also found in *red blood cells*.

(2) *Acetylase*. In the presence of ATP, this enzyme catalyses the formation of acetylcholine from choline and acetic acid in the form of "active acetate", acetyl CoA. "Active acetate" is formed by the action of an enzyme found in large amounts in pigeon liver, *acetylthiokinase*. This synthesis is presented in equation (72).

$$CH_3-\overset{O}{\overset{\|}{C}}OH + CoA \cdot SH \xrightarrow[\text{ATP}]{\text{acetylthiokinase}} CoA \cdot S-\overset{O}{\overset{\|}{C}}-CH_3 + H_2O$$
$$+ \text{choline} \Big\downarrow \text{choline acetylase}$$
$$\text{acetylcholine} + \text{regenerated CoA—SH} \quad (72)$$

It should be mentioned that the left auricle of rat heart, stimulated *in vitro* at a frequency of 2·4 shocks per second, is more strongly inhibited by acetylcholine than when stimulated at 0·1 shock per s. Acetylcholine has little or no action on contractions at a lower frequency (Baumann, Girardier and Posternak, 1960). These phenomena may be related to the rate of enzymic hydrolysis of acetylcholine (cholinesterase being the limiting factor) and it would be of value to examine the effects of eserine and atropine.

Jullien and Ripplinger (1957) showed the presence in the cardiac branch of the vagus of fibres leading to the myocardium of fish which have a negative tonotropic action and of other fibres leading to nodal tissue which produce negative inotropic and dromotropic effects. However, Jullien and Ripplinger were not able to obtain conclusive evidence for the existence of cholinergic processes during cardiac inhibition in fish. The heart of roach contains no acetylcholine even after vagal stimulation. The concept of Loewi would not appear to be universally applicable.

(2) *Cardio-accelerator nerves*

We have just considered the *cholinergic* parasympathetic innervation. Let us now examine *adrenergic* innervation which is responsible for *cardio-acceleration* and *coronary dilation*. This is the *sympathetic* innerva-

19*

tion. Stimulation of cardiac sympathetic nerves accelerates the heart and antagonizes the action of the vagus (notably the work of Cannon).

The effects of sympathetic stimulation are inhibited by *sympatholytics* such as the alkaloid *ergotamine* and 933 F, a sympatholytic preventing synthesis (piperido-3-benzodioxane). *Sympathomimetics* are compounds such as adrenaline, tyramine and more generally, phenolic and non-phenolic aromatic amines (cf. Volume II, catechol amine liberators). Ca^{++} is also a sympathomimetic (Howell).

The sympathetic mediator is a *catechol amine*. However, two compounds must be distinguished:

adrenaline

$$HO-\bigotimes-CHOH-CH_2-NH\,\underline{CH_3}$$

$$HO-\bigotimes-CHOH-CH_2-NH_2$$

and *noradrenaline*
("Nitrogen ohne Radikal" = *nor*)
or *arterenol* (U. S. von Euler).

Adrenaline induces tachycardia in mammals, inhibits tonicity in smooth muscle and has an action on the central nervous system. Adrenaline is identical with *sympathin* I of Cannon and Rosenbluth.

Noradrenaline has a greater effect on maintenance of blood pressure. It is identical with *sympathin* E of Cannon and Rosenbluth.

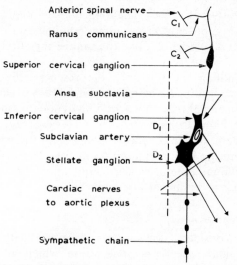

FIG. 82. (After FRANCOIS-FRANCK; from HERMANN)

The cardio-accelerator nerves originating in the medulla emerge *anteriorly* to the spinal cord. They join the ortho-sympathetic chain and converge on the stellate ganglion from where they pass to the aortic plexi and the heart.

Ventricular arhythmia, induced in rat by $CaCl_2$, is inhibited by administration of a sympatholytic and also by medullo-spinal section but bilateral vagotomy or bilateral sympathectomy is without effect (Malinow et al., 1953).

This demonstrates the role of the central nervous system in calcium arhythmia and can probably be extended to rhythmic disorders associated with emotion, anxiety, etc.

The heart of *Myxine glutinosa* is particularly rich in catechol amines (Östlund) but is insensitive both to catechol amines and acetylcholine.

The heart of hemipterous insects contains no cholinesterase and acetylcholine only blocks the heart in diastole at non-physiological concentrations (Mendes, 1957). The function of these hearts does not appear to be associated with acetylcholine.

However, the Japanese opisthobranch mollusc *Dolabella auricula*, studied by Ebara (1955), contains both cardio-inhibitor and cardio-accelerator innervation and the former may be *cholinergic*.

Another aspect of regulation of frequency of cardiac contraction is demonstrated by the experiments of C. Schlieper (1955) on the mussel *Mytilus edulis*. In seawater at 18°C the frequency of cardiac contraction is 40–60/mn but the rate falls to 1–6/mm under the same thermal conditions when the animal closes the valves to air (ecological conditions of low tide). If the concentration of carbon dioxide in seawater is increased the frequency of cardiac contractions decreases and returns to normal when oxygenation is restored. A change from pH 8 to pH 7 also induces bradycardia. It can be concluded that heart rhythm is modified according to metabolic conditions within the mussel and the circulation is thus adapted to metabolic changes.

It should be noted, however, that perfusion of a heart, such as from frog, with a so-called physiological saline results in the slow disappearance of contractile ability. This shows clearly that solutions described as physiological allow the heart to beat only for a few hours whereas the heart *in situ* functions for several years. Hajdu and Leonard (1961) isolated two substances from mammalian blood plasma which have a positive inotropic effect on the frog heart. One of these compounds is a phospholipid called lysolecithin and the other a system of three plasma globulins called *cardioglobulin* A, B and C (the level of these three globulins varies with different cardiac disorders) (cf. p. 259).

III. Sensitive innervation

Normally we possess no *conscious* awareness of our heart. Pathologically, in angina, sensory awareness of the heart is heightened (ischemia lowers the sensitive threshold?).

An unconscious awareness exists as shown by cardiac reflexes.

The sensitive nerves pass from the pericardial, myocardial and subendo-cardial plexi; they function in the *opposite direction to cardio-accelerator nerves* which explains the *importance of the stellate ganglion*, the meeting point of spinal cardiac nerves. Ablation of this ganglion inhibits anginal discomfort and tachycardia as suggested by Francois-Franck and realized by Leriche. It should be noted however, that among teleosts P. Laurent (1956) suggests that *myelinated fibres of the auricles constitute an afferent* (sensitive) *innervation without ganglionic connections*. There are two important reflex mechanisms associated with the heart of mammals, antagonistic reflexes which assure a regulatory balance:

(1) the accelerator reflex or Bainbridge *reflex* from the auricles or great veins during distension due to increased blood pressure;

(2) the *inhibitor reflex* which tends to slow the heart when the volume of blood in the aorta is too great.

Summarily, the accelerator reflex occurs when the output of blood is too small and the inhibitor reflex occurs when the output is too great. We shall see that cardiac inhibition can be brought about by a neuro-circulatory reflex. In this respect we must consider the physiology of blood vessels.

II. Physiology of vessels

A. Arterial circulation

Histologically, two types of arteries can be distinguished:

(1) Elastic arteries.

The larger arteries (aorta, pulmonary artery, brachio-cephalic branch) consist of:

1. an inner coat *(tunica intima)* which is constituted by an endothelium similar to that of capillaries, an endarterial layer of elastic connective tissue and a limiting elastic membrane;

2. an intermediate coat *(tunica media)* which is thin and elastic;

3. an external coat *(tunica externa* or *adventitia)* which is fibrous and characterized by superficial vascularization—the *vasa vasorum* having a nutritive role (the coronary vessels form part of this)—and by the presence of nerve fibres.

(2) Muscular arteries

The media of these vessels contains many smooth muscle fibres. These arteries are both muscular and elastic and are generally situated in the peripheral part of the cardio-vascular network.

The calibre of the arteries decreases along the length of the aorta. The large elastic arteries of the limbs, found always on the flexing surface, undergo deformation during movement and this facilitates blood circulation.

The tonicity of arteries counterbalances the slow variations in the intra-arterial blood volume while arterial contraction adapts the diameter of the vessels to rapid changes in blood volume.

A large part of the energy of the wave of blood which is propelled down the aorta and pulmonary artery is absorbed by the elasticity of the large arteries. This prevents damage to delicate tissues by the force of blood. The blood flow creates a pulse wave which is propagated with decreasing intensity (propagation decrement); the elasticity of the small arteries absorbs the energy of the pulse wave.

The tunica adventitia is richly innervated by a sympathetic plexus; the intermediate coat receives motor innervation. The tunica adventitia and media contain free sensitive basis.

Hemodynamics in general and arterial hemodynamics in particular are characterized by the fact that blood circulates in the vessels at a certain *pressure* with a certain *rate* and *volume*.

Arterial pressure

This depends on numerous factors, the *systolic force*, the *resistance of the vessel walls* and the *volume* and *viscosity* of the blood mass.

(1) Measurement of arterial pressure

(a) *Experimental surgical techniques.* The arterial blood pressure (most often of the carotid) can be measured by inserting a cannula attached to tubing containing physiological fluid and leading to one branch of a U-shaped manometer containing mercury.

As a result of the periodicity of the cardiac cycle the arterial pressure undergoes regular variations. The following elements can be distinguished:

systolic pressure or *"maxima"*, which is greatest and shows most variation;

the *diastolic pressure* or *"minima"* which is constant for a given physiological state (isobar); it increases during muscular exercise.

The difference between the two pressures, the *differential pressure*, gives a rough indication of cardiac energy since this differential depends on numerous other factors including the resistance to blood flow in the vessels.

Variations in arterial pressure can be recorded by relaying movements in the other branch of the manometer to a recording drum via a stylus mounted on a float.

L. Fredericq has divided the periodic variations in arterial pressure into three orders:

1st order: small oscillations of *cardiac origin,* each corresponding to a systole. These are superimposed on oscillations of the:

2nd order: large oscillations of *ventilatory origin,* each corresponding to an inspiration–expiration cycle;

3rd order or Sigmund-Mayer *waves:* by joining the apices of the 2nd order oscillations a wave is derived whose period is long and irregular. These waves may be due to variations in bulbar vasoconstrictor tone (Fig. 83).

FIG. 83.

(b) *Non-surgical techniques* (measurements in man). These are *indirect* techniques termed *sphygmomanometric.* Essentially, they consist of compressing an artery with a known pressure sufficient to stop the circulation. A pneumatic cuff attached to a sensitive manometer is generally used.

When the circulation has been interrupted by compression no oscillations are recorded on the manometer. The pneumatic pressure is then released to the point where the blood circulates at each systole (maxima pressure > parietal pressure). *The value of the pressure in the sleeve is equal to the arterial systolic pressure (maxima).* By further decreasing the pressure, a pressure is obtained at which the amplitude of the observed manometric oscillations is greatest and then begins to decrease: *the pressure is then equal to the diastolic pressure (minima).*

In practice, the passage of blood in the brachial artery can be determined either by digital palpation or by listening to the characteristic sound at the beginning of decompression and the dull sound with increased decompression, corresponding to the minima, or by the *oscillometric* method of Marey-Pachon which utilizes a highly sensitive aneroid assembly (a metal barometer with low mechanical inertia).

In adult man the normal systolic pressure is 130–160 mm Hg and the diastolic pressure is 80–90 mm Hg (measured with Pachon's apparatus).

The values increase with age up to 65 years; after this age cardiac fatigue tends to result in lower values (loss of elasticity — arteriosclerosis — which increases parietal pressure). Blood pressure is dependent on sex, the arterial pressure of females being lower than that of males.

The *arterial pulse* is the expression of the transitory increase in arterial tension at each ventricular systole and can be observed by digital palpation.

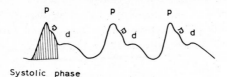

Systolic phase

FIG. 84. Form of sphygmogram. (d) Dicrotic wave (closure of semilunar valves). The rising phase to *P* corresponds with opening of semilunar valves. (s) Ventricular emptying. (GALLAVARDIN, 1921.)

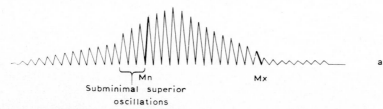

Subminimal superior
oscillations

FIG. 85a. Oscillatory curve of a fairly weak pulse. As a result of the fairly rapid increase in the levels of dilatation, the observer is liable to place the diastolic pressure too low and to take as the great oscillations the group of subminimal superior oscillations, since the true onset of the great oscillations is marked by a change in the rhythm of the pointer rather than by a sharp variation in amplitude. (GALLAVARDIN, 1921.)

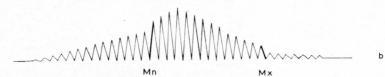

Mn   Mx

FIG. 85b. Oscillatory curve of a moderately strong pulse. As a result of the amplitude of minimal oscillations, the increase and progressive decrease in the great oscillations, clear delimitation of the phase of the great oscillations becomes difficult, but remains possible. (GALLAVARDIN, 1921.)

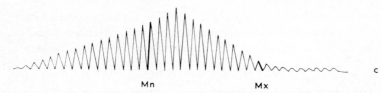

Mn   Mx

FIG. 85c. Oscillatory curve of a very strong pulse. The absence of a sufficiently clear demarcation of the beginning and end of the great oscillations makes it impossible to determine the diastolic blood pres sure by the oscillatory method. (GALLAVARDIN, 1921.)

The time course of the pulse can be followed using a *sphygmograph* which consists of a mechanographic system in which the writing stylus is joined to a spring which rests on the artery (the radial artery in man) such that maximum transmission of arterial deformation at the time of ventricular systole is obtained. The *arterial pulse appears a little later than the beginning of systole* (the "retard essentiel" of Chauveau, measuring approximately 0·05 s in man); the pulse appears with the entry of blood into the aorta when the ventricular myocardium has completed the systolic contraction.

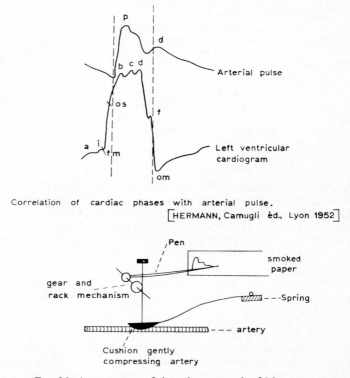

Correlation of cardiac phases with arterial pulse.
[HERMANN, Camugli èd., Lyon 1952]

FIG. 86. Arrangement of the sphygmograph of Marey.

Examination of the pulse has considerable diagnostic importance, the pulse being an easily recorded expression of cardiac frequency and amplitude. It also takes into account the state of the arterial circulation.

(2) Regulation of arterial pressure

Vasomotor tone is considered to result from the action of *adrenaline* or a related substance leading to *vasoconstriction* (of nervous or humoral origin in arterioles) and of *acetylcholine* leading *to vasodilation* and a decrease in systemic vascular tone.

These modifications in diameter of the arterial lumen are of fundamental importance in regulation of arterial pressure by varying the mechanical resistance to flow.

Table 41, compounded by Hermann (1952), provides an excellent summary of the reactions which intervene in the regulation of arterial pressure.

TABLE 41

Factors contributing to arterial pressure	Reactions to hypertension		Reactions to hypotension	
	nature of regulatory reactions	mechanisms of reactions	nature of regulatory reactions	mechanisms of reactions
heart	slowing	increase in bulbar cardio-inhibitor tone; decrease in cardio-accelerator tone	acceleration	decrease or suppression of bulbar cardio-inhibitor tone; maintenance of cardioaccelerator tone; acceleration by adrenaline
resistance of vascular tissue	vasodilation	decrease in bulbar vasoconstrictor tone; reduction in adrenaline secretion (?)	vasoconstriction	increase in vasoconstrictor tone; release of adrenaline controlled by bulbar centre and adrenaline-releasing nerves (splanchnic)
blood volume (increase: plethora)	increase in diuresis	hypertension and renal vasodilation: increase in output of kidney		
blood volume (decrease: hemorrhage)			indraught of tissue water; movement of corpuscles to spleen; oliguria or anuria	drainage of lymph; contraction of spleen due to nervous mechanisms (splanchnic nerves) and adrenaline; hypotension and renal vasoconstriction. Decrease in output of kidney.

Hypotension is related to low cardiac or vascular tone and may be dependent on histamine liberation. Guimarais, after Cecker (1934), has shown that the saliva contains a hypotensive substance which is distinct from the sialagogic substance.

Kallikreins are enzymes which in very low concentrations rapidly produce hypotension when injected intravenously to mammals. They are found in urine, blood serum and in particular, in pancreas from which the name was derived by Kraut, Frey and Werle in 1930 (in Greek *kallikreas* = pancreas). The kallikreins exert their pharmacological effect by an enzymic action on *kallidinogen* (a plasma α_2-globulin) to release a polypeptide called *kallidin* (Werle and Breke, 1948). In 1949, Rocha e Silva, Beraldo and Rosenfeld demonstrated that certain snake venoms (such as from *Bothrops*), wasp and hornet venom and trypsin can liberate a nonapeptide from blood plasma (the structure of the peptide is as follows: arginine–proline–proline–glycine–phenylalanine–serine–proline–phenylalanine–arginine). This peptide which is hypotensive and stimulates smooth muscle is called *bradykinin*. It has been synthesized (Boissonasse, Guttman and Jakenoud, 1960). Kallidin may be the immediate precursor of bradykinin (Elliott, 1963).

Hypertension has several causes:

(1) Cardiovascular: by depriving the arteries of elasticity, *arteriosclerosis* increases the work of the normal heart with an increase in arterial pressure and particularly in diastolic arterial pressure.

(2) Glandular and in particular *adrenal*:

(a) *hyperactivity of the medulla* characterized by increased adrenaline secretion;

(b) *hyperactivity of the cortex* leads to secretion of excess desoxycorticosterone which increases tubular reabsorption of NaCl leading to water retention.

(3) Renal:

(a) by inadequate elimination of urine *(ischuria)*;

(b) by cortical secretion of *renin* during renal *ischemia* (i.e. when the circulation is reduced).

The mechanisms of this regulation are numerous:

(1) *Endocrine. (Post)-hypophyseal antidiuretic hormone* augments water reabsorption (cf. desoxycorticosterone, above).

Renin is a *proteolytic enzyme* which can be liberated from the blood where it combines with a plasma globulin of hepatic origin, *hypertensinogen*. This molecule is split by renin resulting in the formation of a polypeptide: *hypertensin* (or *angiotensin*), a pressor substance which increases the force of cardiac contractions and contracts arterioles. There are two *angiotensins*:

angiotensin I (a decapeptide) and angiotensin II (an octapeptide), the true pressor agent (Bumpus, Smeby and Page, 1961). The kidney contains *hypertensinase* which is able to destroy the hypertensins.

(2) *Nervous.* Vasoconstriction can originate in the central nervous system. The actions of the vasoconstrictor centre and the ventilatory centre, both in the bulbus, are interrelated. With inspiration the activity of the ventilatory centre increases and depresses the tone of the vasoconstrictor centre, the opposite occurring with expiration. As a result, the arterial

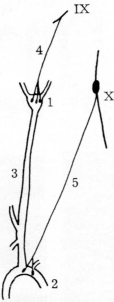

FIG. 87. Pathway of depressor nerves. *1*, carotid sinus; *2*, aortic arch; *3*, primitive carotid; *4*, nerve of Hering; *5*, nerve of Ludwig-Cyon; *IX*, glossopharyngeal; *X*, pneumogastric. (After HERMANN.)

pressure tends to decrease with inspiration and increase with expiration to give the Traube-Hering *oscillations* in arterial pressure which contribute to the production of second order ventilatory oscillations.

The principal regulatory mechanisms involve reflexes arising in the following regions:

(a) *cardio-aortic region* which gives rise to the nerve of Ludwig and Cyon;

(b) *sino-carotid region* which gives rise to the nerve of Hering which follows the path of the glossopharyngeal nerve.

These regions are sensitive to variations in pressure and in the composition of the blood with respect to O_2 and CO_2. *Thus they regulate the circulation of both blood and gases.*

The Reflexogenic Areas

The Ludwig and Cyon nerve and the Hering nerve are centripetal nerves ending in the bulbar centres.

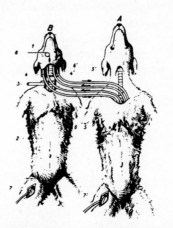

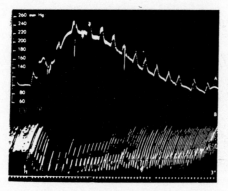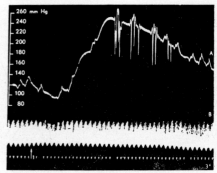

FIG. 88. Technique of isolated head of dog B perfused by dog A. Only the vago-aortic nerves connect head B with body B. Left: A, arterial pressure of donor A and head B; B, cardiac contractions of body B; ↑, intravenous injection of 0·1 mg noradrenaline to dog A; increase in arterial pressure of dog A and head B, bradycardia in body B. Right: Same preparation but nerves of carotid sinus of head B are cut. ↑, intravenous injection of 0·1 mg noradrenaline to dog A; increase in arterial pressure in A and head B but no change in cardiac contraction of body B. (C. HEYMANS and G. R. DE VLEESCHHOUWER, from C. HEYMANS and NEIL, 1958.)

When these nerves are sectioned there is an increase in pressure — the pressure can double — while excitation of the central end of these nerves produces cardio-inhibition, vasodilation and reduction of adrenaline secretion. These facts indicate that these nerves are in a permanently tonic state (cf. p 281, the action of the vagus on the heart).

Experimental evidence for these facts has been provided by cephalic *cross-circulation* and *isolated head* experiments which were derived from a *princeps* experiment of Léon Fredericq (1889). These experiments in which the innervation of the operated animal is intact, have been carried out as follows:

(1) In the experiment of Hédon (1910), two anaesthetized dogs were placed side by side with cephalic cross-circulation, the carotid of one being joined to the carotid of the other and similarly with the jugular veins. *Hypotension in one dog leads to hypertension in the other* and inversely for hypertension (pressure recorded via the femoral artery).

(2) In the experiment of Heymans (1930) the head of one dog was perfused from the vessels (carotid and jugular) of another dog and circulation in the head of the latter was prevented by ligatures and compression. *Any variation in arterial pressure of the donor leads to an inverse change in arterial pressure of the recipient and any increase in pressure leads to slowing of cardiac rhythm.*

C. Heymans demonstrated by perfusion *in situ* that the *carotid sinus* responds to pressure changes by initiating reflex activity and that baroreceptors in the adventitia and Hering nerve are involved. If the carotid sinuses are denervated, hypertension in dog A does not result in changes in dog B. This can also be demonstrated by isolating the carotid sinus with ligatures but conserving the innervation. Injection of a physiological fluid into the sinus leads to hypotension with vasodilation and bradycardia. Heymans *et al.* (1931) using the direct Fick method, have measured CO_2 production and arterial CO_2 in dogs and noted that carotid occlusion increased the cardiac output.

The carotid sinus is not the only blood baroreceptor in mammals. Paintal has produced evidence for the presence of baroreceptors in the cardiac walls which initiate reflex regulation. For example, increased pressure within the heart cavity leads to bradycardia and a decrease in arterial pressure. Baroreceptors are also present in the aortic arch and inititate pressure regulating reflexes via the Ludwig and Cyon nerve. In fact many baroceptors are *voloceptors*.

Intravascular pressure provides the basic stimulus of depressor nerves, called barosensitive; hypotension within the aorta or carotid sinus acts in the same way as cutting the nerves and leads to compensatory hypertension.

It should be mentioned that there is no defined carotid sinus in fish or in birds.

After bilateral adrenalectomy in dog, the arterial pressure falls and denervation of the carotid sinus associated with sectioning the aortic depressor nerves no longer influences the arterial pressure. If these operations are followed by bilateral splanchnectomy the arterial pressure decreases slightly (Guimarais and Rodrigues, 1948).

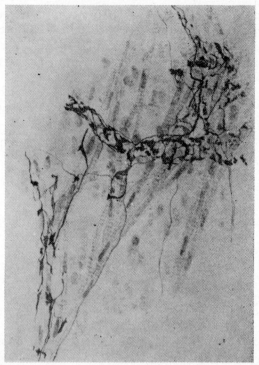

FIG. 89. Capillary from right atrial myocardium (I. A. CHERVENOVA's preparation). Extensive receptor common to capillary and surrounding tissue
(obj. 60 ×, oc. 20×; drawing; reduced in printing to one-half).

It has been known since 1856 that chemical stimulation of the ventilatory tract of rabbit induces pronounced bradycardia (Kölliker). Kish
and Yuceoglu (1955) have shown that inhalation of ether, ammonia or
tobacco smoke in rabbit leads to reflex bradycardia; the mucous membrane
of the nose and larynx being particularly sensitive (reflex action).

In addition, central nervous phenomena are known to produce hypertension. Finally, numerous compounds are hyper- or hypotensive (cf. p.
451, hypophyseal vasopressin). Among the latter compounds is reserpine,
an alkaloid of *Rauwolfia serpentina*, whose prolonged hypotensive action
is of therapeutic interest.

Rate of flow of intra-arterial blood

This must not be confused with *circulation time* which expresses the
time taken for blood to pass from one part of the vascular system to
another and which is estimated by techniques using dyes (Hering technique)
or flow of O_2 by means of arterial and venous PO_2-probes (Rybak).

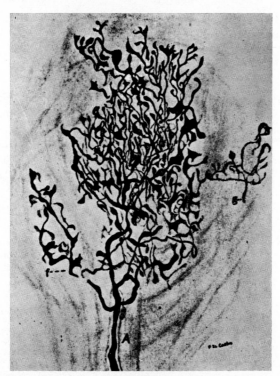

FIG. 90. Type II receptor in wall of carotid sinus of adult man; tangential section ($\times$1050). *A*, large myelinated nerve fibre; *g*, *f*, nerve "terminals". (In DE CASTRO; from C. HEYMANS and NEIL, 1958.)

These techniques give relative values, since it is not possible always to measure the distance between two points even during autopsy because of the numerous ramifications of the vascular tree.

The rate dx/dt is the distance travelled per hour, minute or second and implies exact knowledge of the pathway. It is measured using a *hemodromometer* implanted at one point in an artery.

The Volkmann apparatus consists of a U-shaped tube with graduated side-arms such that the time for arterial blood to pass through the tube can be estimated. The Chauveau apparatus consists of a T-shaped tube (inverted T), the vertical branch of which receives a kind of water-mill paddle immersed in the current of blood. The base of the paddle is attached to a needle which moves in front of a graduated dial: the greater the rate of flow the greater the needle displacement. In dog the rate is 25 cm/s and in rabbit, 15 cm/s but these are mean values because the rate decreases during diastole and the value for blood circulating in vessels some distance from the heart and in veins (cf. p. 299) can be 10–20 times lower than that of arterial blood. Assuming that at any time the output of the left heart is

20

equivalent to the output of the right heart it is implied that the venous mass is greater than the arterial mass.

Arterial blood output

This is a measure of the blood passing each minute through a section of an artery. The method of the "Stromuhr" of Ludwig gives poor results.

One of the best techniques is that of Rein using a "Thermostromuhr" which utilizes a sensitive thermoelectric couple and estimates the loss of heat by the Joule effect of a heated metal element placed against the vessel; in comparison with abacus the technique gives absolute values. Actually new techniques are developed: electromagnetic, etc.

The output increases with local work. The submaxillary salivary gland receives about 70 cm^3 of blood per mn; excitation of the chorda tympani increases the arterial output to 100 cm^3 per mn.

B. Venous circulation

Histology

As in the arteries the veins are constituted by three layers:
1. the inner coat *(tunica intima)* consists of an endothelium, sometimes a thin subendothelial endovein of connective tissue (equivalent to the endo-arterial layer) and a limiting elastic layer;
2. the intermediate coat *(tunica media)* consists of elastic connective tissue and some circular muscle;
3. the outer coat (or *tunica adventitia*) consists of longitudinal elastic connective tissue.

A characteristic in most veins is the presence in the lumen of numerous *semilunar venous valves* which oppose the backward passage of blood (Fabrice d'Aquapedente). If these valves are forced varicose veins develop. The innervation and distribution of the *vasa vasorum* is the same as in arteries. Payard (1926) has classified veins in the following way:

fibrous veins (cerebral, cerebellar veins, etc.) with no muscle fibres and practically no elastic tissue;

fibro-elastic veins (optic, thyroid veins, etc.) with muscular tissue sometimes present;

muscular veins (inferior vena cava, radial veins, etc.) with smooth muscle fibres in the intermediate layer.

Certain veins in the region of the heart, such as the superior vena cava and pulmonary veins, contain *striated* muscle in the media (Baudrimont and de Lachaud, 1937).

Functional considerations

The intracranial fibrous veins and fibro-elastic veins situated above the heart are *passive* conductors in that, in man in particular, the blood circulates under the influence of *weight*.

Veins are muscular only when they have to resist the effect of weight. The blood is thought to be propelled through these veins by rhythmic contractions of the circular and longitudinal muscles in the *media*, associated with contraction of smooth muscle in the visceral organs and of striated muscle as in limbs. The valves maintain the forward movement of blood towards the heart.

Finally, in the propagation of venous blood in mammals, it is necessary to consider *the effect of pressure changes within the heart and the thoracic cavity:*

(1) *The propulsive action* of the heart has little effect on venous blood flow since the force of blood flow is progressively decreased with passage through the arterial and capillary system.

(2) *Drawing effect.* At the moment of ventricular systole *a negative pressure* can be recorded in the jugular region; this results in the flow of blood from the veins to the right heart.

(3) *Intrathoracic pressure is lower than atmospheric pressure* by a few millimeters of mercury and this pressure difference is accentuated during ventricular systole when the volume of the heart is reduced; the reduced pressure tends to cause distension of the walls of the auricles and great veins (mechanical catalysis).

According to Burton-Opitz, the rate of venous blood flow in dog is 62 mm/s in the femoral vein, 85 mm/s in the mesenteric vein and 147 mm/s in the jugular vein. Thus, the rate increases towards the heart and this can be explained by considering that *venous pressure* which is measured with a *water* manometer, decreases *towards the heart* and can even be negative in the jugular vein and vena cava. The venous pressure in dog (after conversion to mm Hg for direct comparison with arterial pressure) is:

+ 5·4 mm Hg in the femoral vein,
+ 14·5 mm Hg in the mesenteric vein,
− 0·1 mm Hg in the jugular vein,
− 2·9 mm Hg in the auricular part of the vena cava.

The venous pressure does not fluctuate in man except in the *portal vein* which has a high resistance to flow and a relatively high blood pressure (up to 20 mm Hg), characteristics of all portal systems; in mammals it shows automatic movements and produces double spike potentials (Funaki and Bohr, 1964). The portal pressure increases when the diaphragm is lowered.

If the venous output is in the region of 50 cm^3/mn at rest it can exceed 140 cm^3/mn during muscular activity.

20*

The *venous pulse* can be observed at the base of the neck and is due to the external jugular vein. The pulse can be recorded *(phlebogram)*. In particular, it is characterized by a deflection occurring at the time of *auricular systole*.

C. Capillary circulation (or microcirculation)

Capillaries were first observed in the lungs by Malpighi in 1660. The capillary system, with the liver, is of great importance in metabolism since it is the site for exchange of food materials and in particular, for *hematosis* (uptake of O_2 and loss of CO_2) and loss of O_2 from the blood and uptake of CO_2. One can consider that there is hematosis in the lungs (or gills or skin as in frogs) and reverse hematosis in tissue regions. In man at rest, the surface for capillary exchange is 2400 m² (Policard). Most of our knowledge of physiology of capillaries is derived from the work of Krogh (1930).

Anatomy:

The capillaries are thin-walled tubes with a diameter of 5–30 μ and supplied with sensitive and motor innervations. They can be divided into: (a) *typical capillaries*, consisting of an endothelium (basis of all vascular tissue, cf. veins and arteries) covered by connective tissue or perithelium in which are contractile cells, the cells of Rouget;

(b) *atypical capillaries* which include:

(1) *capillaries of the embryonic type* in which the endothelium is syncytial: these capillaries are associated with important functions since they are found in intestinal villi and kidney glomeruli.

(2) *sinusoids*, large capillaries with irregular shape, situated in heart, bone marrow and endocrine glands; certain hormones pass into the blood in these regions particularly from the pituitary.

I should like to emphasize that the *basis of all vascularity is the endothelium*, this being the anatomical structure which is common to blood vessels.

Capillaries are arranged in a *network*. In addition to the normal capillary networks there are capillaries which connect two similar vessels such as the *arterial network of the renal glomerulus* and the *hepatic portal network* (cf. also pituitary). In *Myxine* there is a hepatic portal heart, the presence of which does not affect the portal character of this vascular system. In my opinion the existence of this portal heart is related to the fact that this animal leads a relatively inactive life on the muddy bottom which does not favour blood circulation. In amphibia and reptiles and even in birds there is a *pre-renal portal system* which carries blood from the hind part of the body to the urinary tubules.

General considerations on capillary physiology

The capillary walls exhibit a certain amount of tone due to the presence of the cells of Rouget; variations in tone lead to variations in the diameter of the vessels. Inside a vessel of small diameter there is a cylinder of plasma, free of erythrocytes, close to the internal cylinder of the vessel and the red corpuscles are localized in the axial part of the dynamic cylinder of blood; thus the diffusion of gas from the erythrocytes to the interstitial fluid is controlled by the quality and the quantity of the marginal plasma.

The pressure of blood decreases as it passes from arterioles into the small veins. The pressure in the region of capillaries can be measured by estimating the pressure required to cause local ischemia (inhibition of circulation). This is a rough method and according to the area examined gives a value of 20 to 60 mm Hg. A technique by compressing the blood with a catheter seems to give better results.

With H. Cortot and R. Canivenc, I filmed the interior of a spontaneously active heart and demonstrated the *intra-auricular capillary circulation.* In dogfish, after each systole there is a pause in the circulation; the blood appears to be withdrawn slightly before resuming the normal circulation at diastole. This slight backward movement probably allows more efficient release of oxygen and uptake of carbon dioxide by blood.

Krogh (1919) showed that the *number of open capillaries increases with local tissue activity.* In the frog sartorius muscle about 100 capillaries are open at rest and during contraction this number increases to 350. Therefore capillary function involves *recruitment* (cf. Volume II, nerves).

From a study of the number of muscle capillaries in horse, dog and guinea pig Krogh concluded that the smaller the animal the greater the number of capillaries. K. Schmidt-Nielsen and his co-workers (1958–1961) modified this concept by pointing out that such factors as the increase in capillaries by exercise, exposure to cold, altitude and even domestication should be taken into account. It is apparent that density of capillaries is related to metabolic activity per unit weight of mammal and I consider that, in spite of the limitations, the relationship of Krogh contributes to an explanation of the law of sizes and its anomalies (cf. Chapter 2).

After removal of the heart from a frog of which the nervous centres have been destroyed or anaesthetized, capillary circulation in the intestinal mesentery and interdigital membrane of the hind limb continues for several minutes. Ligatures placed around the hepatic portal vein and renal portal vein immediately prevent this circulation which is therefore the result of uptake of blood by the liver and kidney (Rybak and Macouin, 1960).

It should be remembered that the slightly differentiated endothelial cells of capillaries (such as the Küpffer cells of liver) exhibit *phagocytic activity* and thus function in immunity.

The triple response of Lewis

This is a local response which occurs when the *skin* is *scratched*.

For example, if the skin of the forearm is scratched with a sterile needle such that the cornified layer is damaged but there is *no bleeding*, a red flare appears for a distance of up to 2 cm on either side of the scratch after about 1 mn. The intensity of the reaction varies with the individual. After 3 to 5 mn a small pustula develops along the length of the scratch and becomes sur- rounded by a red area.

This represents the triple response: a red line, inflammation and pustula.

(1) The local red area is due to active dilation of capillaries;

(2) The flare is due to arteriolar dilation;

(3) The pustula is a local oedematous area due to transudation of interstitial fluid. The reactions which lead to dilation of capillaries and arterioles are related: the local red line is produced by the direct action of a *histamine-like substance* on the cutaneous capillaries and the flare results from the action of this substance on *vasodilator nerves* in the skin.

The appearance of this reaction is inhibited by antihistamines.

Capillary vasomotor effect

In fact, little is known of capillary innervation; according to Woolard (1926) there is a diffuse plexus. In low doses adrenaline causes vasoconstriction and in high doses vasodilation. Hypophyseal vasopressin induces capillary constriction. Anoxia (acidification) and heat cause dilation while oxygen and cold cause constriction. Under the influence of cold reflex reactions lead to the appearance of a red nose (arteries and capillaries dilate). Blue coloration of the nose during cold results from cyanosis (arterioles contracted, capillaries dilated resulting in blood stasis with hypercapnia). This represents a local stress reaction (cold shock). Hertzman (1959) has stated that the basic function of the cutaneous circulation (L. Fredericq, 1882) is the convective transfer of heat to the overlying epithelium through which heat is conducted to the surface, whence it is lost by radiative, convective and evaporative pathways. Since the temperature of the skin has an important influence in these transfers, one may expect significant correlations between the cutaneous blood flow and temperatures. The convection heat transfer by the blood stream is given by the equation

$$H = F \times k(T_1 - T_2)$$

in which H, F, k, T_1 and T_2 are respectively the heat transfer, blood flow, specific heat of blood and the temperature of the blood before and after heat loss or gain. The analogy with the Fick reaction for cardiac output is obvious. This equation expresses only the numerical resultant of physiological mechanisms but it is still useful. Thus, if one assumes: heat transfer by the cutaneous blood flow at a rate of 45 kcal/m²/h, a constant thermal conductiv-

ity of the epithelium of 9 kcal/m²/h/°C/cm and its effective thickness as
1 mm, and that the venous blood leaves the superficial plexus at a temperature about 0·5°C higher than that of the skin surface, one obtains a family
of curves which shows graphically the required relationships between the
rates of cutaneous blood flow, skin temperatures and inner or core temperatures (Hertzman, 1961).

Acidity dilates capillaries by a direct action on the vascular walls; carbonic acid and phosphorylated compounds such as those from carbohydrate catabolism act in this way. Thus, part of the *regulation* of capillary
diameter according to metabolic requirements is brought about by local

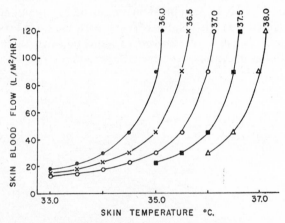

FIG. 91. Theoretical relationships of cutaneous blood flow and temperature
at five different levels of core temperature. (From SENAY *et al.*, *J. Appl.
Physiol.*, 1960; borrowed from HERTZMAN, *Blood Vessels and Circulation*,
ed. by MONTAGNA and ELLIS, 1961.)

metabolic products which include, in addition to acid products, other
compounds such as histamine, serotonin and related compounds (H substance of Lewis). With increased metabolism the capillaries dilate allowing
provision of substrates and elimination of waste products.

This adaptation to local requirements also involves certain *arteriolar
mechanisms which ensure connections between precapillary arterioles and
venous capillaries* by means of arterio-venous anastomoses, the *glomus*
of Masson (cf. p. 188). By increasing blood output, these shunts have an
important role in thermoregulation. In the region of the arterioles the
tissue is muscular and richly innervated.

Vasomotor reflexes:

In higher animals a localized action is generally *metabolic* in nature.
A more diffuse reaction but one localized to a certain region, results from

axonal reflexes which can be secondary to metabolic effects. Finally, the
reaction to more complex or intense stimulus results from a central reflex.

I should like to point out that *all mechanisms of homeostasis are based
on reflexes which can be metabolic (molecular), hormonal or nervous reflexes.*

With respect to vasomotor reactions in general, it is not always possible
to distinguish between a purely metabolic action and a reflex action.

Vasodilation generally results from inhibition of central bulbar or medullary vasomotor tone such as that resulting from ablation of the stellate
ganglion or anaesthesia, particularly by morphine which depresses activity
within the vasoconstrictor centre. This is *passive* vasodilation.

Active vasodilation is peripheral in origin and is due to acetylcholine.

It is considered that vasodilator reflexes, with the exception of thermo-
regulatory reflexes, *are never generalized* indicating that there are no specialized vasodilator centres. However, these reflexes can be produced at

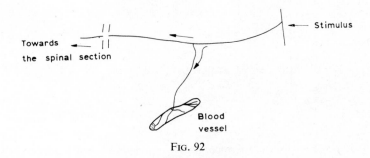

FIG. 92

some distance from the point of stimulation; for example, stimulation of the
superior laryngeal nerves leads to vasodilation in the face. It is known that
emotions such as anger or pleasure lead to flushing of the face and neck
in certain individuals. These are reflex phenomena. The region of the face
and neck is particularly susceptible to vasodilation as shown by the action
of nicotine.

Axonal reflexes (Langley and Anderson, 1900) do not pass through nerve
centres. The diagram in Fig. 92 illustrates the pathway of this short circuit phenomenon which can be demonstrated by cutting the sensory
nerve: before neural degeneration is apparent, the stimulus produces a
normal centripetal impulse but since this cannot reach the spinal cord it
passes along a sensory collateral and excites the blood vessel. The direction of the impulses in the collateral is abnormal and they are said to be
antidromic. The direction is abnormal if one considers the normal destin-
ation of the collateral but from the point of view of nerve physiology we
shall see later (Volume II) that nerve impulses can pass in either direction.

For example, the vasomotor modifications occurring in the hind limbs due to postural changes include the strictly regional vasomotor reflex. This local reflex may be provoked by the stimulus of pressure as shown by rapid injection of 1–2 ml of physiological saline into the femoral artery and is probably due to sudden distension of the vessel walls. The *local pressure reflex* is of the axonal type since it persists after cutting the somatic nerves (sciatic and femoral nerves) and the periarterial sympathetic innervation (Condorelli *et al.*, 1961).

CHAPTER 12

EXCRETION

The products of catabolism are finally eliminated. In this chapter I shall present a unified concept of the various processes of elimination.

General principles of detoxication

Products *which are not retained by the organism* are the *endogenous poisons* formed during the final stages of metabolic chains. Generally, these compounds are eliminated after conversion to less toxic compounds by *detoxication mechanisms* in which structure is modified or metabolic products conjugate with other compounds. The liver which is *an essential organ in detoxication* constitutes an active filter.

The antigen–antibody reaction represents a type of detoxication by conjugation preceded by biosynthesis of a *specific* coupling agent, the *antibody*. *Detoxication is a method by which organisms can maintain the dynamic constancy of the internal environment.*

(1) Detoxication by conjugation:

(a) *with an amino acid:*

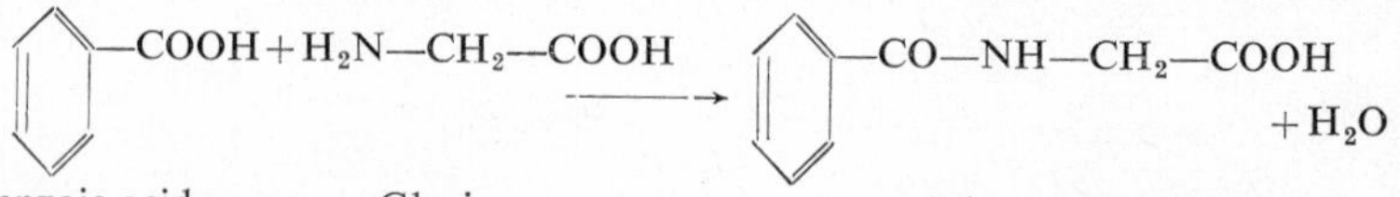

$$\text{Benzoic acid} -COOH + H_2N-CH_2-COOH \longrightarrow \text{(ring)}-CO-NH-CH_2-COOH + H_2O \quad (73)$$

Benzoic acid   Glycine   Hippuric acid

(b) *with a mineral acid:* such as H_2SO_4 to give phenylsulphates, and CH_3COOH to give acetylated derivatives. Conjugation with formation of sulphates, discovered by Baumann (1876), occurs in two stages (Borström and Vestermark, 1961):

(1) *activation of sulphate* (Wilson and Bandurski, 1956):

$$\text{inorganic sulphate} + ATP \xrightarrow[\text{Mg}^{++}]{\text{enzyme}} \text{adenosine-5'-phosphosulphate} + PP$$

$$\text{adenosine-5'-phosphosulphate} + ATP \xrightarrow[\text{Mg}^{++}]{\text{enzyme}}$$

$$\text{3'-phosphoadenosine-5'-phosphosulphate} + ADP$$

(2) *sulphur transfer* (Lipmann and his co-workers, 1958):

ROH + 3'-phosphoadenosine-5'-phosphosulphate $\rightleftharpoons$

$$R—O—\overset{\overset{\displaystyle O}{\|}}{\underset{\underset{\displaystyle O}{\|}}{S}}—O— + \text{3'-phosphosadenosine-5'-phosphate}$$

(c) with glucuronic acid: steroid sex hormones are eliminated after conjugation with glucuronic acid.

(d) with compounds containing SH groups: mercapturic derivatives are formed.

(2) Detoxication by modifications in structure:

(a) by methylation (from methionine, betaines and choline). For example:

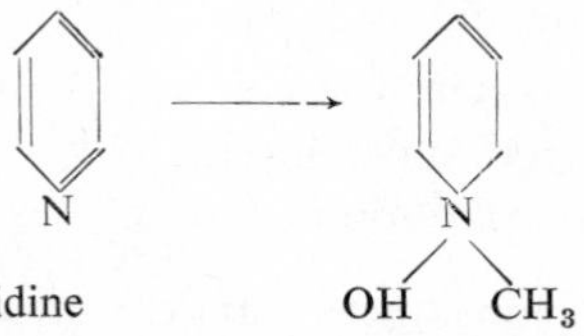

(74)

Pyridine   OH   CH$_3$ Methylhydroxypyridine

(b) *by reduction:* aldehydes and ketones give alcohols.

(75)

(c) *by oxidation:*

There are numerous oxidation processes. For example, reactions leading to the formation of acids from primary alcohols, aldehydes or ketones, production of phenols from aromatic compounds and respiration can be included if one accepts the concept of Wieland (1912) (oxidation = dehydrogenation; this concept was extended by Kossel in 1916 to oxidation by a change in valency and loss of electrons), particularly with reference to water elimination.

Essentially, respiration is a biological phenomenon by which an organism absorbs oxygen and eliminates water and carbon dioxide.

Asphyxia presents many characteristics of poisoning and is essentially a case of poisoning by electrons. In this celebrated work "Sur la respiration des Animaux" produced with Seguin (1790), Lavoisier, the originator of the concept: respiration = combustion, would have considered the lung to be an emunctory, since he had previously remarked on the "considerable quantity of humidity produced by the lung at each expiration" which led him to compare *"cutaneous transpiration"* and *"pulmonary transpiration"*.

In my opinion, the comparison of the lung with an emunctory is justified by the balance which exists between renal function and transcutaneous function in amphibia (bucco-pharyngeal and transcutaneous in certain urodeles without lungs such as the salamander *Aneides lugubris* (Ritter and Miller, 1899), branchial function in certain fish and pulmonary function in birds and mammals. This balance is apparent in mammals, for example during ketonuria when the expired air contains acetone.

Our study of excretion will include therefore the following topics:

— respiratory excretion and in particular the evolutive differentiation of pulmonary ventilation (*hemo-pulmonary* system);
— urinary excretion or renal physiology (*hemo-renal* system);
— salivary excretion and biliary excretion (involving the *hepato-intestinal* system);
— cutaneous excretion.

The media of these excretory pathways are respectively breath, urine, saliva, bile (faeces), sweat. In mammals:

(1) the final product of pulmonary excretion is a vapour–gas mixture;
(2) the final product of renal excretion is a liquid;
(3) the final product of intestinal excretion is a solid.

In the latter two cases, although substances enter the metabolism by the same bucco-pharyngeal route they are eliminated by distinct pathways; in pulmonary function, the detoxicating agent (O_2) and a part of the final products of detoxication (CO_2 and H_2O) enter and leave by the same buccopharyngeal route.

The liver and kidney are of particular importance in the purification of blood. However, the lungs are also organs of detoxication. From the trophic point of view, the circulation begins in the region of the intestinal capillaries and the blood containing both useful products (amino acids, sugars, etc.) and toxic products (ammonia, amines, etc.) must be purified before passing to the different parts of the organism. Similar requirements apply to the embryo in which the vitelline circulation can be considered to be the first stage of the circulation. Detoxication of blood takes place in hepatic, renal and pulmonary tissue and the liver represents an organ of internal detoxication, the kidney and lungs being organs of external detoxication. The liver and kidney possess a specialized capillary system, a portal system.

These capillary networks are situated between two identical vascular structures. Physiologically, in an arterial network the blood which passes through the capillaries remains arterial and in a venous network the blood remains venous. Moreover, from the portal vein to the capillaries of the pulmonary artery, although different vascular structures are involved—a vein and an artery (pulmonary artery) — the blood is always venous blood. Similarly, from the capillaries of the pulmonary vein to the renal capillaries, although the vascular structures are different, these vessels transport arterial blood. The circuit for detoxication can be represented by the following scheme: intestinal capillaries → portal vein → hepatic capillaries and veins → inferior vena cava → right heart → pulmonary artery (venous blood) → pulmonary capillaries (venous blood) → pulmonary capillaries (arterial blood) → pulmonary vein → left heart → renal glomeruli (→ renal vein → inferior vena cava).

Then, from the intestinal capillaries to the pulmonary venous capillaries internal detoxication occurs and from the pulmonary arterial capillaries to the renal glomeruli external detoxication occurs, external pulmonary detoxication being the first and external renal detoxication the second.

Excluding the lymphatic system for purposes of simplification, I consider that the nutrition circuit can be represented as follows:

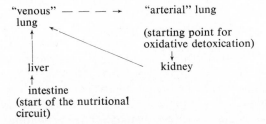

The intestinal capillaries represent the starting point for nutrition and the pulmonary capillaries for oxidative detoxication, between these two regions a system of general detoxication operates (liver).

The concept which I have just presented is an extension of the idea of a specialized capillary network (rete mirabile) and will relate to the physiological aspect of the problem of the circulatory structures being considered.

I. Respiration

A certain number of principles must be examined.

In *unicellular organisms* respiration involves *oxygen dissolved* in the external medium. In multicellular organisms, even in the more highly organized ones, the same applies.

Although this is valid for fish (where hematosis occurs through the gills) and generally valid for amphibia (where hematosis with oxygen transfer occurs across the damp skin), in mammals the ventillatory system of the lungs receives *gaseous* O_2. However, for hematosis in the pulmonary alveoli gaseous oxygen *is dissolved in water*. Not only is there always a certain water vapour tension in the lung but in vascular regions the plasma and interstitial fluid contribute to the dissolution of O_2 in water. *In hematosis of both O_2 and CO_2 there is passage of gases across an air–water interface.* This explains the discomfort associated with being placed in a dry atmosphere.

During ventilation there is not only transfer of gases but also *transfer of heat*.

Finally, since the lung has become differentiated during evolutionary processes and respiration implies a more general phenomenon than active pulmonary respiration, in the latter case one should speak of *ventilation* and in particular, thoracic movements should not be termed "respiratory movements" (which are electron movements occurring at the molecular level) but ventilatory movements. I understand active pulmonary respiration to represent the process characterized by mechanical movements which favour gas exchange. In scorpion, for example, air passes into the lungs by simple diffusion.

Some points concerning comparative physiology

(1) The aerial ventilatory system of insects consists of openings along the side of the body: the spiracles. These openings lead to tubes—the trachea—which penetrate the organs and which subdivide to form smaller and smaller tubes; some branches being less than 1/1000 mm in diameter. How does air circulate in such a system in animals exhibiting no ventilatory movements? P. Portier (1932) studied the tracheal system in butterflies and wrote "the wings of butterflies contain a blood circulation and an air circulation, both being very active during flight. The blood which returns from the wings, circulated by the pulsating organ in the thorax, is oxygenated and bathes the large motor muscles of the wings which are situated in the thorax. During flight, air beats on the striated scales and induces local eddies which result in aspiration in the hollow scales which connect with the trachea. The aspiration is propagated to the tracheal system of the body by the large trachea at the base of the wings and augments circulation of air in the general tracheal system, thus supplementing diffusion which becomes inadequate. CO_2 ... escapes from the scales into the atmosphere". In this example, the essential feature is the presence of air deflectors. These conclusions were reached after experiments in which the wings were varnished. The structure of tracheae, tracheoles and air sacs in different insects has been studied using the electron microscope by Sawaya (1954): the walls of the tubes were found to have a helical structure. Trachea consist mainly of chitin. P. Remy (1925) demonstrated that "white indigo injected to the general cavity of insects is almost immediately oxidized, the reaction being more intense in tracheolar cells around the tracheoles ... indigotin, formed by the oxidation process, is fixed by cytoplasmic granules and other spherical inclusions". Remy considered that the penetration of oxygen was not based on simple diffusion and thus implied the participation of metabolic processes. However, this is not valid for CO_2 transfer, at least in mammals such as mice: Brues and Naranjo (1949)

demonstrated that transfer of $^{14}CO_2$ is the same from both sides of the active lung surface.

(2) The lung of amphibians, reptiles and mammals can be compared with a smooth sac which can be subdivided to form alveolar sacs as appear in turtles and crocodiles. In the bird adapted to aerial life the lung has a different structure (Fig. 93). It is characterized by:

(i) a lung which is traversed by numerous canals between which are found compact regions rich in small canals and capillaries;

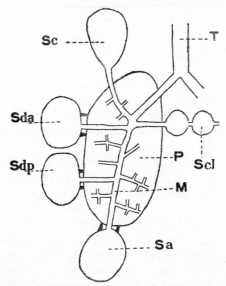

Fig. 93. Diagram of air sacs (after Mouquet). *P*, lung; *M*, mesobronchus; *T*, trachea; *Sc*, cervical sac; *Scl*, clavicular sac fused with the symmetrical sac; *Sda*, anterior diaphragmatic sac; *Sdp*, posterior diaphragmatic sac; *Sd*, abdominal sac. (GUEYSSE-PELLISSIER, 1945.)

(ii) the existence of well-developed *air sacs* which fill the free spaces of the body cavity and penetrate between muscle and *into bone*. These sacs are the extremities of hypertrophied bronchii and the system in some ways resembles that seen in flying insects. During flight, air enters the sacs via the lungs and then passes into the canals. Hematosis occurs only in the lungs but during the inward and the outward passage of air.

When related to body weight the ventilatory rhythm of birds is less rapid than that of mammals. This point is demonstrated by the following values, compiled by C. Richet (1893).

(3) Hemoglobin is not the only blood pigment transporting oxygen and carbon dioxide. However, certain conjectural points should be mentioned concerning hemoglobin and oxygen transport. De Graaf (1957) and Ewer (1959) reported the existence of *Xenopus laevis* which contains no hemoglo-

TABLE 42

Species	Weight in kg	Number of ventilatory movements/mn
cassowary	50	2
pelican	8	4
condor	8	6
cock	2	12
duck	1	18
pigeon	0·3	30
sparrow	0·02	100

bin and a living Myxine has been discovered which also exhibits this unusual character (Rybak, 1960). In the latter case, the absence of an oxygen carrier would not be detrimental if one considers the low activity of Myxine and the low p_{O_2} and temperature which the animal would encounter in its natural medium. The plasma p_{O_2} would be sufficient to supply the oxygen requirements. It would be also the case of fish (Chaenichthyidae) surviving in the Antarctic (θ water 2 °C) without hemoglobin or red cells ("ice fish") (Ruud, 1954).

Although hemoglobin is found in certain invertebrates such as polychete annelids (Bloch-Raphaël, 1939), it is absent from others. The molecular weight of invertebrate hemoglobins is generally higher (about 400,000) than that of the hemoglobin from higher vertebrates (estimated by Svedberg, 1934, using ultracentrifugation). The invertebrate hemoglobins have been called *erythrocruorins*. In most invertebrates (lumbricids, malacostracean crustaceans, etc.) the blood pigment is dissolved in plasma. However, corpuscles which carry pigment are found in echinoderms, lamellibranchs and vertebrates.

In three families of polychetes (including sabellids and serpulids) the "respiratory pigment" (gas carrier) contains iron associated with a different porphyrin to that of hemoglobin. This pigment is *red* in concentrated solution and *green* in dilute solution and is called *chlorocruorin* (H. Munro Fox, 1926) and found dissolved in the plasma.

In sipunculids, in particular, there is another pigment which contains 0·18 per cent iron but *no porphyrin*. This is called *hemerythrin*. It is characterized by a low cysteine content (Holleman and Biserte, 1958); the pigment from *Sipunculus* has a molecular weight of 66,000 (Roche and Roche, 1935).

In cephalopod molluscs, certain gasteropods and *Limulus*, the pigment dissolved in plasma is a protein containing copper but no porphyrin: *hemocyanin*. The molecular weight of hemocyanin from *Helix pomatia* is 6,700,000 as estimated by measuring the sedimentation constant using the ultracentrifugation technique of Svedberg; at pH 6 this molecule is cylin-

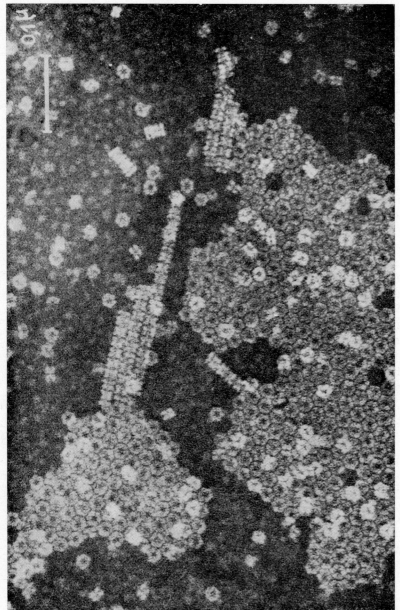

FIG. 94. Hemoglobin molecules of *Arenicola marina* L. obtained from circulating blood (technique of negative staining). The hexagonal form and the arrangement of the six sub-units are visible. (ROCHE, BESSIS, BRETON-GORIUS and STRALIN, 1961.)

21

drical with a diameter of 300 Å and a height of 335 Å (van Bruggen *et al.*, 1962).

Finally, a brown pigment containing *manganese* is found in body fluids of the pelecypod mollusc *Pinna: pinnaglobin*. A pigment containing *vanadium* is present in the body fluid of certain ascidians (Henze, Boeri).

(I) Basic respiration

This is a process which takes place at the *molecular level. The name respiration should be given to the process of acceptance of O_2 and production of H_2O and CO_2 because of its fundamental nature and because it takes place in all cells and tissues of aerobic organisms.*

Water is formed by the combination of O_2 with hydrogen—a detoxication mechanism. CO_2 is not formed by the direct combination of O_2 with carbon but arises principally from the *oxidative decarboxylation* of keto acids. As we shall see, this last reaction is also a method of detoxication. The respiratory phenomenon is expressed by a certain number of *oxidation–reduction* reactions.

Gradual oxidation of metabolites is the principal mechanism for the release of energy. Energy transfers imply oxidation–reduction systems, each one being characterized by an electrical potential. We have seen (p. 9) that a system with a high potential will oxidize a system having a lower potential, resulting in release of energy.

Respiration involves dehydrogenation and the transfer of electrons along a series of molecular systems leading to molecular oxygen which is the hydrogen acceptor in aerobic organisms. In anaerobic organism, hydrogen is accepted by an organic molecule with a suitable E_0' value.

Origin of respiratory water

Dehydrogenases (succinic acid dehydrogenase, lactic acid dehydrogenase, etc.) acting *specifically* on a substrate oxidize these substrates by reacting with mobile hydrogen. In the less numerous cases of aerobic dehydrogenases (L-and D-amino acid oxidases) oxygen is the direct acceptor of liberated hydrogen but most dehydrogenases are anaerobic and intermediate systems couple the dehydrogenases to molecular oxygen. The main intermediate system in respiration consists of the:

Cytochrome system

MacMunn using spectroscopy discovered the presence in muscles and other tissues of compounds which he termed "histohematins" or "myohematins" (1880), but is was not until 1925 that Keilin demonstrated their fundamental significance and 1939 that Theorell obtained a pure preparation of the red pigment, cyctochrome c.

$$\text{Val-Glu-Lys-Cy-Ala-Glu-Cy-His-Th-Val-Glu}$$

$$\begin{array}{ccc} | & & | \\ S & Fe & S \\ | & & | \\ \end{array}$$

$$\llcorner\text{porphyrin}\lrcorner$$

Basic sequence of cytochrome c from ox showing the chain of amino acids (Paléus, 1955)

Keilin (1929) recognized that the cytochrome pigments were associated with particulate material in the cell; today one considers that the enzymes involved in biological oxidations i.e. dehydrogenases (Barnett and Palade, 1957), the cytochrome pigments a, a_3, b and c, are associated with the membranes of mitochondria (Siekevitz and Watson, 1956).

Four types of cytochromes are known. For example, in the heart one finds:

cytochrome a, characterized by absorption bands at 604 mμ and 452 mμ (1 mμ = 1 nm);

cytochrome b, characterized by absorption bands at 564, 530 and 432 mμ;

cytochrome c, characterized by absorption bands at 550, 521 and 415 mμ;

cytochrome a_3, characterized by absorption bands at 600 and 448 mμ and identified with *cytochrome-oxidase*.

Cytochrome c, a heme compound (thus containing iron with variable valency) is widely distributed in aerobic tissues and is the basic factor in respiratory electron transport. Cytochrome c is thermostable. Biosynthesis of cytochrome c is known to be genetically controlled (Ephrussi and Slonimski). *Saccharomyces cerevisiae*, a single haploid cell, synthesizes two distinct forms of cytochrome c (iso-1 and iso-2). This biosynthesis is controlled by several loci which are genetically unrelated (Slonimski and his co-workers, 1964). *Explicative physiology is molecular and its basis is genetics.*

Cytochrome c contains a maximum of 0·46 per cent iron (Paléus) and has an isoelectric point of 10·65. *It does not combine with carbon monoxide, cyanide* or oxygen. At pH 7 the E_0' of cytochrome c is 0·25 V.

Cytochrome a_3 (oxidase) is responsible for the oxidation of cytochrome c by O_2; it contains copper (Cohen and Elvehjem, 1934; Eichel *et al.*, 1950; Okunuki *et al.*, 1957).

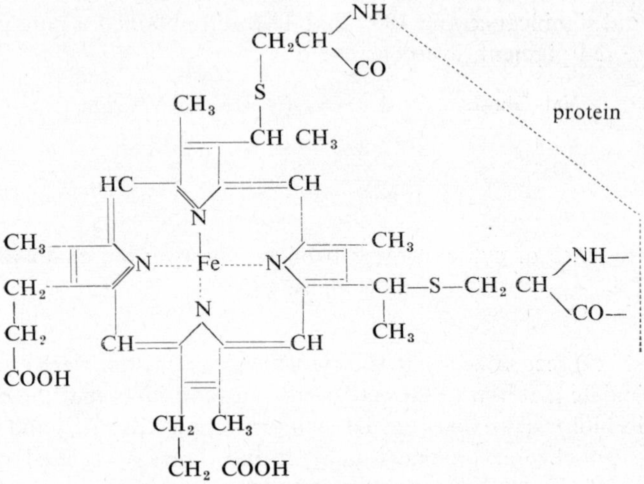

Prosthetic fraction of *cytochrome c*

Cytochrome b (of which there are several types) it thermolabile (as is a_3) and autoxidizable; it does not react with carbon monoxide or cyanide. The E_0' of cytochrome b at pH 7 is 0·04 V.

Cytochrome a differs from a_3 in that the former does not combine with O_2 or CO while a_3 is very sensitive to CO and combines with CO to form a compound with absorption maxima at 589 and 430 mμ. This combination can easily be detected spectrophotometrically.

Warburg (1927) found that cellular respiration is inhibited by CO and that this inhibition can be reversed by illumination. Thus, the a_3-CO complex is destroyed by light *(photolysis)*.

The amount of oxygen consumed by mammalian tissues is *directly proportional* to the concentration of cytochrome c in these tissues.

The fundamental role of the cytochromes in apparently distinct processes is presumed or verified in many conditions of pharmacological interest:

(1) while examining the tetanic responses induced by guanidine derivatives, Banu and Gavrislescu (1933) found that these derivatives considerably reduced the respiration of various rat tissues (notably muscle). An explanation of these phenomena must be found at the level of electron transport systems;

(2) during a study of the effects of *narcotics* (such as phenylurethane, chloral hydrate, morphine, etc.) on the adductor muscle of the Australian oyster *Saxostria commercialis*, G. F. Humphrey (1948) demonstrated that endogenous oxygen consumption is reduced and he concluded that this resulted from an action on the cytochromes;

(3) *potassium cyanide*, the major respiratory poison, inhibits cytochrome a_3. This fact is important not only from the pharmacological point

of view but also from the point of view of advancing our understanding of respiratory phenomena. In fact, *cyanide does not completely inhibit cellular respiration* and this can be explained in two ways:

(i) either this residual respiratory activity is due to cytochrome b (which is insensitive to CO and cyanide and the activity of which is independent of cyanide-sensitive a_3);

(ii) or this residual activity is due to aerobic dehydrogenases which can transport hydrogen to oxygen without passing along the cytochrome system. These reactions involve flavoproteins—which have riboflavin as a prosthetic group (cf. p. 67.) The processes of flavin respiration can be presented in the following way:

$$\begin{array}{l} RH_2 \quad + \quad NAD \leftarrow\! \text{---------} \quad NAD \quad + \quad \text{reduced flavoprotein} \\ \text{substrate} \quad \Big| \quad \text{specific} \qquad\qquad\qquad\qquad\qquad \searrow \\ \qquad\qquad \Big| \quad \text{dehydrogenase} \qquad\qquad\qquad \cdot \qquad\qquad +O_2 \\ \qquad\qquad \downarrow \qquad\qquad\qquad\qquad\qquad\qquad\qquad\qquad \nearrow \\ R \quad + \quad NADH_2 \text{---------}\rightarrow \quad + \quad \text{oxidized flavoprotein} \end{array}$$

These flavoproteins, colourless in the reduced form and yellow in the oxidized form and having an absorption band at 450–470 mμ, are widely distributed in animal tissues and are situated in the electron transport system of normal cells between the cytochromes and NAD (CoI) and NADP (CoII) co-dehydrogenases. Thus the *yellow enzyme* of Warburg catalyses the reoxidation of NADP and the reduced form of this substance can be directly but slowly reoxidized by molecular oxygen. Another flavoprotein reduces cytochrome c: *cytochrome reductase* which accepts hydrogen only from NADP. Table 43 gives the approximate values for the *free energy* of several respiratory agents:

TABLE 43

	$-\Delta F$(cal)
NAD	4000
flavoprotein	15000
cytochrome c	31000
$^1/_2O_2$	56000

The passage of electrons from substrate (up stream) to oxygen (down stream) is determined by a decrease in free energy and the ΔF values determine the direction of the reactions. Generally, electron transport to oxygen is indirect and occurs principally through combination of the substrate with a coenzyme:

$$RH_2 + CoE \rightarrow R + CoEH_2 \tag{76}$$

Each apoenzyme is specific for a substrate and coenzyme (CoE and R are temporarily linked to the apoenzyme at the moment of the reaction).

To summarize these points let us consider the transport of electrons from succinate to molecular oxygen:

succinate $\longrightarrow$ succinic acid dehydrogenase $\longrightarrow$ flavoproteins
(+ Co-dehydrogenase = CoI)

$$H_2O \leftarrow \tfrac{1}{2}O_2 + \xleftarrow{\;e(H)\;} \text{cytochrome } a_3 \xleftarrow{\;e\;} \text{cytochrome } c \xleftarrow{\;e\;} \text{cytochrome } b \leftarrow$$

The oxidation of fatty acids is more complex and requires the *flavoprotein electron carrier* of Green (1956).

It would now appear to be well established that electron transport can proceed not only by oxidation of succinate or from $NADH_2$ by cytochrome c followed by oxidation of reduced cytochrome c but also via a pathway involving a benzoquinone system. In 1957, F.L. Crane (in America) and R. A. Morton (in England) isolated from liver mitochondria a liposoluble compound characterized by a maximum absorption band at 272 mμ. This compound is *ubiquinone* or *coenzyme* Q and resembles α-tocopherylquinone and vitamin K (*n* in the formula of ubiquinone indicates that there are several forms of coenzyme Q; *n* can be 5, 6, 7, 8 or 9).

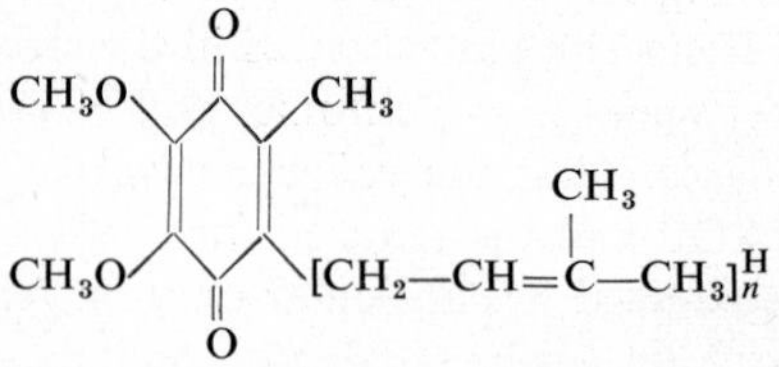

Ubiquinone

According to Green, coenzyme Q is reduced by succinate or $NADH_2$ and reduced coenzyme Q is oxidized by cytochrome c. In the electron transport chain coenzyme Q is situated between the flavoproteins and cytochrome c. These facts provide more evidence for the role of lipoproteins in cellular metabolism.

Dehydrogenases have been remarkably well studied by Thunberg. Generally *methylene blue* (with an E'_0 of 0·01 V at pH 7) is provided as the hydrogen acceptor (in place of O_2) in experimental work. In the presence of numerous tissues this dye is reduced and becomes colourless due to the formation of *leuco-methylene blue*. After washing, these tissues lose their ability to reduce methylene blue but this can be restored by adding various oxidizable substrates (such as succinate, lactate) which donate hydrogen in the presence of specific dehydrogenases.

Oxidative phosphorylation

When a substrate is oxidized by dehydrogenation the energy released can either be stored in the form of high energy chemical bonds, generally phosphate bonds, or can be liberated in the form of work and heat, the latter

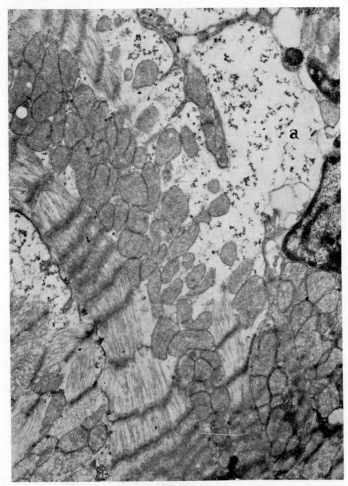

FIG. 95. Section of the ventricle of the normal heart of *Rana esculenta. a*, cytoplasm showing glycogen granules. Normal mitochondria accumulated near the myofibrils. (RYBAK and RUSKA.)

being a degraded form of energy. The liberation of energy from storage forms will appear sooner or later in any case. This energy store ("memory") in the form of high energy phosphate bonds is formed by one bond (or about 12,000 cal) for each transfer of electrons. The coupling of oxidation with phosphorylation to some extent economizes the energy from metabolites. Three atoms of phosphorus are incorporated into high energy bonds ($\sim$P) for each atom of oxygen utilized. However, when electrons are transferred from the substrate to coenzyme I four high energy phosphate bonds (or 48000 cal) are stored per molecule of substrate oxidized. This is the case

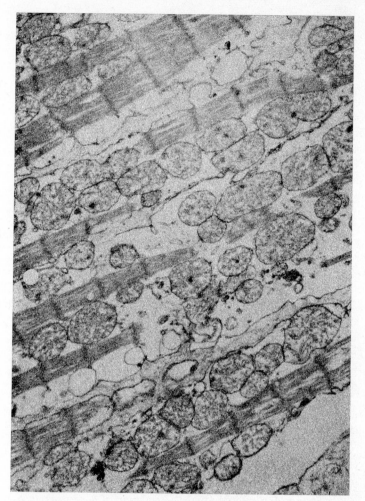

Fig. 96. Section of the ventricle of the heart of *Rana esculenta* rendered electromechanically inert after 17 h of contractions under tension. The emptiness of the cytoplasm is apparent as well as the swelling of mitochondria and changes in the membranes. (Rybak and Ruska.)

in the Krebs cycle of the oxidative transformation of α-ketoglutaric acid to succinic acid. Thus, transport via NAD is an advantageous route for conservation of cellular energy. The conditions for this pathway have been described. An enzyme, *diaphorase*, accepts hydrogen from NAD but cannot reduce cytochrome c. This reduction is carried out by $NADH_2$-cytochrome c reductase which contains iron. The mitochondria are the centres for oxidative phosphorylation (Green, Lipmann). The mitochondria of muscle (sarcosomes) are very sensitive to any change in the function of the cells with which they are associated. This has been demonstrated for frog heart *in situ* (B. Rybak, H. Ruska and C. Ruska, 1964) as shown in Figs. 95 and 96.

The abundance of cristae within the mitochondria of fresh ventricle is related to the important mechanical and electrical work which the ventricle normally performs and also to the importance of oxidative phosphorylation (Siekevitz and Porter, 1955; Palade, 1956). The morphological changes in sarcosomes during ventricular senescence of the frog heart correspond to a decrease in respiratory metabolism (estimated polarographically); this respiratory failure occurring at the same time as glycogen exhaustion. These phenomena must depend on the diappearance of adenosine triphosphate (Lehninger, 1959). Ageing of mitochondria *in situ* might correspond to the ageing of isolated mitochrondria which is characterized by uncoupling of oxidative phosphorylation (Kielley and Kielley, 1951). In addition, we have described a reversible experimental change in the mitochondria of the branchial heart of Myxine during treatment with 2-deoxy-D-glucose (Rybak, Hoffmeister and Ruska, 1962). This change was characterized by an alteration in the shape of mitochondria and by a stumping of cristae related to inhibition of glycogen degradation. Therefore, one must ask whether, under these conditions, the swelling of cardiac mitochrondria *in situ* is an expression of energy exhaustion while the change in the appearance of cristae is associated with energy block.

Our results extend those of Hoffmeister (1961) who found that after prolonged forced flight the alary muscles of *Vespa germanica* show submicroscopical alterations characterized by disruption and even swelling of mitochondria. Mitochondrial swelling in rat is associated with a poor diet (Rouiller, 1960). Thus, *in vivo* the mitochondria are structurally very sensitive to abnormal metabolic conditions and this is apparent particularly during the state of fatigue. These considerations would provide a new way for formulating muscle fatigue which would have the advantage of being related to ultrastructure and being available for direct experimentation.

Certain compounds, such as 2,4-dinitrophenol, disrupt the energy storing process of oxidative phosphorylation and as a result, heat is liberated (uncoupling of oxidative phosphorylation). This can lead to serious disturbances in cellular economy and could restrict "la vie libre" and thus participate in the determination of a parasitic existence.

In order to understand the main processes of oxidative phosphorylation, it is necessary firstly to consider the nature of the phosphorylated compounds involved. Lardy and Wellman (1952) have stated that respiration cannot proceed normally in the absence of the elements of oxidative phosphorylation. With a hexose the reaction is expressed by the following equation:

$$C_6H_{12}O_6 + 6 O_2 + 38 ADP + 38 H_3PO_4 \rightarrow 6 CO_2 + 44 H_2O + 38 ATP \qquad (77)$$

such that ADP which is present in cells in small quantities is the limiting factor in respiration. During cell activity (particularly a muscle cell) ATP is hydrolysed:

$$38 \text{ ATP} + 38 \text{ H}_2\text{O} \rightarrow 38 \text{ ADP} + 38 \text{ H}_3\text{PO}_4 + \text{work-heat} \tag{78}$$

It is remarkable that the *Pasteur effect*—which is inhibition of glycolysis by respiration—can be counterbalanced by the *Crabtree effect*—which is inhibition of respiration by glycolysis. In both reactions there is a competition for ADP and Pi. Thus, if production of ATP by the respiratory process is abolished by the addition of dinitrophenol, the glycolysis rate will increase. Similarly, agents which inhibit glycolysis, such as iodoacetate, will interfere.

Therefore, the first point is that *ADP stimulates respiration* (Lardy; B. Chance; Slater). This phenomenon is particularly well demonstrated by the spectrophotometric technique of Britton Chance which is based on the fluorescence of reduced pyridine nucleotide derivatives (NAD derivatives). The apparatus is a differential spectrofluorimeter (double beam), the fluorescence being excited by light with a wavelength of 366 mμ. Reduction of NAD derivatives in mitochondria is responsible for more than half of the increase in fluorescence observed during anoxia. Nitrogen, sulphide, cyanide and carbon monoxide cause increases in fluorescence to very nearly the same levels.

Then it is in fact the ratio ADP/ATP.Pi (phosphate-potential buffers) which has to be taken in consideration (Klingenberg). Coupling factors are necessary for oxidative phosphorylation in animal tissues. The role of various coupling agents in the assembly of structural units capable of oxidative phosphorylation was studied using the electron microscope (E. Racker; D. F. Parsons and B. Chance); projection units of about 90 A in diameter were noted most frequently in submitochondrial particles that were capable of oxidative phosphorylation. Particles that were depleted of coupling factors by different procedures showed a marked decrease in the number of projection units.

It is possible that mitochondrial phosphoproteins act as intermediates in oxidative phosphorylation by conversion of ADP to ATP with the aid of P $\sim$ proteins (Rabinowitz and Lipmann; Judah).

The oxidation–reduction processes of oxidative phosphorylation are associated with micro-ions. During respiration the K^+ content of mitochondria is maintained but if electron transport is inhibited, K^+ escapes from mitochondria (Bartley; Davies and Gamble). Moreover, isolated respiring mitochondria can accumulate large quantities of Ca^{++} (Vasington and Murphy) and inorganic phosphate—(P_i)—(C. S. Rossi). According to Lehninger (1964) the molar ratio of Ca^{++} and P_i, accumulated in mitochondria, is about 1·67 and equivalent to the ratio in hydroxy-apatite.

Finally, electrons can pass along the respiratory chain in the *reverse* direction when the ratio of (ATP) to (ADP)(P$_i$) is very high (Chance; Klingenberg; Ernster). Lehninger has stated that reversal of the whole respiratory chain is not necessarily a physiological process but the elements of the above ratio have a fundamental role in control of total respiration.

Origin of respiratory carbon dioxide

Carbon dioxide is derived from *oxidative decarboxylations*. Amino acids such as histidine ($\rightarrow$ histamine) and tyrosine ($\rightarrow$ tyramine) are of great pharmacological interest but quantitatively they have little significance as the amount of CO_2 produced from them is negligible.

One important source of CO_2 is the *decarboxylation of dicarboxylic acids* of the Krebs cycle.

$$
\begin{array}{llll}
\text{COOH} & \text{COOH} & \text{COOH} & \text{COOH} \\
| & | & | & | \\
\text{CH}_2 \ \pm 2H & \text{CH} \ \pm H_2O & \text{CH}_2 \ \pm 2H & \text{CH}_2 \\
| \ \underrightarrow{\ \ \ \ } & | \ \underrightarrow{\ \ \ \ } & | \ \underrightarrow{\ \ \ \ } & | \\
\text{CH}_2 & \text{CH} & \text{CHOH} & \text{CO} \\
| & | & | & | \\
\text{COOH} & \text{COOH} & \text{COOH} & \text{COOH} \\
\text{Succinic} & \text{Fumaric} & \text{Malic} & \text{Oxaloacetic} \\
\text{acid} & \text{acid} & \text{acid} & \text{acid}
\end{array}
\tag{79}
$$

For example, malate decarboxylase which is abundant in pigeon liver, simultaneously catalyses the oxidation and decarboxylation of malic acid in the presence of NADP and Mn^{++} (Ochoa):

$$
\begin{array}{l}
\text{COOH} \\
| \\
\text{CH}_2 \quad + \text{NADP} \quad \underset{\longleftarrow}{\overset{Mn^{++}}{\longrightarrow}} \quad
\begin{array}{l} \text{CH}_3 \\ | \\ \text{CO} \\ | \\ \text{COOH} \end{array}
\ + \ \boxed{CO_2} \ + \text{NADP·H}_2 \\
\text{CHOH} \\
| \\
\text{COOH}
\end{array}
\tag{80}
$$

Glyoxylic acid is an inhibitor of respiration (Kleinzeller, 1943). Ruffo has shown this action to be at the level of aconitase (Krebs cycle) and due to the formation during incubation of an inhibitor, from glyoxylic acid and oxaloacetic acid, such that the rate of the Krebs cycle can be regulated by glyoxylic acid.

However, the most general process producing respiratory CO_2 is *oxidative decarboxylation of α-keto acids*, an irreversible process in animal tissues which can be inhibited by arsenites.

For example, with pyruvic acid:
$$
CH_3 \cdot CO \cdot COOH + \tfrac{1}{2}O_2 \rightarrow CH_3COOH + CO_2
\tag{81}
$$

or more exactly:
$$
CH_3 \cdot CO \cdot COOH + H_2O \xrightarrow{\ -2H\ } CH_3COOH + CO_2
\tag{82}
$$

which expresses this reaction in terms of a detoxication process with reference to the physiology of hydrogen (cf. p. 11).

Pigeon brain can metabolize glucose to the stage of pyruvic acid but for
the reaction to proceed further it is necessary to provide cocarboxylase or
thiamine diphosphate (Peters) which explains the fundamental importance
of vitamin B_1 in oxidative decarboxylation. In animal tissues, decarboxylation precedes oxidation. The complete reaction can be written as follows:

$$CH_3CO\ COOH + NAD + CoA \xrightarrow{\quad Mg^{++} \quad} CH_3CO-CoA + NADH_2 + CO_2 \quad (83)$$

Acetyl-CoA is a necessary intermediate in oxidative decarboxylation of
pyruvate.

In general terms, A. Santos-Ruiz (1962) considers that photosynthesis is
a process of reductive carboxylation with rupture of the water molecule and
production of carbohydrates, lipids and proteins while among heterotrophic creatures, the reverse process predominates, i.e. oxidative phosphorylation with liberation of carbon dioxide and water.

(II) Physical chemistry of respiration in mammals

The total mass of the atmosphere is estimated to be $5 \cdot 13 \times 10^{21}$ g. In the
biosphere, "pure" atmospheric air has the following composition by vol-
ume: O_2 $20 \cdot 96$ per cent, CO_2 $0 \cdot 04$ per cent, N_2 79 per cent with traces of rare
gases (Ramsay) and a variable percentage of water vapour.

In towns and biotopes contaminated by the presence of industries, the air
contains variable but significant amounts of CO, SO_3, etc.

In a mixture of gases each gas exerts its own pressure. Thus, the *partial
pressure* (or *tension*) of O_2 at zero altitude represents about 21 per cent of
the total pressure of 760 mm Hg (or *torr*) of dry air, or

$$p_{O_2} = 760 \times 0 \cdot 21 = 159 \cdot 60 \text{ mm Hg} \quad (84)$$

In the pulmonary alveoli the total gaseous pressure — after having subtracted the partial pressure of water vapour (47 mm Hg at 37°C) — is

$$760 - 47 = 713 \text{ mm Hg} \quad (85)$$

The alveolar oxygen content is 14 per cent, thus the partial pressure of oxygen in the lung is:

$$713 \times 0 \cdot 14 = 99 \cdot 82 \text{ mm Hg (approximately 100 mm Hg)} \quad (86)$$

Since the CO_2 content of alveolar air is 5 per cent

$$p_{CO_2} = 713 \times 0 \cdot 05 = 35 \cdot 65 \text{ mm Hg (approximately 36 mm Hg)} \quad (87)$$

Diffusion of gas in lungs

The oxygen in inspired air passes into the blood by diffusion across the humid alveolar membrane. This membrane is the site of pulmonary hematosis. In order to understand the mechanism of hematosis the following points should be considered:

The p_{CO_2} in blood is the pressure (in mm Hg) of dry gas in equilibrium with CO_2 dissolved in blood and similarly for p_{O_2}.

Thus the oxygen tension of alveolar air is 100 mm Hg and in venous blood it is $40-50$ mm Hg.

The difference in pressure of 50–60 mm Hg [100–(40–50)] allows the passage of O_2 from the pulmonary alveoli to venous blood which is thus oxygenated.

The p_{CO_2} in alveolar air is 36 mm Hg and in venous blood approximately 46 mm Hg. This apparently small difference in pressure, about 10 mm Hg, is sufficient for CO_2 to pass from blood into the alveoli, CO_2 diffusing easily across the lung membrane and particularly across the alveolar membrane. A difference in pressure as small as 0·12 mm Hg is sufficient to ensure extraction of venous CO_2.

Under normal atmospheric pressure, the pressure of nitrogen is the same in blood as in the pulmonary alveoli (570 mm Hg).

After gaseous exchange in the lungs the blood becomes arterial from the chemical point of view. Arterial blood has an O_2 tension of 100 mm Hg and a CO_2 tension of 40 mm Hg.

Hemoglobin

With respect to the needs of the mammalian organism, oxygen is relatively insoluble in plasma (about 3 ml per mil in arterial blood) and it is transported by respiratory pigments which are said to have *oxygen-carrying ability* and thus differ from cytochromes which transport electrons.

In mammals this ability is shown by a heme pigment, *hemoglobin*, with a molecular weight of $67,000-68,000$. Hemoglobins (Heidelberger and Landsteiner, 1923)—or precisely globins (Gay and Robertson, 1913)—and hemocyanins (Hooker and Boyd, 1936) show species specificity which can be detected by serological techniques (precipitins). The cytochromes are of primary importance in *electron* transport and pigments such as hemoglobin are of primary importance in transport of *gases*, O_2 and CO_2.

Human blood normally contains about 150 g of hemoglobin per l, associated with the erythrocytes.

The protein of hemoglobin—*globin*—is a basic protein. The prosthetic group—*heme*—is an iron-containing porphyrin, ferrous or *ferroporphyrin*.

Hemoglobin (Hb) can be obtained in a crystalline form.

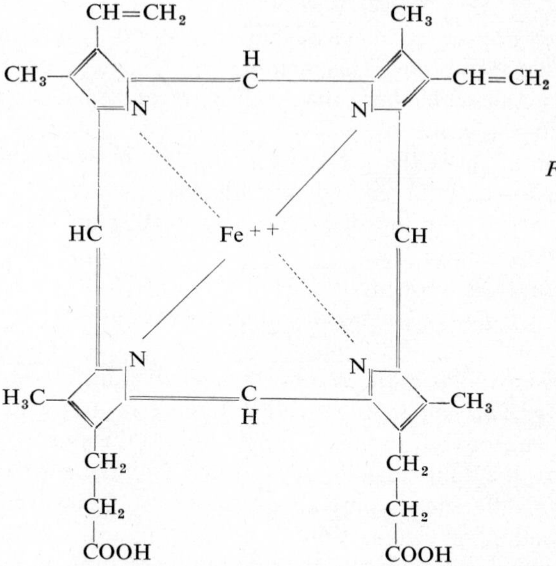

$$CH{=}CH_2 \qquad\qquad CH_3$$

$$CH_3 \qquad\qquad \overset{H}{C} \qquad\qquad CH{=}CH_2$$

Formula of heme

$$HC \qquad Fe^{++} \qquad CH$$

$$H_3C \qquad\qquad \underset{H}{C} \qquad\qquad CH_3$$

$$CH_2 \qquad\qquad CH_2$$

$$CH_2 \qquad\qquad CH_2$$

$$COOH \qquad\qquad COOH$$

The hemoglobin of the most primitive vertebrates at present in existence, the myxinoids, has a molecular weight of the order of 17,000 (Hearst, 1960) which corresponds to the molecular weight of muscle hemoglobin—*myoglobin*. It is of note that myoglobin has a greater affinity for oxygen than mammalian hemoglobin and as indicated by Manwell (1963), it appears that myxinoid hemoglobins are characterized by moderate to very high oxygen affinities. The molecule of myoglobin possesses one atom of iron, one heme molecule and a single peptide chain with 153 animo acid residues which facilitates the study of myoglobin by X-rays (diffraction patterns). Kendrew and his co-workers (1960) have established a three-dimensional model for myoglobin; the same has been done by Perutz *et al.* for hemoglobins (Fig. 97 and 98).

The molecule of hemoglobin is formed by 4 peptide chains joined in two pairs. These chains are different in foetal hemoglobin (F) and adult hemoglobin (A: the principal fraction and A_2: about 2 per cent). These chains, designated by α, β, γ and δ, are distributed such that hemoglobin F contains $\alpha_2\,\gamma_2$, hemoglobin A contains $\alpha_2\beta_2$ and hemoglobin A_2 contains $\alpha_2\delta_2$. (All three hemoglobins have chain α which consists of 141 amino acids while chain β has 146 amino acids). Thus $\alpha_2\beta_2$ consists of 72 alanine residues, 12 arginine, 50 asparagine, 6 cysteine, 32 glutamine, 40 glycine, 38 histidine, 72 leucine, 44 lysine, 6 methionine, 30 phenylalanine, 28 proline, 32 serine, 32 threonine, 6 tryptophan, 12 tyrosine, 62 valine.

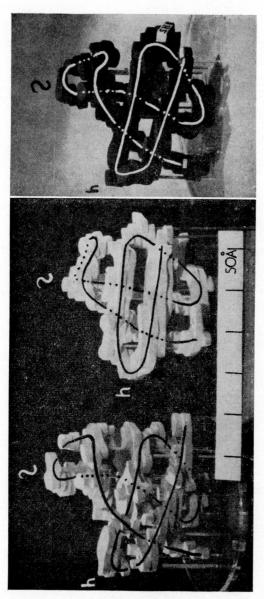

FIG. 97A. Two different polypeptide chains in the asymmetric unit of hemoglobin (left.) The heme groups are at the back of the chains. (PERUTZ, et al., 1960.)

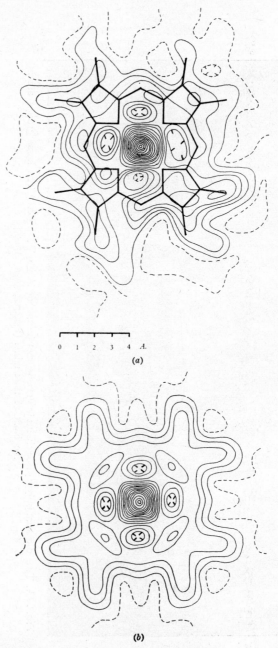

0   1   2   3   4   *A.*

(*a*)

(*b*)

Fig. 97B. (a) Observed electron density distribution in the plane of the heme group, with the atomic arrangement in the group superposed. One contour interval = 0·5 electron/Å. (b) Calculated electron density distribution, computed from the known atomic arrangement. One contour interval = 0·5 electron/Å. (Perutz, *et al.*, 1960. *Nature*.)

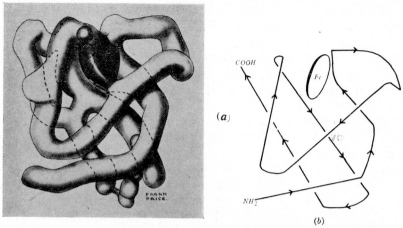

FIG. 98A. (a) Drawing of the tertiary structure of myoglobin as deduced from the 6-Å Fourier synthesis. (b) The course of the polypeptide chain, deduced from 2-Å Fourier synthesis. (KENDREW, *et al.*, 1960, *Nature*.)

FIG. 98B. Partially assembled molecule showing two black chains and one white. (PERUTZ, ROSSMANN, CULLIS, MUIRHEAD WILL & NORTH, 1960, *Nature*.)

Chain α of hemoglobin of gorilla differs from the α chain of human hemoglobin principally in that aspartic acid replaces glutamic acid (Zuckerkandl and Schroeder, 1961).

Each hemoglobin molecule contains 4 heme groups, each capable of combining reversibly with molecular oxygen. However, the 4 iron atoms are not independent and the equilibrium constant of any one of them is influenced by the state of oxygenation of the 3 others. As a result the shape of the oxygen dissociation curve is sigmoid rather than hyperbolic (Adair). However, myoglobin—a single polypeptide chain—shows a hyperbolic oxygen dissociation curve; its Bohr effect (cf. p. 334) is null or negligible while that of hemoglobin is well marked. It appears that oxygenation of hemoglobin leads to a change in its structure (Muirhead and Perutz, 1963).

Fixation of oxygen on the hemoglobin is in some way a function of the iron atom while that of CO_2 is a function of a NH_2 group (cf. p. 337).

If hemoglobin is acidified the two pairs of chains separate and are rejoined in neutral pH. If for example, a mixture of human hemoglobin and canine hemoglobin is acidified, the two pairs of specifically distinct chains separate and recombine haphazardly in neutral solution. In this way one obtains 25 per cent human hemoglobin, 25 per cent canine hemoglobin and 50 per cent hybrid hemoglobin, half of which consists of human α chain and canine β chain and half of the reverse arrangement (Itano and Robertson, 1959). This technique which allows an analysis of abnormal hemoglobins is to some extent similar to that described for the dissociation of the complexes (streptomycin–ribonucleic acid)–(streptomycin–cephalin) and their recombination under the influence of heat and recooling (B. Rybak and F. Gros, 1947).

During gestation maternal blood and foetal blood do not mix. Schapira incubated blood from the human umbilical cord with ^{59}Fe and ^{14}C-glycine and found this blood to be heterogeneous, the specific activity of iron and of the heme content and of glycine isolated from globin being higher for the adult fraction than for the foetal fraction.

The biosynthesis of hemoglobin is controlled genetically and abnormal synthesis can occur:

4 structural genes are responsible for the synthesis of the 4 polypeptide chains of hemoglobin and mutation of one of these leads to the appearance of an abnormal hemoglobin. Moreover, regulating genes control the extent of synthesis of a given chain and other genes control the transformation of hemoglobin F to hemoglobin A; finally, the condensation of the two pairs of chains would be controlled genetically.

The biochemical technique of V.M. Ingram has provided a method for the identification of hemoglobins: trypsin splits the peptide chain in the region of arginine–lysine linkages such that 30 fractions are obtained from hemoglobin. These are subjected to one dimensional electrophoresis on paper followed by chromatography in the other direction. In this way characteristic "finger prints" are obtained. This method has been of value in the analysis of the disease called Mediterranean anaemia or thalassemia in which hemoglobin F persists until an advanced age and in which there are also large amounts of hemoglobin A_2.

The work of Linus Pauling forms the basis of these studies.

According to the simplified concept "one gene → one protein", one can see that biochemical genetics—which is a fundamental aspect of physiology—has been able to develop rapidly. Thorell demonstrated in 1947 that hemoglobinopoiesis starts at the stage of the basophilic erythroblast when the level of cytoplasmic ribose polynucleotides decreases by half compared to the level in the preceeding type of cell (using the spectrophotometric techniques of Casperson).

The *acellular biosynthesis* of hemoglobin, or more exactly of globin, is an example of the biogenesis of proteins which we examined on p. 135. This biosynthesis was realized by Kruh, Rosa, Dreyfus and Schapira (1961) and by Erenstein and Lipmann (1961) using rabbit reticulocytes.

The average life of a human erythrocyte is about 120 days. When cells are broken down the porphyrin fraction of hemoglobin is catabolized and forms bile pigments (cf. p. 105). Shemin and Rittenberg (1946) and later Schapira and his co-workers (1954) have shown that hemoglobin is not renewed until death of the red corpuscle. These experiments were carried out using ^{15}N-labelled glycine or [59]Fe. This fact provides an interesting exception to the "turnover" rule for biological molecules which originated with the work of Schoenheimer (1939).

Transformation of hemoglobin

(1) When blood is treated with substances such as ozone, $KMnO_4$, nitrobenzene, potassium ferricyanide, etc., the ferrous iron (Fe^{++}) of hemoglobin passes to the ferric state (Fe^{+++}) and *methemoglobin* is formed, showing an absorption band in acid solution at 635 mμ. In percutaneous poisoning by certain liniments containing nitrobenzene (as in the case of an infant reported by J. C. Dreyfus in 1955), the clinical symptoms are essentially those of brown cyanosis, localized mainly on the lips and ears. The symptoms disappear when the use of nitrobenzene is prevented. Certain cases of rain water poisoning have been associated with methemoglobinemia (this gives an indication of the possible dangers of atmospheric pollution) but very rarely cases of genetic methemoglobinemia occur.

(2) Carbon monoxide combines *more readily* with hemoglobin than does oxygen and forms carboxyhemoglobin showing two absorption bands at 570 mμ and 542 mμ. Sulphemoglobin is obtained with H_2S and nitrohemoglobin with NO.

22*

(3) Hemoglobin is rapidily denatured in alkaline media:
Hemoglobin + alkali → denatured globin + hematin (= heme with ferric iron or ferriporphyrin).

Heme and hematin have a strong affinity for nitrogenous bases(NH_4OH, nicotine, etc.) and form respectively *hemochromogens* and *parahematins* which are coloured and have a characteristic spectrum which is of value in the detection of small quantities in blood.

(4) One of the most characteristic properties of hemoglobin is the ability to combine with oxygen to give oxyhemoglobin showing three absorption bands at 578 mμ, 542 mμ and 415 mμ. Reduced hemoglobin has an absorption band at 559 mμ.

$$Hb + O_2 \rightleftharpoons HbO_2$$
$$\text{reduced} \qquad \text{oxyhemoglobin}$$

Boeri and Vescia (1949) stated that since there is total reversibility in this reaction there is no hysteresis effect.

The quantity of O_2 associated with Hb is determined by the fact that one atom of iron can combine with one molecule of oxygen. Thus, the *respiratory capacity* of the blood can be defined.

1 Fe combines with 1 O_2 or 56 g Fe combines with 22·4 1 O_2.

Now 1 g Hb contains 3·4 mg Fe and can combine with 1·34 cm³ of O_2. Since 100 cm³ of blood contains an average of 14·5 g of Hb, 100 cm³ of blood can combine with about 19·43 cm³ of O_2 (values for oxygen fixation by erythrocytes; however, the blood of man normally contains about 2 per cent methemoglobin compared with the total amount of hemoglobin and therefore, the oxygen carrying capacity of erythrocytes is lower). To the volume of O_2 combined with Hb must be added the 0·393 cm³ of O_2 dissolved in 100 cm³ arterial plasma at physiological temperatures. Thus, the total respiratory capacity of the blood is about 20 volumes per cent.

Factors involved in regulation of the equilibrium $Hb + O_2 \rightleftharpoons HbO_2$

In 1954 I stated, with reference to sperm motility, that *all living phenomena are dependent on four basic parameters* which represent the physico-

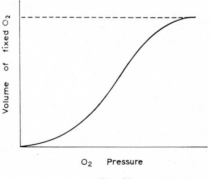

FIG. 99

chemical properties of the medium: rH_2, pH, θ (temperature) and μ (ionic strength). In the present case, these factors are p_{O_2}, p_{CO_2}, θ and μ.

(1) *The quantity of O_2 combined with Hb depends on the partial pressure of O_2 but is not proportional to the partial pressure,* as in the case of dissolved oxygen. *The curve is a sigmoid* (Fig. 99): *with increased pressure the fixation occurs more and more slowly* as the point of saturation is approached. The curve becomes asymptotic to the horizontal at a partial pressure of 100 mm Hg which is normally that of alveolar and arterial oxygen.

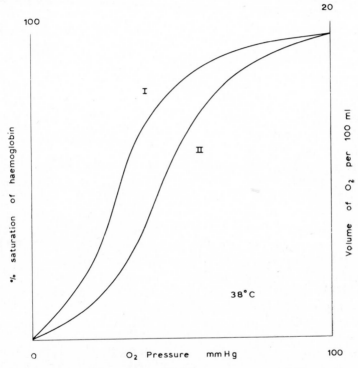

Fig. 100. (I) Arterial blood (40 mm Hg CO_2). (II) Venous blood (45 mm Hg CO_2).

Thus the equilibrium $Hb + O_2 \rightleftharpoons HbO_2$ *is conditioned by the partial pressure of oxygen.* Methods by which the oxygen carrying capacity of stored blood can be maintained for long periods have been sought. The addition of 3277 RP ("Phenergan") to whole human blood does not modify the decrease in oxygen carrying capacity during the first ten days but slows the rate of decrease during the following period (Arnould and Lamarche, 1952). Unfortunately, the administration of "Phenergan" or other antihistamine drugs lowers the resistance of experimental animals to oxygen deficiency (Arnould and Lamarche, 1952).

(2) Another factor determines the equilibrium $Hb + O_2 \rightleftharpoons HbO_2$; this is the *partial pressure of* CO_2. *In arterial blood* the partial pressure of CO_2 is approximately 40 mm Hg and that of O_2 is about 100 mm Hg which corresponds to about $19 \cdot 5$ cm³ of O_2 (96 per cent saturation of Hb). In *venous blood* (of the right heart) the partial pressure of CO_2 approaches 45 mm Hg and that of O_2 is about $40 - 45$ mm Hg which, for 40 torr, corresponds to $14 \cdot 8$ cm³ of O_2 (~ 70 per cent saturation of Hb).

The dissociation curve (Fig. 100) shows the variations in the percentage of oxygen combined with hemoglobin with increasing partial pressure of oxygen at two values of CO_2 tension.

Thus, *as the level of* CO_2 *increases the amount of* HbO_2 *decreases* (increased dissociation of oxyhemoglobin) and inversely, with less CO_2 there is more HbO_2 (reduced dissociation of oxyhemoglobin). The relationship which considers the diminution of the ratio of p_{O_2}/p_{CO_2} is called the BOHR EFFECT; this is an expression of the *acidification* induced by carbon dioxide leading to dissociation of HbO_2.

(3) The equilibrium $HbO_2 \rightarrow Hb + O_2$ is dependent on temperature: *an increase in temperature lowers the degree of saturation of hemoglobin:*

For an O_2 tension of 100 mm Hg, Hb is 93 per cent saturated at 38°C
98 per] cent saturated at 25°C

Thus, in poikilothermic animals (in a cool medium) the resistance to asphyxia is relatively great. This must also apply to hibernating mammals. It should be noted that with decreased temperature a greater amount of oxygen is dissolved in plasma. Haldane demonstrated that rats, in which the carrying capacity of hemoglobin is blocked by carbon monoxide, can survive if subjected to a p_{O_2} of 2 atm which maintains the oxygen content of plasma at a level sufficient to provide tissues with the detoxicating agent. According to Kiese and Bänder (1955), cytochrome oxidase is completely saturated at an oxygen tension of 7 mm Hg, but Lübbers found a lower value of $2 - 3$ torr and Chance less than $0 \cdot 5$ torr. This point has to be reinvestigated.

(4) The *electrolyte* level in plasma also modifies the equilibrium $HbO_2 \rightleftharpoons Hb + O_2$. At low oxygen tensions HbO_2 more easily liberates oxygen in the presence of electrolytes than in their absence. This fact is of great physiological significance in the tissue capillaries. The electrolyte content can be expressed in terms of ionic strength $\mu = \frac{1}{2} Mv^2$ where M is the molecular weight of salt and v is the valency.

The combination of oxygen with other pigments

No Bohr effect is observed in the sipunculid which contains hemerythrin. Portier has remarked that this fact relates to the mode of life of this marine organism which lives in sand containing organic debris where the

CO_2 tension is high; thus the behaviour of hemerythrin is determined in some way by CO_2.

Hemocyanin from the mollusc *Diodora aspera* shows no Bohr effect at pH values between 6·88 and 7·84. At 10°C, the pressure for 50 per cent saturation is 5 mm Hg of O_2. A slightly positive interaction is produced between the sites of combination during oxygenation. The heat of oxygenation determined by calculation is about $-12·6$ kcal/mol O_2 (Redmond, 1963).

Remarks: In the lamprey *(Petromyzon marinus)* the Bohr effect is markedly greater than in humans, and the hemoglobin shows a relatively weak affinity for oxygen. "The lamprey is the only vertebrate known to have a blood hemoglobin with a hyperbolic oxygen equilibrium curve" (Wald and Riggs, 1951). The myxinoid *Eptatretus stoutii* shows a very weak Bohr effect at physiological pH and the hemoglobin has a great affinity for oxygen (Manwell, 1963). This provides a good indication of the diversity of physiological evolution.

Portier (1938) reported that *Limulus,* called the king crab, even shows an inversed Bohr effect, that is to say, CO_2 favours oxygen fixation. It is to be considered here that the addition of oxygen causes a major alteration in the tertiary structure of hemoglobin (Perutz, 1962; Wyman and Allen, 1951). McMenamy (1964) emphasizes that for this alteration to change the pH of the medium, it must alter the electrostatic potential between the macromolecule and the solution; in the present instance, a reduction in electrostatic potential between the molecule and solution would be required for hemoglobin on oxygenation. Thus when hemoglobin carries a net negative charge (as in the alkaline region), the potential changes from the alteration though the addition of oxygen would result in a release of protons to the solution (positive Bohr effect); when hemoglobin is positively charged (as in the acid region) the potential change from the addition of oxygen would result in the uptake of protons from the solution (negative Bohr effect).

In this connection it is noticeable that mammalian erythrocytes contain about 30% of the blood water and that during each phase of oxygenation erythrocytes shrink (chiefly because anionic ionization of Hb increases) while they swell during deoxygenation (T. Hansen, 1961).

Comments on diving mammals

A whale can remain under water for 30 mn or even an hour by suspending ventilation. In terrestrial mammals the two azygos veins situated on either side of the vertebral column collect blood from the intercostal veins. In cetaceans these azygos veins are situated *within* the vertebral column and are protected from pressure changes in the thorax. These veins connect the cerebral and abdominal circulations and cannot develop asphyxia due to compression. Moreover, the basal metabolic rate of a whale weighing 150 tons is 1·7 to 2·25 cal/kg/24 h (Rubner, Barcroft) or about 15 times less than that in man. Portier stated that this value was 10 times less in order to provide a margin of safety and he carried out the following calculation:

Under conditions of basal metabolism, man consumes 4·28 ml O_2 per kg/mn; thus, a whale will consume about 0·4 ml/kg/mn.

"Consider a whale weighing 122 tons. It will consume per kg/mn: $0·4 \times 122000 = 48,800$ ml or, in round figures, 49 l. The blood contains 1700 l of oxygen and the lungs contain 3050 l of air or 610 l of oxygen. If this oxygen is consumed the whale can remain under water for $2310/49 = =47$ mn. Moreover, with air containing 8 per cent CO_2, the oxygen consumption falls to 60 per cent compared with that during ventilation of air at the surface (A. Mayer)". The possibility of increasing the duration of diving is favoured by a remarkable property of the capillaries involved in pulmonary hematosis (cf. p. 342).

However, there is another problem: if a whale dives to a depth of 200 m, where the partial pressure of oxygen in the gas of the lungs is 4 atm, how is it that the animal does not suffer convulsions as described by Paul Bert for laboratory mammals inhaling oxygen in a pressure chamber under similar conditions? This is explained by the fact that when a whale dives, the central nervous system is not saturated with oxygen under high pressure as is the case with a dog completely enclosed in an oxygenated chamber. In the whale only the lung is involved and oxygen is absorbed by the venous blood and rapidly taken up by the tissues from hemoglobin and in particular, from arterial plasma.

The problem remains of decompression during surfacing. This problem also faces frogmen. During compression the blood absorbs nitrogen from the air and this dissolves in plasma and becomes localized in adipose tissue (nitrogen is 6 times more soluble in fats than in water). On surfacing the nitrogen is liberated (Paul Bert) and, if the rate of ascension is rapid, nitrogen is liberated not only from lungs but also in tissues and capillaries. This leads to gaseous embolism with serious consequences and necessitates recompression of the subject followed by slow decompression. The large amount of fat within the whale would be unfavourable in this respect but it is stated that *the amount of nitrogen carried in the lungs is relatively small* compared with body weight. However, this problem should be studied further.

The passage of nitrogen into tissues can occur also in high-altitude aviators with non-pressurized equipment and rapid rates of ascension. However, such accidents are rare nowadays.

Transport of carbon dioxide in the blood

An accumulation of carbon dioxide within the organism would rapidly become toxic; thus, carbon dioxide is eliminated (physical detoxication).

Carbon dioxide is transported by cells and blood plasma although, in the latter case the quantity of dissolved CO_2 is 2·6 volumes in arterial human blood. Unlike dissolved oxygen, the CO_2 content of plasma is of great importance in that any change in CO_2 consumption has an effect on the equilibrium:

$$CO_2 + H_2O \rightleftharpoons H_2CO_3 \rightleftharpoons H^+ + HCO_3^- \tag{88}$$

As the carbon dioxide in *plasma* is carried in the form of sodium or potassium bicarbonate, the carbon dioxide is transported by erythrocytes in combination with hemoglobin.

Carbon dioxide reacts with hemoglobin to give a *carbamino compound:*

$$HbNH_2 + CO_2 \rightleftharpoons HbNHCOOH \tag{89}$$

which represents only 20 per cent of the total carbon dioxide in blood. The importance of this reaction is found in the fact that the exchange of gas is relatively rapid.

In the red corpuscle, the reaction $CO_2 + H_2O \rightleftharpoons H_2CO_3$ is accelerated by the presence of an enzyme: *carbonic anhydrase*. This enzyme will be considered again when we deal with vision and the kidney. For the moment it must be remembered that this enzyme is found not only in erythrocytes but also in kidney tubules, parietal cells of the stomach (cf. p. 92), and pancreas and muscle tissue; it is inhibited by some sulphamides (acetazolamide), sodium sulphite and sodium nitride.

A quantity of carbonic acid is exhaled from the lungs of man each day equivalent to about 30 l of N acid. This acid, which originates in tissues, is transported in the blood after neutralization by alkaline cations. At physiological pH (7·25–7·4) there must be twenty times more bicarbonate than carbonic acid. Any alteration in this ratio leads to acidemia *(acidosis)*—extreme pH ∼ 6·8—or alkalemia *(alkalosis)*—extreme pH 7·9. This leads to a study of buffer systems of blood.

Buffer systems of blood

A buffer system (of pH, rH_2, etc.) represents a basic form of homoiostasis.

This leads us to consider the Henderson–Hasselbach equation:

Henderson (1908) examined the equilibrium $H_2CO_3 \rightleftharpoons H^+ + HCO_3^-$ and deduced that

$$(H^+) = k \frac{[CO_2]}{[HCO_3^-]}$$

where k is the first dissociation constant of carbonic acid which is equivalent to $1-10 \times 10^{-7}$ at $38°C$. Hasselbach (1916), with reference to the concept of pH introduced by Sørensen in 1909, wrote:

$$pH = pK + \log \frac{[HCO_3^-]}{[CO_2]}$$

This equation neglects the second constant ($HCO_3^- \rightarrow H^+ + CO_3^{--}$). The Henderson–Hasselbach equation allows the determination of the concentration of bicarbonate necessary at different temperatures for given values of p_{CO_2} to obtain a certain pH but, as stated by Homer Smith, it is a "monument to human laziness".

(1) The buffer effect of *plasma proteins* is important in that they buffer 10 per cent of total CO_2 due to their amphoteric nature (J. Loeb).

(2) *Phosphates of the erythrocyte* buffer 25 per cent of total CO_2.

(3) *Hemoglobin* is the most important buffer. The isoelectric point of hemoglobin (HHb) is $6·8$ while that of oxyhemoglobin (HHbO$_2$) is $6·65$. Thus, *oxyhemoglobin is a stronger acid than hemoglobin* and a greater quantity of base can combine with HHbO$_2$ than with HHb. In the region of tissues, when HHbO$_2 \rightarrow$ HHb, there is a liberation of base which is utilized in the neutralization of carbonic acid produced in the tissues by decarboxylation.

During pulmonary hematosis the reverse process occurs. As a certain amount of CO_2 escapes from the blood, a corresponding amount of alkali is present in excess but since HHbO$_2$ is formed simultaneously, this excess alkali is immediately utilized to neutralize it. Summarily, *there is neutralization of CO_2 (carbonic acid) in the region of tissues and of HHbO$_2$ in the arterial lung.* According to van Slyke:

If the pressure of O_2 is low (vein):

$$KHb + H_2CO_3 \rightleftharpoons HHb + HKCO_3 \tag{90}$$

carbonic acid—a stronger acid than hemoglobin—displaces hemoglobin from its salts, in this case, potassium hemoglobinate.

If the pressure of CO_2 is low (artery):

$$HHbO_2 + HKCO_3 \rightleftharpoons KHbO_2 + H_2CO_3 \tag{91}$$

oxyhemoglobin tends to displace carbonic acid from its salts.

This expresses the Haldane *effect* which is the inverse of the Bohr effect: *the greater the quantity of oxygen, the less CO_2 fixed on hemoglobin.*

Intervention of chlorides:

85 per cent of the buffering capacity for CO_2 resides in the erythrocytes (60 per cent for Hb and 25 per cent for phosphates). However, because of the bicarbonate content, plasma also has a considerable buffering effect. How are these two buffer systems related?

The wall of the red corpuscle is involved. The membrane of the red corpuscle is normally *impermeable* to hemoglobin, Na^+ (plasma) and K^+ (in erythrocyte) but permeable to $(HCO_3)^-$, Cl^- and H^+.

In plasma one finds:

$$dissolved\ CO_2$$
$$CO_2 + H_2O \rightleftharpoons H_2CO_3$$
$$H_2CO_3 \rightleftharpoons (HCO_3)^- + H^+$$
$$NaCl\ and\ NaHCO_3$$

Within the red corpuscle one finds:

$$dissolved\ CO_2$$
$$CO_2 + HbNH_2 \rightleftharpoons HbNHCOOH$$
$$CO_2 + H_2O \rightleftharpoons H_2CO_3\ (carbonic\ anhydrase)$$
$$H_2CO_3 + KHb \rightleftharpoons HKCO_3 + HHb$$

If the level of CO_2 increases in the blood (venous region) an excess of HCO_3^- is produced, essentially in the red corpuscle which contains carbonic anhydrase. Excess carbonic acid passes into the plasma in the form of bicarbonate and is replaced in the corpuscle by Cl^-.

$$
\begin{array}{c|c|c}
\boxed{Plasma} & & \boxed{Red\ corpuscle} \\
NaCl + HCO_3^- \leftarrow & \text{membrane} & ---HCO_3^- + K^+ \\
\downarrow & & \downarrow \\
NaHCO_3 + Cl^- --- & & \rightarrow Cl^- + K^+ \\
& & \downarrow \\
& & KCl
\end{array}
$$

Thus, the level of Cl^- in plasma decreases and that of bicarbonate increases. The reverse occurs in the arterial system.

The displacement of chloride from the plasma into erythrocytes due to the presence of CO_2 is called the Hamburger effect.

Changes in the oxygen cycle regulate the mechanisms of cyanosis and anoxia and changes in the carbon dioxide cycle lead to modification of the acid–base equilibrium and thus affect micro-ions. As a result certain electrical phenomena of the membrane can be modified, in particular, the phenomenon of amortization (cf. Volume II).

The Haldane effect participates to a greater extent in acid-base regulation than in "respiratory" regulation, bicarbonate being the compound which regulates the formation of $Hb-HbO_2-CO_2$ complexes.

Techniques for measuring acidity:

(1) *The electrical pH meter:* for measuring the pH of total blood or of plasma and serum in order to determine the difference between erythrocytes and the blood medium. Rybak *et al.* (1967) have developed a miniature pH-probe which can be inserted *in situ* into vascular vessels by catheterization. Lübbers and his co-workers (1958) and Severinghaus

used *in vitro* pH electrodes covered by a membrane impermeable to ions but permeable to gas; this method allows measurement of p_{CO_2} in very small volumes of blood but a pH-probe is much more versatile.

(2) *Liberation of CO_2 by an acid:* the volume of CO_2 liberated from x cm^3 of total blood or serum is measured. Pressure, temperature, etc. must be taken into account. If the plasma is equilibrated with air in which the partial pressure of CO_2 is 36 mm Hg (equivalent to alveolar air), the *capacity* for CO_2 can be measured (the blood is removed under oil to prevent exchange with air). The quantity of cations (principally Na$^+$) combined with bicarbonate provides a measure of the *alkaline reserve*. This value can be obtained by considering the ratio of bicarbonates to carbonic acid (20/1) and the capacity for CO_2.

We have now considered the function of the lung as an emunctory and regulator of the pH of the internal medium.

Pulmonary regulation of acid–base balance

Acidosis results either from a decrease in the quantity of bicarbonates (diabetes) or from an increase in the quantity of H_2CO_3 (unsatisfactory elimination of CO_2 from the lungs as occurs in asthma). The first type of acidosis is described as "metabolic" and the second as "respiratory".

Alkalosis can result either from a decrease in the quantity of H_2CO_3 compared with that of bicarbonates ("respiratory" alkalosis of hyperventilation) or from an increase in the quantity of bicarbonates (*metabolic alkalosis* which appears after vomiting producing excessive elimination of HCl and in general terms, when a Cl$^-$ deficiency is produced; this is *hypochloremic alkalosis* which can appear during cortisone treatment). During alkalosis ventilation is slow and shallow.

The importance of the kidney in regulation of the acid–base equilibrium will be studied later (cf. p. 384), but the lung has to be considered now in more detail.

II. Ventilation

The relative pulmonary volume, the pulmonary surface for capillary exchange, must depend on the intensity of oxidative metabolism; among tetrapods (from Proteus to tortoise) there is a progressive increase in these parameters. The importance of this metabolism is also related to the circulation of air and thus, to the frequency and amplitude of ventilatory movements and, as we have seen, this depends notably on the size of the animal. However, since oxidative metabolism implies a distribution of oxygen (and elimination of CO_2), I have deduced that evolution of the lung must have paralleled evolution of the cardiovascular system. These changes must have been contemporary with development of hemopoietic structures and—remembering that we are concerned with metabolism—of the digestive tract. Summarily, I consider that evolution regulated by conditions of ambient oxygenation has been such that, in each phylum, vegetal structures must have evolved *en bloc*.

Anatomy of the lungs:

In man, the lung is surrounded by a *serous* membrane *(the pleura)* and composed of lobes divided into *lobules* a few cm³ in volume and poly-hedral in shape. Each lobule receives a bronchus which divides into *bron-chioles* which terminate in the *alveolar* sacs. The pulmonary artery ac-companies the bronchus into the lobule and forms the *broncho-arterial axis of the lobule*. In man the total number of alveoli is 300 million ± 3 per cent and the total number of air ducts is 14 million ± 11 per cent (Weibel; Liebow).

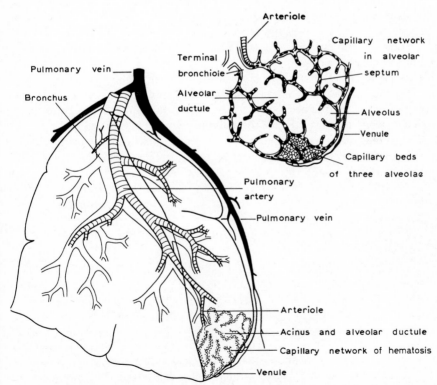

FIG. 101. Diagram of the general circulation of the pulmonary lobule of man.
(DUBREUIL and BAUDRIMONT, 1959.)

The wall of the alveolar sac is very thin; it consists of a network of capillaries (site of hematosis) and the so-called *respiratory* epithelium which forms the surface of the alveoli.

Smooth muscle fibres surround the orifice of each alveolus. G. Dubreuil called these the *plexiform smooth muscle sphincters of the alveolar ducts*. The sympathetic and parasympathetic innervation of the lung terminates in the region of these sphincters.

Pulmonary tissue appears to possess marked mitogenic properties: by endocular homografts of pulmonary tissue in mice, Ganter, Kourilsky and R. M. May (1956) induced *proliferation of the anterior corneal epithelium* at the interior of the pulmonary graft. This action which can also be shown on cutaneous tissue would be specific to the lung.

Pulmonary circulation

It is often forgotten that, although the lung represents an organ which is specialized for gaseous exchange in mammals, it must possess a vascular system which maintains the tissues of the lung. In the same way that the heart of mammals with specialized blood circulation contains a parietal circulation, the coronary vessels, the lung contains a circulation which is not involved in hematosis and which does not enter into contact with the alveolar ducts—the circulation consists of the bronchial vessels.

By the technique of *micro-angiography* (vascular radiography utilizing compounds opaque to X-rays), Morais, Mirabeau-Cruz and de Souza (1957) were able to examine the pulmonary capillary vascularization and discovered long and rectilinear arterio-venous anastomoses in the human pulmonary parenchyma. *Short, tortuous anastomoses* have been described by von Hayer (1953)

The pulmonary circulation of hematosis is the "little circulation" originating from the *right auricle*. In this system arteries carry blood which is *venous* from the chemical point of view and veins carry *arterial blood*.

This circulation carries 25 per cent of the total blood volume and the surface for capillary exchange is 450 m² in man.

In terrestrial mammals the arrangement of capillaries arising from the pulmonary artery is such that one capillary is adjacent to two alveoli, which increases the surface for gaseous exchange. However, in marine mammals, Dubreuil and Baudrimont have pointed out that the alveolar partitions are thick and the capillaries are in contact with alveolar air only on one side. This arrangement decreases the rate of gaseous exchange in air and under water and favours prolonged immersion.

Since the pulmonary circulation is localized within the thorax where the pressure is lower than atmospheric pressure, the vessels of the lung remain open. Moreover, movements associated with inhalation and exhalation induce rhythmic changes in pulmonary hemodynamics, the movement of blood being influenced by the combined actions of the heart and thorax. *The output of pulmonary blood increases during inspiration* (oxygenation phase) due to a lowering of intrapleural pressure and *decreases during expiration* for the opposite reason.

The *systolic pressure in the right ventricle* is lower than the systolic pressure in the left ventricle. By catheterization in man, Cournand has shown that the maximum pressure in the pulmonary artery is 22 mm Hg and that the minimum pressure is 7 mm Hg. This is explained by the lower barometric and vascular resistance to blood flow and the comparative thinness of the right ventricle associated with the smaller amount of work produced.

Using a technique involving impedance measurements allowing a study of the distribution of an electrolyte solution in the blood, i.e. by following the increase in conductivity, Stewart concluded that blood passes through the pulmonary circulation in 3–4 s. Therefore, *passage through the lungs is rapid* as must be hematosis.

Ventilatory innervation

I. Vasomotor innervation

The smooth muscles of the arteries and arterioles contain vasomotor and vasosensorye nerve fibres. The sympathetic nerves produce constriction; these arise in the dorsal roots and pass via the stellate ganglion. The fibres of the vagus induce vasodilation. *Stimulation of the peripheral end of the vagus, the nerve of Ludwig and Cyon or the nerve of Hering leads to pulmonary vasodilation.*

II. Ventilatory centres

Flourens stated that the localization of the inspiratory and expiratory centres is not *cervical:*

(1) Anencephalic foeti of mammals show co-ordinated ventilatory movements.

(2) Dogs in which the cerebral hemispheres have been removed show normal ventilation but Pachon (1893) stated that the ventilatory frequency decreases by half in decerebrate pigeons. In man, voluntary control can be involved (by a trans-thalamic pathway?). In this respect we should consider the Hindu ventilatory technique involved in the practice of yoga. Apart from certain debatable mystical considerations, this technique allows the individual to have remarkable control of the body. F. Lefebure stated that rhythmic ventilation of a certain sequence favoured mental activity through improved oxygenation and thus facilitated meditation. In this respect, we are not far from the peripatetic philosophy since walk-

ing favours blood circulation and good circulation augments cerebral metabolism which is very sensitive to anoxia (cf. Volume II). V. G. Rele (1925) suggested that "Kundalini" of tantrism must be related to vagus nerve activity and by this he explained the control of visceral activities exhibited in yoga, notably that of ventilation and cardiac contraction (Laubry).

Area 13 (according to the system of Brodmann, cf. Volume II) situated on the orbital surface of the frontal lobe induces ventilatory arrest when excited (Bailey and his co-workers, 1938–1940). Intense excitation of a part of area 6 can release this ventilatory arrest. Dusser de Barenne and co-workers (1932–1938) observed that hyperventilation leading to alkalinity of the blood induces hyperexcitability of the cerebral cortex which can cause epileptic crises in certain subjects particularly sensitive to variations in cerebral p_{CO_2} and p_{O_2}.

There are numerous areas of the cortex which can influence ventilation and they are situated principally in the rhinencephalon (cf. Volume II) but the motor areas of the ventilatory muscles (abdomen, diaphragm, thorax) are very restricted. Colle and Massion (1957) discovered, in cat, a cortical area of 2 mm diameter situated in the posterior part of the motor area of the shoulder which when stimulated produced an immediate response in the phrenic nerve (which innervates the diaphragm). This

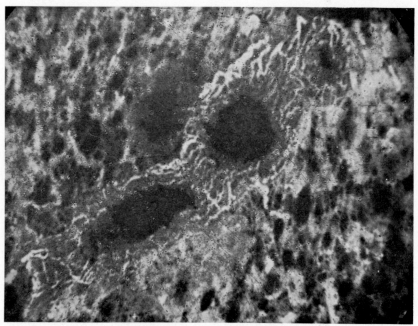

FIG. 102a. Venous plexus in the bronchial wall ($\times$ 45). (BELLO MORAIS, MIRABEAU CRUZ and DE SOUZA, 1957.)

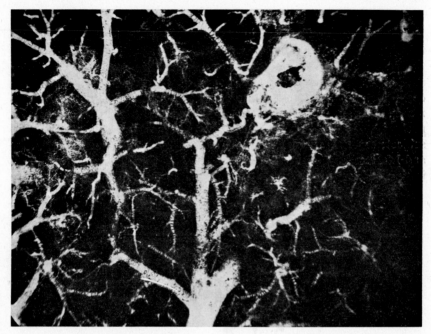

FIG. 102b. Short arterio-venous anastomoses (× 60). (BELLO MORAIS,
MIRABEAU CRUZ and DE SOUZA, 1957.)

discovery led these authors to provide evidence for an extrapyramidal
cortico-phrenic pathway (cf. Volume II) which is independent of the
ventilatory centre. This is a non-specific pathway since cortical stimulation
excites not only the phrenic nerve but also other motor nerves of the same
segments (for example, the sciatic nerve). There is diffuse activation which
probably originates in the nervous reticular substance (cf. Volume II).

Neurones arising in or passing through the hypothalamus form part of
the reverberating system of neurones which includes medullary reticular
neurones facilitating inspiration. Activity in this inspiratory facilitating
system assists initiation of inspiration and drives lower ventilatory system
neurones to produce polypnea and inspiratory apnea. When this inspiration facilitating system is intact, reflex activity induced by tracheal clamping
at the end of expiration is enhanced, while the Hering-Breuer inhibitory
activity evoked by pulmonary stretch receptors (cf. p. 356) is depressed.
On the other hand, when the activity within this system is depressed by
breaking the chain at the hypothalamic level, lower brain stem respiratory
neurones have the capacity to maintain eupnea but reflex responses evoked
by sciatic nerve stimulation and tracheal clamping are depressed and the
electrical threshold of medullary neurones facilitating inspiration is increased (Redgate, 1963).

3

The principal ventilatory centres are bulbar.

When an attempt to destroy the cardiac arteries is unsuccessful, the toreador kills his victim by destroying the bulbar ventilatory centre.

Flourens (1842–1858) localized the co-ordinating centres of ventilation at the point of the *calamus scriptorus* (floor of the fourth ventricle); destruction by puncture leads to immediate arrest of ventilatory movements. This so-called "vital nucleus" should be termed the ventilatory nucleus because destruction of this centre does not result in death of the subject; cardiac activity can be maintained by artificial ventilation. Gad and Marinesco (1892) demonstrated by systematic ablation that the localization of Flourens is too strict: the ventilatory centre is situated in the

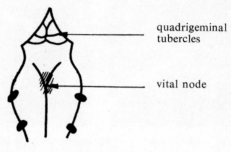

quadrigeminal
tubercles

vital node

Fig. 103

bulbar reticular substance. By faradic stimulation, an *inspiratory centre* and an *expiratory centre*, situated further back in the floor of the fourth ventricle, can be localized.

Simultaneous excitation of the two centres produces inspiration and thus the *inspiratory centre is dominant* (Pitts, 1940). There have been certain objections to these results. Gesel (1940) considers that the neurones forming the inspiratory and expiratory centres are interlaced and, as a result, it is impossible to localize these centres.

However, the bulbar ventilatory centres are paired and *symmetrically placed*, each half maintaining ventilation on one side. Nerve fibres provide connections between the nerves on each side of the medulla.

In young animals, certain authors have described *medullary* ventilatory centres. These are open to question but according to Virchow the young mammal is a *spinal creature* and nervous activity predominates in the bulbo-medullary region.

MOTOR VENTILATORY PATHWAYS. Nerve fibres pass from the ventilatory centres and innervate the muscles involved in ventilation:

(1) Fibres run together in the medullary region to form the *descending motor ventilatory bundle* and join with the spinal motor cells which act

on the intercostal muscles (intercostal nerves) and on the diaphragm (phrenic nerve).

(2) Other efferent fibres pass to the motor nuclei of the following cranial nerves: V (trigeminal), VII (facial), IX (glossopharyngeal); X (vagus), XI (spinal) and XII (hypoglossal) which act on striated muscle of the mouth, tongue, nostrils, pharynx, larynx and smooth muscle of the bronchial tree.

MODE OF ACTION OF THE VENTILATORY CENTRES: A large number of facts support the idea that ventilatory movements are initiated in the *bulbar ventilatory centres*. For example, J. P. and C. Heymans (1927) stated that regular ventilatory movements are maintained in the isolated, perfused head of dog, these movements being seen as movements of the larynx. Winterstein (1911) and Adrian and Bronk (1928) recorded *rhythmic action potentials* in the ventilatory nerves of animals in which spontaneous thoracic movements had been prevented by curarization. If the bulbar ventilatory centres are isolated, rhythmic electrogenic activity disappears. Therefore, these trains of potentials are associated with ventilatory activity. If the cerebral trunk in cat is sectioned below the protuberance without affecting the bulbar region and the pneumogastric nerves are sectioned bilaterally, continuous action potentials can be recorded in the phrenic nerve. At the same time, the animal shows *permanent inspiration* (Pitts), the inspiratory centre being predominant.

For ventilatory rhythm to be maintained there must be regular interruption of this permanent nervous activity. We shall attempt to explain this phenomenon.

There is no doubt that the vagi influence the periodicity of ventilation: in vivo, distension of the lungs at the moment of inspiration stimulates receptors situated in the pulmonary parenchyma which produce inhibitory influx via the sensitory fibres of the pneumogastric nerves. *The frequency of this influx increases proportionally with the increase in the volume of inspired air* (Adrian, 1933), the pulmonary baroreceptors having different thresholds such that the general action is progressive. With the beginning of expiration, activity in the pulmonary baroreceptors decreases and finally disappears and a new ventilatory cycle commences.

Bivagotomy leads to *slowing* of the ventilatory *rhythm* and a marked *increase* in the *amplitude* of ventilatory movements. Bradypnea results from prolonged depolarization of the inspiratory centres, the increase in amplitude being related to an increase in the frequency of the train of potentials originating in the inspiratory centres (Pitts, 1942). Thus, bivagotomy does not lead to a disappearance of ventilatory periodicity and as a result, the vagal nerves are not the only factors contributing to this periodicity. Bradypnea is associated with tachycardia and oxygenation of the blood is maintained at a level which ensures survival. The

23*

ventilatory changes occurring with bivagotomy can be thought to be conditioned by the sympathetic system and/or the activating reticular substance (cf. Volume II).

Ventilatory rhythms are maintained in a mammal in which only the bulbo-protuberal system has been conserved (Rosenthal, 1865). Markwald

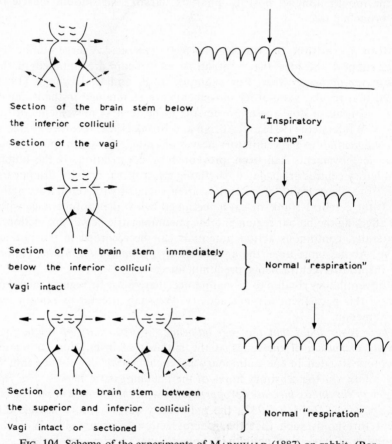

Section of the brain stem below
the inferior colliculi

Section of the vagi

} "Inspiratory cramp"

Section of the brain stem immediately
below the inferior colliculi

Vagi intact

} Normal "respiration"

Section of the brain stem between
the superior and inferior colliculi

Vagi intact or sectioned

} Normal "respiration"

FIG. 104. Schema of the experiments of MARKWALD (1887) on rabbit. (Borrowed from GRANDPIERRE and FRANCK, *Revue médicale de Nancy*, 1950.)

carried out a number of experiments on sectioning in rabbit and the results of these are summarized in Fig. 104.

It can be concluded that *the bulbar ventilatory centres have no proper automatism.* These centres determine a prolonged tonic inspiration (inspriatory cramp), periodic interruption of which produces ventilatory rhythm. This periodic interruption is conditioned by two inhibitory processes which can act independently:

(a) *One process originating in the quadrigeminal tubercles* in rabbit and previously described by Langendorff (1877) and Pachon (in pigeon); (b) the other secondary process associated with centripetal vagal pathways.

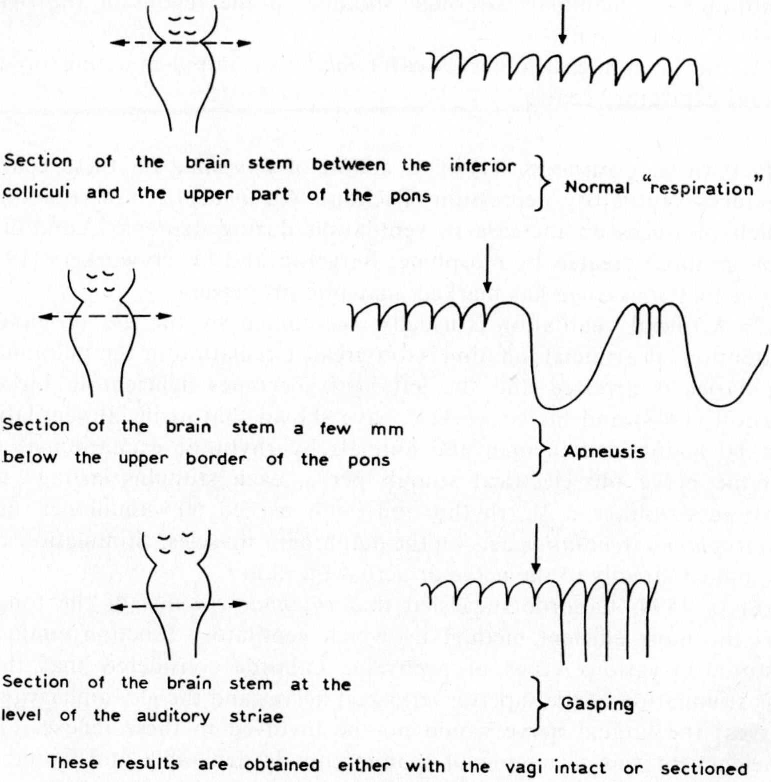

Section of the brain stem between the inferior colliculi and the upper part of the pons } Normal "respiration"

Section of the brain stem a few mm below the upper border of the pons } Apneusis

Section of the brain stem at the level of the auditory striae } Gasping

These results are obtained either with the vagi intact or sectioned

FIG. 105. Schema of the experiments of LUMSDEN (1923), in cat. (Borrowed from GRANDPIERRE and FRANCK, *Revue médicale de Nancy*, 1950.)

These experiments of Markwald have been completed by those of Lumsden (1923) working on cat. The essential results are summarized in Fig. 105.

Lumsden provided a concept of the function of central ventilatory mechanisms which has been modified following further work by Lumsden and by other authors (Pitts, Stella). We can now distinguish four nervous structures involved in ventilation in mammals:

(1) in the bulbar region

(a) an expiratory centre,
(b) an inspiratory centre with prolonged tonicity;

(2) an inhibitory centre (pneumotaxic centre of Lumsden), localized in the superior region of the protuberance and acting on the bulbar expiratory centre;

(3) a primitive centre, the gasping centre, which has no role in normal ventilation of mammals although situated in the region of the "vital nucleus" of Flourens;

(4) the pneumogastric nerves carry *inhibitory* impulses acting on the bulbar expiratory centre.

PRACTICAL COMMENTS. (1) The action of *morphine* on these centres produces ventilatory depression (Pachon). An *analeptic* is a compound which produces an increase in ventilation during depressed conditions such as those created by morphine. Bargeton and his co-workers (1955) found that *strychnine* has marked analeptic properties.

(2) Artificial ventilation is usually maintained by the use of pumps. (Attention: if artificial inflation is too great, circulation in the pulmonary capillaries is arrested and the left heart becomes deficient in blood). Sarnoff (1948) and his co-workers have shown that artificial ventilation can be maintained in man and animals by rhythmic excitation of the phrenic nerve (40 electrical stimuli per s, each stimulus lasting 2 ms, maximum voltage 3 V, rhythm must not exceed 60 stimuli per mn). *Electrophrenic ventilation* acts on the diaphragm muscles. Stimulation can be applied directly to the nerve or across the skin.

(3) In 1894, Laborde suggested that *rhythmic traction* of the tongue was the most efficient method by which ventilatory function could be restored in various types of asphyxia. Laborde considered that there was stimulation of the superior laryngeal nerves and the glossopharyngeal nerves; the lingual nerve would not be involved in these reflexes. The efficiency of this technique of reanimation is debatable and would be non-existent in inhibition of the bulbo-protuberal regions such as occurs with high doses of anaesthetics.

(4) Asthma is a form of dyspnea associated with bronchospasm of vago-sympathetic origin. The "pneumo-bulbar reflex ataxia" described by Moncorgé (1941) led to the following scheme:

Excitation by toxic materials (allergy)	→	Peripheral nervous hyperexcitability	→	Hyperexcit-ability of bulbar centres	→	Asthmatic dyspnea

The liberation of histamine is one of the factors in asthma and anti-histamines are generally active from the therapeutic point of view.

Pulmonary embolism is not due to some reflex action nor to obstruction of the capillaries by particles. As has been clearly demonstrated by Valle

Pereira (1946), it results from a defect of the right ventricle with interruption of the pulmonary circulation and congestion which leads secondarily to vascular obstruction.

Regulation of ventilation

Ventilation depends on nervous activity which is maintained by the circulation of blood and by thoracic movements. Ventilation is controlled and modified by various nervous (afferent nerves) or humoral (blood acting via the neuroglia?) stimuli. We shall examine these factors and later return to a study of certain details.

(1) Chemical factors

The most important chemical factor is carbon dioxide. The action of carbon dioxide was recognized by Hermann in 1870. The facts are as follows:

An increase in the CO_2 content of alveolar air leads to an increase in the amplitude of ventilatory movements while even a small decrease in the CO_2 content of alveolar air produces *hypocapnic* or *acapnic apnea* such as appears notably during hyperventilation (Douglas and Haldane).

Oxygen can also be a factor modifying ventilation.

A decrease in the oxygen content of alveolar air causes an increase in the amplitude of thoracic movements followed by acceleration of the frequency of these movements.

An *increase* in the oxygen content of inhaled air leads to a decrease in ventilation which is generally hardly apparent but which is marked in certain cases (anaesthetized animal) and results in apnea: a paradoxical action of oxygen. This effect is not incompatible with survival. L. Binet, Poutondet and Strumza (1943) maintained an animal in which ventilatory movements had been prevented by anaesthesia, by connecting the trachea to a reservoir containing oxygen. Although there was apnea the animal survived for about 30 mn and the percentage saturation of hemoglobin approached the normal value but there was marked acidosis due particularly to an accumulation of lactic acid (Joels and Samueloff, 1956). This phenomenon is *ventilation by diffusion* and implies ventilation which does not involve pulmonary movements.

Finally, among chemical agents which participate in the regulation of ventilation, it is necessary to mention *barbiturates* which decrease ventilation and *caffeine* which increases ventilation. Moreover, a decrease in blood pH induces augmentation of ventilation while an increase diminishes ventilation. In this respect, it should be noted that CO_2 and lactic acid (produced during metabolism associated with muscular contraction) cause vasodilation. This effect can be compared with the "hormonal" activity of carbon dioxide.

(2) Physical factors

Blood pressure: this acts in the region of the barosensitive vascular zones (cf. p. 356);
Temperature (cf. thermoregulation, Volume II).

(3) Nervous factors acting indirectly on the ventilatory centres

These can be voluntary phenomena leading to hyperpnea or apnea—as in the case of the yogi or pearl fisher—or unconscious phenomena, ventilation being modified by speaking, deglutination, etc.

How is ventilatory regulation accomplished?

(1) Regulation of ventilation by central pathways.

In 1862 Rosenthal established that the chemical changes in blood leading to the appearance of "venous" blood increase ventilation while those associated with "arterial" blood decrease ventilation. This was confirmed in 1890 by L. Fredericq by cross circulation experiments (cf. p. 295) in anaesthetized dogs or rabbits: occlusion of the trachea in one animal caused asphyxia and an increase in the carbon dioxide of the blood which passed in the circulation to the other animal resulting in hyperventilation: inversely, hyperventilation in one animal resulted in hypoventilation in the other (hypocapnic apnea). It has since been demonstrated that CO_2 is the principal agent in determining "venous" blood. Acids other than CO_2 can influence central regulation of ventilation but CO_2 has the most marked stimulant action. This is a selective action which depends on the greater cellular diffusibility of CO_2 compared with the other acids resulting in rapid penetration to the ventilatory neurones. Because of these properties carbon dioxide has been called the "respiratory hormone" which is incorrect in two respects. Firstly, the term "hormone" is incorrect in this context (classically the term is associated with endocrine glands) and secondly, the term "respiratory" is incorrect.

Neither the direct central stimulating action of O_2 deficiency nor the action of an increase in O_2 are constant or marked. It would appear that CO_2 is the limiting factor in determining the venous or arterial reaction, the venous state being considered in ventilatory reactions as a hypercapnic state rather than a state of decreased oxygen. The discovery of Rosenthal demonstrates the way in which regulation is a process which depends on antagonistic states: venous state–arterial state. This combination is simi-

lar to other homoiostatic combinations, parasympathetic–sympathetic,
vasoconstriction–vasodilation, etc. In the case of the venous and arterial
state, the ganglia involved in reflex regulation are the centres of physico-
chemical reflexes. Some reflexes are purely biochemical, others involve
endocrine pathways and other nervous pathways. Homoiostasis results
from the combination of these reflex mechanisms.

When adrenaline is administered during hypocapnic anoxia it has less
effect with respect to vasoconstriction (Hermann and Jourdan, 1941)
and to hyperglycemic effect (Franck, Grandpierre *et al.*, 1943).

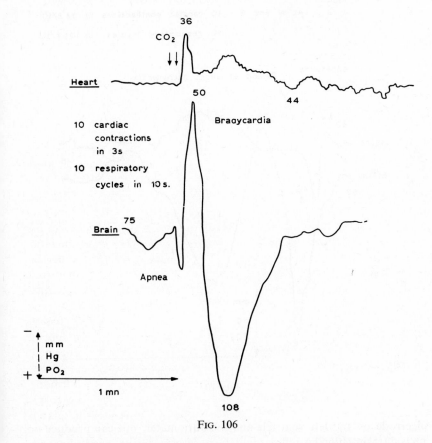

FIG. 106

A reflex involving the lung, heart and vessels and induced by inhalation
of CO_2 or injection of adrenaline or acetylcholine has been demonstrated
using oxygen polarography (Rybak, 1964). A new experimental technique
has been developed which allows prolonged spontaneous ventilation with
the heart exposed. In seeking to establish encephalo-cardiac respiratory
correlations which are particularly interesting, one must either place the

animal under artificial respiration—which can mask the essential regulatory phenomena—or maintain the experimental animal by cross-circulation—which can introduce interference from the associated animals. The new preparation which allows one to study reflexes in the anaesthetized rabbit, is based on the fact that in this mammal there is a mediastinum such that the left thorax can be opened without interruption of unilateral ventilation in the right lung. Under these conditions, the pericardium is cut and the heart exposed and subjected or not to tension by a weight attached to the apex of the ventricle. By placing a polarographic oxygen

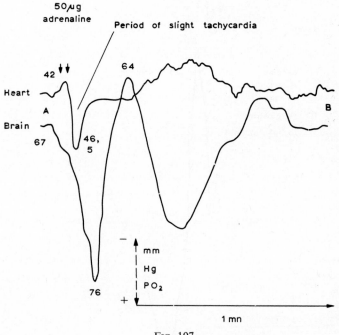

At A and B : 10 cardiac contractions in 3 s 2/10

10 respiratory cycles in 10s 4/10

Fig. 107

electrode on the left ventricle and the dura mater, one can produce polyphasic traces showing reflex regulation of p_{O_2} during conditions which change the normal circulatory function; Figures 106 and 107 show the effects of CO_2 gas and adrenaline.

De Boer (1946) demonstrated that the administration of high doses of sodium succinate induced intense ventilatory stimulation. Lwoff, Monod and Bovet reported that intravenous injection of this compound interrupted apnea induced by hyperventilation. There is an increase in the production of CO_2 and in consumption of O_2 (Heymans *et al.*, 1948). Succinate appears to act as a catalyst of oxidation and to accelerate the utilization of tricarboxylic acids (Krebs cycle).

Santenoise and his co-workers isolated "centropneine" from pancreas. This compound increases the biochemical excitability of ventilatory centres and inhibits carbonic anhydrase *in vitro* (1954–1960).

ROLE OF THE TEMPERATURE OF BLOOD. When the temperature of blood increases there is direct stimulation of the ventilatory centres and a regular increase in the frequency of ventilatory movements. This action, related to central thermoreceptors, in part explains the thermal polypnea and polypnea of fever.

(II) Regulation of ventilation by intercentral pathways.

(a) *Participation of higher nervous centres.* In certain animals, ablation of the brain leads to ventilatory disorders (for example, in pigeon). Moreover, the speech centre modifies ventilation through conditioned reflexes; on the other hand, the role of emotions is well known. The automatic ventilatory rhythm can be controlled voluntarily.

In teleost fish branchial function can be modified by illumination: a light stimulus (6000 lux) in the eye region of an immobile fish kept in the dark, leads to a decrease in the amplitude of opercular movements which can proceed to apnea. At the same time bradycardia is apparent and the degree of bradycardia parallels the extent of the decrease in opercular movements. These reactions disappear with ablation of the eyes. Fish which rely mainly on sight (for example, pike) show less marked but more prolonged reactions than those relying on taste or smell (tench, carp) (Serfaty and Peyraud, 1962).

(b) *Role of the hypothalamus.* Sleep (cf. Volume II) conditions a decrease in ventilatory rhythm.

During sleep of hibernation, mammals (dormouse, marmot) show a very slow rate of ventilation and there is an accumulation of CO_2 within the organism. There is also a decrease in body temperature and Fleming (1954) demonstrated that when the temperature of a dog falls from 38°C to 20°C the p_{CO_2} changes from 45 mm Hg to 130 mm Hg and acidosis occurs.

(c) *Role of the bulbar region.* The deglutition centre controls ingestion and arrest of ventilation (cf. p. 87). The relationship between the ventilatory centres and the cardio-inhibitory and cardio-acceleratory centres must also be considered.

(III) Reflex regulation of ventilation

(1) *Role of vagal nerves.*

Let us examine the effect of excitation:

(a) *Electrical stimuli.* Low intensity excitation of the central end of the vagus leads to acceleration of ventilatory movements while high intensity excitation results in apnea. Low frequency excitation results in inspiration and high frequency excitation (200/s) results in expiration. The exact interpretation of these results is not known but it is possible that, considering the dominance of the inspiratory centre, a lower frequency of excitation is required to release its function.

(b) *Mechanical stimuli.* Hering and Breuer (1868) showed that pulmonary distension in mammals mechanically stimulated the terminations of the sensitive pneumogastric fibres.

During normal ventilation (eupnea), inspiration releases inhibitory impulses which pass along the vagus and which, together with impulses from the pneumotaxic centre, decrease the tone of the inspiratory centre and expiration follows.

During forced ventilation, inspiratory reflexes develop after forced expiration and, inversely, expiratory reflexes develop after forced inspiration. These are the *reflexes of Hering and Breuer* (1865). During normal ventilation, inspiration leads to expiration and during forced ventilation, expiration leads to inspiration. The sensitive pathway of these reflexes is represented by the pneumogastric nerves.

(c) *Chemical stimuli.* Although reflexes induced by alveolar carbon dioxide have been described (Pi Suñer), it is not certain, according to Grandpierre and Franck, that alveolar chemical sensitivity via the vagus is important in normal ventilation. However, arterial reflexes are well established.

(2) *Role of arterial receptors*

Certain areas of the arterial vascular system of mammals contain nerve endings of centripetal fibres which connect with the bulbar region and which participate in regulation of ventilation.

(a) The nerves of Ludwig and Cyon originate in the *aortic arch.* In rabbit these nerves are distinct but in man and dog they are contained in the pneumogastric nerve bundle.

(b) The nerves of Hering originate in the *junction of the common carotid artery* and internal and external carotids and join with the glossopharyngeal nerve.

These regions are of great importance since, in addition to containing *baroreceptors* situated in the walls of the vessels, they contain *chemoreceptors*: the *aortic glomus* and *carotid glomi* situated in the region of the carotid bifurcation.

The role of these two regions in ventilation can be demonstrated using various types of stimulation.

(a) *Electrical stimulation.* François-Frank (1892) produced *reflex dyspnea* by stimulation of the Ludwig and Cyon nerve. Dyspnea can also be obtained by stimulation of the Hering nerve. However, in both cases, certain authors have obtained hyperpnea. This subject requires further study.

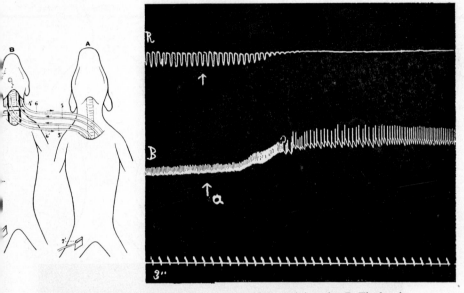

Fig. 108. The donor dog A perfuses the head of the recipient dog B. The head of dog B is connected to the trunk only by the vagus nerves (HEYMANS). R, ventilatory movements of larynx of isolated head; B, systemic pressure of trunk B. Note the hypertension and the bradycardia and reflex apnea to the administration of adrenaline (a). (After C. HEYMANS and A. LADON, from C. HEYMANS and NEIL, 1958.)

(b) *Chemical stimulation.* C. and J. F. Heymans demonstrated that the cardio-aortic and carotid sinus regions are sensitive to the CO_2 level. These results were obtained in the following way:

(i) *Cardio-aortic region*: the activity of ventilatory centres of dog B is followed by *mechanographic recording of laryngeal movements*. The head of B which is attached to the trunk only by the vagi, is perfused by dog A and the heart and aortic arch of B are perfused from dog C. Thus, the head of B is not asphyxial and the response of the larynx to chemical changes produced in the blood of C can be recorded.

(ii) *Carotid sinus region:* the two carotid sinuses of dog B (which are normally innervated or can be denervated for certain experiments) are

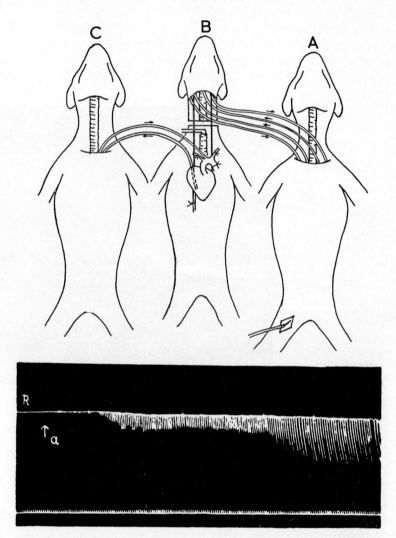

FIG. 109. Upper: The isolated head of dog B is perfused by dog A. Only the vago-aortic nerves connect the head of B to the trunk. Dog C is used to perfuse the cardio-aortic region of dog B: the carotid of dog C provides blood to the ascending aorta of dog B and the blood circulating through the coronary system of dog B is collected in the right ventricle by a cannula connected to the jugular of dog C. Lower: Ventilatory movements of head B: a, asphyxia of dog C; note the stimulation of ventilatory centres of head B by a decrease in pH, in p_{O_2} and an increase in p_{CO_2}. (After J. F. and C. HEYMANNS, from C. HEYMANNS and NEIL, 1958.)

perfused either from dog A or by an artificial circuit containing a perfusion fluid with variable chemical properties.

More simply, a solution can be injected into the carotid artery. The results of these experiments can be summarized as follows:

(1) *Reflex hyperpnea* appears in the perfused dog (B) if ventilation in A is arrested or if the blood of A becomes hypercapnic by inhalation of a mixture of $O_2 + 10$ per cent CO_2.

(2) *Reflex apnea* appears in the perfused dog if dog A is hyperventilated or if the carotid sinus is perfused with a solution buffered at pH 7·4 and having a low CO_2 content.

This effect is prevented by total denervation of the aortic and carotid sinus regions. Thus, the action of CO_2 is on the vascular receptors rather than the bulbar centres.

C. Heymans (1963) described the discovery of the carotid sinus chemoreceptors:

"My father and best teacher always advised his pupils, "Never kill an animal at the end of a planned experiment if the animal may still be used for an experimental purpose, and take profit of this animal to perform any experimental trial, even if it looks foolish but keep yours eyes well open in order to catch any unexpected event". One day, also bearing in mind the statement of Darwin, "I like to perform foolish experiments", we finished a planned experiment on the carotid sinus baroreceptors. One carotid sinus area of the dog was denervated, the other being still innervated. Wondering for what experimental purpose this dog could still be used, we injected into the carotid artery with normal carotid sinus innervation some potassium cyanide which was standing on the laboratory desk. According to expectation marked hyperpnea occurred. When similar amounts of cyanide were next injected into the carotid artery, the carotid sinus of which was denervated, no hyperpnea occurred. The alternate injections were repeated several times and the same very unexpected respiratory responses were observed. Next morning a planned experiment was performed and gave the same results. This primary unplanned experiment, at first sight a very foolish-looking trial, also started a series of planned experiments, performed with several methods, which led to the identification of the chemoreceptors of the carotid body and their functions".

Using the perfused head technique, C. Heymans, J. Jacob and Liljestrand (1947) demonstrated that ventilation in the perfused head was activated during and even after *muscular activity* in the donor. Hyperventilation or inhalation of O_2 by the donor decreases but does not abolish the reaction of the perfused head. Denervation of the carotid sinus does not abolish the stimulating effect of blood during muscular activity but the effect *after muscular activity* is depressed and even abolished. Therefore, O_2 acts through the carotid sinus by a reflex pathway and directly on the ventilatory centres.

Schmidt (1932) considered that a decrease in oxygen to a p_{O_2} of 67 mm Hg (92 per cent saturation of blood or the total oxygen at an altitude of 1200 m) would correspond to the threshold of reflex excitation for oxygen, while a variation of 0·1 of a pH unit or variations of ± 10 mm Hg in

p_{CO_2} with respect to normal would represent the reflex threshold for carbon dioxide.

In any case, I consider that in ventilatory chemoreflexes that which is active is the actual value of pCO_2/pO_2 and not only the value of pCO_2.

Acapnic apnea results from induced alkalosis (Haldane and Priestley, 1935). If this alkalosis is prolonged by extended hyperventilation, *death from asphyxia* can follow. C. Heymans and J. Jacob (1946) showed that *suppression* of the carotid sinus and aortic chemoreceptors *prolongs* acapnic apnea. Moreover, in the normal dog, hyperventilation, as acapnea, does not produce a decrease in arterial pressure. However, if the carotid sinus and aortic region are denervated, arterial hypotension results *during hyperventilation*, while, during *acapnic apnea*, the arterial pressure increases (a secondary decrease in arterial pressure is observed during prolonged acapnic apnea).

Chemoreceptors are present in the ventricles and cisterns of the cerebrospinal system. These receptors, situated notably in the fourth ventricle, are sensitive to changes in pH and pCO_2 and influence the ventilatory centres (Leusen, 1951; Loeschke *et al.*, 1958).

In 1959 G. Nahas recalled an exciting discovery. He wrote:

"In Genesis we learn that our common ancestor, Adam, gained conciousness after the Creator had filled his nostrils with the breath of life. The prophet Elisha was revived by direct insufflation of air into the mouth. However, many thousands of years passed before a doctor could be found who practised and described a method of artificial respiration. In 1743, William Tossach, a medical doctor in a small mining village in Wales, noticed during one of his rounds, a group of men gathered around a miner who had just been brought from the mine shaft in an unconscious state. Doctor Tossach felt the pulse of the miner but could detect no movement. With one hand he closed the nose of the miner, and, applying his lips to the mouth of the unconscious man, he blew into the mouth until the chest was raised. After the fifth inflation, the pulse could be detected and within several minutes the miner began to respire spontaneously. Four hours later he could walk and five days later he resumed his work".

CO_2, acting as a "respiratory hormone," and reflexes induced by pulmonary dilation were responsible for this revival. This method is of great value when a rescuer is ignorant of the method of Schaeffer which involves compression of the thorax and raising the arms. It should be noted that the gaseous mixture of artificial ventilation contains $O_2 + CO_2$.

The participation of carbon dioxide in the ventilatory mechanism is again evident from the pathological case known as the *Cheyne-Stokes rhythm*. When there are alterations in the ventilatory centres the sequence of thoracic movements is very characteristic: at first ventilation is weak, then increases through anoxia and an increase in p_{CO_2} which, at a certain threshold, induces a progressive increase in the amplitude of ventilatory movements. The CO_2 tension then decreases and ventilations weaken. The cycle is disrupted by continuous inhalation of O_2.

(c) *Mechanical stimulation*. C. and J. F. Heymans used the technique of parabiosis in dogs to study this problem. Mechanical stimulation can be performed in several ways: compression of the carotid sinus regions, modification of the perfusion pressure, administration of adrenaline (hypertensive) or acetylcholine (hypotensive). Hypotension in the aortic region stimulates the ventilatory centres and hypertension inhibits these centres. When the perfusion pressure in the carotid sinus region is decreased or the carotids are compressed, the amplitude and frequency of ventilation increases; an increase in perfusion pressure results in apnea.

Thus, *variations in arterial pressure modify the activity of the ventilatory centres* except in cases where the carotid sinus and aortic baroreceptors are denervated.

Other regions which respond to pressure changes by reflex activity are found in mammals, particularly in the left heart (Daly and Verney, 1926; Aviado *et al.*, 1951), in the right heart (Aviado *et al.*, 1951), in the mesenteric region (Gammon and Bronk, 1935).

In considering cardiac structures the reflex of Bezold-Jarisch should be mentioned. Intravenous injection of veratridin or veratrin (a mixture of alkaloids from a Liliaceae, *Schoenocaulon*) produces bradycardia and reflex hypotension by excitation of intra-auriculo-ventricular nervous receptors (Paintal, 1955).

It can be seen that the perfusion techniques of Heymans have contributed much to an understanding of basic cardiopulmonary regulation.

It should be added that in a preparation in which the vagi are intact, increased venous pressure in the right auricle which produces tachycardia (cf. p. 286) also produces hyperpnea (Harrison *et al.*, 1932).

(3) *Role of sensory nerves*

(1) *The trigeminal nerve* is the pathway of impulses of ventilatory structures of the head. Excitation produces reflex inhibition of ventilation (as in the case of the diving duck) or sneezing or the CO_2-apnea reflex of the rabbit (Rybak).

The ventilatory role of sensory nerves involved in smell can be determined: an agreeable smell leads to inspiration and a disagreeable smell to apnea.

(2) *The superior laryngeal nerve* provides the sensory innervation of the laryngeal mucosa but also contains motor nerves. Irritation of these sensitive fibres (which connect with the pneumogastric nerve) induces cough while electrical stimulation leads to apnea.

(3) *The glossopharyngeal nerve* (a motor and sensitive nerve) is involved

in swallowing which produces a transitory apnea and would take part in the Laborde reflex (rhythmic movements of the tongue).

(4) *Cutaneous nerves.* Generally, cutaneous pain accentuates ventilation while an increase in temperature leads to thermal polypnea and intense cold induces immediate inspiration. Thus, there is a relationship between ventilation and the cutaneous nerves.

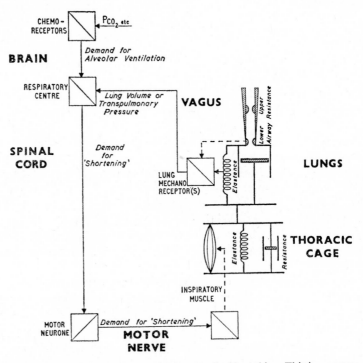

FIG. 110. Conventional schema of the control of breathing. This is a composite diagram in which the neuromuscular components are represented as control functions and the chest and lungs are represented as mechanical components. The inspiratory muscle appears twice, once as a control function receiving a nervous input (demand for shortening) and also, in more conventional guise, exerting tension in parallel with the elastance and resistance of the thoracic cage. The control functions are given the usual anatomical structure labels.

The descriptions applied to the input and output signals of the control components represent the mechanical equivalents of the nervous activity. The usual account of the control of breathing can be obtained by following the diagram anticlockwise starting at the top.

(5) *Muscle nerves.* Muscular activity results in adaptive ventilatory reflexes but the mechanisms of this effect have yet to be fully elucidated. The phrenic and intercostal nerves are of particular importance.

P. Dejours and his co-workers (1956, 1959) found that reflex contractions induced by repetitively striking the tendon of the patella or Achilles tendon are accompanied by hyperventilation. These ventilatory effects

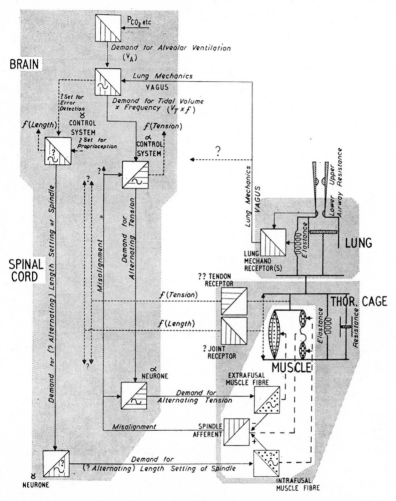

FIG. 111. Suggested schema of the control of breathing incorporating proprioceptive mechanisms in the respiratory muscles and thoracic cage.
This composite diagram uses the same conventions as Fig. 110. The inspiratory muscle (dotted) again appears twice, the translation being represented by boldly dashed lines.
The control functions are only given anatomical or structural labels if the correspondence between functions and structure is fairly certain. Finely dashed lines and question marks indicate less probable functional connexions.

continue even if articulations are blocked. Selective mechanical excitation of receptors in the joint has no effect on ventilation. Without eliminating the possibility of reflex ventilatory action of articular receptors, the results suggest a muscular localization of the proprioreceptors associated with ventilatory changes, which are stimulated by mechanical conditions such as occur in the limb during muscular exercise.

The thoracic muscles are also important. Campbell and Howell demonstrated that there is a constant relationship between the activity of the ventilatory centre, the activity of spinal neurones, the force of contraction or work of ventilatory muscles and ventilation or the work of breathing. They proposed a scheme (Fig. 111) which is more complex than the conventional scheme (Fig. 110) (1962).

In addition, there is visceral regulation of ventilatory movements. Graham showed in 1881 that electrical excitation of the abdominal sympathetic system led to ventilatory reactions and other authors (Gernandt, Siljestrandt and Zottermann, 1946; von Euler and Sjöstrandt, 1947; Tchernigovsky, 1947) demonstrated that mechanical stimulation of the intestine, stomach, etc. induced reflex ventilatory reactions. G. Paulet (1963) stated that distension of the viscera in the upper part of the abdomen of dog (stomach, gall bladder, pelvis of kidney) leads to a diminution of the thoracic component and augmentation of the abdominal component, the overall volume remaining unchanged. Distension of the viscera in the lower abdominal region (small intestine, bladder) results in diminution of the abdominal component (sometimes complete inhibition) and augmentation of the thoracic component, the overall volume showing little variation.

Remarks on pulmonary ventilation in mammals

Pulmonary ventilation can be maintained by a single lung and unilateral ventilation which can be maintained in man and tortoise, allows prolonged survival with compensatory hypertrophy of the functional lung (C. de Fano, 1912; H. Herman, 1921; Addis, 1928).

Inspiration is an active phenomenon involving inspiratory muscles, the diaphragm, intercostal muscles and external thoracic muscles but during forced expiration in man the pectoral and abdominal muscles are more important.

Inspiration is shorter than expiration.

Expiration is normally a passive phenomenon (relaxation of the thoracic cage) but during forced expiration the abdominal muscles act on the diaphragm.

Elastance (Bayliss and Robertson) is expressed by the coefficient of elasticity $\Delta P_{el}/\Delta V$ where P_{el} represents the elastic pressure of the lung exerted on the volume of air which the lung contains. The distension of the lung is important in determining P_{el} and, during normal ventilation, the curve defined by the coordinates elastic pressure–pulmonary volume is rectilinear. The coefficient of elasticity is 4·5 cm of water/l of air in man (Donders). The inverse of the elasticity ratio expresses *compliance* (Mead and Whittenbuerger). Normally, in man, this ratio has a value of 0·22 (Donders) which indicates that when the pressure varies by 1 cm of water, the pulmonary volume varies by 0·22 l. Therefore, certain problems relating to the lung, as to the heart (cf. p. 243) depend on an understanding of the technology of the resistance of materials.

During hyperventilation, the thorax of dog tends to adopt an expiratory position. During the apnea which follows, the position of the thorax returns progressively to that of inspiration (C. Heymans and Jacob, 1948). These rhythmic variations would be related to the level of oxygen.

The maximal tracheal pressure increases when resistance or elastance is placed on the trachea of the unanaesthetized or anaesthetized rabbit at the end of expiration. This phenomenon which results from vegetal reflexes, involves arterial chemoreceptors (hypoxia and hypercapnia) (Dejours, Lefrançois and Gautier, 1962).

The normal ventilatory rhythm in man at rest is higher in the newborn (40 cycles per mn) than in the adult (16 cycles per mn) and this may relate to the dominance of the higher nervous centres. During forward acceleration in man, the ventilatory frequency increases with acceleration in a rectilinear manner up to 12 *G*; the oxygen consumption decreases during acceleration and later increases.

Spirometry

Several pulmonary volumes can be defined:

(1) *Tidal air.* This is the volume of air which enters and leaves the lung at each normal ventilation (about 500 ml in adult man or approximately 8 l/mn; the maximum is 100 l/mn).

(2) *Inspiratory reserve.* This is the volume of air inspired by forced inspiration (about 1500 ml in adult man).

(3) *Expiratory reserve.* This is the volume of air expired by forced expiration (about 1500 ml in adult man).

(4) *Residual air.* This is the volume of air remaining in the lungs after forced expiration or alveolar air+air in the ventilatory passages or the *dead space* (about 1200–1500 ml in adult man).

The residual volume is measured by determining the dilution of an inert gas in the expiratory reserve. Boeri and his co-workers (1949) examined the dead space in man using hydrogen and found that a sudden, deep voluntary inspiration is accompanied by a small increase in dead space.

Boeri suggested that certain parts of the lung which during normal ventilation acted as air passages became ventilatory.

It can be deduced from these volumes that the *vital capacity* = *tidal air* + *inspiratory reserve* + *expiratory reserve*.

Anoxias

Anoxia is a state of oxygen deficiency in tissues; there is a decrease in the detoxicating ability and more and more intense poisoning related, in particular, to disorders of acid–base equilibria.

Anoxias are classified as follows:

(1) *Anoxemic anoxia* results from insufficient saturation of hemoglobin which can be due to lack of oxygen in inspired air (at high altitudes) or to mixing of arterial and venous blood in the heart in the case of the congenital cardiac disorder called "maladie bleue".

(2) *Anemic anoxia* results from hemoglobin deficiency induced either by post-hemorrhage anemia or by poisoning of hemoglobin with compounds such as carbon monoxide.

(3) *Circulatory anoxia* or *congestive anoxia* results from decreased capillary circulation due either to cardiac insufficiency or decreased capillary tone as occurs in states of shock.

(4) *Histotoxic anoxia* results from defective oxygen utilization when respiratory enzymes are inhibited by toxic compounds such as KCN of which the antidotes are sodium thiosulphate—which forms thiocyanate in liver in the presence of rhodanase—or methylene blue, a hydrogen acceptor.

Cyanosis corresponds to the presence in blood of an abnormally high level of reduced hemoglobin.

The organism reacts to a fall in p_{O_2} (hypoxia and anoxia) through chemoreceptors and there is regulation by the ventilatory centres to produce hyperventilation and an increase in cardiac output. There can also be an increase in the number of erythrocytes. Let us consider the conditions which can lead to an increase in the number of erythrocytes.

At zero altitude, p_{O_2} = 160 mm Hg (or in arterial blood approximately 96 per cent Hb saturation); at 3500 m altitude, p_{O_2} = 103 mm Hg (or approximately 85 per cent Hb saturation in arterial blood); at 8000 m, p_{O_2} = 60 mm Hg (or approximately 50 per cent Hb saturation in arterial blood).

Moreover, at zero altitude man contains about 5×10^6 erythrocytes/mm^3; after 24 h at 1800 m altitude the value increases to $5 \cdot 5 \times 10^6$ erythrocytes /mm^3 and to 7×10^6 erythrocytes/mm^3 after 24 h at 4200 m altitude.

This increase in the number of erythrocytes with altitude shows a certain latency; it leads to an increase in the hemoglobin content and oxygen-carrying capacity which compensates for the decrease in total atmospheric oxygen. This is a particular example of ecological adaptation. The increase in erythrocytes is due to three processes:

(1) hemoconcentration which increases during transpiration and is greater therefore, during active climbing than when the subject is transported (for example, by cable car);

(2) mobilization of blood in storage zones of spleen and liver [inverse of the process described after removal of the heart of frog (Rybak and Macouin, cf. p. 301)];

(3) stimulation of erythropoiesis by a mechanism which has yet to be explained fully but which may relate to increased production of *erythropoietin* or its precursor. Hypoxia causes disorders in several endocrine glands such as the pituitary, adrenals, thyroid and gonads (Gordon, Tornetta, D'Angelo and Charipper, 1943) which suggests that stimulation of hemopoiesis by hypoxia results from the action of hormonal mechanisms. Stewart, Greep and Meyer (1935) demonstrated that the hypophysectomized rat does not display an erythropoietic response to lowered barometric pressures corresponding to an altitude of 16000 ft.

These problems lead us to an examination of mountain sickness. Mountain sickness is experienced by man particularly after rapid ascension (as a result of atmospheric depression) or climbing (as a result of transpiration). This disorder is characterized by dizziness which is due principally to a decrease in p_{O_2}; it is associated with hyperpnea.

M. Brooks (1948) prevented mountain sickness by previous administration of tablets containing 0·2 g of methylene blue. This can be explained by the ability of methylene blue to accept hydrogen and thus produce detoxication in oxygen-deficient subjects. However, there is a possibility of methemoglobin being formed.

The bulbar region shows a relatively high resistance to anoxia; moreover, the newborn is a spinal creature (cf. Volume II) and its resistance to anoxia is high. In addition, the respiratory rate of the cerebral hemispheres of a newborn rat is lower than that in the adult (Himwich, Baker and Fazekas, 1939).

Hypercapnea and acapnea

Anoxia and asphyxia must not be confused. Asphyxia represents a pathological state which results from the combined action of anoxia and hypercapnea which is associated with cyanosis.

When the oxygen level in the alveoli is insufficient, or the release of carbon dioxide is impaired, the blood carbon dioxide content increases *(hypercapnea)*. This is anoxia through hypoventilation. An increase in the bicarbonate content of plasma results and there is also a more marked increase in H_2CO_3 leading to acidosis which activates the ventilatory centres and results in compensatory hyperpnea. During hyperventilation there is a decrease in the level of H_2CO_3 and bicarbonates in blood. The decrease in H_2CO_3 is greater than the decrease in bicarbonates and leads to alkalosis followed by compensatory acapnea.

Ventilatory acidosis is characterized by: (1) ventilatory changes (notably rhythmic abnormalities); (2) neuropsychic changes (headaches which are often occipital, trembling, alterations in recent memory); (3) cardiovascular changes (moderate tachycardia, peripheral vasodilation, digestive hemorrhages); (4) increase in acid secretions (notably sweat).

Ventilatory alkalosis is characterized by: tachycardia and arterial hypotension, increase in ventilatory amplitude, increase in urine volume, dizziness.

The implication of changes in acid–base equilibrium will be understood more clearly after we have dealt with the physiology of the kidney.

III. Renal physiology

In mammals, *the renal organs produce a liquid excretory product.* A comparative study of these organs would show the existence of very specialized structures such as the labial organs of Collembola, the urate-containing cells of adipose tissue which form the storage kidneys. The scope of this book is such that we cannot undertake such a study but the above remarks should emphasize the need to avoid a totally anthropological interpretation of physiology and should encourage a study of comparative anatomy and physiology.

Two points must be established before we examine renal function:

(1) Although the excretory and reproductive systems are separated in most invertebrates, in vertebrates they are closely associated. This spatial concentration (association) of vegetal organs in higher creatures appears to be a general rule which relates to an important evolutionary process and of which we shall meet further examples.

(2) *The pulmonary system and the renal system are the principal organs for the regulation of the internal medium* (cf. p. 309) in that they are the parts of the body concerned with the entrance and exit of such compounds as CO_2 which are important in determining the pH of the internal medium of birds and mammals. These two organs are directly involved in the purification of blood which occurs primarily through physical mechanisms: diffusion in the lung, filtration in the kidney and in mammals, concentration by countercurrent. The circulation in the kidney is related to renal function and there is also a nutritive circulation. However, fish, amphibia and birds possess a *renal portal system* which is equivalent to the tubule circulation in provertebrates. In mammals the tubules are nourished by the postglomerular blood. The arterial supply of the renal parenchyma arises from

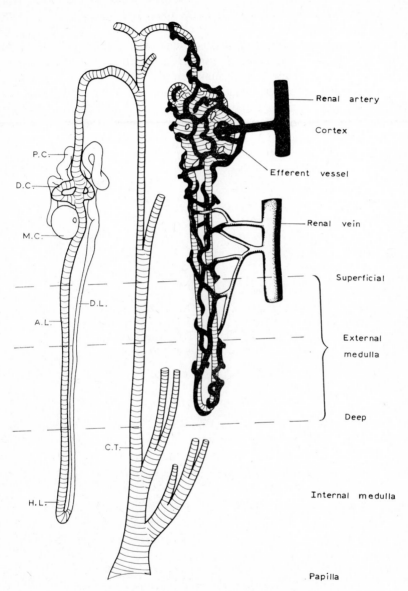

FIG. 112a. On the left, the tubule is drawn according to a diagram of G.C. HUBER. (CUSHNY, 1917.)

the functional arterial system while the veins which pass from the cortex become interlobular and finally form the renal vein which opens into the inferior vena cava.

In mammals, the kidneys consist of tens of thousands of uriniferous tubules which form the urine. Figure 112a illustrates the anatomy of a uriniferous tubule or nephron in mammals.

Urine is initially formed from blood circulating in the glomerulus situated in the cortex which also contains the proximal and distal convoluted tubules. The loops of Henle with parallel branches (the importance of

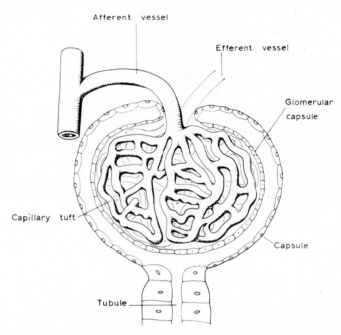

FIG. 112b. Diagram of a Malpighian corpuscle. (CUSHNY, 1917.)

which we shall appreciate later) are situated in the subcortical medullary region of the kidney. The glomerulus is formed from a network of arterial capillaries with an afferent arteriole and an efferent arteriole. The network is enclosed in the Bowman *capsule*.

The formation of urine in mammals takes place in three stages: glomerular filtration and selective secretion and reabsorption in the tubules. The modern concept (since 1951) considers that the processes of concentration must be explained in terms of a countercurrent. We shall examine the classical concepts first of all:

(1) Filtration

Filtration is passive and the quantity of glomerular filtrate depends on blood output and filtration pressure.

From 1 to 1·3 1 of blood pass through the kidneys of man each minute. The renal artery is a branch of the aorta and the hydrostatic pressure of blood in the afferent glomerular artery is about 75 mm Hg which is a high value for capillaries. However, this hydrostatic pressure is opposed by the osmotic pressure of plasma proteins—about 25 to 30 mm Hg—the interstitial pressure acting on the capillaries—about 10 mm Hg—the pressure required for movement of fluid in the tubules, about 10 mm Hg. The final filtration pressure is 25–30 mm Hg. During an arterial hypotension when the pressure in the glomerular capillaries is 45–55 mm Hg the filtration is prevented. This condition is anuria or oliguria which occurs when the intracapillary pressure is lower than the normal pressure.

In man, the normal output of glomerular filtration, as determined by dyes, is 120–125 ml/mn.

Normally, the fluid in the glomerular sac (primary urine) contains no proteins but urea, uric acid, creatinine, phosphates, chlorides, sodium bicarbonate and glucose (albuminuria, hematuria are indications of glomerular defects). The composition of this fluid was determined by intraglomerular puncture in frog followed by analysis of the chemical composition of the extracted fluid (Wearn and Richards, 1924).

(2) Tubular function

The controversy between the theories of *filtration–secretion* (Bowman, 1842; Heidenhain, 1874) and *filtration–reabsorption* (Ludwig, 1844; Cushny, 1917) is of little interest today and I shall consider the current ideas.

(a) Secretion

Many types of compounds are secreted and among the normal constituents of the blood, potassium is of particular importance. Secretion constitutes another form (tubular) of renal excretion. However, secretion is of particular interest as it relates to foreign compounds such as penicillin, iodinated compounds used in urography since they are opaque to X-rays, phenolphthalein which is employed in a test of renal function, etc.

Compared with glomerular filtration, this form of excretion which can reach 50 mg/mn for certain iodinated radiographic compounds, is an *active phenomenon*—a process of *active transport*—which derives energy from the oxidation of succinate. Inhibitors of succinic acid dehydrogenase decrease the excretion of phenolphthalein, for example; thus, succinic acid dehydrogenase acts as an *enzyme of permeation*.

(b) Reabsorption

The concept of reabsorption is derived from quantitative and qualitative information. The composition of urine is different from that of the glomerular filtrate (for example, the urea and glucose content) and the volume of urine is less than that expected if the glomerular filtrate was excreted without modification.

Reabsorption takes place in the tubules. Reabsorption is selective in the sense that not all micromolecules in the glomerular filtrate are reabsorbed.

Normally 100 per cent of glucose is reabsorbed and 99·5 per cent of water and 40–50 per cent of urea.

However, the thresholds of reabsorption must be considered. Although certain compounds are totally reabsorbed when their concentration in blood is normal, when the blood level of these compounds increases, they are not totally reabsorbed (saturation of permeation). For example, glycosuria appears when the level of glucose in plasma reached 2 g/l; this is the *renal threshold for glucose* of Claude Bernard.

Glycosuria

In man the normal concentration of glucose in blood plasma is 1 g/l and the glomerular filtration rate is approximately 120 ml/mn. Therefore, 120–160 mg of glucose pass into the glomerular filtrate per mn. However, the urine contains no glucose. The glucose is reabsorbed by the *proximal convoluted tubules.* The mechanism of reabsorption—*active transport*—involves *hexokinase* (which is identical with that in the intestine) which catalyses the phosphorylation of glucose to give glucose−6-phosphate (ATP is required). In this case, the enzyme for permeation is hexokinase. The phosphorylated ester of glucose is transported across the tubule cells to the capillaries where it is dephosphorylated by glucose-6-phosphatase with the liberation of glucose. Inhibition of phosphorylation by phlorhizin prevents the reabsorption of glucose and produces *phlorhizin diabetes* or *phlorhizin glycosuria.* The ability of tubular hexose-phosphorylation is limited such that hyperglycemia "restrains" the reabsorption of glucose (threshold of C. Bernard). It must be emphasized that in renal diabetes there is no hyperglycemia but tubular function is affected (a congenital disorder or by phlorhizin). F. S. Sjöstrand (1944) has shown that renal proximal tubule cells of several mammalian species contain autofluorescent granules which are segmentally distributed. Maunsbach (1964) has identified these granules as ultrastructures possessing acid phosphatase activity (cf. p. 373).

Reinecke (1943) stated that glycemia in the renal vein is greater than in the renal artery and concluded that there is biosynthesis of glucose in the kidney. This would appear to have been confirmed by Baïsset and his co-workers (1950).

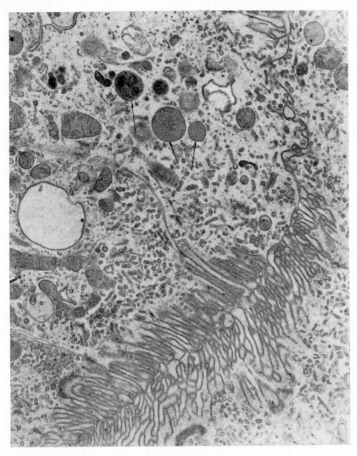

FIG. 113. Kidney of cat. Electron microscope. Original magnification:
×5000. Final enlargement: ×1500. Cells of initial segment: apical part.
Note the characteristic structure of this segment of the nephron such as the
brush border, mitochondria, cytosomes, vacuoles, numerous vesicles and
ribosomes. The arrows indicate the structures which are probably the autofluorescent "dense granules" of Sjöstrand–Maunsbach. (Figure kindly
provided by H. RUSKA.)

Hydruria

The main difference between the classical theory of H. W. Smith and
that of Wirz, Hargitay and Kuhn (1951) essentially concerns this aspect of
renal function.

(I) Classical concept

Under normal conditions the glomerulus passively filters from 120 to 130 ml of water per mn. However:

1. *In the proximal tubule,* electrolytes are partially reabsorbed and glucose is completely reabsorbed as well as 85 per cent of water *("obligatory reabsorption").* No reabsorption would occur in the loop of Henle through which passes 20 ml of water;

2. *In the distal tubule* (according to Smith), 10–12 ml of water are reabsorbed through the action of hypophyseal *antidiuretic hormone ("facultative reabsorption")* and there is also reabsorption of electrolytes.

3. *At the junction of the distal tubule and the collecting duct,* 7 ml of water are reabsorbed due to antidiuretic hormone.

Therefore, 1 ml of concentrated urine passes into the bladder each minute.

170–190 l of glomerular filtrate per day in man finally provides about 1·5 l of urine.

(II) New concept

It should be noted that mammals produce a hypertonic urine with respect to plasma. The deep regions of the kidney medulla contain the loops of Henle which consist of three segments: the *pars recta* leading from the proximal tubule, the smooth segment which generally forms a loop within the *medulla interna* and the ascending large segment in the *medulla externa* which connects with the distal tubule.

A comparative study of kidney sections from mammals shows that the *medulla interna* and *papilla* are more developed in animals from a dry habitat or ingesting dry food (Peter, 1909; Sperber, 1944). As Morel and Guinnebault (1961) have stated, it would appear that excretion of concentrated urine is associated with elongation of the smooth and large segments of the loop of Henle. Using the methods of micropuncture (developed by Richards) and microcryoscopy, Wirz, Hargitay and Kuhn have shown that the osmotic pressure of intratubular urine increases during the passage from the cortex, where it is equivalent to that of plasma, to the papilla where it is equivalent to that of urine due to accumulation of urea, Na^+ and Cl^-. The blood circulating in the extremities of the vascular loops is hypertonic with plasma and isotonic with urine. Following a suggestion by Kuhn, a concept of renal function was developed based on the principle of molecular concentration by countercurrent.

Consider a tube with the shape of a hair pin in which the central wall is formed by a membrane which is permeable only to water (Fig. 114). An electrolyte solution circulating in a certain direction encounters a straight tubular section and in this region the hydrostatic pressure is rapidly reduced such that a certain amount of water crosses the membrane and short circuits the narrow section, while salts become concentrated in the region of the loop. In this way the liquid develops a high osmotic pressure. Kuhn and his co-workers consider that such a mechanism is to be found in the loop of Henle. Moreover, the walls of the collecting tubules are considered to be highly permeable to water, at least in the presence of antidiuretic hormone (Morel *et al.*, 1959). The osmotic pressure of the fluid passing along these ducts is continuously equilibrated with that of neighbouring

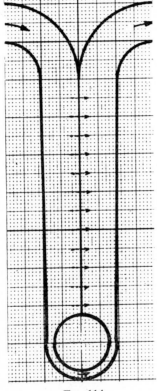

FIG. 114

tissues and is progressively increased by passive extraction of water. The urine reaches its maximum concentration at the point where it passes into the papilla. It is possible to study the renal distribution of tritiated water. Morel, Guinnebault and Amiel (1960) used this method to examine the kidney of the golden hamster *(Mesocricetus auratus)* kept on a diet containing no water. At intervals varying from $1/2$ mn to 10 mn after intravenous injection, the concentration of the tracer in renal tissue fragments was determined by scintillation in liquid medium of the β radioactivity of tritium. The distribution of ^{22}Na injected simultaneously served as a reference and was measured by γ scintillation in samples placed in a hollow crystal of NaI. These authors found a uniform distribution of ^{22}Na in different regions of the kidney in less than 2 mn while the renewal of water in the deeper regions of the kidney was not obtained until 10 mn after injection. These results can be explained if the walls of the vascular and urinary loops are more permeable to water than to sodium such that there is a countercurrent exchange of water between the ascending and descending branches.

The antidiuretic hormone is found in the (posterior) hypophysis and in *the supra-optic nuclei*, the (posterior) hypophysis being a storage organ (cf. p. 452). Secretion of this hormone is stimulated by a decrease in the water content of blood: hemoconcentration (for example, of altitude), hemorrhage. The sensitive region—a cellular "osmoreceptor" (Verney)—is found in a part of the hypothalamus measuring about 100 mm^3 in dog (Jewell and Verney, 1957). This hormone is partially destroyed in the liver.

Partial and progressive stenoses of the common carotid in the rabbit give rise to variations of diuresis, of plasma electrolyte level and to significant bioelectrical responses of the mesencephalic tegmentum and of the habenula. Constriction of the *vena cava inferior* produces a significant drop in urinary Na$^+$ excretion while arterial pressure and diuresis remain stable. This situation is suppressed by destruction of the habenula; destruction of the central tegmental grey substance and of the tegmentum of the ponto-mesencephalic boundary, on the contrary, enhances it (J. Faure *et al.*, 1962–1965). According to Faure the habenula-epithalamic system stimulates the secretion of aldosterone. This is to be correlated with the results of Farrell *et al.* (1959–1962) showing that stretching of the right auricle decreases the secretion of aldosterone while the emptying of the right auricle following partial occlusion of the *inferior vena cava* induces the liberation of aldosterone (Bartter, 1960).

The classical theory of *diabetes insipidus* considers that it results from the absence of antidiuretic hormone in the *distal* tubule alone. There is no reabsorption of water and water elimination increases at least by 10 such that 15 l of *dilute urine* can be produced each day.

This polyuria cannot be compared exactly with that induced by ingestion of large volumes of water. In the latter case, dilution of the blood leads to a decrease in plasma osmotic pressure and inhibition of secretion of hormone from the supra-optic cells resulting in polyuria.

On the other hand, according to the classical theory, the mechanism for the appearance of polyuria associated with sugar diabetes is as follows:

The glucose which is not reabsorbed increases the osmotic pressure of tubular urine and decreases the reabsorption of water in the proximal tubule such that 25 ml of water pass into the loop of Henle compared with 20 ml under normal conditions. This leads to the daily production of 3–6 l of urine with a high glucose content. These facts lead us to an examination of *diuretics*.

Diuretics

As we have just seen, intravenous injection of glucose leads to diuresis. Since the mechanism of activation of diuresis in this case is osmotic, other sugars can produce the same effect and also hypertonic salt solutions.

Other compounds act as diuretics either through an increase in cardiac work, for example caffeine, (V. Schroeder, 1887) or through a toxic action on the tubules (for example, mercurials such as calomel which acts on the distal tubule.)

Administration of desoxycorticosterone acetate leads to polyuria (Harned and Nelson, 1943). On the other hand, high concentrations of adrenaline have been reported to be antidiuretic (Rydin and Verney) while low doses of adrenaline would increase the volume of the kidney and the amount of urine. In the first case, vasoconstriction would be the predominant effect and in the latter case, augmentation of arterial pressure. Excitation of the splanchnic nerves (part of the sympathetic system) decreases urine secretion. Moreover, Claude Bernard demonstrated that polyuria can be induced by puncture of the floor of the IVth ventricle (osmoreceptor zone or stimulation of the reticular formation?).

Following a study of induced polyuria, A. Mayer and F. Rathery stated that the number of glomeruli open to excretion varies according to the degree of diuretic work performed, the majority of glomeruli being normally at rest. (These experiments involved the use of dyes such as Janus green and capillaroscopy). These results can be compared with those obtained by Krogh with intramuscular capillaries (cf. p. 301). This renal work "according to the demand" proceeds in the following way: during mild polyuria, the lumen of a certain number of convoluted tubules is approximately round while the other tubules remain at rest; in severe polyuria, most of the tubules are functional and show an increase in diameter of the lumen. Out of a total of 2×10^6 nephrons in man, only 50 per cent are functional under normal conditions.

Lamy and A. Mayer (1906–1936) visualized the glomerulus as a *"kind of peripheral heart"* which, by continuously beating at the end of the uriniferous tubule, ensured the passage of the filtrate by overcoming friction and capillary attraction within the tubule. However, aglomerular nephrons exist such as the nephridia of oligochetes and the nephrons of certain marine animals *(Opsanus tau)*; this anatomical arrangement appears to be related to reduced elimination of urine. It is a secondary adaptation to the environment and the statement of Lamy and Mayer would appear to remain valid in cases in which renal work is important such as in mammals. However, mention should be made, with respect to the large volume of urine eliminated, that the crab-eating frog, *Rana cancrivora* (from southeast Asia) can live in sea water and, in such a concentrated medium, its blood is slightly hypertonic to the medium. A considerable part of the

25

osmotic concentration is due to urea, the excretion of which is greatly reduced. This phenomenon is not due to active tubular reabsorption of urea but primarily to a *low urine flow* caused by increased tubular reabsorption of water and reduced glomerular filtration. Osmoregulation of the crab-eating frog in sea water resembles that of elasmobranchs (cf. p. 155) except that there is no evidence of active tubular reabsorption of urea in the frog (K. Schmidt-Nielsen and Ping-Lee, 1962).

Extra-renal purification

This is a therapeutic procedure which is of value during a limited period of renal insufficiency. It involves either the replacement of blood by transfusion (cf. p. 204) or the use of semipermeable membranes.

As the blood purifies the tissues, the kidney purifies the blood. In both cases the method is that of a capillary consisting of a membrane with limited permeability, in that non-renal capillaries allow macromolecules and even cells (diapedesis) to pass through the walls. Extrarenal purification consists of *dialysis*. The methods include:

(a) *Peritoneal dialysis* (Gantner, 1923; Putnan, 1923), in which exchange takes place across the peritoneum between circulating blood and a liquid perfused through the peritoneal cavity.

(b) *Intestinal dialysis* in which exchange takes place across the digestive wall between blood and a dialysis fluid rapidly circulating through the intestine. Derot (1958) considers that there is a modification of the composition of the dialysis fluid. However, this method is not often used.

(c) The "artificial kidney": this is a misnomer in that an artificial membrane hemodialyser is incapable of carrying out reabsorption. The blood circulates outside the organism in a tube made of a semipermeable membrane (generally cellophane) immersed in a dialysis bath. At the most, this apparatus could be called an "artificial glomerulus", forming a kind of primary urine. The "kidney" designed by Kolff (1940) is mainly used.

Clearances

Clearances express the purification of plasma, that is, the amount of blood plasma cleared by the kidney of a known micromolecular compound in one minute (Van Slyke). Here again we find a differential, before–after system.

Determination of clearance values

If U is the concentration of a *micro*molecule in *urine* and V (ml) the volume eliminated by the kidney in 1 mn, then $U \times V =$ total amount of the compound eliminated in 1 mn.

The concentration C_u of the substance u per ml of blood *plasma* can be determined and therefore, $U \times V/C_u$ gives the maximum clearance.

In order to measure these volumes it is necessary that:

(1) the substances introduced are not metabolized and their concentration can be estimated easily and accurately;

(2) time allowed for the experiment is such that a sufficient volume of urine is obtained in order to decrease the error;

(3) measurement of the concentration in plasma is carried out when the plasma concentration of the compound is stable.

There are essentially, as we have seen above, three processes in the nephron: (1) filtration; (2) reabsorption; (3) secretion.

The method of clearances allows these to be studied separately and is of great value in studying renal function. Three examples can be given in which the clearance of a substance is equal to, higher, or lower than glomerular clearance.

(a) The clearance of a substance which is only filtered (by the glomerulus) will be equal to the volume of the glomerular filtrate or about 120 ml per mn (for example, inulin or sodium thiosulphate).

(b) The clearance of a substance which is excreted by the glomerulus and the tubule represents the sum of glomerular filtration and tubular excretion (for example, phenolphthalein, creatinine).

In this case, the renal blood flow can be measured by catheterization and application of the Fick principle. If tubular excretion has removed all trace of phenolphthalein from blood passing through the tubular artery, one can conclude that the quantity of plasma cleared by the kidney is equal to the quantity of plasma which is effectively passed through the kidney in 1 mn.

(c) If the substance is partially reabsorbed by the tubule, the difference between its clearance and glomerular clearance gives the degree of reabsorption. For example, the clearance of *urea* in man is 70 ml/mn. However, tubular reabsorption can be measured using other compounds, notably mannitol, and ions (particularly ^{24}Na according to Gemell and his coworkers, 1957).

The segments of the nephron in which the processes of tubular reabsorption and excretion occur can be defined by using the technique of "interrupted diuresis" (stop–flow technique of Malvin, Wilde and Sullivan, 1958). The principle of this method is as follows: urine flow is stopped mechanically for several minutes during osmotic diuresis (of mannitol, for example) such that the hydrostatic pressure of urine in the tubule increases to a point where glomerular filtration ceases. Under these conditions, the composition of the static urine is modified according to the physiological activity of each tubular region. When the urinary flow is reestablished, the first fraction to appear corresponds to the urine of the collecting ducts and distal tubule, the intermediate fractions correspond to the different distal regions and the last fraction corresponds to normal urine. Morel and his co-workers (1961) have used this procedure in conjunction with radioactive tracer ions, Na^+, K^+ and Cl^-, and glomerular indicators and have shown that in rabbits: (1) the distal reabsorption area exhibits high permeability to potassium, low permeability to chloride and no permeability to sodium ions; (2) the immediately more proximal segment is by contrast, highly permeable to sodium and chloride ions; (3) glomerular indicators always occur in samples collected later; filtration continues slowly during the stop–flow run. This filtration does not affect the distal urine samples.

There are two points to be made, one concerning urea and the other, Na^+ and K^+:

(1) In man, the concentration of urea in blood is about 0·30 g/l which indicates a very low urea content; however, the concentration in urine is 25 g urea/l.

This results from the almost complete reabsorption of water and the partial reabsorption of urea such that urea is concentrated in the urine of the collecting ducts. This urinary concentration depends on diuresis: when diuresis increases urinary elimination increases. It should be noted that, during polyuria, the concentration of urinary urea falls but urea excretion is greater than during normal elimination.

The variations in urea output are expressed by the *laws of* Ambard:

First law: when the kidney produces urine containing urea at a constant concentration (*C*), the output (*D*) varies proportionally with the square of the blood urea concentration.

Second law: when the urea concentration in blood is constant, the urea output is inversely proportional to the square root of the urea concentration in urine.

The Ambard *constant* (*K*) is an empirical relation expressing the constancy and fluctuations in urea excretion:

$$K = \frac{\text{concentration of urea in blood (g/l)}}{\sqrt{\left(D \times \dfrac{70}{\text{wt. of subject in kg}} \times \dfrac{\sqrt{C}}{5}\right)}} = 0.07$$

normally in man.

(2) The sodium ion of blood plasma filters easily through the glomerulus and the glomerular urine contains sodium ions in the same concentration as in plasma. 85 per cent of the sodium ions is reabsorbed in *the proximal tubule* (reabsorption of chloride, phosphate, bicarbonate and water occurring at the same time). It would be interesting to know if transfer of these micro-ions is associated with changes in the electrical potential of cell membranes (cf. p. 388 for comparison with the results from the salt gland of herring gull).

A second sodium fraction is reabsorbed in *the distal tubule*. This reabsorption is related to the requirements of the organism for sodium and bases. Thus, distal reabsorption is facultative.

There is also marked reabsorption of sodium in the loops of Henle (Wirz).

A total of 99·55 per cent of sodium ion is reabsorbed.

According to the type of food, 3–5 g are eliminated daily in man.

Potassium ion is reabsorbed in the *proximal* tubule at the same time as sodium ion. However, in the *distal* tubule, there is excretion of potassium ions in exchange for sodium ions. When sodium ion in urine decreases, potassium ion increases. The concentration of K^+ increases in alkaline urine.

The essential facts of renal function regarding these alkaline ions are that there is distal reabsorption of sodium and distal excretion of potassium. These processes are modified by hormones notably aldosterone (cf. p. 428). There is an increase in the elimination of sodium ions in urine when adrenal secretion is low (aldosterone favours Na^+ retention).

Let us now consider the consequences of renal function with respect to the fate of renal products by considering their elimination on the one hand and their re-utilization on the other or the product of excretion, urine, and the effects of urine production on the ionic equilibrium of blood. It is to be noted that a mammal can survive with a single kidney but this kidney performs supplementary work. There is compensatory hypertrophy.

Human urine

(1) Excretion

(a) *The need for excretion.* The end products of catabolism constitute a useless, toxic fraction. Thus, accumulation of high levels of urea in mammals leads to a uremic coma; the same would apply for uric acid retention (during nephrites or disorders of the nephron). The toxicity of urine has been shown directly: 100 mg of human urine kills an adult rabbit after intravenous injection. The toxicity is due principally to potassium salts, polypeptides and to ill-defined hypertensive and neurotropic substances.

(b) *The route of excretion.* Urine passes to the bladder along the ureters which undergo rhythmic peristaltic contractions. The bladder, composed of smooth muscle fibres, distends and the orifice of the ureter is closed such that urine does not re-enter the ureter. Urine flow is prevented principally by the urethral sphincter tone but urine can also be retained voluntarily. The striated fibres of the prostatic gland and the membraneous part of the urethra are important in voluntary retention. These muscles are contracted or inhibited by fibres of the pudendal nerves.

In mammals, the bladder is a passive vessel and has no connection with the rest of the organism across the bladder wall. However, certain amphibia with a predominantly terrestrial habitat—and exposed to severe dehydration as in the case of *Cyclorana* of Australia—show the ability to reabsorb water from the bladder (Dawson, 1951).

Peachey and Rasmussen (1961) described the amphibian bladder as the phylogenic and functional analogue of the collecting tubule of mammalian kidney. This organ becomes highly permeable to water and urea in the presence of oxytocin and arginine—or lysine–vasopressin (cf. p. 451) (Maffly *et al.*; Rasmussen *et al.*; Sawyer, 1960).

(c) *Mechanism of excretion.* When the pressure within the bladder increases due to the accumulation of a certain volume of urine (about 400–500 ml in man), some of this urine passes across the cervix of the bladder and excites the innervation of the prostatic part of the urethra. If urine is not retained voluntarily, micturition occurs consisting of relaxation of sphincters and evacuation of the bladder. The bladder returns to the relaxed state while urine is finally eliminated by contractions of the *bulbus cavernosum*.

The following nerves take part in micturition:

The *hypogastric nerves* of the sympathetic system principally innervate the bladder sphincter; these nerves originate in the dorso-lumbar region of the spinal cord.

The *erector nerves* of Eckard are parasympathetic and innervate the longitudinal fibres of the bladder; these nerves originate in the sacral region of the spinal cord.

The *hypogastric plexus* receives branches from these two types of nerves. Excitation of the nerves produces contraction of the bladder.

Emotions are known to influence bladder contraction and micturition. The reflex centres of micturition are found mainly in the hypothalamus and spinal cord (Barrington).

(2) Composition

(a) *General properties.* Normally the pH of human urine is about 6·0 (from 4·7 to 8). On standing, urine becomes alkaline due to the formation of NH_3 from urea. Urine has an aromatic odour but after eating asparagus, for example, urine acquires a special odour (due to methylmercaptan?). Acetone can be detected during ketonuria.

No sugar is found in urine under normal conditions (but following a large meal, milligram quantities of glucose can be detected per l of urine). During diabetes, the urine can contain 100 g of glucose per day.

50 mg of ascorbic acid are found in adult urine each day.

Urochrome gives urine a yellow coloration and urine also contains small amounts of urobilin and hematoporphyrin. Normally urine contains no proteins ("albumin" is detected by heating. However, during this experiment, phosphate precipitation can occur and therefore, it is necessary to add a few drops of acetic acid after cooling. If the precipitate persists, albumin is almost certainly present). A special globulin, called the Bence Jones *protein*, appears during certain diseases such as lymphosarcoma. This protein exhibits an unusual property in that it is precipitated when urine is heated to 60°C and redissolves at 100°C. It is not precipitated by cooling.

(b) *Principal constituents.* (1) *Urea.* The urea content increases with protein catabolism. The concentration of urea is estimated using *urease* or hypobromite (volumetric estimation). In the latter case the following reaction takes place (93):

$$CO(NH_2)_2 + 3NaOBr \rightarrow 3NaBr + N_2 + CO_2 + H_2O \qquad (93)$$

(2) *Ammonia.* Fresh urine normally contains very little ammonia. When large quantities are detected these are derived generally from urea rather than being preformed.

(3) *Creatinine* and *Creatine.* The amount of creatinine eliminated is fairly constant (cf. p. 144) and for this reason has been used as an index. The concentrations of creatinine are:

20–26 mg/kg body weight/day in normal man

14–22 mg/kg body weight/day in normal woman.

Thus, the urinary level of creatinine constitutes a secondary sexual characteristic which is dependent on muscle mass. Then urinary creatinine could be related to the law of surfaces (cf. p. 26).

Creatine is found mainly in the urine of infants and increases during gestation. The urinary content decreases during hypothyroidism.

(4) *Uric Acid.* An average concentration of 500 mg/l is found in urine of man.

(5) *Amino Acids.* The principal amino acid of urine is glycine followed by histidine and cystine.

Adult man eliminates about 200 mg of amino acids per day in urine. Thoai and Roche (1957) also reported the presence of monosubstituted derivatives of guanidine in urine of rats and man.

(6) *Steroid hormones* and *gonadotropins.*

(7) *Ions*

Cl^- (mainly in the form of NaCl);

sulphur in the mineral form or conjugated (sulphur derivatives of detoxication) and also in the neutral form (cystine, taurine, thiocyanate);

PO_4^{---} (in combination with Na^+, K^+, Ca^{++} and Mg^{++}).

Role of the kidney in ionic equilibria of the internal environment

(1) Osmotic regulation

The kidney regulates both water and mineral salts. With respect to mineral salts, renal work is concerned principally with NaCl. We shall return to this topic when we consider the hormonal regulation of water and mineral balance (cf. p. 456).

(2) Acid–base regulation

Amino acids and proteins play some part in acid–base regulation. However, proteins do not normally filter into the tubule whereas micro-molecules such as amino acids pass through. In so far as the kidney affects amino acids it conditions the role of protein derivatives in acid–base regulation.

It is important to understand that while oxidation–reduction potentials correspond to electron transfer, acidity potentials correspond to proton transfer.

A proton consists of the nucleus of a hydrogen atom, i.e., a hydrogen atom which has lost an electron.

According to Brönsted's concept:

acids are compounds which tend to lose protons,
bases are compounds which tend to gain protons.

$$\text{acid} \rightleftharpoons \text{base} + p \qquad (94)$$

Thus

$$CH_3CO_2H \rightleftharpoons CH_3CO_2^- + p \qquad (95)$$

similarly for water: in the oxidized form

$$H_2O + p \rightleftharpoons H_3O^+ \qquad (96)$$

From left to right, water acts as a base and from right to left, the hydronium ion H_3O^+ (or hydrated hydrogen ion) acts as an acid; in the reduced form

$$H_2O \rightleftharpoons HO^- + p \qquad (97)$$

By combining (96) and (97):

$$2H_2O \rightleftharpoons HO^- + H_3O^+ \qquad (98)$$

(protons no longer appear).

By applying the law of mass action:

$$\frac{[HO^-][H_3O^+]}{[H_2O]^2} = K \text{ (for a given temperature)} \qquad (99)$$

The dissociated fraction is always small compared with the total number of water molecules:

$$[H_3O^+][HO^-] = K_{H_2O} \leftrightarrows 10^{-14} \text{ at } 25°C \qquad (100)$$

K_{H_2O} defines the acidity constant of the acid–base complex. It is more convenient to use $pK = -\log K$ or, in this case, 14. A scale of pH from 0–14 is obtained.

All ionizable compounds have one or many pK values according to the number of their ionizable groups. Thus, the buffering ability of cysteine is dependent on the ammonium and sulphydryl groups.

In strongly acid solutions, cysteine cation has the following structure:

SH
|
CH_2 which gives three distinct stages of ionization, three pK
|
$HC-NH_3{}^+$ values, with the three acids groups SH, $NH_3{}^+$, COOH.
|
COOH

At the isoelectric point cysteine has the following structure:

SH
|
CH_2
|
$HC-NH_3{}^+$
|
COO^-

Thus, the kidney modifies the buffering capacity of the internal medium by the elimination of amino acids. The kidney also modifies the acid–base balance of the internal medium by eliminating non-volatile acids: lactic acid, sulphuric acid, phosphoric acid, etc.

It is through the regulation of the CO_2–bicarbonate system that the kidney has the greatest influence on the acid–base balance. We must consider the elimination of hydrogen ions by exchange with sodium or ammonium ions of bicarbonates, phosphate acid and chloride, exchanges which take place in different regions:

(1) *In the proximal tubule:* sodium ion of *bicarbonate* is mobilized. *Carbonic anhydrase*, discovered by Meldrum and Roughton in 1932, catalyses the production of carbonic acid. Figure 115a illustrates the exchange reactions for bicarbonate in the system: blood–proximal tubule cells–tubular filtrate.

(2) The filtrate from which all bicarbonate has been reabsorbed passes into *the distal tubule* where H ions are exchanged with sodium ions of Na_2HPO_4. These processes are summarized in Fig. 115b.

(3) Cells of the *distal tubule* produce ammonia, by deamination of amino acids, which combines with H ions to give ammonium derivatives. These reactions are summarized in Fig. 115c.

It should be noted that the blood and intracellular mechanisms of these processes of H^+ elimination are the same, except in the case of ammonia production. The mechanisms mainly differ in the tubule according to the salt. I feel there is one point which can be made which unifies the processes of excretion: *hydrogen elements (electrons in respiration and protons in acid–base balance) are the principal toxic products of highly evolved aerobic organisms.*

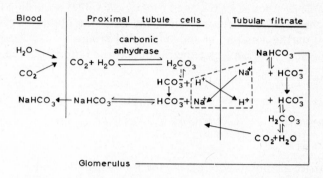

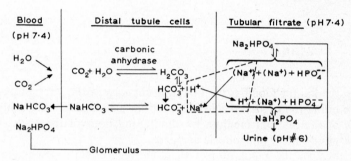

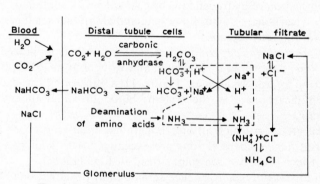

FIG. 115. (From PITTS, 1950; and H. A. HARPER, 1957.)

I could not develop here this point of view but, in general terms, I agree with V. M. Goldschmidt (1952) that the evolution of living creatures occurred in increasingly oxidative ambient conditions. This fact helps to explain a certain number of evolutionary phenomena and particularly, the toxicity of hydrogen elements for mammals and man in particular. The first living creature must have been anaerobic.

When the p_{CO_2} of the internal medium increases, the production of carbonic acid and thus, of hydrogen ions in the tubular cells, is augmented. This facilitates the compensatory reabsorption of bicarbonate such as occurs in ventilatory acidosis. During *ventilatory alkalosis* the reverse process occurs with *elimination* of bicarbonate to a point where the normal ratio (1/20) of carbonic acid to bicarbonates is re-established.

Carbonic anhydrase is inhibited by sulphamide derivatives such as *acetazolamide*. Administration of this compound to mammal or teleost (Maetz, 1956) results in the appearance of alkaline urine and an increased amount of sodium bicarbonate. This provides a method for elimination of Na^+ in chronic hypertension. Moreover, Mölleström and his co-workers demonstrated an increase in diuresis and urinary glucose excretion.

Stolkowski (1950) found that carbonic anhydrase is transformed by dilution and is inhibited by NaCl. This can be related to the formation of calcite in shells of molluscs or, in other cases, of aragonite (two calcareous products cf. p. 43).

Other aspects of kidney physiology

(1) Renin (cf. p. 292) is normally associated in the kidney of rabbit with intracellular granules of cells localized in the vascular part of the glomeruli. These cells appear to be the principal sites of renin storage (Cook and Pickering, 1963). These granules disappear when renin activity is reduced by acute experimental nephrites in rat (Vikhert and Serebrovskaja, 1963).

(2) Kidney extracts stimulate erythropoiesis (kidney factor of Osnes, 1960). However, Baciu and his co-workers (1963) concluded from studies in dogs and rats that the kidneys do not produce erythropoietin. The debate is open.

(3) The kidney is particularly sensitive to any change in the general condition and Reilly has shown that irritation of the autonomic system induces renal disorders. Although the kidney adapts favourably to conditions constituting the gravid state, the kidney is the first organ to react (albuminria, etc.) to any disorder during gestation. The determination of kidney function is of interest, therefore, during experiments on mammalian embryology. Moreover, we have seen that glucose solutions augment diuresis when injected intravenously. On some occasions parenteral administration of glucose leads to swelling of the proximal part of the nephron and rapidly fatal oliguria. This sugar nephrosis is particularly serious when glucose administration is accompanied by saline depletion (notably by peritoneal dialysis). This is a particular reaction of the kidney to sugar and not a general reaction to hypertonicity. Rouiller and Modjtabai (1958) studied the alterations during sugar nephrosis in cells with striated borders

using electron microscopy. Normally, the cells contain rod-shaped granules called Heidenhain rods. During sugar nephrosis, the rods degenerate to form small granules and then thin mitochondria which are eliminated and not regenerated. There is tumefaction with the formation of cavities between intra-cytoplasmic membranes.

(4) *Kidney injury* has been studied in rabbit by J. J. Bounhiol (1935) and causes death in young specimens (about 1-year old) and malignant growth in older specimens (3 years). Initially there is hematuria and albuminuria followed by renal insufficiency and hyperkalemia and finally anuria.

IV. Other excretory pathways

Apart from phaneres (cf. p. 48) and the gills of fish—equivalent of lungs—which have an excretory role, four excretory pathways can be mentioned: lacrymal excretion, salivary excretion, biliary excretion and excretion by sweat glands.

I do not wish to discuss the human lacrymal secretion which essentially contains sodium chloride but a related secretion. In marine animals such as birds and oceanic reptiles (for example, turtles), the ingestion of seawater produces no toxic effect. K. Schmidt-Nielsen (1958) demonstrated

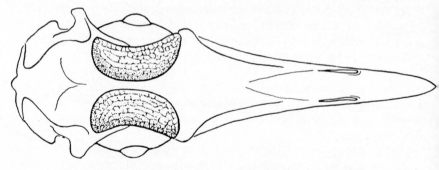

FIG. 116. Situation of the salt glands. (SCHMIDT-NIELSEN, 1958.)

that the ability to absorb such a concentrated solution is related to a particular anatomical feature. The nasal glands on each side of the head maintain efficient extra-renal elimination of salts (particularly NaCl). Two-thirds of absorbed NaCl are eliminated by this nasal pathway. The droplets which are very hypertonic to blood give rise to the characteristic behaviour of such animals: gulls and penguins are known to carry out rapid movements of the head. Schmidt-Nielsen has demonstrated that shaking of the head corresponds with elimination of nasal secretions. The nasal glands are controlled by the parasympathetic system in *Larus* (R. Fänge, Schmidt-Nielsen and Robinson, 1958). This system forms part of a reflex involving osmoreceptors and augmentation of blood osmotic pressure by NaCl solutions or sucrose leads to secretion by the nasal glands.

When the salt gland of the herring gull excretes sodium chloride the duct of the gland becomes positive relative to the blood. Strophanthin, an inhibitor of active sodium transport, prevents the establishment of the positive potential and also blocks gland secretion. The findings suggest that an active transport of sodium from the blood to the gland lumen may be a primary secretory mechanism (Thesleff and K. Schmidt-Nielsen, 1962). Thus, electrophysiology is not only of interest with respect to excitable structures.

Let us now consider the other three types of excretory pathway.

I. Saliva

Saliva does not serve only as a lubricant but also provides an excretory pathway for certain virus (e.g. rabies), compounds such as morphine, ethanol, acetone and mineral compounds: K^+, Ca^{++}, HCO_3^-, I_2, SCN^-. Thiocyanates are found in the three types of saliva. Rhodanase (K. Lang, 1933) which is present in saliva is a detoxication enzyme (thiosulphate is an antidote of HCN):

$$HCN + Na_2S_2O_3 \xrightarrow{\text{rhodanase}} HCNS + Na_2SO_3 \qquad (101)$$

Certain hormones are found in saliva, for example, insulin (Scott), pituitary and ovarian hormones in the pregnant female (Trancu), blood group substances.

II. Bile

Bile eliminates pigments formed by the destruction of erythrocytes in the reticulo-endothelial system (bilirubin), various toxins, Cu^{++}, Zn^{++}, Hg^+ and cholesterol which sometimes precipitates to give calculi which are not necessarily formed with Ca^{++}.

Since the phenomena involved are of considerable physiological importance, I shall consider certain points concerning urobilin derivatives. These compounds are formed by rupture of the porphyrin ring and have the following general formula:

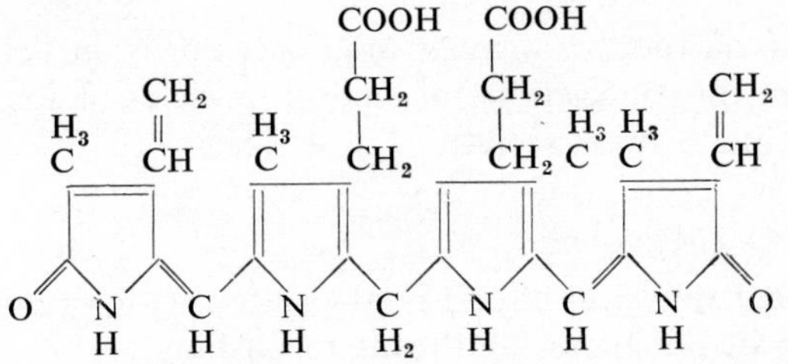

Urobilins are derived from urobilinogens and pass along the entero-hepatic circulation (von Müller, 1887). Yamaoka and Kosaka (1951) demonstrated that the bile pigment ("direct bilirubin") is an ester of bilirubin. In the liver, bilirubin conjugates with glucuronic acid and Malloy and Lowenstein (1940) suggested that jaundice could be due to difficulty in the conversion of bilirubin to "direct bilirubin" in the liver. In certain regions of South Africa the plant *Tribulus terrestria* is responsible annually for the death of large numbers of sheep suffering from jaundice and becoming photosensitive. Rimington and Quinn (1934) identified the agent responsible for photosensitivity as phylloerythrin, a chlorophyll derivative. Rimington, Quinn and Roets (1937) isolated a triterpenic acid from another plant, *Lippia rehmanii*. This compound which they called "icterogenin" reduced biliary secretion to a small volume of white bile after administration *per os* in a dose of 1–4 g.

III. Sweat

Sweat is produced continuously over the whole area of human skin and constitutes *insensible perspiration* produced by the *sweat glands*.

The pH of human sweat is acid but in cat and horse the pH is alkaline (sweat is not produced by dog or rabbit).

The odour of sweat is associated with the presence of volatile fatty acids: butyric acid, caproic acid and formic acid.

The production of sweat must not be confused with cutaneous evaporation which does not involve specialized glands and represents a simple physico-chemical phenomenon.

In temperate climates, 1 kg of sweat is eliminated each day by man while 5 kg/day can be produced under conditions of heat or physical activity.

Sweat contains: urea, the total of which varies between 1 and 5 g/l of sweat depending on renal work, 1 to 10 g of lactic acid/l (depending on muscular activity), traces of proteins in man while the sweat of horse contains detectable amounts, 3·5 g/l NaCl and finally KCl, calcium and magnesium phosphates (0·3 g/l).

Mechanisms of sweat production

Sweat passes via the "*vis a tergo*" and accessorily in the sweat glands of the axilla, by the contractions of smooth muscles of the glands; local depressions facilitate sweat secretion.

(1) Action of the circulation

(a) *warm sweat* (physical effort) is associated with vasodilation of the vascularization of the glands and leads to flushing,

(b) *cold sweat* (agony, fear, etc.) is associated with constriction of cuta-neous capillaries and paling. The existence of cold sweat clearly suggests that sweating is not necessarily associated with the importance of the local circulation. Direct participation of the nervous system seems established.

(2) Action of the nervous system

(a) *Nerves*. Fibres innervating the sweat glands are found in the sciatic nerve, the facial nerve, the trigeminal nerve, etc. After sectioning the sciatic nerve to the paw of cat, no sweat is secreted by that paw when the animal is placed in a heated chamber although secretion occurs in other regions of the body.

These fibres originate in the dorso-lumbar region of the spinal cord and form the white *rami communicantes* which join the post-ganglionic fibres of the gray *rami communicantes* in the ganglia of the sympathetic chain. These post-ganglionic sympathetic fibres are cholinergic in man and cat. Therefore, it is not possible to generalize as to the type of nerve involved or as to the type of sweat produced. Acetylcholine or parasympathomimetics stimulate the production of alkaline sweat in cat while in horse, adrenaline stimulates the secretion of alkaline sweat. In man, acidic sweat is produced by the action of acetylcholine or parasympathomimetics.

(b) *Higher centres*. These are localized in the hypothalamus, bulbus and medulla. They are stimulated directly by CO_2 (asphyxia induces sweating in cat) and by *reflex* pathways during changes in ambient temperature. Sudation associated with fever results from the direct participation of the higher centres.

Sweat is a factor in the cooling of homoiotherms and is also a medium for excretion: thus, the sweat glands have a similar function to that of the kidneys. The excretory function of the sweat glands leads to hemoconcen-tration which explains the therapeutic value of sweating.

SOMATIC AND SEXUAL ENDOCRINOLOGY

INTRODUCTION

The concept of endocrine glands extends from anatomical considerations to physico-chemical consequences.

Many glands of the animal organism release secretions into specific ducts. These are the exocrine glands. Other glands produce substances or *hormones* which are released generally into the blood circulation *(hemocrine)* or within nervous tissues *(neurocrine)* which constitute the pathways for transport in the organism. These are the *endocrine glands*. Endocrine glands can be vesicular (thyroid), reticular (adrenal medulla), tangled (parathyroid) or diffuse (cells in connective tissue such as the interstitial gland of the testis). Hormones are necessarily non-toxic products in the concentrations found in the organism; however, exocrine secretions, as shown by the fact that they are eliminated, are products which cannot be tolerated by the organism (cf. p. 14).

In his *Leçons de Physiologie expérimentale* of 1855, Claude Bernard clearly explained the distinction between exocrine and endocrine secretions. He wrote: "It must now be well established that there are two secretory functions in the liver. One, the exocrine secretion, produces bile which flows to the exterior while the other, the endocrine secretion, produces sugar which immediately enters the blood of the general circulation". However, Claude Bernard recognized only the *metabolic* function of internal secretions and not their function as catalysts in the utilization of metabolites. Today this function is being understood more fully but Brown-Sequard and d'Arsonval (1891) were the first to provide a more general hypothesis based on the results obtained from injections of *extracts* of *glands*—including an injection of a testicular extract to Brown-Sequard: "We accept that each tissue or more generally, each cell of the organism secretes products or special ferments which are released into the blood and which, by the intermediary of this fluid, affect all other cells which are united by a mechanism other than that of the nervous system". The idea of a humoral pathway was presented (confirmed later with the discovery of antibodies) and the fundamental role of substances which are not necessarily metabolites was formulated. However, it was not until 1902 that the concept of a *hormone* began to have a more precise significance. Bayliss and Starling demonstrated that acidic extracts of duodenal mucosa stimulated the production of the external secretion of the pancreas by way of the blood stream. They gave the name "chemical messenger" (1903) to this

394

type of biological substance. In 1905, the term *hormone* (from the Greek: I excite) was proposed by W. B. Hardy and adopted by Starling.

At the present time we can use the definition of Selye: "Hormones are physiological compounds of organic structure which are formed by certain cells and whose role is to direct, regulate and co-ordinate the activities of the same organism at sites far away from the site of production". Sainton, Simonnet and Brouha added (1950): "These substances act in very low concentration".

A hormone can be defined in morpho-physiological terms and as an example of such a definition I include that of P. J. Gaillard (1960): "Hormones can be defined as substances which are normally produced by groups of cells and which are transported to other parts of the organism in such a way as to allow sensitive tissues to undergo morphological and/or functional changes of a specific character".

The concept of a hormone should not be restricted to that of a vegeta catalyst. Neuro-hormonal relationships suggest that hormones are involved together with nervous influx in transmitting information.

This cybernetic aspect leads to a consideration of the function of hormones in integration. Endocrinal integration constitutes a stage in the organization of homoiostasis. One can consider there to be several codes. I would draw attention to the concept of coding presented in 1955 with respect to memory. The transference of information depends on a nucleic code (desoxyribonucleic acids, messenger RNA), a hormonal code and lastly a neuro-hormonal code (the manifestations becoming progressively less chemical and more physical), the whole directing the metabolic code which corresponds to interrelationships between free or "structural" molecules. Thus, each living organism presents a spatio-temporal focus in which all information is interconvertible. The concept of reflexes, which has fundamental significance with respect to homoiostasis, includes reflexes which modify different codes and which pass along an information circuit from up stream to down stream, from peripheral to central regions and inversely. In this section we shall be concerned with coupling of reflexes, as neuro-hormonal reflexes particularly when we consider the relationships between the hypophyso-pituito-hypothalamic system and other endocrine glands. In the second volume we shall consider a system of greater complexity, the integrating system controlling biotic rhythm.

This system of codes and decoding implies specificity, that is, affinity. Specificity relates to the selective aspect of homoiostatic organization. In the case of hormonal molecules, their effect will be determined by the number of receptor molecules within the organism. These receptors can be localized or widely distributed such that many organs can respond to varying degrees—that is to say, at different thresholds—to a single hormone ("functional ambivalence"). Specificity can be described in terms of a key-lock linkage and implies certain conditions: existence (affinity),

26*

quantity (efficacy) and distribution (topography determining accessibility). This analogy provides an indication of the way in which the concept of the evolution of internal secretions should be formulated to include the substances with only nutritive capacity (the trophins) and the messenger compounds, that is, information carriers with catalytic affinities (the tropins).

The concept of endocrine glands and hormones is of significance not only with respect to mammals but can be extended to include vertebrates in general and invertebrates. Endocrinology of invertebrates which is beginning to be understood in the arthropods will not be included systematically in this book which will expand the concept of "chemical messengers" essentially in terms of vertebrate endocrinology.

ENDOCRINE GLANDS AND METABOLIC EQUILIBRIA

CHAPTER 13

THE THYROID GLAND

THE important dates in the history of thyroid physiology are:

1895–1896 Baumann showed the presence in the thyroid of iodine in combination with organic compounds and demonstrated the formation of iodinated proteins in the gland.

1915 Kendall isolated *thyroxine* in crystalline form.

1926 Harrington and his co-workers provided the chemical structure of thyroxine and di-iodotyrosine; Harrington and Barger (1927) synthesized thyroxine.

1927 Smith described the relationship between the regulation of the thyroid and the pituitary.

1929 Marine and his co-workers discovered goitrogenic dietary factors with antithyroid activity; this discovery led to the therapeutic use of synthetic antithyroid compounds (Astwood).

1935 Collip and Anderson isolated pituitary *thyrostimulin*.

1939 Hertz, Roberts and Evans used *radioactive iodine* in a study of thyroid function.

1952 Identification of 3, 5, 3'-tri-iodothyronine in the thyroid gland (Gross and Pitt-Rivers; Roche, Lissitzky and Michel).

Origin and structure of the thyroid

The thyroid arises from an *endodermal* outgrowth from a medial bud situated in the posterior part of the pharyngeal floor.

In adult man the thyroid consists of two lobes joined by the isthmus and situated symmetrically with respect to the trachea in the neck (average weight $= 30$ g).

The thyroid consists of a fibrous capsule and internal parenchyma. The physiological entity of the thyroid is the follicle. The gland contains numerous follicles of unequal size, separated by connective tissue containing numerous capillaries. The thyroid is one of the most highly vascularized organs: in terms of unit weight the thyroid receives more blood than the kidney. At branching points of small arteries of the connective tissue near the thyroid gland, cushions of the intima projecting into the lumen are to be seen; the cushions belong to a sphincter (Reale and Luciano, 1966)

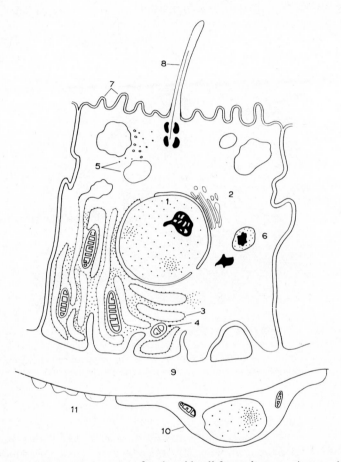

FIG. 117. Schematic structure of a thyroid cell from electron micrographs. (Modified from EKHOLM and SJÖSTRAND, 1957 by STOLL, MARAUD, LACOURT, *Ann. d'Endocrinol.*, 1961.)
1, nucleus; *2*, Golgi apparatus; *3*, α-cytomembranes consisting of an outer wall of Palade granules and surrounding a central canalicular cavity; *4*, mitochondria associated with α-cytomembranes; *5*, various granules with nonparticulate contents; *6*, granules with particles; *7*, microvilli; *8*, filament projecting in vesicular colloid.
Note in addition the presence of Palade granules in the cytoplasm situated between the α-cytomembranes. The cell is separated from the capillary by a basal membrane or periendothelial space (*9*) formed by 2 osmophilic layers surrounding a central clearer layer. The endothelial cell (*10*) limits the capillary (*11*) and shows in places thinner regions where the plasmic membranes are united.

which, very probably (Rotter, 1952) serves to regulate the blood circulation. Each follicle consists of a flat epithelium which secretes the colloid containing the thyroid hormone or hormones.

Activity of the follicles is accompanied by structural changes (Aron) consisting of hypertrophy of the follicular epithelium, nuclei and Golgi apparatus and vacuolization of the colloid mass indicating resorption of hormonal products (or deficiency). When the two lobes of the thyroid,

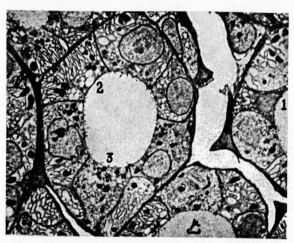

FIG. 118. Thyroid of a newly hatched chick.
Note the varied cell images which are related to the state [of the cytomembranes which are shown as dark areas (*1*), clear areas (*2*) or associated with colloid (*3*). Note on the right the presence of a capillary with a discontinuous wall of the sinusoid type ($\times$ 3500). (STOLL *et al.*, 1958.)

which is formed initially from a compact cellular mass, acquire a filamentous structure, the α-cytomembranes—which are ultrastructural lamellae associated with mitochondria and limited by a wall lined with ribonucleic acid granules (Palade granules)—are filled with a substance which Stoll and Maraud (1961–1962) described as the polysaccharide fraction of colloid substance (Fig. 117). Synthesis of the protein fraction must depend on ribosomal activity.

Methods of study

Although these methods are described for the thyroid, they are, in essence, applicable to *endocrinological systems in general*.

(1) Total or partial ablation of the gland (puncture, biopsy) or grafting.

(2) Administration of extracts prepared from homogenates by the use of various solvents; isolation and purification by chemical or physico-chemical fractionation of the hormone (or hormones) and intermediate products

of synthesis and degradation notably by the use of radioactive tracers (in this case, ^{131}I which emits γ-rays); finally, the receptor molecules must be isolated in order to establish the mode of action of the hormone.

(3) A study of the fixation of ^{131}I by the thyroid (detected by a scintillation counter or *in situ* autoradiography).

(4) Histophysiological studies on different animals at different ages (staining, phase contrast microscopy of tissue slices, autohistoradiography, electron microscopy particularly in studying the chondriome).

Often, abnormal or pathological conditions are induced in order to understand the physiology of an organ. Abnormalities contribute to an understanding of the normal condition.

The important results

The important results are physiological, chemical and biochemical.

I. Physiological results

Thyroidectomy leads to characteristic disorders particularly in *young* mammals, these disorders appearing after a *latent period* (10–20 days). They are characterized by a slow rate of growth, delayed ossification (principally of the long bones), muscular atrophy which can affect the myocardium, lack of development of genital organs and obesity, a decrease in the liver content of cytochromes and amino acids oxidase, inhibition of the transformation of carotene to vitamin A leading to visual disorders, changes in nitrogen metabolism with a reduction in basal metabolism, there being a possible reduction of 30 per cent in O_2 consumption per kg body weight and of 20 per cent in CO_2 production of the rabbit.

In adults, thyroidectomy leads to degeneration of Nissl granules (cf. Volume II) associated with degeneration of cerebral neurones and loss of mental vigour (the tone of the cerebration depending therefore not only on nervous stimuli). There is general apathy with hypotension and hypothermia together with loss of sexual libido.

These disorders can be overcome by grafts or administration of hormones. *A graft is successful only in conditions of marked thyroid insufficiency* which indicates that the organism can tolerate only a certain quantity of active thyroid hormone and that saturation results in rejection of excess active products. Simon and Morel (1960) provided rats with water containing radioactive iodine and established that the thyroid secretes 1 μg hormonal iodine each day in animals receiving 5 μg dietary iodine each day.

This result leads us to a consideration of the nature of the internal secretions of the thyroid gland.

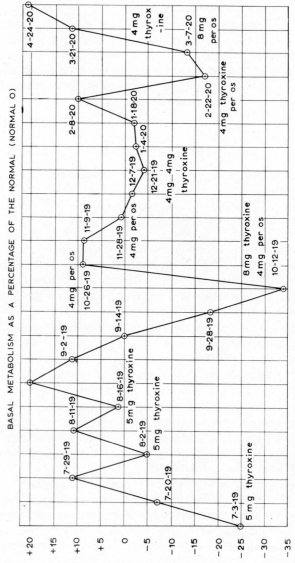

FIG. 119. Effect of thyroxine administration on a cretin. (After SNELL, FORD and ROWNTREE, borrowed from TERRONE and ZUNZ, 1925.)

II. Chemical and biochemical results

The colloid substance includes an active *protein, thyroglobulin* (molecular weight: 700,000) which contains thyroxine, an iodinated amino acid derivative of tyrosine, which functions, to some extent, as the prosthetic group of the glycoprotein. By ammonium sulphate precipitation, Derrien, Michel and Roche (1948) purified thyroglobulin and indicated that the level of iodine in thyroglobulin from normal thyroids shows considerable variation such that the iodine content cannot be used as a criterion for the purity of thyroglobulin. During the course of biosynthesis of thyroglobulin other iodinated proteins are formed. These can be detected by the use of tritiated tyrosine (Nunez and Roche, 1962). The thyroid gland of vertebrates contains 20 to 300 mg I/100 g (fresh weight) while the adrenals contain only 0·2 to 0·4 mg/100 g and the gonads 0·1 to 0·2 mg/100 g. Since the definition of a hormone is related to the efficacy of the *natural* product under consideration, it is difficult to state whether there is one or several thyroid hormones. *Synthetic iodinated proteins* can have similar biological actions (thyroxinomimetics) to those of thyroxine. With respect to the mode of action, it has been suggested that thyroxine acts on *oxidative phosphorylation* of mitochondria partly by *uncoupling*—in the manner of dinitrophenol—the respiratory system and the phosphorylating system (Martius and Hess) associated with swelling of mitochondria (Lehninger, 1960). However, the problem is complicated by the fact, as stated by O. Lindberg (1963) that it is difficult to reproduce experiments on the action of thyroid hormones on respiration and phosphorylation.

Four derivatives of the amino acid *thyronine* have been identified in the thyroid and blood plasma.

HO—⟨ring: $3'$ $2'$ / $4'$ $5'$ $6'$ $1'$⟩—O—⟨ring: 3 2 / 4 5 6 1⟩CH_2—CH—$COOH$
   |
  NH_2
Thyronine (T_0)

HO—⟨ring: $3'$ (I) ⟩—O—⟨ring: 3 (I) ⟩CH_2—CH—$COOH$
   |
  NH_2
3,3'-Di-iodothyronine (T_2')
(J. Roche *et al.*, 1955).

HO—⟨ring: $3'$ (I) ⟩—O—⟨ring: 3 5 (I, I) ⟩CH_2—CH—$COOH$
   |
  NH_2
3,5,3'-Tri-iodothyronine (T_3)
(G. Ross *et al.*; Roche *et al.*, 1952).

HO—⟨ring: $3'$ $5'$ (I, I) ⟩—O—⟨ring: 3 (I) ⟩CH_2—CH—$COOH$
   |
  NH_2
3,3',5'-Tri-iodothyronine (T_3')
(Roche *et al.*, 1954)

HO—⟨ring: $3'$ $5'$ (I, I) ⟩—O—⟨ring: 3 5 (I, I) ⟩CH_2—CH—$COOH$
   |
  NH_2
3,5,3',5'-Tetra-iodothyronine (T_4)
(Kendall, 1915; Harrington, 1926),
or thyroxine.

By the method of isotopic equilibrium, Simon and Morel (1960) have found that the thyroids of rats receiving 5 μg iodine per day renew hormonal iodine at a rate equal to the average rate of renewal of the total pool of functional iodine. However, the thyroids of rats receiving 50 μg iodine per day renew hormonal iodine at a higher rate than the average rate of renewal of the total thyroid pool. The principle of the method of isotopic equilibrium is based on the fact that a stationary pool reaches a dynamic equilibrium, that is, the loss and gain of a metabolite are equal. In this case, the stationary state is obtained by placing the animals (rats) on a basic diet deficient in iodine and administering by way of the drinking water, a certain daily dose of iodine in the form of KI. After two months (steady state), the iodine is replaced by an equal dose of stable iodine labelled with ^{131}I with a chosen specific radioactivity. Under these conditions, the iodine pool is continuously renewed from the labelled iodine, the specific radioactivity increasing progressively until the same ratio of isotope as in the dietary precursor is attained. The isotopic equilibrium quantitatively defines the total pool of renewable iodine.

The *first stage* in the biosynthesis of thyroxine is the formation of *mono-iodotyrosine* in the presence of Cu^{++}; iodide is oxidized to iodine and spontaneous iodination of tyrosine *(in position 3)* occurs *within the gland.* The concentration of iodide by the thyroid is an active phenomenon: the "iodide pump". [Mammary tissue can concentrate iodine such that a high milk: plasma ratio of ^{131}I is maintained (Reineke, 1963)]. *Antithyroid compounds* (such as thiourea) *inhibit iodination* in the thyroid

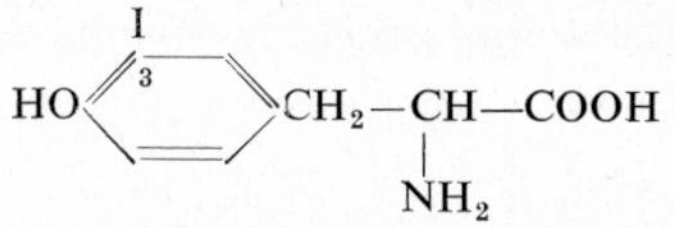

$$HO \underset{\diagdown \underline{} \diagup}{\overset{I}{\overset{\diagup \overline{} \diagdown}{}}} \overset{3}{} CH_2 - \underset{\underset{NH_2}{|}}{CH} - COOH$$

Mono-iodotyrosine

Mono-iodotyrosine is iodinated to give 3,5-di-iodotyrosine which is the direct precursor of L-thyroxine. Thyroxine is formed by the condensation of two molecules of 3,5-di-iodotyrosine with the loss of one alanyl residue. Approximately 75 per cent of hormonal iodine is present in thyroxine. However, *3,5,3′-tri-iodothyronine is considered to be the active form of thyroid hormone.* Thyroxine represents a reservoir which provides 3,5,3′-tri-iodothyronine by an enzymic process of *deiodination* in the receptor tissues, liver (Sprott and McLagan) and pancreas (Lissitzky, Michel, Roche and Roques). Summarily, one finds:

In the gland: *thyroglobulin* (the storage form) which *is normally never found in blood.* During secretion, thyroxine is liberated by cathepsic *proteolysis* (de Robertis, 1941); the addition of hyaluronidase prolongs the action of cathepsins (Stoll, Maraud, Mounier and Blanquet, 1959).

In plasma: *thyroxine* and very little 3,5,3′-tri-iodothyronine: the active circulating form being fixed in man to a serum α-globulin (a glycoprotein;

Freinkel *et al.*, 1955; Tata, 1959). In the salmon significant amounts of thyroxine are not linked to plasma proteins (Leloup, 1958–1961).

In receptor tissues: only 3,5,3'-tri-iodothyronine. Thyroxine-deiodinase only catalyses the deiodination of *free* thyroid hormones, the hormones being protected by combination with certain proteins (Lissitsky, 1960). The intracellular action of the hormone would depend on the degree of protein fixation (Tata, 1960). According to Roche and his co-workers (1963) the deiodination of thyroid hormones occurs in two phases: (1) the hormone is fixed to protein, probably enzymic; (2) this complex partially liberates iodine and retains the organic skeleton, iodothyronine. The coenzyme for dehalogenation of T_4 is a flavin (Lissitsky, Bénévent and Roques, 1961).

3,5,3'-tri-iodothyronine enters cells more rapidly than thyroxine; it is also more easily degraded. The thyroid produces about 0·35 mg thyroxine per day, the production being regulated by *pituitary thyrotropic hormone* (TSH). Total hypophysectomy *is accompanied by a decrease in thyroid iodine*. TSH does not act only on the *fixation* of iodide but also on the incorporation of iodine to organic derivatives. Moreover, thyrotropic hormone leads to *mobilization* of thyroxine in the circulation. Proteolysis of thyroglobulin may be regulated by thyrotropic hormone. *This provides a good example of the action of hormones on enzyme systems through a coenzyme-type reaction.*

Using radioiodine it has been shown that, during activity, the *ovary* is an important centre for the storage of iodine and that the histophysiological differentiation of the thyroid of *chick embryo* depends in part on the quantity of iodine present in the *vitellus* of the egg (Blanquet, Stoll, Maraud, Meyniel, Mounier). By radiochromatography the existence of thyroid hormones has been shown in the chick from the tenth day of incubation. *The vitellus represents the site of concentration and distribution of hormonogenic iodides. The function precedes the organ (the definitive organ).*

Thyroid hormones and, to a greater extent the acetic derivatives, can be fixed by leucocytes (Roche *et al.*, 1962) and nucleated erythrocytes of ducks, dogfish and sipunculid (T_3 being fixed more rapidly than T_4) but iodotyrosines and iodides do not penetrate the erythrocytes.

Di-iodotyrosine has been identified in the skeleton of gorgonians (Drechsel, 1896) and sponges (Wheeler and Mendel, 1909). It has also been found in the byssus and periostracum (scleroproteins) of the mussel and in the wall of zooids of the bryozoan *Bugula neritina* (Roche, André and Covelly, 1962, 1963). Roche (1960) has stated that the widespread natural distribution of iodotyrosines does not correspond with the existence of rudimentary thyroid growth; endocrine function of the thyroid type can be assumed only by the presence of iodothyronines. In the scleroproteins (*gorgonin* of gorgonians, *spongin* of sponges, *antipathin* of antipatharians) there are not only iodinated derivatives of tyrosine but also brominated derivatives. From his own work and that of Barrington (1957), Jean Roche concludes that thyroid hormogenesis is first realized in the tunicates (*Ciona intestinalis, Salpa maxima*, etc.) and no biochemically differentiated thyroid rudiment appears to exist in invertebrates with the exception of the protochordates.

What is the fate of thyroid hormones in the adult organism?

Thyroid hormones are not bound by the thyroid or by the pituitary (Courrier *et al.*) which may seem surprising in that the pituitary secretion of thyrotropic hormone appears to be regulated by thyroid hormones. However, the pituitary is stimulated by low doses of unbound thyroid hormones.

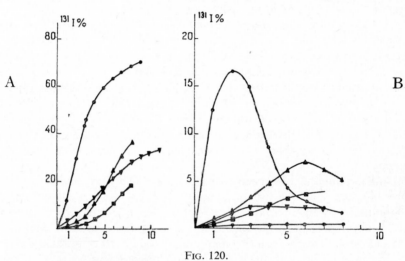

FIG. 120.

Part A: total biliary excretion of ^{131}I after injection of 1–2 mg of DL-thyroxine (■); DL-3,5,3'-tri-iodothyronine (▲); DL-3,5',3'-tri-iodothyronine (▼); DL-3,3'-diiodo-thyronine (○) and iodides (●) to thyroidectomized rats (200 g). Abscissa: time after injection (h). Ordinate: per cent injected ^{131}I recovered in bile.

Part B: Hourly biliary excretion of ^{131}I calculated from the experimental results shown in the curves of part A. Abscissa: time after injection (h). Ordinate: per cent injected ^{131}I eliminated per h in bile.

(After ROCHE, MICHEL, ETLING, NUNEZ, *Biochim. Biophys. Acta*, 1956.)

The hormones are localized principally in the *liver* and, to a lesser degree, in the digestive tract and kidney. In thyroidectomized rats, hepatic metabolism of hormones labelled with ^{131}I results in the appearance in *bile* of a mixture of radioactive products including sulphur derivatives of 3,5,3'-tri-iodothyronine and *glucuronic acid derivatives* of the hormones (Taurog, Briggs and Chaikoff, 1951; Roche *et al.*). Steroid hormones are also eliminated in the form of glucuronic acid derivatives (cf. p. 430). Free iodothyronines are found principally in fecal material but they can be absorbed by the intestine and pass into the entero-hepatic circulation.

3,5,3'-tri-iodothyronamine is found in the intestinal contents of rat after administration of 3,5,3'-tri-iodothyronine and is formed most probably by bacterial decarboxylation. T_4 gives thyroxamine (Roche, Michel and Stutzel, 1961, 1962). These amines are of physiological interest since they potentiate the inhibitory action of adrenaline on intestinal motility (Lachaze and Thibault, 1952). Moreover, in the presence of polyphenol oxidase from potatoes, the carbon skeleton—L-thyronine or T_0—is degraded by splitting the oxygen bond joining the two benzene rings (Lissitzky and Bouchilloux, 1955).

Inhibition of thyroid function by synthetic antithyroid compounds

A diet rich in cabbage, peas and soya beans leads to the development of goiter characterized by enlargment of the thyroid. Thiocyanates, certain sulphonamides, thiourea and thiouracil are *antigoitrogens*. Recent studies using ^{131}I have shown that *thiocyanate blocks the penetration of iodine to the thyroid while thiourea and its derivatives inhibit the conversion of mineral iodine to hormonal iodine.*

$$\begin{array}{cc} HN\!-\!C\!=\!O & \\ | \quad\; | & \\ S\!=\!C \quad C\!-\!H & \\ | \quad\; \| & \\ HN\!-\!CH & \end{array}$$

Thiouracil   Thiourea

Cellular and luminal potentials of rat and guinea-pig thyroid glands have been measured with glass micro-electrodes. The luminal potential was zero with respect to the interstitial fluid. Cellular potentials average about 50 mV, inside negative with respect to the interstitial fluid. Propylthiouracil treatment and TSH administration lowered the size of cellular potentials. The potential measurements permit the definite conclusion that the uptake of iodine by the thyroid is *active*. If iodine is taken up as I^-, then no electrical work is required to transport it into the lumen. On the other hand, a considerable amount of electrical work must be done if I^- is concentrated inside the cells (D. M. and J. W. Woodbury, 1963).

We shall now examine the *anomalies* of *thyroid function* (in man):

A. *Hypothyroidism* produces *myxoedema* characterized by sleepiness, inattention, dry skin (interference with avitaminosis A ?), obesity, imbecility, general debility. It is prevalent in certain regions and can be *congenital* or *acquired after birth*. The thyroid is often *atrophied*. Treatment consists of thyroid therapy which increases basal metabolism and body temperature. This provides an example of the influence of non-nervous factors on cerebral tone.

B. *Hyperthyroidism* produces *goiter* which can be a simple increase in the volume of the gland (*without* hyperactivity) or an increase in the volume of the gland *with hyperactivity*.

In the simple form, the dietary supply of iodine is inadequate; in the intense form which has serious consequences, there is excessive production of thyroid hormone. The thyroid of rats deficient in proteins shows histological signs of hypoactivity and the pituitary of these animals contains only small amounts of thyrotropic hormone (Aschkenasy *et al.*, 1952–1961).

Exophthalmic goiter (or Basedow's disease) is the final expression of hyperthyroidism with hyperactivity. The main symptoms are:

(1) *Tachycardia* (180 beats per mn in man at rest); antithyroid compounds reduce then the heart rate (Ruskin, 1951).

(2) *Goiter*, a variable characteristic.

(3) An increase in basal metabolism of up to 75 per cent above normal with an increase in oxidative processes and *loss of weight*.

(4) *Exophthalmia* (70 per cent of subjects) which is not always bilateral. Protruding eyes are characteristic and are due to enlargement of the palpebral slit.

(5) *Psychological disorders:* violence, irritability and trembling are characteristic.

Summarily, the characteristics, as formulated by Peter, are: *large neck, large heart, large eyes*.

Goiter is treated surgically, by antithyroids, by iodine solutions (such as that of **Lugol**) which decrease the activity of the gland and by ^{131}I which becomes concentrated in the gland and subjects the gland to β and γ irradiations. The last form of treatment is of particular value in the case of thyroid tumours.

The thyroid and other endocrine glands

Hormones produced by an endocrine gland pass in the blood to an *effector* which can be an apparently *unrelated* organ or *another endocrine gland* which, in its turn, influences other organs and particularly, the gland which produced the original secretion. These interactions (reflexes) contribute to the regulation of the internal environment by a chain reaction of the following type:

$$\text{endocrine} \rightarrow \text{hormone} \left\langle \begin{array}{l} \text{effector} \rightarrow x \\ \text{effector} \rightarrow \text{endocrine} \end{array} \right. \rightarrow \text{hormone} \left\langle \begin{array}{l} \text{effector} \rightarrow x' \ldots \\ \text{effector} \rightarrow y' \ldots \end{array} \right.$$

In certain respects, this organic chain is similar to the molecular chain of metabolic cycles such as the Krebs cycle. We shall now consider briefly these functional relationships in the case of the thyroid gland.

Thyroid and hypophyso-pituitary system

Smith (1927) discovered that total hypophysectomy resulted in atrophy of the thyroid. *Thyrotropic hormone* (a glycoprotein) increases iodine fixation by the thyroid and accelerates the incorporation of iodine to thyroxine (Chaikoff and Taurog). Fontaine (1957) found that thyrotropic hormones differed according to the zoological origin.

Thyroxine *inhibits the production of thyrotropic hormone.* Administration of thyrotropic hormone induces hypertrophy and thyroid hyperactivity. This is the mechanism of *exophthalmus* since *thyroxine cannot release this syndrome which is produced by thyrotropic hormone. Thyroxine which is inhibitory can even cause regression of exophthalmus* except in the case of Basedow's disease which is characterized by insensitivity of the pituitary.

Nervous control of the thyroid would appear to be important since experimental lesions of the diencephalon inhibit the appearance of goiter during administration of antithyroid compounds (R. Greer, 1952); (cf. p. 441 concerning thyrotropic hormone).

Thyroid and gonads

In this case, the observations are not always in agreement. However:

(1) Slight hyperthyroidism stimulates sexual development (notably in mice).

(2) Continuous administration of thyroxine to young rabbits increases sexual libido and stimulates spermiogenesis. This has also been demonstrated in mountain finches *(Fringilla montifringilla)* by L. and M. Vaugien (1958). The quality of spermatozoids is improved. This fact is of interest with respect to *artificial insemination.*

(3) According to J. Leloup (1959), the iodine content of the thyroid of sexually immature yellow eels is high and the content decreases progressively during the development of genitalia to reach a very low level in the sexually mature golden eel; this could be explained by increased thyroid activity in the golden eel.

(4) Thyroidectomy or administration of antithyroid compounds arrests sexual maturation in prepubescent animals and in the adult hen for example, results in a decrease in the number and weight of eggs.

Comment. In gallinaceans, thyroidectomy also causes an increase in the length of feathers and development of fringes while melanin is replaced by a reddish-brown pigment. However, a chick produced from a cock and a hen which have received repeated injections of thyroid hormones shows white plumage from birth and for the first three months (Sainton, Simonet and Brouha). These facts, which indicate the great sensitivity of ectodermal structures to thyroid hormones, appear to me to lead us to examine the direct or indirect mobilization of thyroid hormones by desoxyribonucleic acid or derivatives in the curious genetic experiments carried out by Benoit and his co-workers on ducks.

Hypothyroidism is a cause of sterility and of troubles of character in the female. During menopause there is accentuation of hypothyroidism in hypothyroid subjects and of hyperthyroidism in hyperthyroid subjects. Oestrogens have an inhibitory action on the peripheral effects of thyroid hormones (Sherwood): follicle stimulating hormone decreases cellular oxidation and therefore has the opposite metabolic effect to thyroid hormones.

In general, the action of hormones does not relate only to functions but also to structures which are the basis of these functions; thus, hormones regulate function and morphogenesis. A typical example is provided by the action of thyroid hormones on metamorphosis in amphibia.

Gudernatsch (1912) demonstrated that tadpoles fed on thyroid gland rapidly metamorphosed, that is, developed quickly into frogs. From the anatomical point of view this is obviously a very important effect but ethologically it is no less important: under the influence of thyroid hormones a vegetarian aquatic creature with branchial "respiration" is transformed to a carnivorous amphibian creature with pulmocutaneous "respiration"; also, tadpole hemoglobin binds oxygen more firmly than does frog hemoglobin; tadpole hemoglobin is adapted for binding the maximum amount of oxygen from water and frog hemoglobin is adapted for releasing oxygen easily to meet metabolic needs—E. Frieden. It would be of value to make a comparative study of the "primitive" hemoglobin of *Myxine* and that of the frog tadpole. When raised in a solution containing thiouracil (an antithyroid compound), tadpoles of *Discoglossus* reach the stage of metamorphosis only after 5 months compared with metamorphosis after 30 days in normal individuals (Delsol, 1952). The animals which show delayed metamorphosis continue to grow resulting in the appearance of *giant tadpoles* showing *normal morphological development of gonads and renal system.* Certain giant tadpoles contain testes rich in spermatozoids *(neoteny).* Delsol considered that in these giant tadpoles the pituitary growth hormone (producing larval gigantism) and gonadotropic hormone (producing sexual maturity) were functioning normally. Ablation of the hypophyseal complex of tadpoles inhibits metamorphosis.

Among the batracians, neoteny (or paedogenesis)—that is, *the development of reproductive organs and reproduction in organisms retaining larval somatic characteristics*—is marked, particularly in axolotl, a kind of urodele with external gills. Thyroid treatment results in the transformation of this animal to another whose identity was not recognized until 1875: *Amblystoma.* Axolotl is a hypothyroid form since, although a thyroid is present, it is unable to influence the tissue receptors and produce metamorphosis. In such a case, the hormonal threshold is not reached.

Toivonen and his co-workers (1956, 1957) carried out a quantitative histological study of the pituitary and thyroid during normal and abnormal development of *Xenopus laevis* and stated that there is an increase in the size of the nuclei and number of granules in pituitary cells *until metamorphosis*. At metamorphosis there is nuclear and chondriosomal disintegration in the pituitary and a pronounced decrease in the quantity of ^{131}I in the thyroid. During neoteny, the activity of pituitary and thyroid cells is low. Thus, *thyroid–pituitary interrelationships appear to be well established in batracians and express an important modification of the internal environment with metamorphosis, which can be compared with the passage from uterine life to aerial life in mammals.*

In *teleost fish*, the thyroid gland has no precise location and thyroid function cannot be studied by thyroidectomy. However, antithyroid compounds which selectively inhibit the various islets of the diffuse organ allow a study of the gland. (These animals provide another example of the concentration of vegetal structures as a function of evolution). In surface dwelling marine fish, the volume of the thyroid islets indicates *greater* thyroid activity than in deep sea fish, bathypelagic fish (Bougis and Buser, 1952). During metamorphosis of the sardine, Buser and Ruivo (1954) noted a rapid and marked increase in the volume of the thyroid follicles (three times greater than the increase in weight of the animal); after metamorphosis the ratio of the volume of the thyroid to body weight decreased. Metamorphosis of the larval form to the young fry coincides with a *significant change in ecological behaviour*. The larvae which have led a pelagic life in the ocean, migrate towards the warm coastal waters and sometimes, to less salt water. They appear to follow an electrolytic gradient (migration being a search for a new biotic equilibrium). Therefore, the thyroid would function in the regulation of the internal ionic equilibrium. It would appear that migration of cyclostomes and pond-dwelling fish (towards fresh water) is also partly dependent on thyroid hyperactivity (Fontaine, 1943) corresponding with an increased sensitivity to high salinities. Therefore, the appearance of osmoregulatory disorders related to the thyroid and genetic repercussions is of considerable importance with respect to the maintenance of the species. Arvy, Fontaine and Gabe (1956) found that experimental hyperthyroidism in trout is accompanied by:

(1) cytological changes associated with an *increase in secretion by the cells of the preoptic nucleus* (cf. p. 452);

(2) *movement of neurosecretory products* towards the *hypophysis* and movement of granules and droplets resembling that occurring in salmon at the time of catadromic migration (cf. pp. 463, 465).

Thus, *the thyroid is important in determining migration of sardines at the time of metamorphosis and reproductive migration in the salmon.*

Therefore *the thyroid gland regulates metabolic, sexual and ethological activities.* Among the vertebrates, *thyroid hyperactivity* is accompanied by hypophyso-pituitary hypothalamic hyperactivity and adreno-cortical hyperactivity during reproduction and, it would seem, during any change in biotic conditions.

Thyroid and adrenal cortex

I. Synergism (action of the thyroid on the adrenal cortex)

In high doses, thyroxine leads to hypertrophy of the adrenal cortex and during myxoedema there is atrophy of the adrenal cortex. Stimulation mainly affects the production of 11-oxycorticoids which are concerned with the metabolism of proteins and carbohydrates.

Mention should be made of the Donaggio reaction (with thionine) which has been developed for quantitative estimations by Tayeau, Jensen and Reiss (1960). This reaction shows the presence of urinary mucoproteins. The elimination of mucoproteins is related principally to metabolism of the basic material of supporting tissues (Tayeau). Stoll, Sparfel and Maraud (1961, 1962) found that the thyroid modifies the Donaggio reaction through an indirect mechanism by influencing adrenal cortex function which is responsible for urinary elimination of mucoproteins.

McGavack considers that the increase in thermogenesis during thyroid hyperactivity is related to the action of the adrenal cortex on protein metabolism. Adrenalectomy prevents the calorigenic response (cf. p. 30 SDA) and the increase in basal metabolism associated with thyroxine administration (Verzar and Wirz).

According to Mansfeld, a "cooling hormone" may exist: an animal subjected to heat would liberate a hormone which would decrease ventilatory exchange in another animal; this hormone would not appear in a thyroidectomized animal.

The thyroid is important in hibernating mammals. Peiser (1906) demonstrated that there is a flattening of the glandular epithelium in the thyroid of bats during winter. Lachiver and Petrovic (1960) stated that thyroid activity, measured *in vivo* after injection of ^{131}I, increases in *Eliomys quercinus* and rat but decreases in *Citellus citellus* following a short exposure to cold in summer. These authors interpret these results by considering the *Citellus* to be a diurnal, terrestrial rodent possessing a burrow in which it can protect itself from external aggression. This animal would have a higher limiting temperature than the nocturnal, tree-dwelling *Eliomys* which builds temporary hides in summer. Thus, the limiting temperature at which the thyroid reaction to cold is inhibited or reversed varies with species and various factors such as previous thermal adaptation which is probably related to the thermal characteristics of the biotope of each species.

II. Antagonism (action of the adrenal cortex on the thyroid)

In high doses, extracts of adrenal cortex inhibit the increase in basal metabolism produced by administration of thyroxine or thyrotropic hormone. Therefore, these extracts are beneficial in the treatment of Basedow's disease (Kinsell *et al.*). *The action of adrenal cortex extracts could be explained by an inhibition of succinic acid dehydrogenase and cytochrome oxidase.*

These differential correlations contribute to the regulation of thyroid and adreno-cortical activity, *regulation resulting always from a "balance of power". When one process is predominant it is moderated by the opposite process; an aspect of Le Chatelier's principle of moderation.*

Thyroid and adrenal medulla

As in the latter case, synergism and antagonism participate in the reciprocal regulation of the function of these glands.

I. *Synergism:* during hyperthyroidism, there is an increase in the production of adrenaline.

II. *Antagonism:* adrenaline decreases the fixation of ^{131}I by the thyroid. Abelin and Goldstein (1956) demonstrated a *direct correlation between thyroxine and adrenaline.* These two hormones produce a similar series of reactions but differing in speed:

increase in general metabolism;
increase in basal metabolism and production of calories from fats;
increase in production of lactic acid by muscle;
increase in the volume of the thyroid by adrenaline;
increase in the volume of the adrenal gland by thyroxine;
thiouracil inhibits the production of thyroxine and adrenaline;
increase in the production of thyrotropic hormone by adrenaline;
increase in the production of ACTH by thyroxine.

The latter two points indicate that the correlation extends to the pituitary.

In conclusion, the thyroid influences diverse phenomena (behaviour including migration, metamorphosis, shock, etc.) and provides an example of the functional ubiquity of hormones.

CHAPTER 14

THE PARATHYROID GLANDS

THE important dates in the history of the parathyroid glands are:
1880 Discovery of parathyroids by Sandström.
1891 Demonstration of the rôle of parathyroids in rabbit by Gley.
1925 Extraction of *parathormone* or *parathyrin* by Collip.

Anatomy and origin

In general, the glands are paired—two external and two internal—but there can be one or even six pairs.

In man the glands are about 6 mm long and 1·5 mm thick and weigh approximately 50 mg.

Embryologically, the glands have a double origin: the external parathyroids are derived from *endoderm* of the ventral surface of the third branchial pouch and the internal parathyroids are derived from the ventral face of the fourth branchial pouch.

The parathyroids are surrounded by adipose tissue in the adult and situated beneath the thyroid.

Parathyroids do not appear to be present in invertebrates in which calcium metabolism is less important than in vertebrates. Parathyroids appear in the amphibia and, according to Waggener (1930) must have some correlation with the method of ventilation in terrestrial vertebrates. Greep (1963) remarked that the parathyroids appear with the disappearance of the gills. Thus, a function related to mineral metabolism appears in the form of a differentiated gland when the processes of gaseous transfer become less dependent on water. This example indicates the way in which comparative anatomy and physiology can provide a method of reasoning in problems of evolution of organized creatures.

The parathyroid consists of an external *fibrous capsule* containing a *glandular parenchyma* formed of anastomosed epithelial cords.

Between the parathyroid cells arranged in lobules are capillary sinuses. The parathyroid of man contains principally two types of cells:

the principal cells with chromophobic cytoplasm;

the chromophilic or acidophilic cells with numerous cytoplasmic granules.

414

Important aspects of physiology

The parathyroid glands are essential for the life of mammals: total ablation results in death preceded by the following symptoms *("parathyreoprive" tetany)*:

tonic and clonic contractions (convulsive movements with contractions and relaxations);
increase in central temperature;
tachycardia, arterial hypertension and hypopnea.

During the intervals between convulsions, neuromuscular hyperexcitability and exaggeration of reflexes are apparent.

Tetany is related to a decrease in blood calcium which is eliminated through the intestine (calcium diarrhoea).

Chemical physiology

(1) *The hormone.* The hormone is extracted from ox parathyroid by dilute hydrochloric acid and precipitation with acetone. It is thought to be a polypeptide but has yet to be isolated in a pure form. A *lipid* fraction of the parathyroid may produce hypercalcemia (Prelot, 1958).

(2) *The effects of the hormone:*

(a) parathormone increases the total serum calcium and decreases the serum phosphate level but increases the urinary elimination of calcium;

(b) parathormone decalcifies bones particularly if the dietary level of Ca^{++} is insufficient;

(c) parathormone inhibits hypocalcemic tetany. Ringer and Loeb demonstrated that excitability of tissues is determined by the ratio $(K^+ + Na^+)/(Ca^{++} + Mg^{++})$. A decrease in the Ca^{++} total results in increased excitability, a fact wich has been verified by Heilbrunn.

We have seen that the parathyroid consists of *two types of cells* and it appears that it contains at least *two hormones* (Munson, 1955):

(1) *one factor mobilizing calcium,*

(2) another, called the *phosphaturic factor,* which lowers tubular reabsorption of mineral phosphates and regulates the blood and urinary distribution of phosphate. This factor is found in crude extracts of parathyroids.

One of these factors must be responsible for the radioprotective role of parathyroid extracts administered sub-cutaneously to rats subjected to X-rays (Rixon *et al.*).

Marnay and Raoul (1959) stated that the parathyroid "hormone" only shows hypercalcemic activity in young male rats which are not deficient in vitamin D. But thyrocalcitonin, from thyroid, (McIntyre), has to be considered.

Anomalies of parathyroid function

(1) *Hypoparathyroidism*. This leads to muscular weakness, tetany with hallucinations and aggressiveness, irritability, anxiety (Gley, 1891). The psychic disorders are particularly characteristic (cf. Volume II, the role of calcium in nervous excitability). Hypoparathyroidism can be treated by assimilable calcium salts such as calcium gluconate. Calciferol (vitamin D_2) increases intestinal absorption of Ca^{++} and excretion of phosphate by the kidney. During tetany the condition is improved by injection of parathormone which, however, is short-acting possibly because of the formation of an anti-hormone.

(2) *Hyperparathyroidism*. It results generally from a tumour of the gland. This condition is characterized by decalcification of bones leading to deformity and spontaneous fractures, anorexia (loss of appetite)—Ca^{++} possibly having a direct action on the hypothalamus—nausea and polyuria.

This disorder is treated surgically by partial ablation of the parathyroids.

A psycho-hormonal link will again be noted with respect to parathyroid function.

CHAPTER 15

THE ENDOCRINE PANCREAS

THE important dates in the history of the elucidation of the physiology of the endocrine pancreas are:

1889 Mering and Minkowski demonstrated that ablation of the pancreas results in severe sugar diabetes.

1892 Hédon overcame this diabetes by sub-cutaneous implantation of pancreatic fragments.

1893 The internal secretion of the pancreas arises in the islets of Langerhans (Laguesse) and the name *insulin* is given to the product of this secretion (Meyer, 1905).

1922 Isolation of insulin by Banting and Best using 80 per cent ethanol (ethanol extracts insulin and inhibits enzymic proteolysis). Treatment of sugar diabetes.

1923 Kimball and Murlin isolated a hyperglycemic factor from pancreas: *glucagon*.

1942 Clinical observation by Janbon of the anti-diabetic action of certain sulphonamides. Analysis of the phenomenon by Loubatières.

Origin and structure of endocrine pancreas

In this section we shall consider the natural history of the endocrine pancreas.

During the study of digestion, we considered the role of the exocrine pancreas in the degradation of food (trypsin, amylase, etc.). The pancreas has a functional bivalence in that it also acts as an endocrine gland. Moreover, the pancreas exhibits a latent cytological bivalence, to some extent resembling that of bone marrow which produces both red and white cells, since the exocrine parenchyma is derived from the same duodenal buds (of the primitive intestine) as the endocrine islets. During the life of the pancreas there is a balance (Laguesse) between the islets and the acini (metaplasia). In adult man, the pancreas is situated transversely in the region of the first lumbar vertebra; the weight varies from 60–160 g. *There is no typical pancreas in invertebrates.*

The *exocrine* pancreas consists of acini and is similar to salivary glands while the endocrine pancreas consists of *large cells* distributed through the parenchyma of the exocrine pancreas.

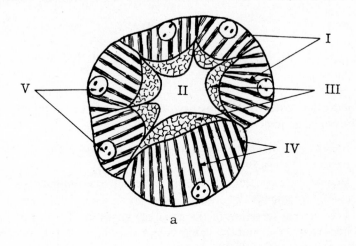

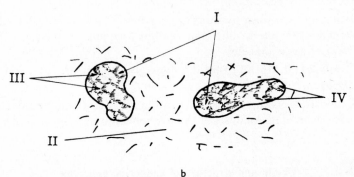

FIG. 121. (a) Schematic representation of a pancreatic acinus: *I*, secretory pyramidal cells; *II*, lumen; *III*, granular zone; *IV*, basophilic zone with chondriosomal striations; *V*, basal nucleus. (b) *I*, islets of Langerhans; *II*, exocrine pancreas; *III*, mass of clear α cells; *IV*, mass of β cells.

In the islets of Langerhans there are two types of cells:

α cells, the largest cells containing secretory granules but no visible mitochondria; none in *Myxine glutinosa* (Falkmer, 1966);

β cells, more numerous than α cells but smaller and exhibiting less intense staining. They contain mitochondria.

This cytological duality suggests, as in the case of the parathyroid, the existence of two hormones (D, E and X cells have been described).

Histophysiology of endocrine function of the vertebrate pancreas

This subject has developed in the following way:

(1) Lancereaux noted the existence of pancreatic lesions during pancreatic diabetes.

(2) Following the work of Mering, Minkovski and Hédon, Laguesse showed that the islets of Langerhans produced an internal secretion which regulated sugar metabolism.

(3) In certain teleosts *(Lophius piscatorius* and *Cottus scorpius*—the daddy sculpin—) the endocrine pancreas consists of large cell masses ("principal islets"), distinct from the acini, which can be selectively removed. Thus, Banting and Best were able to obtain preparations rich in the active constituent—insulin. *These facts support again the principle of evolutionary concentration of vegetal structures: in lower vertebrates these organs are separate but there is a tendency towards unification in higher animals* (in this case, combination of the exocrine and endocrine pancreas).

(4) Laguesse ligatured and sectioned the pancreatic excretory duct of rabbits. These animals survived for several years if the deficiency in acinar secretion was provided for. However, these animals showed no changes in glycemia and although the exocrine pancreas had degenerated, the islets of Langerhans were present in the mass of fibrous tissue.

(5) By injection of *alloxan* (a pyrimidine derivative) to dogs, Shaw, Dunn and Polson (1942) produced a form of sugar diabetes called *alloxan diabetes* (which must not be confused with phlorizin diabetes) in which there is selective destruction of β cells. These cells can regenerate spontaneously.

$$
\begin{array}{cc}
\text{HN—C=O} & \\
| & | \\
\text{O=C} & \text{C=O} \\
| & | \\
\text{NH—C=O} & \quad \text{Alloxan}
\end{array}
$$

According to Lazarow (1946, 1949), alloxan produces necrosis of the pancreatic β cells by destruction of SH^- compounds particularly of reduced glutathione (GSH). In *Cottus scorpius* alloxan produced a marked lowering of the GSH content of isolated islet tissue and erythrocytes but not of myocardium and brain (S. Falkmer, 1961).

The β cells of the islets of Langerhans produce insulin. The α cells of the islets produce glucagon.

Pancreatic hormones (insulin, glucagon and vagotonin)

I. Insulin

Insulin is a protein with an acid isolectric point. It has been obtained in crystalline form and the *chemical composition* was determined by Sanger (1951) using his technique of end-group analysis following partial hydrolysis. Insulin consists of 4 peptide chains, 2 with glycine as the free amino acid of the chain and 2 with phenylalanine. The chains are linked by disulphide bridges. Each insulin unit has a molecular weight of 12,000. The *activity of insulin is related to the integrity of the S—S bonds and hydroxyl groups in the molecule.* Insulin is destroyed by proteases and therefore, insulin cannot be administered *orally*. The protamine–zinc–insulin complex possesses a more prolonged hypoglycemic action than insulin alone.

A unit of insulin is the quantity of hormone needed to reduce in 24 h the blood sugar content of normal adult rabbit weighing 2 kg from 120 to 45 mg/100 ml.

Injection of high doses of insulin to normal subjects results in hypoglycemia accompanied by weakness, hunger, irritability, convulsions. These symptoms disappear with cessation of treatment or by administration of glucose.

Insulin has an action on sugar metabolism.

Goldstein and his co-workers (1953) consider the "role" of insulin to be the mobilization of glucose of extracellular fluid and the penetration of glucose into cells of certain tissues, thus augmenting the oxidation of glucose and also of galactose. Therefore, insulin seems to modify the *permeation* of sugars.

Insulin antagonizes certain actions of the pituitary and adrenal cortex. In the pathways of sugar degradation (cf. p. 121) the first reaction is the phosphorylation of glucose to give glucose-6-phosphate by the enzyme *glucokinase* (hexokinase). Price, Cori and Colowick demonstrated that *glucokinase is inhibited by pituitary extracts*, particularly in the presence of adrenal cortex extracts. *This inhibition is prevented by insulin.*

Lipogenesis from sugars (cf. p. 134) is augmented by insulin which acts on glycolysis and thus decreases glycemia. Through its action on energy metabolism, insulin produces a supplementary consumption of blood amino acids which leads to an increase in proteins. Thus, *insulin regulates glycemia, lipogenesis and indirectly, proteinogenesis.* However, insulin can also influence proteinogenesis by increasing the synthesis of ribonucleic acid (I. R. Wool, 1962).

The activity of insulin is *destroyed* by incubation with slices or extracts of liver. Two factors appear to be important: (a) an inactivating enzyme, *insulinase*; (b) an inhibitor of this enzyme, the inhibitor being thermostable and non-protein.

Hypoglycemic sulphonamides such as 2254 RP (Loubatières) *potentiate* the action of insulin, at least in diabetic subjects. These sulphonamides act, according to Loubatières, by stimulating the formation of species-specific insulin by normal cells and by producing neoformation of β cells. Thus, sulphonamides have a direct metabolic action as well as a structural action. The arylsulphonamides open the door to prevention and prophylaxis of human diabetes.

$$H_2N-\!\!\!\left\langle\bigcirc\right\rangle\!\!\!-\overset{\displaystyle O}{\underset{\displaystyle O}{\overset{\|}{\underset{\|}{S}}}}-\overset{H}{\overset{|}{N}}-\overset{S-C-CH\!\!<^{CH_3}_{CH_3}}{\underset{N}{\underset{\diagdown\!\diagup}{C}}}N$$

2254 RP
(Thiodiazole derivative; isopropyl)

The *anomalies of insulin function* are related to the deficiency or excess production of this hormone.

(1) DEFICIENCY. Diabetes *mellitus* is characterized by abnormal glucose metabolism with *hyperglycemia* and *glycosuria*. This leads to disorders of lipid metabolism with *ketonemia, ketonuria* and *acidosis* (cf. p. 132). Diabetes is characterized also by dehydration because of the amount of water required for the urinary elimination of excess glucose. These disorders can result in coma.

It has not been shown that *all* sugar diabetes result from insulin deficiency due to dysfunction of the β cells. *Excessive destruction of insulin by hepatic insulinase could be another cause of sugar diabetes.* In fact, insulin treatment is ineffective in certain subjects.

In general terms, the simple presence of a compound does not necessarily indicate that it is active. In the present case, it is necessary to distinguish between *circulating glucose* and *effective glucose.*

Pathological glucose metabolism resulting from an absolute or relative deficiency in insulin can be attributed to decreased oxidation of glucose. The increase in circulating glucose results in part from hepatic glycogenolysis and glycogenic amino acids.

In considering *the causal physiology which relates phenomena to molecular and genetic factors,* it is interesting to note that sugar diabetes is partially dependent on genetic factors (Levit; Pincus; Harris): DD being the genotype of normal subjects—D being an autosomal gene—Dd individuals would show delayed diabetes while dd individuals would show diabetes early in life.

Diabetes is treated by suitable diets and insulin (which however is antigenic) or special sulphonamides administrations.

(2) OVERPRODUCTION. This often results from a tumour but a tumour is not the systematic cause. This state is associated with hypoglycemia and emotional disorders, the hormone being of primary importance in determining character: depression, anxiety, confusion, delirium and sometimes convulsions and death.

II. Glucagon

Following the discovery of Banting and Best, McLeod as well as Bürger reported that certain pancreatic extracts produced *hyperglycemia*. This pancreatic extract was given the name *glucagon* by Kimball and Murlin (1923). Glucagon has been obtained in crystalline form (molecular weight approximately 4200). It contains 15 amino acids arranged in a straight chain. Glucagon differs from insulin in that it contains no cystine, proline or isoleucine but contains large amounts of methionine and tryptophan which are not found in insulin. (Commercial insulin contains about 10 per cent of glucagon.)

Glucagon is produced by the α cells of the pancreatic islets and probably from cells resembling the α cells which are situated in the gastric mucosa of dog and rabbit (but not of pig and sheep) and the mucosa of the duodenum and ileum of dog.

These cells can be selectively destroyed by subcutaneous injection of cobalt chloride (van Campenhout and Cornelis, 1951). Total hypophysectomy results, after a period of time, in the partial or complete disappearance of α cells while injection of pituitary extracts to dogs, in doses sufficient to produce Young's diabetes (cf. p. 490), increases the level of glucagon in the pancreas (Dorner and Stahl, 1957).

Glucagon stimulates hepatic glycogenolysis leading to an increase in glycemia. There is still much to be learnt about insulin–glucagon relationships. Some authors state that secretion of insulin and glucagon would occur alternately. Others suggest that they are secreted simultaneously. Glucagon augments synthesis of *hepatic phosphorylase* (cf. also adrenaline) which converts glycogen to glucose-1-phosphate (E. W. Sutherland, 1956) and activates protein catabolism, an effect which is opposed by insulin (L. L. Miller, 1960).

III. Vagotonin

In 1927, Santenoise isolated a protein from pancreas which had a weaker though more prolonged hypoglycemic action than insulin. *This compound increased the tone of the vagus by inhibiting cholinesterase.* This action is similar to that of eserine or physostigmine—whose anticholinesterase properties were described by Loewi and Navratil (1930). This compound is a hypotensive agent and can be considered to be an anti-hypertensive "hormone" rather than a hypoglycemic "hormone".

CHAPTER 16

THE ADRENAL GLAND

I. The adrenal medulla

From the endocrinological point of view, there is only one important date, 1901, when Takamine and Aldrich isolated the principal hormone, *adrenaline*.

One of the interesting problems concerning the adrenal medulla is that of the *chromaffin paraganglia* (Kohn, 1898, 1903). In man, the two adrenal capsules are situated above the two kidneys. However, the adrenal medulla and the adrenal cortex arise from different embryological structures. In lower vertebrates the two regions are separated spatially: *the cortical region constituting the inter-renal organ and the medullary region the suprarenal organ.*

In birds and higher vertebrates the two organs are fused. *These facts provide another example of the concentration of vegetal structures during evolution.*

The *adrenal cortex* arises from *coelomic mesoderm* while the *adrenal medulla* arises from *neural ectoderm*. The adrenal medulla is formed in the following way:

(1) After the neural tube has closed, ectoderm forms *ganglionic crests* in various parts of the tube.

(2) When these crests develop into the buds of spinal ganglia, certain cells, the *sympathogonia*, separate and become situated on either side of the large sagittal vessels and the vertebral column.

(3) The sympathogonia give rise to two types of cell: the sympathetic cells and the paraganglion cells, so-called because they are grouped in *paraganglia*, the most important being the adrenal medulla. However, paraganglia are also found around the descending aorta and at the *bifurcation of the internal and external carotid*, forming the *carotid body*—a chemoreceptor region. Chromaffin cells of the paraganglion type are found also in the heart of Myxine (cf. p. 235) and I suggest that they are significant as pacemaker molecules, in Myxine at least (Table 40). It will be noted that these cells are intermediate between purely endocrine elements and nervous elements and provide a particular example of neuro-hormonal reflex activity.

The paraganglia are described as chromaffin since they show a marked affinity for chromium salts which produce a brown coloration, Henle's reaction.

The adrenal medulla consists of large polyhedral cells arranged in anastomosing cords. The chondriome consists of *adrenalogenic granules* which give characteristic colour reactions: Henle reaction, Vulpian reaction with ammoniacal iron perchloride (green), iodate reaction of Hillarp and Hökfelt (1953) which is selective for noradrenaline. With the electron microscope, granules of catechol amines (Sjöstrand and Wetzstein, 1956) appear as black circular areas surrounded by a clear border (cf. p. 235). During functional activity of the adrenal medulla, these granules pass into the sinusoidal capillaries (Gottscham); the storage granules probably split into granules which are four times smaller by rupture of ATP linkage. During excitation of the splanchnic nerve, the adrenalinogenic granules disappear to be reformed at rest. Using the Vulpian reaction, evidence can be obtained for the presence of catechol amines (adrenaline and related compounds) in adrenal venous blood.

Innervation of the adrenal gland is important but there are only a few nerve cells in the adrenal medulla of adult man while they are absent from guinea pig and primitive in cat and rat.

The physiology of the adrenal medulla has been better understood since it was shown that in man, 80 per cent of the secretion consists of adrenaline and 20 per cent of the demethylated derivative of adrenaline: *noradrenaline* or *sympathin* N (E according to Bacq). These two compounds have been found in the chromaffin tissue of dogfish (Olivereau, 1959). Noradrenaline is the precursor of adrenaline in the adrenal medulla of dog (Peyrin and Cier, 1961).

In dog, *chlorpromazine* in concentrations higher than 1 mg per kg *reduces the secretion of adrenaline* produced by excitation of the splanchnic nerve or by injection of nicotine. Chlorpromazine acts as an adrenolytic, notably in the region of the adrenals (Brunaud and Decourt, 1953).

Administration of *adrenaline* produces *vasodilation* of the vessels of cardiac muscle (coronary vessels), *vasoconstriction* of arterioles of skin and splanchnic viscera with an increase in blood pressure and *stimulation* of frequency, force and blood output of the heart. Most of these effects were observed after administration of adrenal extracts to dogs by Oliver and Schäfer in 1894.

Noradrenaline has less effect on cardiac output than adrenaline but produces more marked *vasoconstriction* and therefore, a more pronounced rise in blood pressure. Noradrenaline is administered in cases of shock since it does not produce tachycardia.

Adrenaline produces *relaxation* of the stomach, intestine and bladder and *contraction* of the sphincters of stomach and bladder.

Adrenaline increases glycemia by augmenting glycogenolysis. Adrenaline *augments* the synthesis of active phosphorylase which converts glycogen to glucose-1-phosphate, an increase which is also produced by glucagon (Cori, Sutherland, 1956). This explains the hyperglycemic action of adrenaline.

Finally, another important property of catechol amines: adrenaline–noradrenaline constitute the hormone of fight, anger and strong muscular effort. Bradykinin, kallidin, has a direct excitatory effect on the adrenal medulla (J. Lecomte and his co-workers, 1961, 1963).

II. The adrenal cortex

The historical dates in the elucidation of the physiology of this gland are:
1927 Stewart and Rogoff and later Hartmann obtained a physiologically active extract from the adrenal cortex.
1929 Extraction by organic solvents of a fraction containing active components, *cortin*, by Pfeiffner and Swingle.
1934 Kendall and later Reichstein recognized the *steroid nature* of the hormones and fractionated the active extracts.
1949 Application of compound E of Kendall or *17-hydroxy-11-dehydro-corticosterone* or *cortisone* to the treatment of rheumatism (Hench).

Structure

Figure 122 illustrates the distribution of various cortical and medullary layers in the human adrenal. Let us examine the structure, starting with the external regions.

The *capsule* is *fibrous;* we shall not consider this region in any detail.
The cortex consists of:
(a) The *zona glomerulosa* consisting of small cells containing basophilic cytoplasm rich in mitochondria.
(b) The *zona fasciculata* consisting of anastomosing cords of cells. The cytoplasm of these cells is vacuolated (spongiocytes).

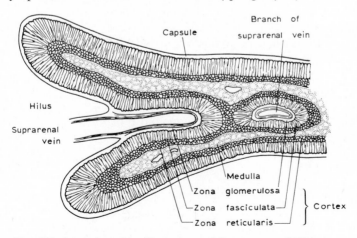

FIG. 122. Adrenal capsule. (DUBREUIL and BAUDRIMONT, 1959.)

(c) The *zona reticularis* forming the inner region of the *zona fasciculata* with anastomosing cellular cords and clearly defined vessels. The cytoplasm of these cells contains *lipochrome granules* with deep yellow pigmentation.

In a brephoplastic graft of adrenal glands beneath the albuguinia of mice, the *zona reticularis* disappears, the medulla degenerates to some extent but the *zona glomerulosa* and *zona fasciculata* persist (R. M. May and J. Chadefaut, 1963).

Between the adrenal cortex and medulla is a layer of cells called *zone X* which persists until puberty in female mice (Howard–Miller, 1927). This zone is present in man, rabbit and cat but *involutes during senescence*. Zone X is thought to produce one or several hormones which have an action resembling that of testicular testosterone *(adrenocortical androgens)*. The possibility of zone X being an androgenic region has been disputed by Delost (1956).

Main aspects of adrenocortical physiology

In teleosts, such as salmon, the homologue of the adrenal cortex, the dispersed *inter-renal* gland (and also the corpuscles of Stannius), contains corticosteroids (hydrocortisone and corticosterone) (Fontaine and Leloup-Hatey, 1959). Imbalance of the internal medium produced by changes in salinity produces an increase in the plasma content of 17-hydroxycorticosteroids of the carp. This increase can be produced by motor agitation of the carp.

About *thirty* sterols—the basic structure of which is found in cholesterol *(cyclopentanophenanthrene)*—have been isolated in crystalline form from the adrenal cortex but only *five* of these normally exist in significant quantities in man:

(a) *Those which are soluble in benzene* (non-polar solvent):
    (1) dehydrocorticosterone (compound A of Kendall);
    (2) corticosterone (compound B of Kendall);
    (3) desoxycorticosterone (compound D of Kendall).
(b) *Those which are slightly soluble in water:*
    (1) dehydro-hydroxycorticosterone (compound E of Kendall);
    (2) hydroxycorticosterone (compound F of Kendall).

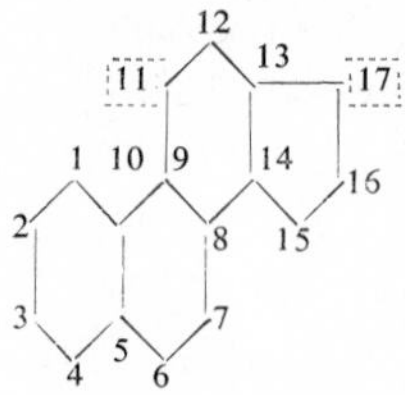

Cyclopentanophenanthrene

To this list must be added a special compound: *aldosterone* which brings the number of adrenal steroids to be examined to six.

Most corticosteroids possess:

two methyl groups in position 10 and 13 (except aldosterone);

a lateral chain in position 17 $(CO-CH_2OH)$;

a double bond between 4 and 5.

TABLE 44

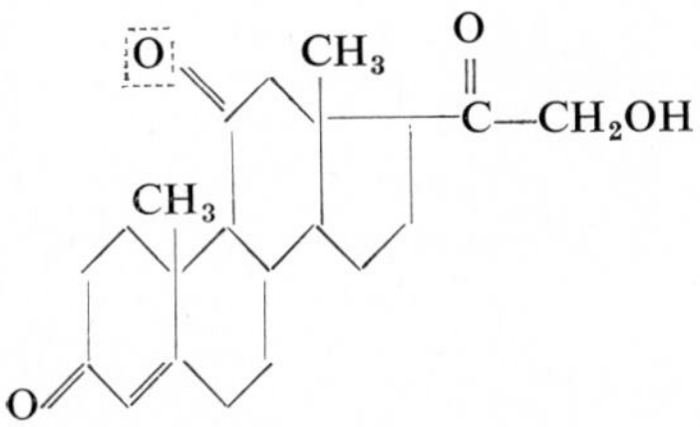

11-Dehydrocorticosterone
(compound A of Kendall)

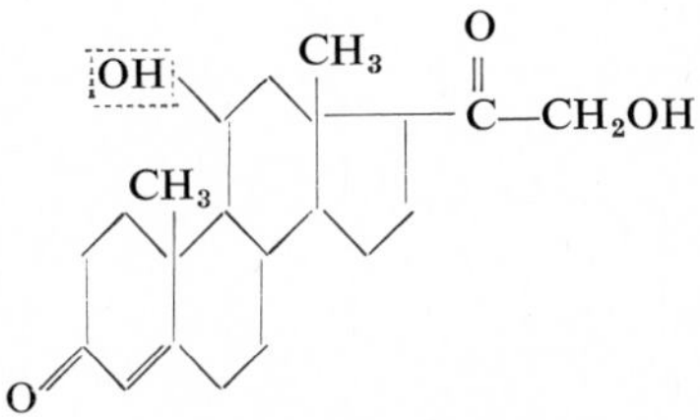

Corticosterone
(compound B of Kendall)

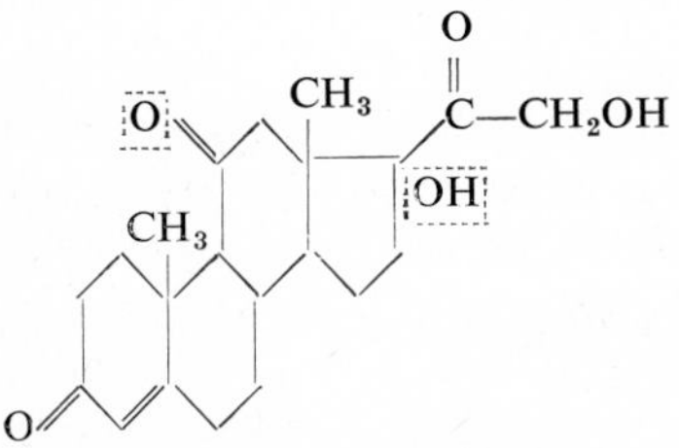

17-Hydroxy-11-dehydrocorticosterone
(compound E of Kendall)

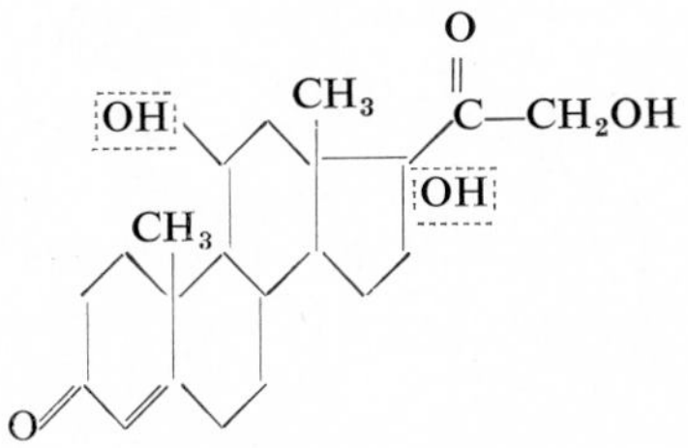

17-Hydrocorticosterone
(compound F of Kendall)

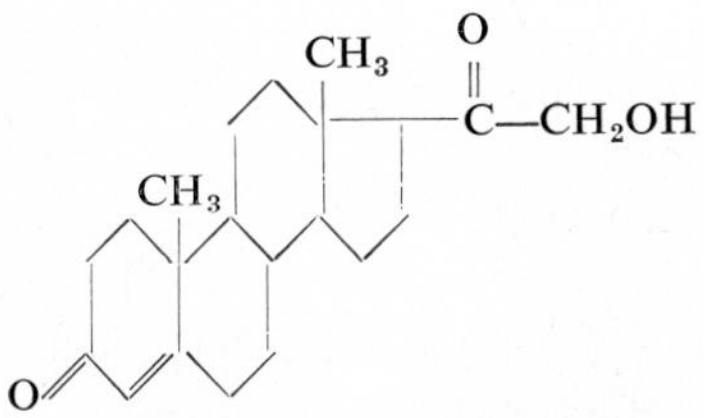

Desoxycorticosterone
(compound D of Kendall)

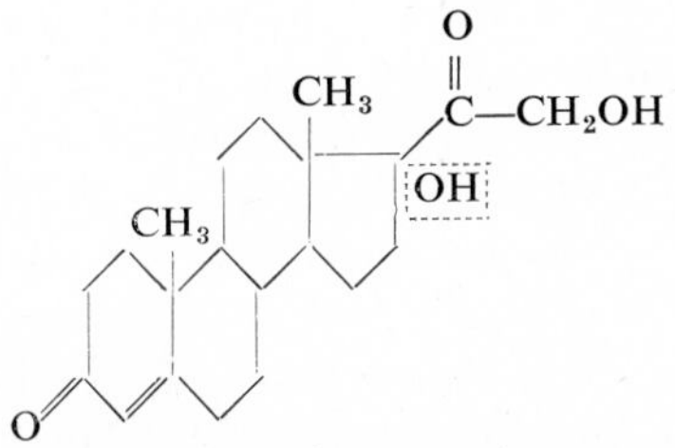

17-Hydroxy-11-desoxycorticosterone
(compound S of Reichstein)

One can distinguish between:

(a) *11-oxysteroids (or glucocorticosteroids):*

(1) Compound A of Kendall;

(2) Compound B of Kendall;

(3) Compound E of Kendall (cortisone);

(4) Compound F of Kendall;

(b) *11-desoxysteroids (or mineralocorticosteroids)*:

(1) Compound D of Kendall;

(2) 17-Hydroxy-11-desoxycorticosterone or desoxycortisone or *compound S* of Reichstein.

In considering somatic endocrinology we must include *aldosterone* or *electrocortin*. This compound can be extracted from urine and can be detected in blood plasma.

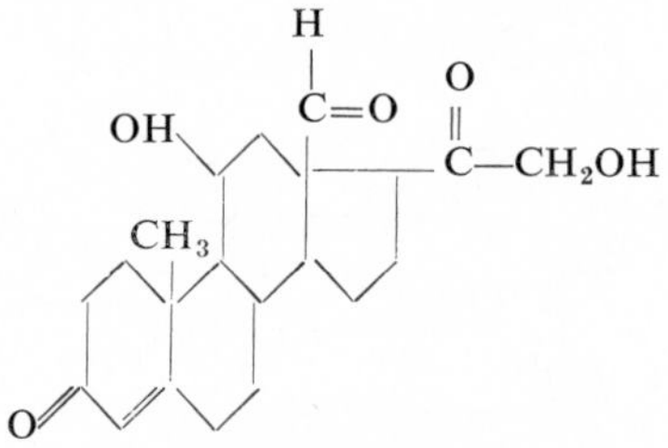

Aldosterone
(aldehyde form)

In addition, compounds can be obtained from the adrenal cortex which are not specific to the cortex and which show sexual activity: oestrone, progesterone and androgens but it should be noted that typical adrenocortical steroids can be formed from progesterone by the action of 17α-hydroxylase and of 21-hydroxylase.

These compounds can be obtained in the laboratory by partial synthesis from desoxycholic acid. However, since 1949, they have been prepared from plant sources using sapogenins which include sarmentogenin extracted from certain *Strophantus* of equatorial Africa.

In considering the biosynthesis of these compounds, it is important to note that total hypophysectomy leads to atrophy of the adrenal cortex while pituitary hyperactivity leads to hypertrophy. The pituitary secretes *adrenocorticotropic hormone* (ACTH). At rest, the *adrenal cortex* contains large amounts of *cholesterol* and *ascorbic acid* (Szent-Györgyi, 1932) but the level of these compounds falls markedly during stress or through the action of ACTH. The cortical hormones can be synthesized directly either from preformed cholesterol or from "active acetate". The pathway from "active acetate" is as follows: acetate → mevalonic acid → squalene → cholesterol (Prokhorova and Toumanova, 1963). The oxidation of Δ^5–3β–hydroxysteroids to Δ^4–3–ketosteroids by a specific dehydrogenase system (which consists of two different enzymes: an isomerase and a 3β–hydroxysteroid–dehydrogenase) occurs early in the only well-known pathway of biosynthesis of steroid hormones in vertebrates for corticosteroids and also for oestrogens, androgens, progestins. G. Pincus and his co-workers

(1951) perfused the *adrenals of cow* (and man) *in vitro* with oxygenated blood containing antibiotics. By this method endocrine function was maintained and augmented (but only augmented, showing the relative functional independence of the adrenal cortex) by *the addition of ACTH* to the perfusion fluid. Under these conditions, two products were liberated by the gland: *corticosterone* and *17-hydrocorticosterone*. Moreover, by this perfusion technique it is possible to analyse the fate of different steroid hormones. Thus:

(1) with *desoxycorticosterone* in the perfusate, corticosterone was released;

(2) *corticosterone* was not modified and, therefore, is a final product of adrenocortical metabolism;

(3) perfusion with *pregnenolone* (related to cholesterol) results in the formation of progesterone and, after prolonged perfusion, progesterone is converted to corticosterone.

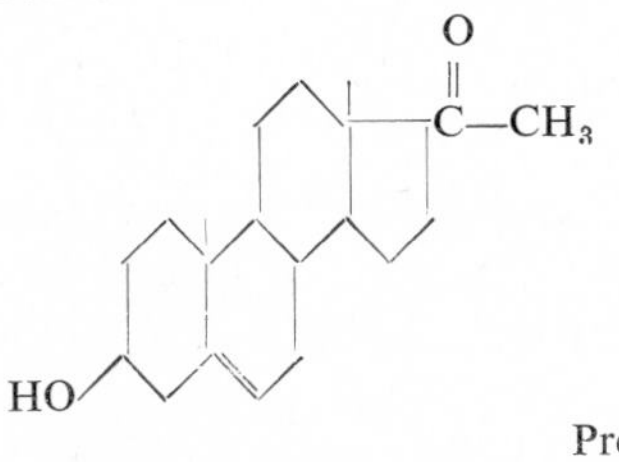

Pregnenolone

The results of the numerous experiments of Pincus can be summarized by the following scheme:

$$\text{cholesterol} \xrightarrow{\text{ACTH?}} \text{pregnenolone} \to \text{progesterone} \nearrow^{\text{corticosterone}}_{\searrow \text{17-hydroxycorticosterone}}$$

Metabolism of specific hormones

About half of the corticoids are transported in serum in combination with protein-carriers. To give some idea of the concentration in plasma, let us say that plasma contains an average of 100 mg/l hydrocortisone, the corticoid found in the highest concentrations in man.

In the liver, 17-hydroxycorticoids are inactivated by *conjugation* with glucuronic acid. Table 45 summarizes the pathways for the elimination of corticosteroids.

The *placenta* exhibits biosynthetic activity (but not storage) of the adrenocortical type such that the condition of females showing dysfunction of the adrenal cortex (for example, Addison's disease) is improved during gestation. Foetal steroid synthesis seems to be concerned mainly with the formation of a variety of "primitive" forms, i.e. different 3β–hydroxy–Δ^5–steroids; the conversion of which into α, β–unsaturated ketones takes

TABLE 45

	In liver	In intestine	In kidney
Hydrocortisone-cortisone	conjugated with glucuronic acid	excreted by bile in a free or conjugated form and partially reabsorbed by the intestine (entero-hepatic cycle)	few free corticoids in urine. Most found in inactive conjugated form
Aldosterone	mostly in the conjugated form as above		little free aldoster-one in urine
17-Ketosteroids	as above and also with sulphuric acid		in urine, in conjugated forms

place predominantly in the placenta. There is an extensive aromatization of C-19 neutral steroids in the foeto-placental unit (Diczfalusy, Pion and Schwers, 1965).

Functional effects of specific hormones

Total (bilateral) adrenalectomy is always fatal. It can be well performed in rats in which the adrenals are separated from the kidneys. Gestation or administration of progesterone prolongs the duration of survival of these animals.

In *elasmobranchs*, the "adrenal cortex" is separated from the "adrenal medulla" such that by selective ablation of each of these structures, one can easily study the function of each part. Ablation of the "medulla" leads to minor disorders while ablation of the inter-renal gland (equivalent to the cortex) resembles total adrenalectomy.

Bilateral adrenalectomy in mammals leads to a certain number of functional disorders which can be summarized as follows:

(1) decrease in arterial pressure due to removal of the medulla;

(2) decrease in basal metabolism but in rodents, this is seen only if the thyroids have been removed;

(3) death with hypothermia following pronounced *dehydration* associated with *hypochloremia, hyponatremia* and *hyperkalemia*.

These effects can be overcome either by grafts (intrarenal in dogs) or by *administration of corticoid hormones*.

Let us examine firstly the *effects of these hormones on water and mineral metabolism* (role of mineralocorticoids: 11-desoxysteroids).

Corticosteroids augment renal reabsorption of sodium and chloride ions and lower excretion of these ions by the digestive tract as well as by salivary glands and sweat glands. The retention of Na $^+$ is accompanied by increased excretion of K $^+$ in the distal renal tubule. The ability to increase reabsorption varies with the compound being considered. Hydrocortisone is the least active while aldosterone is a thousand times more active than hydrocortisone.

Na $^+$ retention is related to *an increase in the volume of extracellular fluid* and an increase in blood and urinary mass. These changes are the reverse of those produced by bilateral adrenalectomy.

Aldosterone is more specifically involved in regulation of water – mineral balance than other corticoids (hence, the name *electrocortin*). Liberation of aldosterone is less influenced by ACTH than other corticoids. However, urinary excretion—and thus the production?– of aldosterone is augmented by blood deficient in sodium ions or containing excess potassium ions.

Aldosterone is involved in certain forms of pathological water retention via sodium retention (oedema syndrome). J. A. Cella (1956–1959) synthesized a compound with a steroid structure, a *spirolactone*, which antagonizes desoxycorticosterone (cortexone) and aldosterone in the adrenalectomized rats; this inhibition occurs in the kidney region through an effect on sodium reabsorption.

The mode of regulation is still not well known. However, the pineal gland could control the aldosterone secretion. Moreover, Mulrow, Ganong and Borycska (1961–1963) consider that the renin–angiotensin system stimulates aldosterone secretion in dog; angiotensin inhibits renin secretion in dog and man (Vander and Geelhoed, 1965; Genest *et al.*, 1965). This suggests some *renal* regulation of metabolism of certain essential electrolytes.

Let us now consider the effects of corticoids on organic metabolism (role of glucocorticoids). F. Gros, Bonfils and Macheboeuf (1951) found that administration of cortisone to rabbits lowers the hepatic desoxyribonucleic acid content (by at least 30 per cent in two days). Vitamin B_1 shows antagonistic activity, that is, it increases the hepatic desoxyribonucleic acid content. It is to be noted that cortisone also shows another effect on the liver: there is an increase in the glycogen store and an increase in lipids (production of artificial fatty liver). Under certain conditions, thiamine has the opposite effect: thiamine produces regression of certain fatty livers except in the case of a choline-deficient diet when thiamine increases the deposition of fat. It can been seen that the problem of fat accumulation in the liver is particularly complex.

Although the thymus gland is rich in desoxyribonucleic acid, cortisone produces degeneration of the thymus (cf. Chapter 10). This phenomenon may be related to involution of the gland at puberty. It must be added that numerous studies with ACTH and corticosteroids have demonstrated that these compounds cause eosinophilia (an increase of up to 25 per cent of the normal level) and decrease the ratio of immature myeloid cells/adult granulocytes (Aschkenasy). In rat and monkey, cortisone acetate produces hyperleucocytosis with neutrophilic polynucleosis, lymphocytosis and eosinopenia (Girod, 1959).

Moreover, these oxygenated sterols produce a series of effects on sugar, lipid and protein metabolism in mammals. Administration of 11-oxy-steroids leads to:

(1) hyperglycemia: they antagonize the action of insulin;

(2) an increase in arginase activity (urea formation) and a decrease in alkaline phosphatase activity of liver;

(3) an increase in protein catabolism and an increase in secretion of HCl, gastric pepsinogen and pancreatic trypsinogen. This explains the aperient and digestive properties of cortisone.

Cortisone and hydrocortisone rapidly mobilize the ribonucleic acid precursors of the liver of rat (Jervell). Moreover, hydrocortisone modifies the biosynthesis of an inducible enzyme: tryptophan-pyrrolase (which oxidizes tryptophan to give α-hydroxytryptophan and kynurenine) can be increased in the liver of adrenalectomized rats by administration of substrate (L-tryptophan)—or a non-substrate analogue such as α-methyl-tryptophan—and administration of hydrocortisone results in an increase in total enzyme formed (Tanaka and Knox, 1959). Molecular complexes are formed between purine derivatives and steroid hormones such as desoxycorticosterone and also testosterone and progesterone. These complexes are characterized by the fact that the hormones are more soluble in complexed form (J. F. Scott and L. L. Engel). Steroid hormones also combine with proteins, notably, blood serum proteins (Brunelli, 1934; Westphal). Ribonucleic acid is related to potassium ion and cortisone administration results in a decrease in the potassium content of cells (Bachrach, Reinberg and Stolkovsky, 1953; Trémolières, 1954). F. Shapira has indicated that administration of desoxycorticosterone leads to an increase in plasma aldolase.

Among the *general biological effects* of corticoids should be included the antimitotic property of cortisone (Green), the anti-inflammatory property and the mobilization of corticoids during stress. Cortisone and ACTH induce *thermoanalgesia* in rat (Szerb and Jacob, 1951). With respect to thermal effects, it should be mentioned that an increase in potassium ion produces an increase in resistance to raised temperatures and, inversely, a decrease in potassium ions reduces resistance to cold (Bachrach *et al.*, 1941; Reinberg, 1952). These facts must be related to the report concerning the RNA–K^+ complex and corticosteroids. Hypothermia induced by cooling or hypercapnic hypoxia produces adrenocortical hyperfunction characterized by a marked increase in the plasma cortico-sterone level (Boulouard and Buzalkov, 1963).

The adrenal cortex produces hormones which are similar to sex hormones (androgenic hormones, 17-ketosteroids and oestrogenic hormones) and therefore a direct action of the adrenal cortex on genital organs must be considered. For example, adrenosterone isolated from the adrenal cortex of ox, shows androgenic activity corresponding to about one fifth of the activity of androsterone.

Isosteres of cortisone and hydrocortisone have been obtained by the introduction of a double bond between carbons 1 and 2. These compounds lose some of the activity of natural hormones, for example the action upon mineral salts, but retain anti-inflammatory activity which is of therapeutic value.

Pathological physiology of the adrenal cortex

(1) Adrenal hypofunction.

We have already examined the effects of ablation.

Addison's disease reproduces certain characteristics of the reaction following ablation by producing adrenocortical hypoactivity in man. Addison's disease which is tubercular or traumatic in origin, results in hypotension, muscular hypotonus, hypothermia, hypoglycemia, a decrease in resistance to stress and a characteristic symptom, progressive brown pigmentation. With the exception of pigmentation and hypoglycemia, these symptoms result from a deficiency in mineralocorticoids.

Metopirone blocks the $11-\beta$ hydroxylation of corticosteroids and, in dog, 2,2-bis (parachlorophenyl)-1,1-dichloroethane causes adrenal cortical atrophy (Nelson and Woodard, 1948; Nichols et al., 1957).

(2) Adrenal hyperfunction.

Hyperfunction can result from the presence of a tumour but can also be produced by administration of ACTH or adrenocorticoid hormones. It is characterized by the inverse effects to those of hypofunction: hyperglycemia and glycosuria, oedema (retention of Na^+ and H_2O) with an increase in blood volume leading to hypertension, hypokalemic alkalosis and virilism (an androgenic effect) which constitutes the adreno-genital syndrome characterized in the female by a male voice, growth of the clitoris and development of hair.

Cushing's disease is characterized by hyperactivity of the pituitary (ACTH) due to the presence of a tumour. There is deposition of fat, insulin-resistant hyperglycemia and hypertension. It occurs more frequently in females than males. This is an indirect disease of the adrenals (thus, differing from Addison's disease), the adrenals showing the effect of pituitary dysfunction. However, Cushing's disease can also be produced directly by cortisone.

CHAPTER 17

THE HYPOTHALAMO-HYPOPHYSO-PITUITARY SYSTEM

ALTHOUGH the adrenal is a complex gland, there is another system of greater complexity which could be described as the "brain" of the endocrine system. This "brain" consists of the hypophyseal-pituitary system and the hypothalamus, two structurally distinct entities, each of which shows structural and functional complexity but which have a *direct* functional relationship. It should be noted that endocrine systems are associated generally with indirect functional relationships. The hypothalamic-hypophyseal-pituitary complex is situated close to the brain and consists of a *nervous* formation (the hypothalamus), a *neuroglial* formation (the hypophysis) and an *endocrine secretory* formation (the pituitary).

The hypophyseal-pituitary complex

The structure which is generally termed the "hypophysis" comprises *two embryologically distinct units:*

the anterior lobe (=ante-hypophysis=*pituitary gland*) of *cutaneous ectodermal* origin (the pouch of Rathke, an invagination of the primitive bud or *stomodeum*);
the posterior lobe (=neurohypophysis = post-hypophysis = *hypophysis*) of *neural ectodermal* origin (a downward outgrowth of the diencephalon making a pair with epiphysis).

I shall use the term *pituitary* for the anterior hypophysis and *hypophysis* for the posterior hypophysis.

The anatomical and physiological complex is suspended in the *sella turcica* beneath the floor of the third ventricle. The human gland weighs about 600 mg.

A *fibrous capsule* attaches the organ to the *dura mater* of the *sella turcica* and the hypothalamic stalk attaches the hypophysis to the floor of the third ventricle.

The pituitary gland consists of three regions:

(1) the large *anterior lobe;*

(2) the thin *posterior lamina,* attached in the adult to the membrane which separates the pituitary and hypophysis;

(3) the *pars tuberalis,* a linguiform prolongation which extends along the length of the pituitary stalk and into the membranes of the *pia mater.*

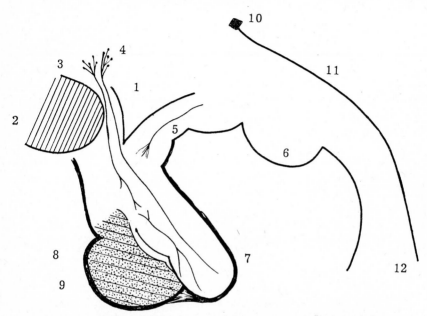

FIG. 123. Schematic topography of a sagittal section of the hypothalamo-hypophyso-pituitary zone. *1,* IIIrd ventricle; *2,* optic chiasma; *3,* supraoptic nucleus; *4,* paraventricular nucleus; *5,* tubero-hypophyseal tract; *6,* mamillary body; *7,* "post"-hypophysis (neuroglial and neural hypophysis, posterior lobe); *8,* pituitary (anterior lobe); *9,* fibrous capsule; *10,* foramen of Monro; *11,* thalamic zone; *12,* cerebral peduncle.

Histology

I. Pituitary

As was stated in 1963 (N. S. Halmi; J. Racadot), our knowledge of pituitary histophysiology is confused and, according to A. R. Currie, it appears that the fluorescent antibody technique offers the most immediate hope of clearing the confusion surrounding pituitary cell nomenclature and function. J. M. Marshall (1957) successfully used this technique in the localization of cells containing ACTH in dog while Rennels and Wolf (1963) used the method in the localization of cells containing FSH and LH in sheep.

(a) The *anterior lobe* is an endocrine gland of the *reticular* type consisting of anastomosing epithelial filaments interlaced with a network of sinusoidal capillaries. The staining method of Mann (acidic methylene blue and eosin) distinguishes:

> *chromophobic cells*—small cells with poorly developed chondriome; *chromophilic cells*—large cells which are rich in ascorbic acid (Giroud and Leblond, 1934), some being *eosinophilic* (or acidophilic), containing numerous mitochondria and showing a Golgi apparatus and eosinophilic cytoplasmic granules, others being *cyanophilic* (or basophilic) with several large granules and fatty inclusions.

Since the pioneer work of Romeis (1940; Fig. 124), the situation has been a little confused (a situation which was described by Halmi as the "terminological jungle"). A basis for the classification of pituitary cells in mammals is provided by Racadot and is based principally on the work of Herlant:

> *cells containing glycoprotein granules* (beta, delta and gamma cells of Romeis—the so-called basophilic cells—the beta cells containing FSH, the gamma cells, LH and the delta cells, thyrotropic hormone); *cells with non-glycoprotein granules* (the alpha, epsilon and eta cells— the "acidophilic" cells—the alpha cells secreting growth hormone, the epsilon cell, corticotropin while the eta cells would secrete prolactin and thus correspond to the "pregnancy cells" of Erdheim and Stumme, 1909).

In the pituitary of birds which have no *pars intermedia* there is another type of cell which has been provisionally termed the kappa cell (A. Tixier-Vidal, 1963).

All pituitary cells are rich in *acid phosphatases* (Abolins, 1949). In man, the basophilic cells associated with ACTH contain a peptidase (cathepsin) (Pearse and Van Noorden, 1963).

(b) The *posterior lamina* is of the reticulo-vesicular type. Certain vesicles contain a colloid substance.

(c) The *pars tuberalis* which to some extent resembles the posterior lamina is reduced in man.

II. The hypophysis

This structure consists essentially of *neuroglial cells and fibres* and contains numerous unmyelinated nerve fibres which originate in the region of the hypothalamus.

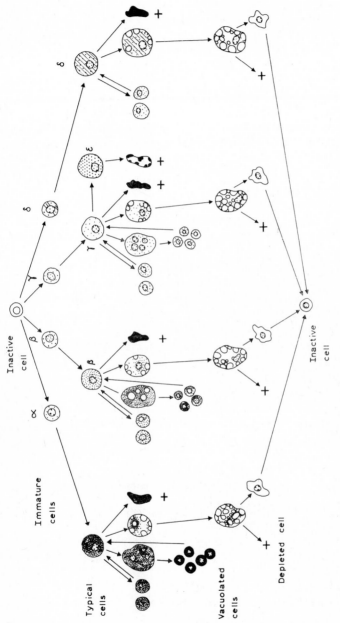

Fig. 124. The secretory cycle of pituitary cells of man. (From B. Romeis, 1940.)

III. The hypothalamus

The hypothalamus is separated from the *thalamus* (the *optic layer)* by the hypothalamic sulcus. The hypothalamus forms the floor and part of the wall of the third ventricle.

The hypothalamus is a differentiation of the ventral wall of the dien-cephalon, the dorsal wall providing the thalamus.

In urodeles, the hypothalamus consists only of a periventricular cellular zone surrounded by a zone rich in nerve fibres.

The hypothalamus of higher vertebrates consists of three areas of gray matter: the periventricular zone, the zone containing several secondary cell groups and the zone containing numerous cell groups including the *supra-optic nucleus* and the *paraventricular nucleus* which have great functional significance. In lower vertebrates (fishes and batracians) these two nuclei form a single preoptic group.

It is appropriate at this stage to formulate the following concept:

I have already expressed the principle of concentration and spatial unification of vegetal organs as a function of evolution. In considering the hypothalamic nuclei, we find an example of the complementary principle. During evolutionary processes, there is a progressive diversi-fication and separation of neural structures associated with an increase in the complexity of neural architecture in higher animals. Thus, *anatomo-physiological evolution in vertebrates appears to be characterized by spatial economy of vegetal structures and spatial expansion of neural elements* as if, by economizing on certain elements, the organism is capable of building structures of higher quality.

The cells of the two hypothalamic nuclei are large and binucleated with Nissl granules and numerous *vesicles. These cells are penetrated by capillaries.* This is exceptional in that nerve cells are generally surrounded but not invaded by capillary networks. Finally, efferent fibres from the supra-optic and paraventricular nuclei innervate the hypophysis.

Important dates in the elucidation of hypophyso-pituitary physiology

1909 A *hypopituitarism* syndrome (hypothermia, arrest of growth, atrophy of genital organs and obesity) was obtained after more or less com-plete ablation of the hypophyso-pituitary complex (Cushing).

1911 Ablation of the hypophysis shown to produce prolonged polyuria (Cushing).

1913 Camus and Roussy produced *diabetes insipidus* by lesions in the infundibulum. Velden and also Farini showed that hypophyseal extract temporarily suppressed spontaneous *diabetes insipidus.*

1916 The *thyrotropic action* of the pituitary was discovered (P. E. Smith).

1921 Evans and Long obtained giant rats after administration of a saline extract of the hypophyso-pituitary complex. Discovery of the *somatotropic action.*

1923 Localization of growth hormone in lateral acidophilic cells of the pituitary (P. E. Smith).

1928 Evidence was obtained for the lactogenic properties of pituitary by Stricker and Greutner.

1937 Localization of thyrotropic hormone in central basophilic cells of the pituitary (Voitkewitch).

1940 Discovery of the corticotropic action (on the adrenal cortex). Bates, Li, Simpson and Evans isolated the luteotropic hormone.

1941 Localization of gonadotropic follicle-stimulating hormone in the basophilic cells of the pituitary (Friedmann and Hall).

1953 Evidence obtained for the chemical structure of *oxytocin* (hypophysis) by Ressler. Synthesis of oxytocin by du Vigneaud and his co-workers.

Physiology of the hypophyso-pituitary hypothalamic system

A. Pituitary

The suffix *trophic* indicates a nutritive function and *tropic* an action directed towards a certain object. Therefore, in the scope of biological cybernetics, the suffix tropic is more appropriate since it is associated with the concepts of *target organ* and of *molecular receptors.*

Pituitary hormones are *antigenic* such that parenteral administration of the heterospecific hormone can lead to the formation of antibodies which inhibit the hormonal action of the antigen. This antigenicity is of value in that it allows a detailed analysis of the structural specificity of the hormones being considered (Kabak and Stulova, 1937 and Hayashida, 1962, for prolactin; S. C. Werner, 1961, for thyrotropin, etc.).

We shall consider the action of pituitary hormones on exocrine glands and their action on endocrine glands.

(1) Control of an exocrine effector

LACTOGENIC HORMONE (PROLACTIN or LUTEOTROPIN)

It is not a glycoprotein but an open peptidic chain with 211 amino-acid residues (Li).

The lactogenic hormone stimulates the formation of "crop milk" by the crop glands of the pigeon and this action is utilized in the assay of this hormone.

However, the *main* action of this hormone is seen *in mammals.* During gestation, the placenta secretes a hormone which inhibits the production of prolactin by the pituitary. At the time of parturition, after elimination of the placenta, prolactin is released and results in milk production.

(2) Control of endocrine effectors.

THE TROPINS OR STIMULINS

(a) SOMATOTROPIN (or growth hormone)

Somatotropin is a protein of which the amino acid composition is known (Li and Chung, 1956).

Somatotropins from man and monkeys are identical but differ from that of ruminants.

The hormone from ruminants has a molecular weight of about 46,000 corresponding to 396 amino acid residues while that of primates has a molecular weight of 25,400 (monkey) to 27,100 (man) corresponding to 242–246 amino acid residues.

Moreover, the molecular configuration differs. The protein from ruminants is in the form of a branched chain and that from primates is a straight chain. This introduces the concept of *iso-hormones* which can be compared to iso-enzymes.

Chemical differences lead to differences in some specificity of action. The hormone from ruminants can stimulate growth in rat but not in man while the hormone from primates can stimulate growth in rat.

Controlled hydrolysis of the hormone from ruminants by chymotrypsin does not destroy biological activity. This suggests that some fraction of the hormone would be the true hormonal agent.

This hormone *stimulates the growth* of soft regions and of long bones in the area of the epiphyses.

Somatotropin acts by stimulating degradation of fats and by inhibiting catabolism of amino acids and thus would function as a hormone for the *conservation of amino nitrogen*. Growth hormone also shows diabetogenic activity. Total hypophysectomy in the young mammal leads to arrest of growth and reduces protein biosynthesis by reducing the number of polysomes (Korner), an effect which is explained (Korner, 1966) as resulting from a direct action of growth hormone on the gene. Growth and increasing activity of ribosomes are restored by prolonged administration of somatotropin. In the normal animal, hypersomatotropism leads to *gigantism*.

(b) GONADOTROPINS

These hormones act on the testis and ovary, the gonads being both endocrine glands (production of hormones) and exocrine glands (production of gametes).

(1) *Follicle stimulating hormone* (= prolan A = FSH) which has a molecular weight of 25,000–30,000 in mammals (Steelman, 1959). Acetylation of the amine groups by ketene and the action of cysteine or versene (a metal chelating agent) decrease the activity of FSH.

FSH augments the production of oestrogens by the Graafian follicles *in the female* and augments growth of the testes and spermiogenesis *in the male.*

(2) *Luteinizing hormone* (= prolan B = LH = progestation hormone). This hormone has a molecular weight of 30,000 in sheep (Squire and Li) and 100,000 in pig.

LH induces ovulation and the development of the *corpus luteum.*

Luteotropic hormone (LTH, prolactin) produces permanent secretion of *progesterone* by the *corpus luteum.*

In the male, LH stimulates the production of testicular *testosterone* which explains the indirect action of LH on spermatogenesis and the development of auxiliary sex organs such as the prostate and seminal vesicles. LH is also called *interstitial cell stimulating hormone* (ICSH).

In a totally hypophysectomized mouse, an atrophied testis which has been inactive for several months regains specific activity after it has been grafted beneath the kidney capsule of a bilaterally castrated male adult mouse of the same strain (R. M. May, 1958). Thus, total hypophysectomy leads not to degeneration of the testes but to the production of a dormant state which can easily be reversed.

Gonadotropic hormones are found in the *placenta* (chorionic gonadotropic hormone). These hormones are glycoproteins. They act in a similar manner to LH-ICSH (Gurin). However, a highly purified preparation of human chorionic gonadotropic hormone with a molecular weight of 30,000 and an isoelectric point of 2·95 shows no FSH activity (Got and Bourrillon, 1959–1961).

According to Moszkowska (1955), the epiphysis would have an inhibitory action on the activity of the genital system. Thus, epiphysectomy leads to precocious puberty, hypertrophy of genital glands, etc. (Foa; Simonnet and Thiéblot; Kitay and Altschule). The pituitary of rat can secrete gonadotropic hormones *in vitro* and this secretion is decreased in the presence of epiphyses (1958). This is an example of a different type of anti-hormonal action to that which develops with antigenic hormones.

(c) THYROTROPIN

Thyrotropin = thyrotropic hormone = thyroid stimulating hormone = TSH. The precise chemical composition of this hormone is not known. However, TSH from the eel has been purified by Y. A. Fontaine and Condliffe (1963) and is rich in aspartic and glutamic acid but contains no galactosamine. The hormone is inactive in mammals in concentrations which produce marked stimulation of the thyroid of teleosts. Purified preparations of mammalian TSH show typical thyrotropic activity in both mammals and teleosts. It would appear that there is a zoological specificity of phylogenetic significance associated with thyrotropic hormones (M. and Y. A. Fontaine). In fact a factor is present in mammalian pituitaries (heterothyrotropic factor = HTF) which is different from TSH:

it strongly stimulates teleostean thyroid but is 100–200 times less effective on mammalian thyroid (M. and Y. A. Fontaine).

Pituitary deficiency leads to thyroid atrophy. Administration of TSH produces hyperthyroidism and administration of thyroxine (or iodine) decreases the production of TSH.

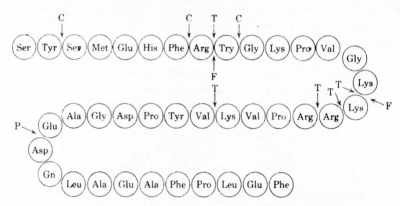

FIG. 125. The structure of porcine α-corticotropin. Points of attack of chymotrypsin (C), trypsin (T), pepsin (P) and fibrinolysin (F) are indicated by the arrows. (From ANFINSEN, *The Molecular Basis of Evolution*.)

(d) ADRENOCORTICOTROPIN (ACTH)

Following enzymic hydrolysis or acid and alkaline hydrolysis of the extracted protein, the products still exhibit biological activity. ACTH is thermostable at neutral pH (C. H. Li). The *active peptides* (Li and Pedersen, 1950) have a molecular weight of 1200 (7–9 amino acid residues) but the most active product—β ACTH—has a molecular weight of 4566 (39 amino acid residues). This provides another example of active fractions from a parent molecule (cf. p. 440).

At the molecular level ACTH augments the synthesis and excretion of corticoids with the apparent exception of aldosterone. ACTH *augments also the incorporation of acetate to adrenal tissues* (as shown by administration of ^{14}C-acetate). Acetate is the precursor of steroid hormones.

Administration of ACTH to man produces the following effects:

(1) augmentation of nitrogen excretion (antagonism of growth hormone) and excretion of potassium ion and orthophosphate;

(2) retention of chloride and sodium ions and thus, water retention;

(3) hyperglycemia and glycosuria;

(4) hypercholesterolemia;

(5) changes in composition of blood.

In teleosts, the anterior inter-renal gland is under the control of the pituitary (Pickford; Chester-Jones, 1957) in the eel, hydrocortisone is the principal glucocorticoid of plasma (Fontaine and Hatey, 1954; Hatey, 1961).

The regulation of ACTH secretion is dependent on a balance between stimulation and inhibition. Corticoids such as cortisone depress pituitary secretion and during psychic, thermal or traumatic stress there is an initial liberation of ACTH which stimulates the adrenal cortex and, thus, contributes to the development of protection against the immediate effects of aggression. *This type of reaction is similar to a reflex.* Stimulation of the hypothalamus by adrenaline and (or) histamine via the blood stream modifies the ACTH-producing basophilic cells of the pituitary. Delost and Chirvan-Nia (1959) have shown that prolonged cortico-stimulating activity of adrenaline is modified according to the age of mice. There is a positive effect in animals of 55–70 days and no effect in animals of 43–48 days.

Finally, in this examination of pituitary tropins, the possible existence of another hormone, the *pancreotropic hormone*, should be indicated.

The pituitary exhibits diabetogenic activity, that is, a substance in the pituitary exhibits the opposite action to that of insulin. Prolonged injection of pituitary extract (to dog but not to cat) leads to sugar diabetes by a change in the insulin producing tissue of the pancreas. Moreover, *diabetes produced by ablation of the pancreas can be improved by ablation of the pituitary* (the Houssay effect). Since totally hypophysectomized animals become particularly sensitive to insulin, a functional relationship between the activity of the endocrine pancreas and pituitary is suggested.

It is widely recognized that sugar diabetes is not necessarily related to an *absolute* deficiency in insulin and diabetes may also result from *insulin deficiency related to pituitary hyperfunction.*

The pituitary *diabetogenic hormone* has not been isolated. The administration of TSH or ACTH or even *growth hormone* can induce a paradiabetic state. The difficulty of the problem is related to the possible dual or multiple action of pituitary hormones. A similar problem is associated with the possible existence of a *ketogenic pituitary hormone*. It has been stated that pituitary extracts are ketogenic. However, ACTH is ketogenic too and ketonemia associated with ketonuria could result from hyperglycemia.

To end this discussion, one has to mention a hormonal factor from the pituitary which would regulate the mobilization of fats in the organism and their deposition in the liver. This factor is called *adipokinin* (Levin and Farber, 1952). Large doses of cortisone produce fat deposition in the liver (Gros *et al.*). ACTH is involved in this phenomenon. However, adipokinin would be another factor. The problem of fat deposition and fat elimination is far from solved but it has been suggested that vitamin B_1 (which produces fat deposition in liver when given in high doses, cf. p. 431), pituitary, adrenocortical and genital factors are important. *With the solution of this problem, definitive information would be obtained*

*which could be correlated with the concept of terrain which is complementary
to the concept of the internal environment.* The work of Birk and Li (1964)
has contributed to the solution of the problem. They isolated from the
pituitary a peptide containing 54 amino acids and with an isoelectric
point at pH 6·9. This peptide, *lipotropin* or LPH, mobilizes fats *in vivo.*

Concepts of pathological physiology of the pituitary

(A) HYPERPITUITARISM

Excessive production of ACTH (basophilic tumour) results in the development of Cushing's disease. Excess growth hormone (eosinophilic
tumour) results either in *gigantism* which is associated with pituitary
hyperactivity *during* growth (Gulliver at Brobdingnac) or *acromegaly*
associated with pituitary hyperactivity *after cessation of growth.* The
acromegaly is characterized by deformities: enlarged chin and nose, swelling of limbs. This localized gigantism indicates a latent disposition for
growth in these structures.

(B) HYPOPITUITARISM

This is generally associated with atrophy or infarctus of the gland.
Hypopituitarism is characterized by *dwarfism* corresponding to deficiency
during growth (Gulliver at Lilliput) or by *pituitary myxoedema* (TSH
deficiency) with idiocy and cachexia.

B. Intermediate lobe

This lobe secretes *intermedin.* Intermedin (= MSH = melanophore
stimulating hormone) modifies the state of pigmentation in skin of lower
vertebrates and the cutaneous storage of melanin by melanophores of
man. Melanin is formed by polymerization, the initial stage being the
oxidation of tyrosine in the presence of *tyrosinase* (an enzyme containing
copper).

A study of the pituitary of ungulates shows that the *pars intermedia*
is well developed in those animals which can tolerate water deprivation
(camel). However, this lobe is reduced in semi-aquatic ungulates or those
dependent on water (rhinoceros). The neurosecretory content of the
(neuro-)hypophysis of camels deprived of water for 25 days is only slightly
diminished as compared to the normal camel (E. and M. Legait and
Charnot, 1963).

(1) Chemistry of intermedin

Intermedin has been isolated in a pure form and the structure is known
(Li *et al.*, 1956). This hormone is a peptide containing 18 amino acids and
with a molecular weight of about 2177. *It is interesting to note that MSH*

contains a pattern of 7 amino acids which is also found in ACTH and that *ACTH shows weak MSH activity.* However, MSH does not exhibit ACTH activity. There is a unidirectional ambivalence.

One can consider that there are two types of ambivalence:

(1) A functional ambivalence due to the presence of receptors on the target organs, some of which are more sensitive since they contain more receptors or the receptors are better situated. (This would explain the action of a hormone such as growth hormone which does not act on a single target organ but on many organs.)

(2) Functional ambivalence due to a similarily in part of the hormonal molecule. This ambivalence which can be considered to be central in origin as compared with the former type which is peripheral, does not exclude—even includes—the selectivity of peripheral receptors which are themselves molecules.

(2) Physiology of intermedin

Cortisone and hydrocortisone (which inhibit the secretion of ACTH) *inhibit secretion of MSH. Adrenaline and noradrenaline inhibit the action of MSH.* In Addison's disease the skin develops a brown pigmentation and this disease is associated with a decrease in the production of corticoids. Under these conditions, synthesis of melanin is augmented and the characteristic pigmentation appears.

In considering the action of MSH on *chromatophores*, the main factor of interest is the chromatic adaptation to light which is important in camouflage and, then, in natural selection.

The structure and physiology of the cutaneous receptor, the chromatophore, varies according to the type of animal. In cephalopods, the chromatophores change form by the action of radial smooth muscle fibres while in batracians, the chromatophore is a cell or syncytium in the shape of the network in which pigment is distributed by the action of intracytoplasmic currents. The pigment can be melanin (black), carotenoids (yellow and red), guanine derivatives (white).

The most important factor in colour change is the *stimulus of light* which acts either directly on the chromatophores (primary response) or through the eyes (secondary response). In the latter case, the response is determined by the ratio:

$$\frac{\text{quantity of light striking the eye}}{\text{quantity of light reflected by the background}}$$

On an illuminated black background, this ratio is high and the animal darkens while on an illuminated white background, this ratio is low and the animal pales. There would be only a primary response in larval batracians. In the adult, the action on the melanophores is humoral. Transfusion of blood from a dark frog to a light frog results in darkening of the recipient. Hypophyso-pituitary ablation in the larva or adult produces permanently pale individuals (G. M. Smith, 1916).

Allen (1920) implanted an intermediate lobe beneath the skin of a hypophysectomized tadpole and the animal darkened.

Torpedo marmorata has two well separated lobes and the intermedio-posterior lobe is almost exclusively represented by the intermediate lobe. Following hypophysectomy of *Torpedo*, C. Veil and R. M. May (1937) concluded that the *pars intermedia* is of considerable importance in the regulation of pigmentation.

R. Collin and Drouet (1933) found that injection into the lymph sac of frog of urine from an individual showing hypophysopituitary hyperfunction produced darkening of the skin. Many substances other than intermedin cause dilation of melanophores (for example, nicotine) while adrenaline produces constriction. The test of Collin and Drouet gives variable results and cannot be used as an index of hypophyso-pituitary activity but it is of value to report this test in that it draws attention to the possible interferences in endocrinology.

Illumination acts by way of the hypothalamus (cf. p. 531, the effect of light on sexual maturation):

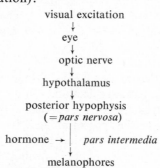

visual excitation
↓
eye
↓
optic nerve
↓
hypothalamus
↓
posterior hypophysis
(=*pars nervosa*)

hormone → | *pars intermedia*
↓
melanophores

The results obtained by Hogben and Slome (1936) for *Xenopus* suggest that two factors are secreted:

(1) one causing dispersal of melanin and darkening: substance B;
(2) the other being an antagonist and producing lightening: substance W.

The production of these substances would depend on the area of the retina stimulated:

(1) basal elements of the retina are associated with the formation of substance B in the *pars intermedia;*
(2) peripheral elements are associated with the formation of substance W by the *pars tuberalis.*

Substance B appears under conditions of intense illumination and the secretion of substance W occurs more slowly than that of substance B.

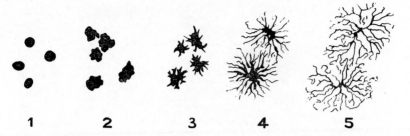

FIG. 126. Melanophore index of HOGBEN and SLOME (frog skin). (1) Maximum retraction of cell (after hypophysectomy or injection of adrenaline). (5) Extreme expansion of melanocyte showing long, thin branches (35 mn after frog placed on black background or following injection of intermedin). (BENTEJAC, 1938.)

When placed in a humid, dark environment, a frog from which the pituitary only had been removed is dark while a totally hypophysectomized frog is pale.

Kleinholz (1940) detected intermedin in the hypophysis of larval *Rana pipiens* measuring 4–7 mm in length. This *intermedin appeared before morphological differentiation*. This latter point should be emphasized. In this case, as in the case of thyroid or the appearance of antigens in sea urchin embryo, it appears that, although the function does not create

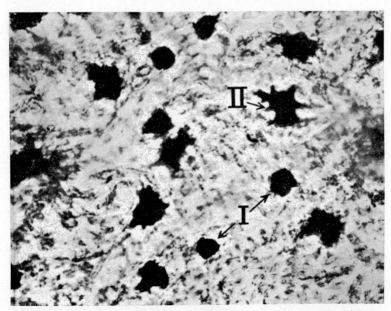

FIG. 127a. Group I melanocytes (complete retraction) and group II melanocytes. ×250. (BENTEJAC, 1938.)

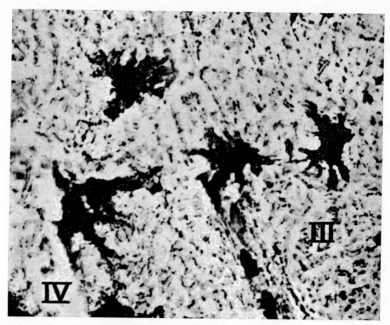

FIG. 127b. Group III and IV melanocytes. (BENTEJAC, 1938.)

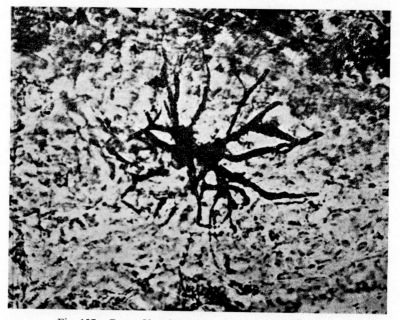

Fig. 127c. Group V melanocytes. (BENTEJAC, 1938.)

the organ as in the lamarkian concept, *the function does precede the organ* and this fact expresses an essential aspect of early molecular life.

The presence of intermedin in a young animal which is unable to protect itself by change in colour suggests that intermedin has a more general function. Collin and M. Verrain (1953) reported that intermedin injected to rabbit caused *hypoglycemia* which would resemble *insulin hypoglycemia*. If this result is confirmed:

(1) regulation of glycemia would involve a supplementary pituitary factor;

(2) there would be another example of direct or indirect functional ambivalence of certain hormones.

The latter point emphasizes that causal physiology explains or tends to explain at the most fundamental level, the molecular level, the macroscopic biological effects which were apparent to early observers and experimentalists. *Physiology is concerned with the discovery of significant biological effects which are detectable at the molecular level. This aspect of physiology can be termed molecular physiology which is distinct from molecular biology or chemical genetics.*

May and Thillard (1957) found that injection of epiphyses from mice resulted in the appearance of longitudinal bands of pigmentation in the lizard *Anolis roquet*. Therefore, there may be an epiphyseal factor which remains to be isolated. Lerner and his associates (1958) have isolated from pineal gland a substance, *melatonin*, which, in fact, is anti-MSH.

Colour adaptation in the invertebrates has been studied principally in Crustacea. Pouchet (1872) demonstrated that ablation of eye stalks of prawn produced a permanently dark animal. When *Crangon* is placed on a light background it pales and on a dark background it darkens (Koller, 1925). Injection of blood from a dark prawn to a pale prawn results in darkening of the recipient. However, the reverse change cannot be produced. Lightening appears to be a passive phenomenon and the existence of a darkening factor is suggested. Hanström (1923) demonstrated that the eye stalk contains a small gland which is responsible for colour changes and which is called the sinus gland since it is situated close to a blood sinus. The situation of the gland suggested that one or more substances diffused from it into the blood and it is known that extracts of sinus gland can produce colour changes. Brown (1933), working with animals from which the eye stalks had been removed, found that electrical stimulation of the optic nerve resulted in chromatic responses. Therefore, the sinus gland is not secretory but stores and releases chromactivating substances produced by specialized neurones. This complex resembles the hypothalamo-hypophyseal complex. The neurones associated with the sinus gland contain neurosecretory structures. Knowles (1953) showed the presence in *Leander serratus* of a post-commisure organ situated near the oesophagus and concerned with the development of nocturnal coloration. The sinus gland is involved in the development of diurnal coloration. In recent years, an important contribution has been made by Kleinholz in collaboration with Josefsson and others (1961, 1964). Kleinholz localized several hormones from the eye stalk of Crustacea: (1) a hormone effecting light-adaptation of the distal retinal pigment (DRPLH): (2) a hormone producing an increase in blood glucose concentration (HGH). The properties suggest that it may be a polypeptide. The hyperglycemic hormone from eye stalk extracts of *Libinia*, *Pacifastacus* and *Pandalus* is inactivated by brief heating at 100°C. On the basis of physiological and specificity tests, the light-adapting retinal pigment hormone and the hormone which concentrates pigment of chromatophores appear to be separate entities.

C. Hypophysis and hypothalamus

Two hormones are found in this region: *vasopressin* and *oxytocin*.

Chemical structure

The *synthesis* of hypophyseal hormones is one of the most outstanding achievements of contemporary biological chemistry. The synthesis was carried out by Du Vigneaud, Ressler and Trippet (1953).

The structure of oxytocin shows it to be a small *peptide* containing 8 amino acids and having a molecular weight of approximately 1000.

The structure of *vasopressin* is very similar to that of oxytocin, suggesting the existence of similar ribonucleic templates.

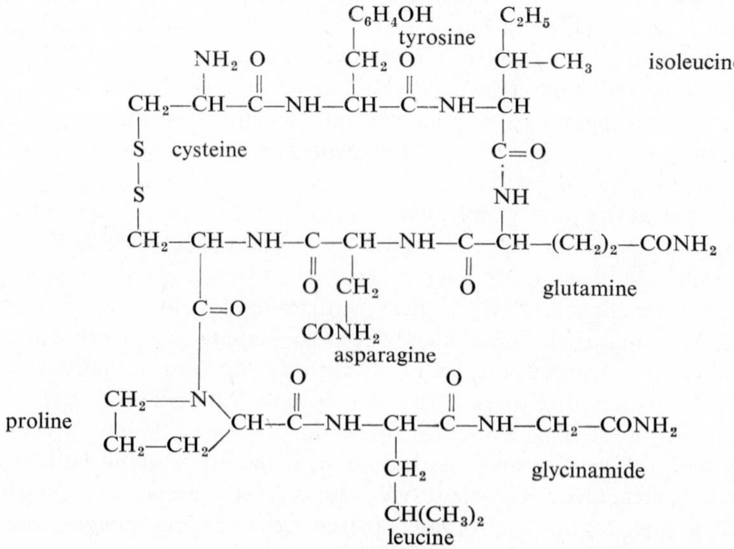

Oxytocin

Isoleucine of oxytocin is replaced by phenylalanine in vasopressin; *leucine* of oxytocin is replaced, *in vasopressin of ox*, by arginine and in *vasopressin of pig* by lysine — two more basic amino-acids.

Three compounds have been found in chicken and frog. The three active molecules from chicken have been purified. In addition to oxytocin and arginine-vasopressin which are similar to those from mammals, there is a third hormone with an intermediate chemical structure and possessing both oxytocic and vasopressor activity. The name "Natriferin" was given to this compound since it has an action on active transport of sodium in batracians (R. Acher *et al.*, 1960).

Function

A. OXYTOCIN

This compound, which is present in both males and females, produces uterine contractions *in vivo* and *in vitro* and this property is utilized in obstetrics. Oxytocin produces an increase in the tone of smooth muscle fibres of many organs: mouth, stomach, gall bladder, urethra. However, there are species variations. Mary Pickford (1961) reported that oxytocin produces vasodilation of the skin and muscle of man and woman with only a transient fall in blood pressure; oxytocin produces vasodilation in rat but there is no fall in blood pressure. (Oxytocin causes a rise in blood pressure during the latter half of pregnancy in rat.) It should be noted that a certain number of pharmacologically active compounds exhibit oxytocic properties (uterine stimulating compounds) such as dihydroergotamine, D- and L-sparteines (Sandberg *et al.*, 1961) and also angiotensin II-Val$_5$-octapeptide (Renson, Barac and Bacq, 1959).

Oxytocin is important in milk ejection during lactation (Ott and Scott).

B. VASOPRESSIN. Vasopressin increases arterial pressure by constriction of peripheral blood vessels and is used in the treatment of shock. Vasopressin produces coronary vasoconstriction; the spasmogenic action is inhibited by papaverine (an alkaloid derivative of isoquinoline).

Vasopressin has an *antidiuretic effect* and is called the *antidiuretic hormone* (ADH). ADH acts in the region of the kidney tubules (cf. p. 374). In the absence of this hormone, *diabetes insipidus*, that is, increased diuresis, is apparent.

Significance

The most important problem concerning the hypophysis is perhaps, not that of the nature of the hormones, but that of the *origin of the hormones. Are the hormones produced and localized in the hypophysis in significant amounts or are they formed elsewhere to be stored in the hypophysis?*

Let us examine some of the facts which will help in answering this question.

Aschner in 1911 and Camus and Roussy in 1913 found that a lesion of the *hypothalamus* (but not of the *neurohypophysis*) resulted in *diabetes insipidus*. Moreover, in 1924, R. Collin demonstrated histologically the existence of granular colloidal secretions which would pass from the pituitary to the hypothalamic nuclei via the pituitary stalk. He termed this phenomenon the *interstitial neurocrine process*. However, Roussy and Mosinger described a process of neurosecretion by the hypothalamic nuclei of man which they termed a *neuricrine* process.

The terms neurocrine and neuricrine must not be confused: it is the existence of the neuricrine process which has been confirmed and corresponds to the *neurosecretion* of great importance particularly with reference to hypothalamo-hypophyseal relationships.

The hormonal effects of the neuricrine process are dependent finally on passage of hormones via the blood stream *(hemocriny)*.

The colloidal granules and droplets which vary in size and number from cell to cell of the paraventricular and supraoptic nuclei, are found along the whole *length of the neurone* and penetrate the narrowest pericapillary nervous branch in the posterior lobe. *A characteristic of secretory neurones is the presence in the fibres of granules which are particularly rich in secretory material.* The largest granules are the *Hering bodies*. The findings of E. Scharrer (since 1932) in fishes and batracians led to the recognition of the glandular properties of certain neurovegetal cells in the diencephalon. However, Kopec (1917) discovered the role of cerebral ganglia in the metamorphosis of insects. The neurosecretory material of these ganglia was found to pass into the *corpora cardiaca* (Bounhiol, Arvy and Gabe). But, let us return to the vertebrates for a more detailed study of neurosecretion.

In order to study the β cells of the islets of Langerhans, the American histochemist, Gomori, developed a staining technique using chromic hematoxylin—phloxin. Bargmann applied this technique to the study of diencephalo-hypophyseal neurosecretion and found that neurosecretory granules could be stained. This reaction could be called the Bargmann–Gomori *reaction*. However, it is usual to speak of the "Gomori reaction" or of "Gomori-positive" granules.

Diencephalo-hypophyseal neurosecretion has been established and it is now certain that the direction of flow of neurosecretory material is from the *diencephalon* to the *hypophysis*. This has been shown by:

(1) section of the hypophyseal stalk of batracians (Hild, 1950) and in rats (Stutinsky, 1952),

(2) the presence of vasopressor and oxytocic principles in the supraoptic and paraventricular nuclei and the fact that *the quantity of neurosecretory material detected histologically* (Bargmann–Gomori positive material)

appears to be directly proportional to the level of hormones in the corresponding regions (Barry, 1954). The hypothalamic nuclei contain 25 per cent ot the total anti-diuretic hormone which is secreted (Melville and Hare, 1945).

(3) depletion of the store of neurosecretory material which coincides with a decrease in the level of antidiuretic hormone.

However, *these facts do not suggest that the hormones* (vasopressin and oxytocin) *are represented by the secretory droplets.* We have seen that the hormones localized in the pituitary are relatively small molecules (cf. du Vigneaud). Therefore, the droplets which are detected histologically can only be the carriers of the hormones and not the hormones themselves. A protein called *neurophysin* (with a molecular weight of 25,000 according to Ginsburg and Ireland, 1963) is able to bind oxytocin and arginine-8-vasopressin (Stouffer, Hope and du Vigneaud).

With reference to the neurosecretory granules of the hypophyseal hypothalamic complex and elsewhere [for example, the X glands of crustacea and certain vertebrate ganglion cells such as the neurosecretory cells of the caudal ganglion of teleosts and elasmobranchs which were described by Dahlgren (1914) and studied by Spiedel (1919) and Enami (1945)], it is necessary to consider not only the production of hormones but also the pathways of diffusion of the hormones.

The blood is not the only vehicle for the transport of "chemical messengers". The blood is the *principal* pathway for hormonal transport in mammals in which the blood system shows a widespread distribution but the *humoral pathway* includes *all body fluids* and in particular, the lymph. This can be exemplified in the case of hypothalamic neurosecretion.

Histological evidence for the passage of various materials (granules, pigments, etc.) in cerebrospinal fluid was provided by R. Collin in 1926 and Cow (1915) detected oxytocic activity in cerebrospinal fluid. Hild found that droplets from the pericaryon of certain cells in the preoptic nucleus passed into the cerebrospinal fluid of *Bufo vulgaris.* These droplets were Bargmann–Gomori positive. However, hypothalamo-hypophyso-pituitary relationships are not confined to the hypothalamo-neurohypophyseal system since it appears that *other regions are directly related. Neurosecretory fibres enter the epithelial tissue of the intermediate lobe.* This lobe contains numerous neuroglial cells such that, in certain animals, it cannot be distinguished from the posterior lobe.

The fibres of the supraoptic-hypophyseal tract which carry the secretion are in direct contact with the capillaries of the wall of the infundibulum. Blood from these capillaries passes to the pituitary and Bargmann (1958) stated that "one can envisage the possibility of a *direct diencephalic influence on the anterior hypophysis* through the active principles". It has been suggested that the secretion of ACTH is controlled by a hypothalamic stimulus (Sayers, 1948; de Groot and G.W. Harris; Saffran *et al.,* 1958).

This stimulus is provided by pituitropic hypothalamic mediators such as the corticotropin releasing factor or CRF (Saffran, Schally and Benley, 1955). This compound is not identical with vasopressin although vasopressin can induce ACTH secretion *in vivo* and *in vitro* (Guillemin, 1955). Guillemin and his co-workers isolated two peptides (α_1-CRF and α_2-CRF) by their coefficients of retention on carboxymethylcellulose. The amino acid composition of α_1-CRF is similar to that of α-MSH but also includes leucine, threonine and alanine. The amino acid sequence of α_2-CRF is as follows

$$R—Ser·Tyr·Ser·Met·Glu·His·Phe·Arg·Try·Gly·Lys·Pro·Val—NH_2$$

(Schally, Lipscomb and Guillemin, 1963). This is the structure of α-MSH with a terminal serine residue without acetyl. Guillemin (1963) suggested that the α-CRF could be the precursors of a complete molecule of corticotropin. However, a β-CRF has been isolated which is similar in structure to vasopressin and which is very active in releasing ACTH (Guillemin).

The concentration of corticoids in the organisms or certain regions of the organism controls the liberation of ACTH. Smelick and Sawyer (1964) demonstrated by implantation experiments with crystalline hydrocortisone acetate that the inhibitory action is not only on the pituitary but also on certain areas of the hypothalamus. These areas are situated in the anterior part of the median eminence and the postoptic region. From the work of Selye, *non-specific reactions* such as injection of a substance or manipulation are known to produce stress and the release of ACTH. Therefore, in these experiments, it is essential to use animals in which the non-specific adrenocorticotropic responses are suppressed. Dell and Dumont (1960) have shown that the rhinencephalon and reticular formation of the brain stem can function in the regulation of adrenocortical activity.

There are other "releasing factors" acting on the pituitary which are less well known than CRF, for example, TRF (thyrotropic releasing factor) and LRF (luteotropic releasing factor).

In the case of TRF, Shibusawa (since 1956) has demonstrated that administration to intact dogs of an aqueous extract of hypothalamus produces an increase in liberation of thyrotropic hormone. The liberation of thyrotropic hormone is measured by the fixation of ^{131}I by the thyroid of guinea pigs. Faure, Croizet, Blanquet and Meyniel (1963) found that electrolytic destruction of a region of the hypothalamus resulted in a significant reduction in thyroid function. This region, in the anterior hypothalamus, was situated in the median and paramedian zone, ventral to the paraventricular nuclei and extending from the anterior part of the arcuate nucleus and the suprachiasmatic nucleus to the dorso-ventral part of the preoptic nucleus (Fig. 128). The reduction in thyroid activity was as great as 92 per cent as shown by the degree of ^{131}I fixation. The possible relationship between inhibition of morphogenesis in the tadpole and suppression of thyrotropic activity of the diencephalon may be questioned (this inhibition results in the appearance of "salamandroformes tetards" in *Bufo, Rana* and *Alytes:* Voitkevitch, 1962; J. J. Bounhiol and C. Remy 1962). In young virgin female mice, high concentrations of adrenaline inhibit thyrotropic activity of the pituitary (Delost and Carteret, 1958)

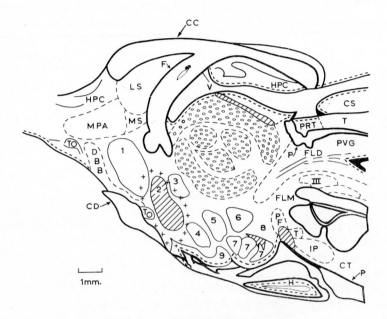

Fig. 128. Diencephalon and associated structures in white rat (sagittal section at 0·2 mm from median line of De Groot).

In black: fibres and bundles. In clear circles: thalamic nuclei. Marked with numbers: hypothalamic nuclei; *1*, preoptic region; *2*, anterior hypothalamic region; *3*, paraventricular nucleus; *4*, ventromedian nucleus; *5*, dorsomedian nucleus; *6*, posterior hypothalamus; *7*, mamillary group; *8*, supramamillary region; *9*, arcuate nucleus; *10*, suprachiasmatic nucleus.

Area circled by crosses = zone in which destruction leads to marked decrease in thyroid metabolism, containing a hatched area (the anterior hypothalamic region) which appears to be the critical region.

Abbreviations: *CC*, = corpus callosum; *HPC*, hippocampus; *MPA*, median parolfactory region; *TO*, olfactory tubercle; *DBB*, diagonal band of Broca; *CO*, optic chiasma; *LS*, lateral nucleus of septum; *MS*, median nucleus of septum; *A*, anterior commissure; *F*, fornix; *V*, ventricle; *HM*, median nucleus of habenula; *SM*, medullary furrow; *PRT*, pretectal area; *SC*, superior colliculus; *T*, tectum; *P*, posterior commissure; *PVG*, periventricular gray substance; *FLD*, dorsal longitudinal bundle of Shutz; *FLM*, median longitudinal bundle; *III*, oculomotor nerve; *PF*, parafascicular nucleus; *T*, ventral nucleus of the mesencephalic tegmentum of Tsai; *PI*, interpeduncular nucleus; *CT*, central nucleus of the tegmentum of Bechterev; *H*, hypophysis; *P*, protuberance (Faure, Croiset, Blanquet and Meyniel, 1963).

and reserpine inhibits transmission between the hypothalamus and pituitary (G. Mayer, 1962). Thus, "external" hormones such as adrenaline
or 5-hydroxytryptamine could act directly or indirectly in the liberation
of pituitary hormones by the hypothalamus.

TRF is being purified by Guillemin and his co-workers.

LRF (McCann; Harris; Campbell *et al.*; 1960–1961) is a small thermostable polypeptide which can produce ovulation and formation of *corpora
lutea* after administration to animals in which spontaneous ovulation is
blocked by stereotaxic lesions of the ventral hypothalamus (Schiavi and
Guillemin). Another compound, FRF, controls the secretion of FSH
(Mc Cann; Moszkowska and Kordon; Jutisz). Hypothalamic control of
luteotropic hormone (prolactin, LTH) is considered to be by *inhibition*
(J. W. Everett, 1954; Alloiteau, 1959) in that certain regions of the
hypothalamus inhibit the secretion of prolactin and in the absence of
hypothalamic inhibition, the pituitary secretes prolactin (cf. placental
function, p. 549).

Regulation of hypophyseal secretion

The hypophysis is under direct nervous control and the active agent
may be acetylcholine. Nervous impulses from the region of the supra-
optic and paraventricular nuclei pass to the hypophysis where they release
hormones from this storage organ. Evidence for the nervous stimulation
of hypophyseal secretion—a good example of a neuro-hormonal reflex—
is provided by a number of experiments:

Huang (1938) found that the efferent blood from the isolated head
of dog induced, after stimulation of the base of the infundibulum, *hypertension* in a second dog connected to the head by cross-perfusion. This
effect is not observed *after hypophysectomy*.

Anderson (1951) produced *milk secretion* in ewes and goats during
lactation by electrical stimulation of the *supraoptic region*. This increased
milk secretion, coincident with the production of concentrated urine,
can be explained by the liberation of oxytocin and antidiuretic hormone.
This effect was observed *after sacral anaesthesia and denervation of the
mammary gland*. Thus, electrical stimulation of the hypothalamus does
not stimulate the mammary gland by a direct nervous pathway but acts
by nervous stimulation of the hypophysis which liberates the active
hormones.

Furthermore, peripheral nervous stimuli modify neurosecretory reflexes.
Verney (1947) explained the action of "osmoreceptors" situated in the
carotid in this way. The secretion of *antidiuretic hormone* is inhibited by
cold and augmented by heat. These effects involve a hypothalamic reflex
system. The hypothalamic-hypophyseal complex is capable of rapid
responses and therefore, has an important role in adaptation and in the
reaction to stress.

CHAPTER 18

REGULATION OF WATER METABOLISM

"DURING the life of a creature... the internal fluid medium formed by the circulating organic fluid surrounds and bathes all anatomical elements of tissues; this fluid is the lymph or plasma, the fluid part of blood which penetrates all tissues and forms all interstitial fluids and provides a means of nutrition and is the source and site of all elementary exchanges. A complex organism must be considered to be a complex of simple creatures which are the anatomical elements and which live in the internal liquid medium" (Claude Bernard, 1878).

In man, the water content of the body is about 70 per cent of which:

> 50 per cent is cellular water,
> 15 per cent is interstitial water,
> 5 per cent is plasma water.

This water can be either of *endogenous* origin (metabolic water such as produced from fats specially in the camel) or of *exogenous* origin (absorption).

A body compartment is defined by the diffusion space of a tracer (various dyes and isotopes). These tracers are used in dilution methods which determine the kinetics of exchange in each compartment. When the results obtained are related to total body weight, the cellular compartment is found to be larger in men than in women while the extracellular compartments are not significantly different. The variations disappear if the results are related to lean body mass. The lean body mass of men and women is identical in subjects aged between 30 and 50 years. Subjects of 60 to 90 years show a decrease in lean body mass which is characterized by a progressive decrease in total water (ageing being associated with dehydration). In order to express the experimental results of lean body mass per unit body weight (kg), Haxhe (1964) calculated regression equations based on $y = a+bx$ where y represents a constituent/body weight and x corresponds to total water/body weight. The equations are resolved for an empirical value of $x (x=732)$ which corresponds to the amount of water (in ml) per kilogram lean body mass. This method is of interest but it is apparent that, since the metabolism of man and woman depends on the hormonal characteristics of the individual (these hormones acting mainly on lipids), this individuality will influence any evaluation of lean body mass. However, the difference between lean body mass and total body mass would provide a good indication of the distribution of fat in obesity and in the sexes.

The water content of an individual is regulated and is fairly constant under normal conditions. Neuro-endocrine mechanisms are involved in such regulation in terrestrial mammals.

Let us examine firstly certain aspects of the comparative physiology of water regulation.

In protozoans such as *Vorticella* (a peritriche), the contractile vacuole is osmoregulatory, the total vacuolar fluid ejected being inversely proportional to the salinity of the medium (Chadwick, 1960).

Among the invertebrates we must distinguish between those with an aquatic life (marine or freshwater) and those with a terrestrial life.

Most marine invertebrates are *poikilosmotic* and the internal medium rapidly equilibrates with natural or experimental changes in concentration of the external medium. Among marine invertebrates, some are highly permeable to water and less permeable to salts and were described by M. Lafon (1953) as behaving as osmometers (for example, many polychete annelids and molluscs such as the mussel); others are more permeable to mineral salts and react to changes in the medium by a *transference of water* in the reverse direction to salt transference such that the volume changes until osmotic equilibrium is reached (for example, *Aplysia*, starfish).

The internal medium of *freshwater invertebrates* is generally *hypertonic* with respect to the external medium. This hypertonicity results from the retention of salts by reabsorption and *elimination of water*. Examples of this group are crayfish and terrestrial oligochetes which "in spite of the name earthworms are animals with typical aquatic physiology which utilize the interstitial water of arable soil; the urinary output of earthworms is considerable, being 2 to 5 per cent of body weight per h (Bahl, 1947)" (M. Lafon).

Invertebrates in a terrestrial environment retain the maximum amount of water. However, the capacity for water retention varies from one insect species to another. The mechanisms by which water is retained are of three types:

(1) *Storage of metabolic water:* this occurs in *Tenebrio* larvae which do not drink and consume dry food, water being obtained from the oxidation of food.

(2) *Reabsorption of water from the excretory organs:* urine produced in the Malpighian tubules is dilute and contains urates in solution. Water is reabsorbed in the rectal gland region and solid matter excreted (Wigglesworth).

(3) *Control of cutaneous evaporation:* this occurs principally in the region of intersegmentary membranes. Cutaneous evaporation depends on rela-

tive humidity and environmental temperature. Over a certain range of temperature, evaporation is regulated by the waxes which cover the integument.

The water regulatory mechanisms of lower vertebrates are similar but less complex than those of mammals.

Movements of water occur freely in amphibians. *In adults water enters through the skin* (Przylecki) while *in larvae* (as shown by ligature of mouth and anus), *water enters both through skin and digestive tract* (Rey, 1937). Cambar (1947) found that, at 14°C, the maximum water penetration occurred in frog larvae during the first days of free swimming (45 per cent through the skin) and water penetration decreased progressively to reach a level of 30 per cent (passing equally through skin and intestine) by the time of metamorphosis. In the adult, the level of water penetration was 15 per cent, water passing only through the skin. The urine of adult frog is more concentrated than that of larvae which are exclusively aquatic.

Hypophyseal hormones augment tubular water reabsorption and the rate of cutaneous water movements in the frog and toad (Ussing and Zerahn, 1951; Leaf and Frazier, 1962). Antidiuretic hormone could affect the size of water-filled channels (pores) but, according to experiments carried out on frog skin subjected to high pressure (up to 1000 kg/cm^2), it seems that solvents as well as solutes cross the lining membrane by interacting with the same component of the barrier (Schoffeniels, 1963). These facts lead to a study of the *endocrine factors in water regulation in mammals:*

(1) Lesions of the supraoptic-hypophyseal tract or removal of the hypophysis or selective ablation of supraoptic neurosecretory cells by electrical methods induce *diabetes insipidus* in mammals (Ranson and Magoun, 1939).

(2) Neuroglial cells of the hypophysis (the pituicytes) do not produce antidiuretic hormone when cultured *in vitro* and the hormonal content of the cells disappears completely during culture (Hild, 1954).

(3) Extracts of neurosecretory cells isolated from the diencephalon contain antidiuretic hormone (Hild and Zetler, 1953). Water-starved animals show a decrease in the volume of neurosecretion.

Therefore, antidiuretic hormone appears to be produced in the hypothalamus and stored in the hypophysis from where it is released when required. The antidiuretic activity of the hypophysis can be demonstrated using the isolated perfused kidney (Verney and Starling). Dilute and abundant urine is produced by the kidney but addition of a hypophyseal extract to the perfusate inhibits diuresis and augments chloride excretion, without a significant effect on vascular tissue.

The secretion of antidiuretic hormone is regulated by an osmotic mechanism. Injection of hypertonic solutions to the carotid results in renal water reabsorption as does an injection of hypophyseal hormone. No increase in water reabsorption occurs in hypophysectomized animals or animals in which the supraoptic-hypophyseal tract is sectioned.

30*

$$\text{Hypertonicity} \frac{\text{hypothalamus}}{\text{(osmoreceptors of Verney)}} \rightarrow \text{Secretion of antidiuretic hormone}$$

Regulation implies both *excitatory and inhibitory reactions* leading to a *dynamic equilibrium* and therefore, we must examine the *antagonists of antidiuretic hormone.* Thyroxine decreases the antidiuretic action of ADH on the renal tubules. However, *adrenocortical hormones are the most active antagonists. Desoxycorticosterone acetate* (cf. p. 377) *produces polyuria* with retention of chloride and sodium, loss of potassium ion and an increase in extracellular water volume (Harned and Nelson). *Adrenocortical insufficiency has the reverse effect*, producing hypochloremia and hyponatremia, hyperkalemia, a decrease in plasma and extracellular fluid volumes and an increase in intracellular fluid volume.

These results can be summarized as follows:

Firstly, (1) an increase in plasma osmotic pressure liberates antidiuretic hormone; (2) a decrease in plasma osmotic pressure inhibits the hypophysis (Verney);

Secondly, (1) a decrease in the Na^+/K^+ ratio (hyperkalemia) leads to liberation of corticoids (aldosterone which, according to Müller and Mach, is dependent on ACTH and according to Johnson and his co-workers, is independent of ACTH); (2) an increase in extracellular volume inhibits the adrenal cortex.

These effects depend on stimulation of the hypothalamic centres by changes in intracranial pressure (a mechano-hormonal reflex) associated with a more specific adrenocortical action in the case of Na^+ and K^+.

Lysine-vasotocin has a specific antidiuretic action in *Rana esculenta.* Oxytocin (which is inactive) suppresses the action of lysine-vasotocin. The oxytocin molecule differs from that of lysine-vasotocin in the substitution of one amino acid of the chain and differs from that of vasopressin by the substitution of one amino acid of the ring. Inhibition of lysine-vasotocin by oxytocin suggests competitive fixation on the receptor by the common part of the ring of both molecules (Jard and Morel, 1961).

A remark of H. and E. Legait (1951) is of particular significance: the *pars intermedia* is absent from mammals living near water and from marine mammals while it is well developed in mammals adapted to arid regions. According to these authors, intermedin would act on the hypothalamus and hypophysis and in higher vertebrates, the role of intermedin in water regulation would be the principal action with the effect on cutaneous pigmentation being secondary. The use of tritium and deuterium oxides in conjunction with scintillation spectrometry would be of considerable value in these problems and particularly in the study of water distribution (Leibman *et al.*, 1960).

Petit-Dutaillis and Bernard-Weil (1958) stated that cranial damage following cellular dehydration is improved by treatment with hypophyseal extracts. Thus, the hypophysis is important in water regulation but certain aspects must be examined further. The kidney of infant rats and other infant mammals is structurally and functionally different from that of adult mammals. The kidney of infant rats has a limited glomerular filtration rate, a not fully functioning countercurrent system and does not react to vasopressin with a sufficiently pronounced increase in tubular water reabsorption. Yet it can concentrate urine under certain conditions, without vasopressin being demonstrable in the plasma (Krecek and Heller, 1962). Antidiuretic hormone does not act only on the kidney. From experiments based on the increase in weight of batracians in water and on the changes in water distribution of mammals with kidneys removed, Montastruc (1962) concluded that antidiuretic hormone has an extra-renal action on water and ionic exchange and modifies intestinal secretion of water and salts, salivary and gastric secretion, insensible perspiration and sweat secretion. In all regions antidiuretic hormone tended to bring about internal water conservation. *In vitro*, vasopressin has been shown to act on red corpuscles, cerebral tissue, intestinal wall, muscle, skin and bladder of certain batracians. In addition to these physiological effects, vasopressin produces more pharmacological effects such as contraction of intestinal muscle, hyperglycemia and stimulation of adrenal cortex.

CHAPTER 19

REGULATION OF NA$^+$ AND K$^+$ METABOLISM

THE regulation of Na^+ and K^+ follows, to some extent, the regulation of water and therefore, we shall examine only certain aspects of this problem.

We have seen (p. 385) that proximal tubular reabsorption of Na^+ parallels that of Cl^- and this process depends on the action of aldosterone which also acts on K^+ excretion. Na^+ regulation differs from that of K^+. Furthermore, the production of aldosterone increases with *hyponatremia* but decreases with *hypokalemia*, and inversely.

These are the essential facts concerning regulation of natremia and kalemia but it should be noted that *desoxycorticosterone* acetate is also very active in Na^+/K^+ regulation. In addition, there are aldosterone antagonists such as spirolactones (synthetic) and progesterone. The total body water—measured with deuterium oxide—is increased in normal pregnancy compared with the total in normal non-pregnant females and is further increased in pre-eclampsia. Exchangeable sodium and chloride ions are increased in normal pregnancy and the amount of exchangeable potassium is slightly increased in normal (and pre-eclamptic) pregnancies but the amount per kg body weight is decreased in both according to Macgillivray (1960) (cf. lean body mass?).

The selectivity of the adrenal cortex in its response to ions as chemically similar as Na^+ and K^+ is remarkable and, as yet, the mechanism is unknown (direct or indirect route via the pituitary or to the epiphysis?). However, one point concerning the distribution of alkaline ions throws some light on this problem: K^+ is essentially *intra*cellular and a large fraction is non-diffusible while Na^+ is essentially *extra*cellular. Thus, hyperkalemia is more dramatic than hypernatremia. It is conceivable that the adrenal cortex would be more sensitive to $K^+(Na^+/K^+)$ since the pathological conditions associated with increased K^+ are particularly serious. The metabolisms of K^+ and Na^+ are related but the production or secretion of corticoids would depend on imbalance of an enzyme system appearing at the threshold level of K^+. In chronic *sugar diabetes*, interruption of insulin treatment results in *hyperkalemia* with dehydration and with renewed administration of insulin (hormone of permeation), there is *hypokalemia* followed by rehydration (R. S. Mach, 1954). The possible role of phosphorylated glucose derivatives (such as glucose-6-phosphate) in the regulation of kalemia should be studied.

Nerve and muscle activity leads to liberation of K$^+$ but the capacity of the kidney for K$^+$ elimination is such that this transitory hyperkalemia does not affect the general state except in cases of renal disease. Strophanthin (a glycoside) appears to inhibit specifically the active transport of Na$^+$ and K$^+$ (Schatzman, 1953; Glynn, 1957; Koefoed-Johnson, 1957; Whittam, 1958).

Water is associated with the equilibrium of sodium and potassium ions and the regulation of the water–mineral equilibrium is a complex problem. As in frogs and toads, vasotocin can be shown to promote active sodium movements across the skin of the urodele *Triturus* (H. Heller, 1965). F. Morel stated that hypophyseal and adrenocortical hormones have only a *regulatory* function on water–mineral balance. Using ^{24}Na, Morel found that, in rats, "ablation of the hypophysis (producing *diabetes insipidus*) or adrenal cortex does not affect the maintenance of normal water–mineral balance provided that dietary factors compensate for the permanent excretory disorders. A diabetic rat is dependent for existence on a continuous supply of water and the adrenalectomized rat on a continous supply of electrolytes such that the independence of these animals from changes in external medium is greatly reduced". These facts indicate the importance of endocrine systems in the liberation of animals from ecological constraint such that, by regulation of the *internal* environment, organisms become more independent of the *external* environment. In lower animals, for example in batracians, endocrine regulation of water–mineral balance is more limited (Sawyer, 1954): water–mineral regulation in fish is dependent on external conditions: Arvy and Gabe (1954) found, in two species of teleosts, that immersion in hypertonic saline produces, after 30 mn, a marked decrease in the production of neurosecretory material in the cells of the preoptic nucleus and congestion in the hypophysis with a decrease in neurosecretory products. After return to normal seawater, the endocrine processes are rapidly restored (within half an hour). Immersion in hypotonic solutions leads to an accumulation of neurosecretory products. Therefore, *the preoptic region and the hypophysis are very sensitive to osmotic changes.* Following this work, Arvy, Gabe and Fontaine (1955) studied the hypothalamo-hypophyseal neurosecretory changes during the amphibiotic life cycle of the salmon.

The natural history of the salmon can be summarized as follows: "The salmon starts life in freshwater and the vitelline sac is reabsorbed during the larval stage of about two and a half months. The following stage is called a *parr*. It remains in freshwater for one, two or three years and during this period a large percentage of males reach sexual maturity. At the end of one, two or three years, the *parr* undergoes certain morphological and physiological changes which herald catadromous migration, these changes being termed *smolting*. *Smolts* migrate to the sea and live there for one or many years during which time, they increase in size and weight. After this period of active feeding, the salmon migrate back to freshwater (anadromous migration) where they remain for several months without feeding before spawning. After spawning, certain salmon die while others undergo morphological changes among which the development of silver coloration resembles that in smolting and begin another catadromous migration. This last stage of the developmental cycle is termed the "*kelt (mended)*."

Thus, the salmon undergoes several changes, each stage constituting an allomorphosis which is prepared by the previous stage.

The euryhalinity of the *smolt* is greater than that of the *parr* (Fontaine and Baraduc, 1954) and the level of Cl^- in muscle of *smolt* is decreased (Fontaine, 1951). This phenomenon could be explained either by the action of thyroxine (Baraduc, 1952) or by the action of the pituitary. Euryhalinity of the trout (holobiotic Salmonidae) increases after treatment with somatotropic hormone (D. W. Smith, 1956). Smolting is certainly associated with thyroid hyperactivity which affects the "iodine pump" of Vanderlaan and Caplan (1954) (cf. p. 404) as shown by a change in the ratio of concentration of iodide in the thyroid and in the plasma as measured by [131]I (Fontaine and Leloup, 1960). In the trout living in a rapid water current, the level of muscular Na^+ and Cl^- is decreased compared with that of a trout living in calm water (M. Fontaine and Chartier-Baraduc). A comparable decrease occurs in salmon from the sedentary *parr* stage to the migratory *smolt* stage but there is no decrease in sedentary smolts (Chartier-Baraduc, 1960). The increase in the K^+/Na^+ ratio in brain and muscle of the smolt is corrected (as emphasized by M. Fontaine (1960), in a medium such as seawater in which there is a high Na^+ content and a relatively low K^+ content. During smolting marked endocrine changes occur which result in a functional change and in particular, a modification of nervous activity. Histological changes in corpuscles of Stannius indicate increased activity in these glands when a young salmon in fresh water undergoes *parr* to *smolt* transformation (M. Fontaine and M. Lopez, 1965). In the eel, one month after removal of the corpuscles of Stannius the natremia declines while kalemia remains within the normal range (Chester Jones *et al.*, 1965). M. Fontaine and his co-workers have demonstrated that the inter-renal organ is hyperactive in the *smolt*. Moreover, motor activity in the carp stimulates this organ and the plasma level of 17-hydrocorticosteroids is ten times higher than normal after continuous swimming (J. Leloup-Hatey, 1958). Motor activity resulting from swimming against a current is associated in the trout with thyroid hyperactivity (Fontaine and Leloup, 1959). *Protopterus* (a dipnoan)—which is able to live in a cocoon made of hardened mucous during periods in which the natural medium is dehydrated—undergoes changes in thyroid activity and transference to an aquatic medium results in a rapid increase in thyroid activity, characterized by release of hormone into the internal medium (J. Leloup, 1963). On the other hand in the male salmon at the time of spawning, the levels of oestriol in blood are similar to those in the female while in human species, oestriol has been found only in females (Cedard, Fontaine and Nomura, 1961). Furthermore, during anadromous migration of salmon, the level of ascorbic acid in the hypophyso-pituitary system decreases and this must correspond to an increase in the ascorbic acid requirements of the gonads (Fontaine and Leloup-Hatey, 1959).

In considering the regulation of water–mineral balance in migratory fish, Arvy, Fontaine and Gabe were unable to establish a direct correlation between changes in hypothalamo-hypophyseal neurosecretion and the passage from freshwater to seawater. However, hypothalamo-hypophyseal changes precede the change in habitat. Thus, preoptic neurosecretion is abundant in the *parr* (its hypophysis is poor in neurosecretion) while the hypophysis of the *smolt* contained large amounts of neurosecretory material and the hypothalamus contained little material.

In the returning salmon, the distribution of neurosecretion is the same as in the *parr* (high in hypothalamus, low in hypophysis) and this distribution is at a maximum during spawning. However, transformation to the *kelt* is coincident with changes similar to those of smolting. Osmoregulation leads to a decrease in neurosecretory material of the hypothalamo-hypophyseal complex. The reaction to an osmotic stimulus increases as the animal becomes less euryhaline. *Comments:* The *parr* stage and the spawning stage both correspond to periods of sexual activity and similar neurosecretory activity is apparent in both cases.

Ionic and water regulation in the lamprey which shows anadromous migration for reproducing in freshwater (where it lives for six months) has been studied by Bentley and Follett (1963): during its anadromous migration the lamprey takes up water osmotically and excretes it through the kidneys at a rate of 200–400 ml/kg/24 h. The urine is markedly hypotonic, but even so, losses of sodium and potassium are appreciable, amounting to about 8·7 per cent of the total body sodium and 0·9 per cent of potassium per day. The urinary concentration of sodium and potassium decreases with an increase in urine volume, which suggests physiological regulation of the tubular reabsorption and/or secretion of electrolytes. Most of a large injected sodium load is, however, excreted extrarenally. In contrast to the regulation of normal Na$^+$ and K$^+$ excretion, renal tubular processes play little part in the control of urinary water loss. Arginine-vasotocin is present in the neurohypophysis of lampreys (Follett and Heller, unpublished observations) but the injection of large amounts of this hormone did not alter the urine volume; arginine-vasotocin increases urinary sodium loss in the lamprey. Arginine-vasotocin, estimated by natriferic assays on the frog bladder, was only present in small amounts in the lamprey hypophysis, and immersion of the fish in a hypertonic solution of seawater (45 per cent) to produce an osmotic stress did not alter the amounts stored. Neither aldosterone nor cortisol affected the urinary losses of electrolytes. Aldosterone did, however, reduce the rate of sodium loss from lampreys and this action would therefore appear to be an extrarenal one. Thus, despite the importance of the kidney for water excretion and electrolyte conservation, it was not possible to demonstrate a physiological role of neurohypophyseal or adrenocortical hormones in the control of these processes by the lamprey kidney.

It is possible that mineral regulation in migratory fish would involve the Bohr effect (cf. p. 334) related to the state of oxygenation of animals in different biotopes. Vanstone, Roberts and Tsuyuki (1964) have found in all cases that the *parr–smolt* transformation in Pacific salmon is associated with changes in the electrophoretic pattern of hemoglobins: "The synthesis of high oxygen affinity adult hemoglobin by salmon and low oxygen affinity adult hemoglobin by other vertebrates is understandable when the ecological changes are considered. After smolting, salmon leave their well aerated natal waters to live in the ocean or lake where several factors may lower the oxygen tensions considerably. By contrast, after termination of the embryonic period or metamorphosis from the larval stage, the other vertebrates enter an environment that contains higher, less variable, oxygen tensions than did their previous habitat."

CHAPTER 20

REGULATION OF GLYCEMIA

GLYCEMIA in birds is maintained at a higher level than in mammals: an average of 1·8 g/l compared with 1 g/l. However, in both cases, glycemia is maintained at a constant level. Table 46 gives the glycemia constants for different species and significant variations are apparent.

In the human species, biochemical anthropology has provided evidence for different blood glucose levels in different races according to the following scheme (G. A. Heuse, 1956):

Southern Mongolian < African Negroes and White Euramericans < Indians

To explain these differences one thinks, in the first instance, of nutrition but for the same carbohydrate diet, Hindus and Indochinese show the most extreme levels of the human species and African Negroes, with a predominantly carbohydrate diet, show average or slightly higher than average levels. A climatic theory is equally unacceptable since, for example, Hindus living in London maintain high blood sugar levels. Parasitism may be a determining factor, the hypoglycemic action of trypanosomes—specifically African—being characteristic, but it is difficult to establish that all group differences result from parasitism. In fact, glycemia seems to be related to genetic factors. The blood glucose level of mice is transmitted genetically (Cammidge and Howard, 1930). However, comparative anthropology of endocrine glands shows the adrenal glands of Negroes to be small and the thyroid of Indochinese to be small. Furthermore, in the frog, there is seasonal hyperglycemia associated with thyroid activity. Pathological anthropology has shown that diabetes is generally rare and benign in yellow races and that their reaction to insulin is marked while diabetes occurs as frequently in African Negroes as in white races and Hindus.

The example of glycemia shows that at the finest level physiological studies are orientated towards physiological genetics.

Following a heavy meal glycemia increases and excess sugar is stored in tissue spaces. When glycemia is decreased by renal elimination (glycosuria) and glucose utilization, the tissue glucose is mobilized and passes to the liver where it is converted to glycogen which can be reconverted to glucose when required.

Blood glucose is maintained at a constant level by hormonal and nervous mechanisms.

TABLE 46. *Blood glucose levels in various animal species during fasting [from* H. H. DUKES *(1942)* Tabulae Biologicae *and* FLORKIN *(1936)]*

Species	Glucose levels in mg/100 ml of total blood
Mammals	
Man: newborn	90–130
Man: adult	70–100
Dog	100–140
Cat	120–140
Rat	100–120
Mouse	90–110
Horse	70–110
Guinea pig	120–140
Rabbit	90–140
Hibernating marmot	20–30
Pig	40–200
Ruminants: Cow	40–60
Goat	45–65
Sheep	30–60
Birds	
Pigeon	120–140
Chicken	150–200
Duck	150–200
Goose	150–200
Turkey	175–210
Amphibians	
Frog	30–45
Fish	100–120
Decapod crustaceans (Roche and Dumazert; Florkin)	(values/100 ml blood plasma)
Carcinus maenas	16·5–12·5
Homarus vulgaris	10
Maia squinado	9–7·5–7
Palinurus vulgaris	7·5–11·7

The endocrine glands involved are the pancreas, adrenal cortex and medulla, pituitary, hypophysis and hypothalamus. We shall examine briefly these effectors and the nervous regulations of their internal secretions.

Physiological release of insulin from the pancreas has been demonstrated by Zunz and La Barre (1933) by anastomosing the pancreatic vein of a donor dog to the jugular vein of a recipient (Fig. 129).

When the recipient has been rendered diabetic by removal of the pancreas, the pancreatic-jugular vein anastomosis allows the normal blood glucose level to be maintained for several hours. Using this technique, Zunz and La Barre examined insulin secretion during hyperglycemia:

the donor receives an injection of 12–15 g of glucose;
the recipient is adrenalectomized and, as a result, is more sensitive to the hypoglycemic action of insulin.

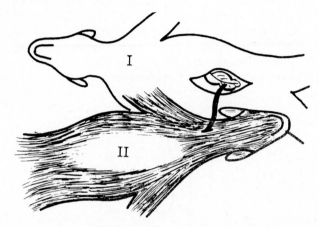

FIG. 129. Pancreatico-jugular anastomosis of ZUNZ and LA BARRE (1933).
(I) Donor dog; (II) Recipient dog.

After glucose administration, the glucose balance of the animal is restored by released insulin as shown by hypoglycemia in the recipient. For this post-hyperglycemic hyperinsulinemia, the parasympathetic and nervous centres must be intact. The reaction is not seen after vagotomy and by combining the isolated head technique of Heymans and the pancreatic-jugular vein anastomosis, it is found that localized hyperglycemia of the nervous centres produces compensatory hyperinsulinemia through stimulation of nervous centres and transmission of information of the pancreas via the pneumogastric nerves.

However, two facts must be considered: (1) Geiger (1928) found that a hypoglycemic action is not specific to glucose; injection of a hypertonic solution of NaCl into the pancreatic vessels also produced hypoglycemia; (2) perfusion of the isolated pancreas with blood containing high glucose concentrations does not result in supplementary release of insulin (La Barre). However, the secretory function of the endocrine pancreas is maintained after transplantation. Houssay and his co-workers (1939) found that, following subcutaneous transplantation of the denervated pancreas, insulin secretion was dependent on the blood glucose level in the gland. However, a nervous action has been demonstrated by other experiments of Zunz and La Barre in which they examined insulin secretion during hypoglycemia (Fig. 130).

Perfusion of the nervous centres with blood from a donor, rendered hypoglycemic by insulin administration or hepatectomy, results in a decrease in insulinemia. Thus, there is post-hypoglycemic inhibition of insulin secretion, this inhibition being controlled by the central nervous system. Furthermore, the islets of Langerhans are innervated by the pneumogastric system and, in rabbit and dog, the right vagus sends numerous fibres to the pancreas. Electrical excitation of the vagus results in insulin secretion as shown by Zunz and La Barre using their original technique. Insulin secretion is markedly decreased after vagotomy. Insulin secretion "tone" would be maintained by the vagus through the activity of higher centres. This process provides an example of a chemico-neuro-endocrine reflex system.

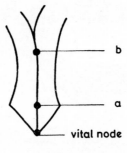

FIG. 130

We must consider the sensitory centres (Fig. 130):

As described previously, puncture at the point of the *calamus scriptorus* (floor of the fourth ventricle) of mammals arrests ventilatory movements (cf. p. 346). Puncture above this "vital node" at *a* produces glycosuria in rabbits (C. Bernard, 1849) while puncture at *b* causes polyuria without glycosuria *(diabetes insipidus)* and above *b*, albuminuria. Thus, *the calamus is an extremely important metabolic regulatory centre.*

The vagus nerve stimulates insulin secretion and by vagotomy, it can be verified that the puncture described by C. Bernard still induces diabetes. Furthermore, stimulation of the adrenergic splanchnic nerve results in hyperglycemia. Thus, parasympathetic nerve excitation produces hypoglycemia (the anterior hypothalamus would be the regulatory centre) and sympathetic nerve excitation (catecholamines) produces hyperglycemia, the sympathetic system representing the pathway from the centre of C. Bernard (posthypothalamus?).

Many glycoregulatory centres are currently known or suspected (activating reticular formation) but their modes of action and precise interrelationships remain conjectural.

Excitation of centres in the *medulla oblongata* produces hypoglycemia (Brugsch *et al.*) and these regions could be important in the development of the hunger syndrome (cf. Volume II). Lesions of the *medulla oblongata* in frog lead to glycosuria (Schiff).

Paraventricular nuclei of the hypothalamus are concerned with glycoregulation and, on the other hand, lesions in the tuberal zone induce sugar diabetes. This hypothalamic centre regulates glycemia through:

(1) Nervous pathways

    (a) sympathetic innervation of the adrenal medulla producing adrenaline secretion with hyperglycemia;

    (b) parasympathetic innervation of the pancreas producing insulin with hypoglycemia.

(2) Humoral pathways, the agents being pituitary growth hormone and ACTH which liberates a hyperglycemic hormone from the adrenal cortex. Thus, the adrenal gland is in any case hyperglycemic.

Furthermore, a reflex hypothalamic reaction can be induced by emotional factors. Emotion produces hyperglycemia, the hypothalamus being linked with the cerebral cortex. Similarly, as we shall see, the hypothalamus is concerned with compensatory hyperglycemia to cold (cf. Volume II).

These facts are summarized in the scheme of Weill and Bernfeld (Fig. 131).

However, this scheme does not take into account all the regulatory mechanisms of glycemia:

(1) Claude Bernard demonstrated that curarization leads to transitory hyperglycemia by:

    (a) a decrease in glucose metabolism of curarized muscle,

    (b) reduction in pulmonary ventilation leading to post-asphyxia hyperadrenalinemia;

    (c) mobilization of glycogenic stores.

(2) The thyroid gland regulates glycemia, subcutaneous injection of thyroxine producing hyperglycemia. Thyroxine antagonizes the action of insulin (Bodansky, 1923; Houssay *et al.*, 1948). Hyperthyroidism is associated with a marked decrease in total NADP (mainly $NADPH_2$) in liver (Dickens, Glock and McLean, 1959). NAD also decreases and these processes could be related according to these authors to decreased mitochondrial synthesis of ATP due to uncoupling of oxidative phosphorylation.

(3) The pituitary contains a diabetogenic factor described by Houssay in 1930. Administration of pituitary extracts (thyrotropic ?) leads to hyperglycemia. Hepatic glycogenesis is inhibited by total hypophysectomy.

The pituitary contains numerous factors influencing glucose metabolism:

an adrenotropic factor (Kepinov, 1940) which sensitizes the liver to the

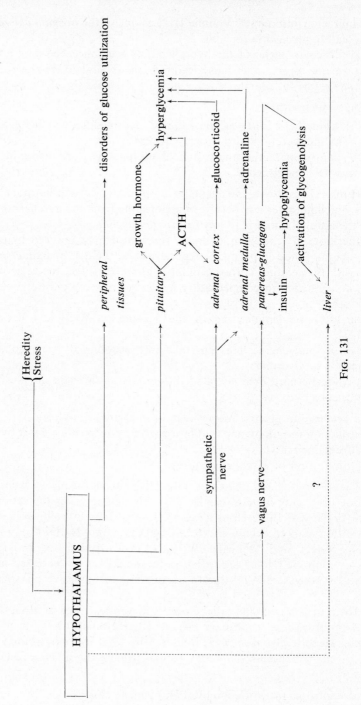

FIG. 131

glycogenolytic action of adrenaline;
a glycotropic or anti-insulin factor (Young, 1937), etc.

These facts have been included in order to demonstrate that biological phenomena are not easily isolated as functional entities.

(4) Finally, one has to emphasize that the adrenal gland plays an important role in regulation of glycemia. The adrenal cortex contains a diabetogenic factor, evidence for which was provided by Houssay and his co-workers (1935):

(a) Administration of cortical extracts leads to hyperglycemia coincident with an increase in glycogen reserves of liver and muscle.

(b) Cortisectomy, as total hypophysectomy, improves the symptoms of diabetes produced by ablation of pancreas in mammals.

The hyperglycemic agent of the adrenal medulla is adrenaline, producing glycogenolysis. Adrenaline secretion is under reflex control (Cannon et al., 1921). Insulin hypoglycemia stimulates the bulbar centres with the release of adrenaline which stimulates hepatic glycogenolysis and restores normal glycemia. Houssay and Molinelli (1925) produced an adreno-jugular anastomosis between two dogs. Injection of insulin to the donor produces hyperglycemia in the recipient except following denervation of the anastomosed adrenal. The reflex pathway includes the splanchnic nerves. It is of interest that injection of growth hormone to arterial blood passing to the pancreas of dog, liberates large quantities of catechol amines localized in this gland (Galansino et al., 1963; Loubatières et al., 1963, 1964). The aetiology of Young's diabetes should be considered in this context.

CHAPTER 21

CALCIUM CONTENT, OSSIFICATION AND GROWTH

I. Regulation of calcemia

THE Ca^{++} content of blood is remarkably constant in man at the value of about 100 mg/l (Guillaumin, 1931). This constancy is found in other animals but in the rabbit, for example, calcemia shows considerable variation.

Factors modifying the calcium ion content of blood

(1) Dietary:

A calcium deficient diet reduces calcemia to the stage in which tetany is produced; furthermore, anti-ricket vitamin deficiency also produces hypocalcemia.

(2) Endocrine:

The specificity of these factors is variable. Insulin increases the calcium content of blood but this action is secondary to a decrease in inorganic phosphate, the metabolisms of Ca^{++} and phosphorus being linked (cf. p. 42).

(a) OESTROGENS. During the period of laying, the level of total and non-diffusible Ca^{++} in plasma of fowl increases. This is particularly marked during the pre-ovulatory period. Injection of oestrogens to male pigeons produces the same effect.

(b) PITUITARY. Certain authors suggest that the pituitary modifies calcemia but there is no evidence to support this view.

(c) ADRENAL CORTEX. Administration of cortisone results in a marked increase in calcium ion excretion in urine and faeces but has little effect on calcemia.

(d) ADRENAL MEDULLA. Adrenaline increases calcemia.

(e) PARATHYROID. This organ is of considerable importance in the regulation of blood calcium levels.

McCallum and Voegtlin (1909) found that removal of the parathyroid glands resulted in a fall in calcemia. There is an inverse relationship between calcemia and the volume of parathyroids (Stoerk and Carnes, 1945).

The mode of action of parathormone is open to debate, very little being known on this subject.

At present, there are *two main theories:* one theory suggests that the hormone acts on bone; the other suggests that the hormone acts on renal excretion (Albright).

In fact, as is often the case, both mechanisms appear to be involved, with the action on bone being predominant. Parathormone decreases tubular reabsorption of phosphorus and augments urinary excretion of Ca^{++} but, using ^{45}Ca and ^{32}P, Talmage and his co-workers (1955) found that this is a secondary effect, parathormone acting principally on bone. Furthermore, Albright has admitted that his own theory is impossible to maintain: in experimental animals deprived of kidneys, injection of para-thyroid extracts induces osteoclastic resorption of bone. However, the possibility remains that "parathormone" is a complex of two or more distinct substances.

II. Ossification and growth

The hormone, or the parathormone complex, does not only control the calcium ion and mineral phosphate equilibrium of blood but also their metabolism in bone.

Before examining the mechanisms of ossification, the physiological properties of parathormone will be described.

The skeleton contains 99 per cent of the total calcium of the organism. Under the influence of parathormone, calcium can be partially mobilized and the bone undergoes decalcification. Histologically, this phenomenon is characterized by an increase in activity of osteoclasts, the giant cells concerned with bone resorption. These changes are seen only with high doses of parathormone. The reverse occurs with low doses of parathormone, that is, instead of stimulation of resorption by osteoclasts, there is stimulation of osteogenic activity with hypercalcification. These processes show us a new method of regulation. *Ossification does not appear to depend on the function of two antagonistic substances and we find a situation in which regulation is through a single substance whose action is dependent on dose.* In the case of vitamins, as of hormones, it is not only the qualitative aspect that is important but also the quantitative factors.

So far we have considered the role of added parathormone. We shall now examine the effects of parathormone deficiency although the effects of parathyroidectomy are difficult to determine since this operation rapidly leads to death of the subject.

In order to appreciate the importance of the parathyroids, we must first consider the factors determining the importance of skeleton.

The skeleton supports the whole of the vertebrate organism and the presence of an internal skeleton constitutes a significant taxonomic criterion in the definition of a vertebrate compared with an invertebrate. The acquisition of an upright position in man is related to skeletal modifications. In vertebrates, bone presents a particular aspect of Ca^{++} metabolism.

Finally, bones are palaeontological documents and almost all palaeontology—and thus the understanding of evolution of species—is based on an examination of these relics. In particular, endocranial dimensions allow an interpretation of the state of the brain in ancient vertebrates (archeo-psychology). In this respect, it should be noted that fossilized bones are heavier than fresh bones.

The actual processes which affect the different lines of vertebrates can be derived from a study of bone. The meristic characteristics of vertebrae allow one to recognize varieties of animals from different geographical locations (for example, the work of M. Ruivo (1957) on sardine). In man, bones show secondary sexual characteristics (cf. p. 564). Certain bones, through the presence of marrow, act as glands of internal secretion (secretion of cellular elements of blood, cf. p. 168).

In considering analogous functions, certain vital processes show broadly similar characteristics, the differences being due mainly, apart from genetic constitution, to evolutionary adaptation or to particular modes of life. Thus, in a ciliated protozoan *(Coleps hirtus)*, the tegument is an armoured plate consisting of rigid and ornamented lamellae which contain a large proportion of calcium (Fauré-Fremiet and Hamard, 1944). Benesch and his co-workers (1945) found that calcification of bone in vertebrates is inhibited by sulphonamides acting on phosphatase or carbonic anhydrase. In *Coleps*, Fauré-Fremiet, Stolkowski and Ducornet (1948) inhibited calcification of the tegument with benzene-sulphonamide. Since this animal contains no carbonic anhydrase but does contain alkaline phosphatase, inhibition of calcification results from inhibition of phosphatase by sulphonamide. The tegument consists of a mixture of bihydrated and anhydrous dicalcium phosphate. In addition, shells of molluscs provide excellent palaeontological records.

In order to show the historical continuity of scientific research and the fortuitous nature of many discoveries of facts and techniques, I shall quote a part of the "Quinzième Leçon" of P. Flourens which was given at the Muséum d'Histoire naturelle in 1855 and is published in his *Ontologie naturelle* of 1861.

"Belchier, a London surgeon, while dining with a cloth dyer, remarked that the bones of a young pig were red. He was curious to know the reason for this strange coloration. He was told that the animal had been fed with bran containing an infusion of madder used by the dyer.

Belchier carried out certain experiments (1736): he mixed powdered madder root with food given to a cock. After sixteen days the cock died. *All the bones were found to be red, but only the bones.* Muscles, membranes and cartilage and all other parts had retained their normal coloration.

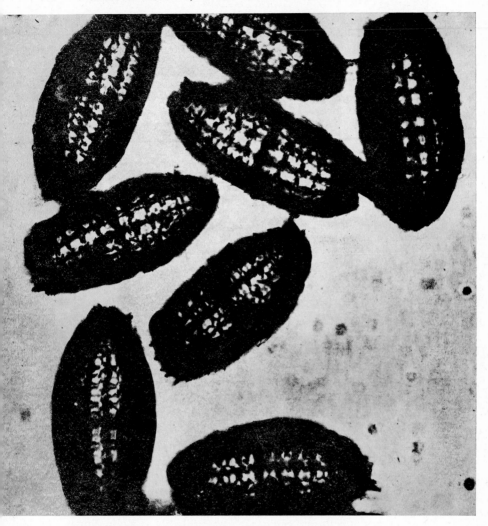

FIG. 132. *Coleps hirtus*. Silver impregnation of calcified armour. (From FAURÉ-FREMIET *et al.*, *Exp. Cell. Res.*, 1948.)

When Duhamel heard of Belchier's work he immediately repeated the experiments on chickens, pigeons and pigs (1739). He observed that madder constantly coloured bones and only bones.

Duhamel performed these experiments in the following way. He returned to a normal diet a pig whose bones had become red following a diet of madder; six weeks later he killed the animal and having sectioned the bones, found that the red layer of bone was surrounded by a white layer.

Another pig which Duhamel had given alternately a diet containing madder and a normal diet, showed alternating red and white layers of bone. From his experiments Duhamel came to a fundamental and accurate conclusion. Bone increases in size by successive superimposed layers.

But this is not the only process occurring during bone growth. As the bone increases in size by addition of external layers, the medullary canal increases by resorption of internal layers. I obtained evidence for this resorption which no one had suspected.

I provided an animal with a mixed diet containing madder. The layer of bone formed during this diet is red. I returned the animal to a normal diet. The new layers formed are white and surround the red layers; then this red layer became the internal layer and the white layers which it surrounded had disappeared; later the red layer disappeared.

Another procedure gave me the same result. I wound a platinum wire around a bone of a young animal. After a certain period the ring of platinum wire which initially was around the bone, was surrounded by bone and contained in the medullary canal.

As the bone becomes covered with new layers on the external surface, on that region called the external periosteum, it loses other layers on the internal surface, the internal periosteum. This double process of external growth and internal resorption is the mechanism by which bones increase in size.

I was also able to demonstrate experimentally that, just as the bones increase in size by superimposed layers, they increase in length by juxtaposed layers.

Thus, bone is formed in layers; it is resorbed in layers. But which organ is concerned with this formation and resorption?

This organ is the periosteum."

General mechanisms of ossification

Bone is formed either from a cartilaginous matrix which is resorbed and replaced (metaplasia) by connective tissue—as in the case of ribs, limb bones, etc.—or by ossification of connective tissue.

In all cases, ossification results from two opposing mechanisms which are balanced during successive modifications of developing bone:

(1) A *building mechanism* involving osteoblasts;

(2) *A resorption mechanism* involving other mesenchymatous cells, the osteoclasts which exhibit lytic activity.

According to this process one has to emphasize that prolonged exposure to 96 per cent oxygen at atmospheric pressure or to hyperbaric oxygen produces toxic effects in animals and cells in culture. According to Allison (1963) hyperoxia induces enzyme release from lysosomes into cytoplasm and he supposes that hyperoxia increases the permeability of lipoprotein membranes of cells and cell organelles, probably as a consequence of lipid peroxidation; Tappel *et al.* have already demonstrated release of lysosomal enzymes from rabbit liver homogenates by lipid peroxidation damage. Now Johnson (1953) has demonstrated that exposure of skeletal tissues in culture to elevated partial pressures of oxygen causes resorption and may affect the differentiation of the tissue. Sledge and Dingle (1965) explain that high partial pressures of oxygen act like vitamin A, that is, by alteration of the lipoprotein lysosomal membrane, allowing escape of hydrolytic enzymes and consequent degradation of the ground-substance. This phenomenon has important physiological implications. Cartilage is normally an avascular tissue. Invasion by blood vessels leads to its resorption and replacement by a tissue capable of existing in an oxygenated environment;

this is the normal sequence of events in the epiphyseal plate. Sledge and Dingle suggest that a local rise in oxygen tension may cause an increased release of lysosomal enzymes from the cartilage cells. These enzymes include an acid protease which degrades the protein–polysaccharide complex of the ground-substance and facilitates invasion by vascular tissue.

The first stage of ossification is the transformation of the basic material of connective tissue into ossein by the osteoblasts. Ossein is a collagen-like

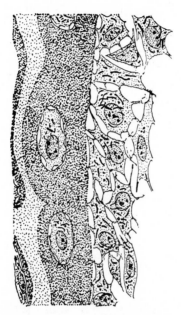

FIG. 133. Osteoblasts. The agents in building processes in growth of endochondral bone. From optical microphotography. High magnification. The richly anastomosing osteoblasts are arranged in epitheloid order along the length of the layers of ossein (in black), the formation of which they induce directly or indirectly at the surface of the cartilaginous matrix (in gray). Connective tissue cells and vessels between matrix. (DUBREUIL and BAUDRIMONT, 1959.)

substance. A glycoprotein, osseomucoid, is also present. The mineral salts of bone are mainly calcium phosphate and carbonate. X-ray studies have shown these compounds to be related to hydroxyapatite:

$$Ca(OH)_2 \cdot 3Ca(PO_4)_2$$

About 1 per cent citrate and also fluorine and strontium can be detected in bone.

As in all tissues of the body, the constituents of bone are in a dynamic state, exchange taking place between bone and plasma.

Ossification mainly implies the precipitation of Ca^{++}, HPO_4^{--} and

PO_4^{---} in the protein matrix of bone. Alkaline phosphatase liberates phosphate from organic phosphoric esters and this mineral phosphate precipitates with calcium. It should be noted that alkaline phosphatase is not found in the protein matrix but in the osteoblasts of developing bone. However, Sevastikoglou (1957), studying endochondral osteogenesis in chick embryos cultured *in vitro* (Fig. 134), found that calcification (hydroxyapatite formation) occurred only 50 to 60 days after explantation, at a

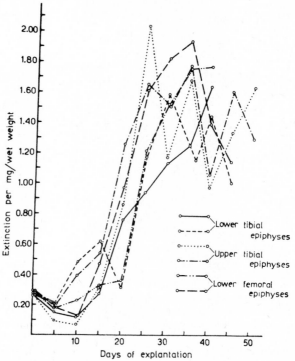

FIG. 134. Alkaline phosphatase activity during growth *in vitro* in a solid medium. (SEVASTIKOGLOU, *Exp. Cell. Res.*, 1957.)

time when alkaline phosphatase activity was declining. Enzymic activity is optimum at the time of ossein formation (40th day).

Lacroix and Ponlot (1957), in experiments utilizing radiocalcium, demonstrated that the young skeleton fixes proportionally more radiocalcium than the adult bone but retains radiocalcium for a shorter period due to more rapid mobilization.

Cartier (1959) found that during bone growth the ratio pyrophosphoric $P (H_4P_2O_7)$/orthophosphoric $P (H_3PO_4)$ did not vary significantly with the degree of mineralization of bone, while the ratio pyrophosphoric P/total nitrogen *increased progressively during growth* and became stabilized in the adult when mineralization was constant. Thus, calcification is paralleled by a fall in the level of organic material. *The maximum ratio is attained most rapidly in the diaphyses.*

According to Cartier (1952), the pyrophosphate is derived from ATP (adenylpyrophosphoric acid) after hydrolysis by ATPase. It then forms $Ca_2P_2O_7$. ATP would be regenerated by the process of anaerobic glycogenolysis similar to that which exists in striated muscle during contraction. To summarize, in this case ATP would have more of a structural, morphogenic role than an energetic role; ATP would act essentially as a donor of pyrophosphate.

Endocrine factors in ossification

The presence of vitamin D in the diet favours ossification since vitamin D augments Ca^{++} absorption by the intestine. Ossification also requires a stable acid–base equilibrium since fixation of Ca^{++} decreases with acidosis. Vitamin A seems to modify skeletal tissue by augmenting the activity of certain enzymic systems, one of which resembles in its action the protease of papain (H. B. Fell, 1960).

Hormones have an important role in ossification.

Sex hormones: hypergenitalism, precocious puberty, produces arrest of skeletal growth. Furthermore, *folliculin induces bone neoformation* (J. Benoit and Clavert). On the other hand, eunuchs are generally tall; testosterone inhibits cellular proliferation of hyaline cartilage.

Thyroid ablation in young animals arrests growth and in the young rat at birth, slows maturation of the skeleton (Scow and Simpson, 1945). *Endochondral ossification* is retarded through the absence of proliferation of cartilage, the cells of which are hyperplastic and the ground substance swollen. However, the basic structure of bone is not modified. The absence of growth mainly affects the long bones such that the limbs remain short (basset). However, periosteal ossification continues and bones are of normal thickness. *Congenital thyroid insufficiency in man leads to dwarfism with myxoedema and idiocy.* Thyrotropic hormone has been shown to affect the growth of bone (Silberger). However, the role of the thyroid in ossification has not been fully elucidated. From a study of propylthiouracil on rat, Jost, Moreau and Fournier (1960) concluded that this antithyroid did not have a specific action on the primary centres of ossification but it retarded general development of the foetus. Thus, thyroid hormones do not appear to be necessary in the initial formation of skeleton but would modify differentiation and maturation.

Hypophysectomy results in arrest of all endochondral processes; a specific example of this action is the "shrinking" of bones seen during pituitary hypoactivity of old age. Administration of growth hormone stimulates endochondral ossification but ossification of epiphyses does not occur. Growth hormone can be assayed by measuring the thickness of the proximal epiphyseal cartilage of the tibia of young totally hypophysectomized rats. Growth hormone is administered by intraperitoneal injection (Evans *et al.*, 1943). Epiphyseal stimulation can also be seen in a totally

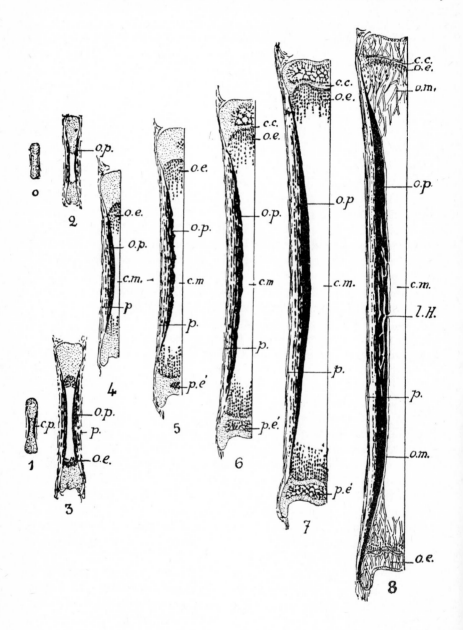

hypophysectomized rat in which a pituitary gland is grafted near to the epiphyseal cartilage (Vaillant and Jost, 1963).

Endocrine interrelationships are apparent. Testosterone counteracts to some extent the effects of total hypophysectomy (cf. p. 440). *Adrenocorticotropic hormone* (ACTH) *retards osteogenesis in the normal rat* (Becks et al.; Evans and Li, 1944). Cortisone administered to young rats leads to the formation, at the extremities of growing bone, of a region of trabeculae of calcified cartilage resulting from inhibition of osteolytic activity (Follis, 1957). However, such an action has not been observed following the administration of cortisone to mice, rabbits or guinea pigs. This example of the variations between species emphasizes the need to avoid premature generalizations which are so often made.

All endocrine glands are concerned with processes of ossification but the most important gland is certainly the parathyroid which affects the metabolism of calcium and mineral phosphorus.

During *hyperparathyroidism* the destructive, osteoclastic action predominates. Furthermore, the fibrous tissue of the Haversian canals undergoes proliferation resulting in poor nutrition of bone.

It is of interest to note that using his technique of joint cultures, P. J. Gaillard (1955) found that in the presence of newborn human parathyroid tissue fragments of parietal bone from embryos of preparturient mice show resorption, although no typical osteoclasts are present, after four days culture *in vitro*. These results suggest that certain products diffusing from the parathyroid explant have a direct action on bone.

The problem of the mode of action of parathormone is centered around the determination of the metabolic site of action. Parathormone may act on a

---

Fig. 135. Development of a long bone (tibia) during the period of growth. Semi-schematic according to G. Dubreuil.

o.p., periosteal bone; *o.e.*, endochondral bone; *o.m.*, medullary bone; *c.m.*, medullary canal; *c.c.*, cartilage plate; *p.*, periosteum; *p.é.*, epiphyseal point.

O, Primitive cartilage model. *1,2*, Perichondral bone collar, elongation, appearance of external periosteal bone lamellae, formation of primitive medullary canal. *3,4*, Endochondral bone forms at the extremities of medullary canal, new lamellae of periosteal bone thicken the diaphysis, the internal layers of this bone thicken and become more dense, formation of cartilage model. *5*, Continuation of these processes, but, the two epiphyseal points appear and enlarge medullary canal by resorption of internal layer of dense periosteal bone. *6,7*, The same processes continue, enlargement and elongation of diaphysis, cartilage plate is limited to region between diaphyseal and epiphyseal endochondral bone; the medullary canal reaches its ultimate size. *8*, The epiphyses are formed, the cartilage disappears allowing connection between epiphyseal medullary bone and diaphyseal medullary bone (which substitutes primitive endochondral bone), the epiphyseal bony plate which has limited extension of epiphyseal bone towards cartilage will persist for a long time. Modification of dense periosteal bone by lacunae of Howship. Formation of basic internal system. *8*, corresponds with *7* of following figure.

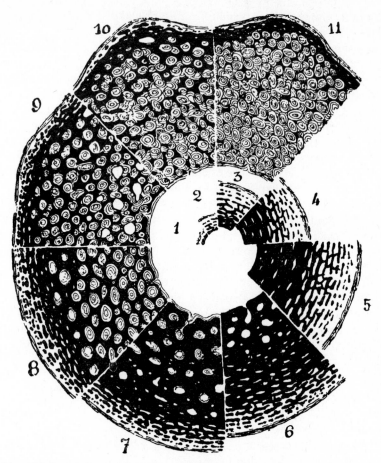

FIG. 136. Successive stages in development of the diaphysis of a long bone. Semi-schematic, from G. DUBREUIL (Vigot). At the periphery, the periosteum. In black, the periosteal bone lamellae. In concentric circles, the Haversian system seen in transverse or oblique section. *1*, Stage of perichondral bone collar. *2*, Periosteal bone lamellae. *3,4,5*, Thickening of internal layers of periosteal bone, addition of external lamellae. *6*, Appearance of lacunae of Howship in thickened periosteal bone and appearance of primary Haversian system. *9*, Extension of processes towards periphery, formation of boney crest at surface, appearance of secondary lacunae of Howship in periosteal and internal Haversian bone. *10, 11*, Thickening of external periosteal bone (basic external system), accentuation of bony crest, extension of substitution of periosteal bone by Haversian bone, intermediate Haversian and periosteal systems persisting, the latter becoming less and less numerous. *1–6*, increase in medullary canal by resorption of internal periosteal bone by osteoclasts. *7–11*, fundamental internal system formed around medullary canal which reaches its ultimate dimensions. Modification of bone tissue continues throughout life, at a reduced rate or at an increased rate (osteoporosis).

FIG. 137. Wall of diaphysis of a long bone. (DUBREUIL and BAUDRIMONT, 1959.)

Ca^{++} acceptor which is thought to be a mucopolysaccharide, chondroitin sulphate (Boyd and Neuman, 1951) or on *anaerobic glycogenolysis.* Glycolysis is a key reaction in ossification (Gutman *et al.,* 1949; Shorr *et al.,* 1953). Glycogen accumulates in cartilage cells and disappears during calcification. Also calcification does not occur in the presence of glycolysis inhibitors.

Another possible site of action of parathormone is the mobilization of calcium ions.

A concept including many mechanisms was proposed by Neuman (1958):

Vitamin D and parathormone act on bone at different stages of glucose metabolism . . . let us consider that parathormone blocks oxidative cycles and augments glycolysis and the accumulation of citric, lactic and pyruvic acid . . ., if vitamin D acts only on condensation of pyruvate to give citrate and not on glycolytic reactions, then parathormone would be capable of augmenting the production of lactic and pyruvic acid and leading to mobilization of mineral salts.

However, other processes must be taken into consideration.

While studying morphologically and autoradiographically the influence of parathyroid extract (PTE) on radius rudiments of 15-day-old mouse embryos in tissue culture, Gaillard (1960, 1961) observed, among other things, a distinct effect on protein (collagen) metabolism. Applying radiometric methods, De Voogd Van Der Straaten (1962) confirmed these findings and related this effect to a decreased ribose synthesis. Approaching the problem enzymologically, Hekkelman (1962) related the influence of PTE on bone tissue to a decrease in the NADP available in bone cells which modified the NADP mediated reactions in the phosphogluconic shunt and the Krebs cycle. The result supports the idea of a PTE-induced increase in NADP activity.

III. Regulation of growth

In considering the whole process of growth it should be apparent by this stage that regulation will depend on limiting factors such as the level of nutritive materials and also the presence or absence of congeners or predators defining the *allelostatic* and *allelocatalytic* relationships. These facts are particularly important when considering a practical problem which is so fundamental in food production. P. E. Victor (1964) regarded the biomass as the capital and the intensity of reproduction within the biomass as the interest on capital. Antarctic marine animals show particular characteristics in this respect, since, despite the constant cold, the absence of light and other unfavourable factors, a great variety of bottom fauna is found in this region, a fact which could be attributed to the presence of plankton (Tressler, 1964). Surveys carried out by Wohlschlag (1964) at the McMurdo Sound Station (Antarctica) showed that fish living in a habitat with a temperature of $-1 \cdot 9°C$ at a depth of 600 metres have a similar metabolic rate and total growth to some fish from temperate regions. Compared with fish from temperate or tropical zones the Antarctic fish show similar metabolism/weight ratios while the metabolism/temperature ratios and metabolism/swimming ratios are higher. These characteristics could be explained in terms of steroid hormone content and the lipotropic action of cold.

We shall confine this study of hormonal factors to the endocrinology of somatic growth, genital growth being considered in the chapters devoted to sex endocrinology.

Growth can be studied in the whole organism, in an organ in culture or tissue cultures but an index of growth can also be obtained from tissue regeneration or regeneration of an organ or, in lower creatures, of part of the organism.

Regeneration of fins in teleost fish

The following account is based mainly on the work of J. Buser (1953). Regeneration of fins is an example of bone regeneration and the description of this process will complete our study of ossification. *Regenerative ossification* of pelvic fins, particularly in marine teleosts such as *Gambusia* and *Lebistes, is initiated by pituitary and thyroid factors:*

(1) regeneration occurs when the animal is subjected to *continuous illumination* but is suppressed by total hypophysectomy (later we shall consider the relationships between light and pituitary hormones in the case of gonadotropic activity; p. 530).

(2) *pregneninolone* (or *ethinyltestosterone*) possesses antithyroid activity on the thyroid of *Discoglossus* larvae (Gallien, 1949) and this substance inhibits regeneration of fins. This action of pregneninolone is through the intermediary of the pituitary.

Ethinyltestosterone

Thus, not only does the thyroid affect growth such as in metamorphosis (batracians) but shows a direct action on volume and weight of organs and parts of the body.

Growth of tibiotarsals in culture in vitro

The most significant example is that of the Creeper mutation (short feet) in chicks (W. Landauer, 1927, 1934) since it establishes a direct link between genetic phenomena and biochemical and morphological manifestations. In the lethal homozygote form, CpCp, the individuals generally die before the third day of incubation but some embryos survive up to the eighteenth day and exhibit phocomelia, parrot beak, etc. The heterozygotes Cp $^+$ are characterized by achondroplasia (viable basset mutant). E. Wolff and K. Haffen (1952) added chick plasma to the medium of an organ culture and found that development of the normal cartilaginous limb buds varied according to the nature of the embryonic extract in the culture. An extract of Creeper embryo was less favourable for linear growth of tibiotarsals than extracts of normal embryos. A Creeper inhibitory substance is present which is thermolabile (100°C) and precipitated by cold acetone (M. Kieny, 1962). This Creeper inhibitory factor is not only localized in long bones but is found in viscera and thus in parts of the embryo not normally affected by the Creeper mutation. As stated by Wolff and Kieny (1963), the extracts contain a diffusible substance with a heterospecific action and important macroscopic effects on growth. Since *p*-aminobenzoic acid improves development of embryonic tibiae (M. Kieny, 1958), it is possible that the Creeper factor is an inhibitor of this acid.

From these considerations we can conclude that the regulation of growth of organs is more complex than the regulation of growth of cell populations in which facilitation *(allelocatalysis)* and inhibition *(allelostasis)* of cells, one by the other, regulates the modes of growth and even allows synchronization of cell division. E. Wolff found that to obtain *organotype cultures* the medium must possess certain characteristics:

(1) the medium should contain little nutritive material such that explosive proliferation of cells of the organ is prevented;

(2) favourable respiratory conditions must be maintained and dessication prevented;

(3) there should be close adherence between the medium and explant to allow nutritive exchange;

(4) a favourable substratum must be provided (L. Marin, 1960): for example, one can obtain simultaneously with the same explant an organotype development on agar and histiotype culture at the edge of a glass plate (the wettability of surfaces by the explants would be an important factor).

Aspects of growth, moult and metamorphosis in invertebrates

Destruction of the brain of insects (Kopec, 1916) inhibits moult and metamorphosis. This inhibition would be related to neurosecretory factors. On the other hand, the cephalo-thoracic gland (Lyonnet, 1762) disappears in the imago and directly controls nymphal and imaginal moult through stimulation from the brain. The hormone in this case is, according to Scharrer, a growth and imaginal differentiation factor. The *corpora allata* liberate factors which delay development of the imago. In silkworms, following removal of the *corpora allata* after the third moult (the fourth being the final moult), the animal produces a cocoon and pupates. The fourth moult has been inhibited but this does not prevent the development of a dwarfed but characteristic moth (Bounhiol, 1936, 1937). This anti-imaginal or juvenile action of *corpora allata* is also shown by the experiments of Ishikawa (1939) and C. M. Williams (1939) on lepidopterous larvae. The caterpillars were decerebrated, one sample being left as such and the other receiving an implant of corpora allata from adults. The latter group moulted but it was a larval moult not an imaginal moult. Compounds with auxotropic activity have been isolated from numerous arthropod organs and organisms:

(1) "ecdysone" of Butenandt and Karlson, a steroid with the following formula

In maggots ecdysone modifies tyrosine metabolism and the formation of the chrysalid (Karlson); but, ecdysone shows chiefly important genetic properties as demonstrated by Clever and Karlson (1960): injection of ecdysone into mature larvae of the midge *Chironomus tendans* induced puffing of certain loci on the giant chromosomes; since puffing is a correlate of gene activity, the hormone can be visualized to induce the DNA-dependent synthesis of messenger RNA and, in this way, the *de novo* formation of certain proteins. Thus it is demonstrated that hormones can act directly upon genes.

(2) *"neotenin"* of Williams, Schneidermann and Gilbert.

As regards the biochemical mechanisms, secretion of chitinolytic and proteolytic enzymes is a general property of arthropod epidermis at the time of moulting. Chitinase activity is always found in crustaceans but the ability of the cells to secrete this enzyme is increased or realized at the time of moulting (Jeauniaux, 1960). On the other hand, chitinase activity in the silkworm is a cyclic phenomenon and in larvae no chitinase is found during intermoult (Jeauniaux, 1957). Digestion of chitin of the procuticle and free proteins liberates acetylglucosamine and amino acids which are reabsorbed to a great extent across the hypodermis. This hydrolysis is associated with moulting. During moulting in crustaceans there is marked absorption of water which produces high water pressures within the stomach and filtration under pressure from the gastric cavity to the fluid cavity (Drach, 1939).

It should be noted that ablation of the *corpora allata* inhibits biosynthesis of proteins and amino acids accumulate in tissues (L'Helias, 1957, in *Dixippus*).

Thus, we see that these complex growth phenomena in arthropods involve mainly proteogenic and endocrine processes.

We shall now examine the endocrine glands which are active on growth of vertebrates. The term "active on growth" is used in the sense that these glands modify growth which is a property of cells. Hormones, as vitamins, are the cofactors of enzymic reactions, the *catalysts* which *augment* reactions which would occur spontaneously although slowly and often in a disordered manner. This should always be remembered when considering hormones particularly in the case of pituitary hormones. Thus, elimination of the hypophyso-pituitary complex during the earliest stages of life does not affect body growth when the complex is destroyed by *direct ablation* immediately after birth (in rat, Fugo, 1940) or by the extreme method of *decapitation* (in chick embryo, Ancel *et al.*, 1938). This technique of Ancel has been successfully applied in experiments by Jost on mammalian sex endocrinology (cf. p. 504). At this stage we shall not examine the role of somatic endocrine glands on genital growth.

Action of endocrine glands on somatic growth

Eunuchism leads to increased height as we have already noted. Castration of young males results in man in an increase in length of the posterior limbs. The epiphyso-diaphyseal cartilages are the most sensitive. *Testosterone* has an inhibitory action on growth. Aron considers this to explain the differences in size between males and females. The male is smaller than the female in batracians, fish and certain birds.

However, the pituitary through the secretion of growth hormone has the greatest effect on somatic growth.

Acromegaly is characterized by hypertrophy of limbs, bones and certain soft regions of the face such that these regions appear to show a greater sensitivity to growth than other parts of the body. This syndrome was first described by Pierre Marie in 1886. Later studies have shown this disease to be produced by a tumour of pituitary acidophilic cells. Furthermore, it is also known—mainly from clinical studies since little experimental work has been done in this respect—that *lesions of the hypothalamus can produce acromegaly.* In the case of stature it is possible that the hypothalamus acts either through excitation of somatotropic eosinophilic cells or more directly, on the tissues through the autonomic nervous system as in the case of any "direct" action of the hypothalamus on a receptor. However, there is no definitive evidence for either pathway. The only definite fact is that *growth hormone is the essential factor in regulation of stature.* It is of importance to determine the duration of activity of the hormone within the mammalian organism. A known quantity of hormone is injected into a totally hypophysectomized animal and by estimating the hormone content of blood samples removed at regular intervals, the rate of disappearance can be estimated. In rat, *50 per cent of injected hormone is destroyed within 26 mn.*

If protein anabolism is an expression of growth, STH is certainly a growth hormone since it leads to a marked accumulation of proteins in the organism and within several hours after injection produces a fall in plasma amino acids (Li and Evans, 1948); could it act as a permease?

On the other hand, since the work of Young (1945, 1953), *STH is thought only to stimulate growth when the organism is subjected to a certain quantity of insulin.* Moreover, insulin can augment protein synthesis from free amino acids. Li and Evans (1950) formulated a synoptic hypothesis on the endocrine mechanism of proteogenesis control: *the action of STH would be supplemented by that of TSH and ACTH. TSH stimulates lipid catabolism and augments the liberation of carbon chains, thus increasing proteogenesis. ACTH inhibits protein anabolism and thus opposes the action of STH.* This antagonism is the basis of regulation and the synergism with TSH could be the origin of eventual substitution. However, the hypothesis of Li and Evans does not consider the role of insulin (Young).

Young who has studied this problem since 1936, induced sugar diabetes in cat with a purified preparation of STH (1945). Young considers that the somatotropic and diabetogenic action of STH would be differentiated by the endocrine pancreas which would either produce or not produce the amount of insulin required for the anabolic action. STH would stimulate growth or have a diabetogenic action (this may be a question of the amount of insulin liberated by STH).

These facts indicate that dysfunction of an essential metabolic chain (in this case of sugars) can interrupt growth. *In multicellular creatures and particularly in mammals, molecular events are found to be regulated by other endogenous molecules, the hormones.*

It should be added that Borstein and Foa, following the work of Young on cat, concluded that STH can stimulate secretion of glucagon, the pancreatic hyperglycemic hormone.

Young's diabetes develops in three phases:

(1) Administration of STH leads to compensatory hypersecretion of insulin with transient hypoglycemia;

(2) continued administration of STH results in a decrease in insulin secretion by the β cells of the pancreas with hyperglycemia and glycosuria which disappear with discontinuation of STH administration. The pancreatic cells which are still active produce glucagon. It should be noted that acromegaly is associated with sugar diabetes with polyphagia and glycosuria (characteristic of sugar "intolerance"); acromegaly and Young's diabetes represent two aspects of somatotropic action;

(3) further administration of STH leads to loss of insulin production. Thus, STH, according to the conditions of administration, can either have an *anti-insulin* action or an *insulinomimetic* action (cf. p. 420).

Before ending this brief examination of the somatic endocrine glands, a few examples must be included of the effect of hormones on "terrain" which expresses a continuously regulated internal environment.

For example, *adrenalectomy* considerably reduces the resistance to infection and excess cortisone favours infection (for example, by intestinal bacilli). The thyroid with an action on cellular oxidation and basal metabolism conditions the "terrain" as much by its effect on metabolism as by its thermogenic ability, the constancy of central temperature of homoiotherms being a characteristic of the internal environment.

Thus, homoiostatic control mechanisms such as hormonal regulation and neurohormonal regulation which will be examined later (thermoregulation, hunger, etc.) demonstrate the existence of servomechanisms. We have seen that the hypothalamo-hypophyso-pituitary complex could be considered to be at the centre of somatic endocrine control mechanisms, sending information by way of hormonal messengers to the peripheral receptors of target organs which are mainly endocrine glands. The target organs in their turn influence numerous receptors and, in particular, the hypothalamo-hypophyso-pituitary complex. These facts extend the concept of chemoreceptors which we considered earlier with respect to the regulation of ventilation. The hypothalamo-hypophyso-pituitary centre is linked also with sexual endocrinology. The different regions of the hypothalamo-hypophyso-pituitary complex which respond to messengers from the endocrine glands can be compared with the *projection areas* of the cerebral cortex. In the nervous system, information is transmitted along differentiated structures, the nerves, but the messengers of the endocrine system are the hormones which follow humoral pathways. Thus, *the hypothalamo-hypophyso-pituitary hormones are the products of the endocrine motor areas*. These hormones are *signals* equivalent to the action potentials of nerves. This dialectic cybernetic approach expresses the characteristics of regulation of the internal environment: regulation consists of *endocrine sensitory-motor circuits* with hormonal messengers, the endocrine reflexes (cf. p. 395) which are associated with nervous reflexes in which the messengers are the action potentials.

REPRODUCTIVE FUNCTIONS

REPRODUCTIVE endocrinology provides one of the best examples of the gene → hormone → metabolism → morphology relationship.

In higher animals this aspect of endocrinology cannot be separated from somatic endocrinology but it is convenient to differentiate these subjects in order to understand the unity of reproductive functions.

CHAPTER 22

PHYSIOLOGY OF GENITAL ORGANS

The notion of sex is, in some ways, contingent as shown by the studies on the extreme caryotypes in human (for example the *syndrome of Klinefelter* where, in the variety XXY, there is a male hypogonadism, or the *syndrome of Turner*, XO, where there is a female hypogonadism). However, the *statistical fact of normality* will determine our scope.

I. Physiology of the male genital organs

Anatomy

The *testis* is the *principal* functional unit of the male genital apparatus. The testis, as the pancreas, is both an *exocrine gland* (in that it produces reproductive cells rather than simple molecules) and *an endocrine gland* (producing *androgens*).

Spermatozoa are produced by the *seminiferous tubules* and the hormones are formed in the *interstitial cells*.

The production of spermatozoa depends on the following processes:

Proliferation of germinal epithelium
↓
Testicular bud   ←———   Cell cords *containing*
with reduced cortical region   *all germinal cells*
and developed medullary region
↓
Seminiferous tubules ———→ Spermatozoa in the pubescent

Between the seminiferous tubules are the *interstitial cells* of Leydig which constitute the *interstitial gland*, a *disseminated* endocrine gland. The interstitial cells are derived from certain small cells of the germinal epithelium.

The cells of Leydig *must not be confused with the syncytial* Sertoli *cells* which are nutritive cells intercalated between the seminal cells. However, the origins of these two types of cell are related.

The paired ducts of Wolff which receive convoluted efferent ducts (= epididymis), form the pathway for elimination of sperm. The initial part of the duct of Wolff is the *epididymal duct* and the terminal part is the *vas deferens* which receives the seminal vesicles. The canals of Müller are atrophied in the male.

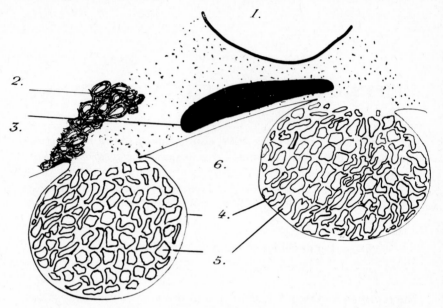

FIG. 138. Section of an embryonic testis of chick (18 days). *1*, aorta; *2*, renal bud; *3*, vein; *4*, cortex; *5*, seminiferous tubules; *6*, coelom.

The genital ducts and the external genital organs of the male regress after castration but injection of testicular extracts to castrated animals prevents the effects of castration.

Histologically the interstitial islets of Leydig, with their marked vascularization, resemble an endocrine gland as was recognized by Bouin and Ancel in 1903.

The following facts provide evidence for the production of testicular hormones by the interstitial cells:

(1) When the testes fail to descend into the scrotum at puberty and are retained in the abdominal cavity *(cryptorchidism)*, the germinal cells disappear but the secondary sexual characters remain. Thus, the exocrine function of the testes is lost but the endocrine function is retained. Injection of extracts of cryptorchid testes to castrated males such as the capon (Pézard) leads to the appearance of secondary sexual characters. One concludes from these experiments that the *germinal elements are not necessary for the formation of the endocrine factors.*

(2) *In certain cases of eunuchism* (in rabbit, cock, etc.) normal spermatogenesis occurs but the interstitial cells degenerate and the subjects exhibit the characteristics of castration.

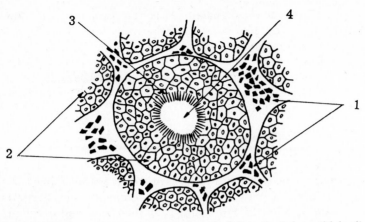

FIG. 139. Diagram of a transverse section of adult testis. *1*, interstitial cell;
2, seminiferous tubule; *3*, groups of spermatozoids; *4*, central lumen.

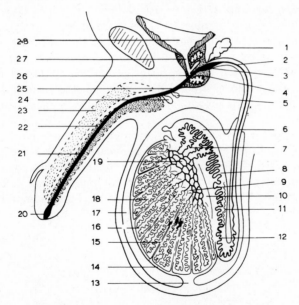

FIG. 140. Diagram of male genital apparatus of man. Seminiferous tubules,
genital ducts and glands. *1*, seminal vesicle; *2*, ductus deferens; *3*, ampulla
of ductus deferens; *4*, prostate; *5*. urethral glands (of Littré); *6*, head of
epididymis; *7*, cone of ductuli efferentes; *8*, duct of epididymis; *9*, tubuli
recti; *10*, mediastinum testis; *11*, vas aberrans of Haller; *12*, tail of epididymis; *13*, gubernaculum; *14*, tunica albuginea; *15*, interlobular septum;
16, lobule of testis; *17*, cavity within tunica vaginalis; *18*, seminiferous
tubule; *19*, rete testis; *20*, urethra of glans; *21*, penile urethra; *22*, corpus
spongiosum; *23*, corpus cavernosum; *24*, bulbo-urethral glands (of Cowper); *25*, menbraneous urethra; *26*, common ejaculatory duct; *27*, prostatic
urethra; *28*, bladder. (DUBREUIL and BAUDRIMONT, 1959.)

(3) In the triton and salamander there are no permanent interstitial cells (Aron). They appear *seasonally* (March–April) and coincide with the appearance of secondary sexual characters. During May–June the interstitial cells regress at the same time as the male characters. *Androgens are formed in the cells* of Leydig but androgens and oestrogens can be detected in all *adipose tissue*, at least in mammals. These regions are probably storage sites (Vague and Garrigues, 1957).

Androgens

Many androgenic hormones have been isolated from *testes* and *urine*. Their physiological activity is determined by examining the effects of injection to castrated animals. For example, in the capon androgens produce growth of the comb. The principal male hormone, *testosterone*, is synthesized by the Leydig cells from active acetate and cholesterol (via *pregnenolone, progesterone* and *17-hydroxyprogesterone* which is converted to a 17-ketosteroid, *4-androsterone-3,17-dione*, the direct precursor of testosterone). 4-Androstenedione has been found in the spermatic vein of dog.

Among the steroids isolated from *testicular* extracts are *oestrogens*. Zondek (1933) reported that the urine of *stallion* contained oestrogenic activity. Oestrone and oestradiol have been isolated from this urine and their testicular origin is shown by the absence of oestrogens from the urine of geldings.

The presence of oestrogens and androgens in testes is not surprising when the similarities between their chemical structures are considered. The biosynthesis of oestrogens from androstenedione depends essentially on aromatization of the ring containing carbons 1, 2, 3, 4, 5, 10 according to the following scheme (Ryan, 1959; Longchampt *et al.*, 1960; Hayano, 1962).

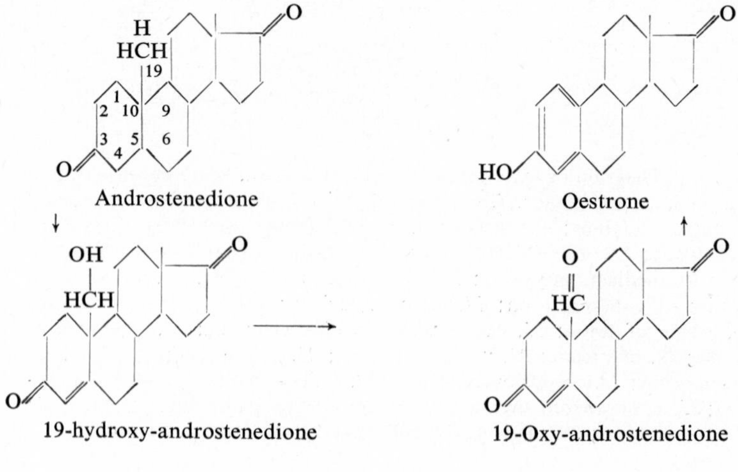

Androstenedione

19-hydroxy-androstenedione

Oestrone

19-Oxy-androstenedione

Prenenolone is the precursor of: (a) adrenocortical hormones, (b) testo-sterone, and (c) progesterone.

It is important to know the *fate of testosterone*. When administered orally testosterone is 100 times less active than when administered by injection. Testosterone is *denatured* during passage through the *hepatic* circulation although this is not the case with all sterols for example cor-tisone which is as active orally as by injection.

Urinary excretion of 17-ketosteroids is used as a test of testicular hormone production. Excretion of 17-ketosteroids occurs from the age of 8 years in man and can reach 2·5 mg per day. This factor is of great importance in that it allows detection of the true *prepubescent state* which is manifested not only by testicular activity but by general changes and in particular, by adrenocortical activity. The coloured reaction (of Zimmerman) between 17-keto group and *m*-dinitrobenzene is used for the determination of 17-ketosteroids.

CH$_3$

—OH

CH$_3$

O

OH$_3$

=O

CH$_3$

OH

H

Testosterone (Laqueur, 1935)

Androsterone (Butenandt, 1931; synthesized by Ruzicka)

CH$_3$

=O

CH$_3$

HO

Dehydroepiandrosterone (principal 17-ketosteriod of blood)

From 8 years to puberty the excretion of 17-ketosteroids can increase to reach 10 mg per day *in both boys and girls*. It is not until after puberty that there are variations in excretion according to sex: about 10 mg per day in females and 15 mg per day in males.

The level of urinary ketosteroids is not a characteristic of eunuchism since certain of the products eliminated originate in the adrenal cortex but in cases of testicular *tumours* the excretion of 17-ketosteroids is augmented.

The effects of testosterone have been attributed to its *proteogenic activ-ity*. However, an ethyl 17-hydroxyl sterol derivative has been synthesized which shows anabolic activity but no androgenic activity. It should be noted that although STH is the principal anabolic hormone of infancy, the androgens are the principal anabolic hormones in the adult. Anabol-ism could be characterized roughly by retention of nitrogen and potassium.

The composition of mammalian semen

Ejaculated semen consists of two fractions: *spermatozoa* and *seminal fluid*. Seminal fluid contains the secretions of the epididymis, prostate gland, seminal vesicles and Cowper glands. These are not endocrine glands and the secretions provide a more or less nutritive medium for the spermatozoa. However, seminal fluid produces contraction of the uterus (Kurzrok and Lieb, 1930) and jejunum (Euler, 1934). These properties are attributed to the presence of *prostaglandin* (Euler, 1935) which behaves as a mixture of relatively stable unsaturated lipid soluble acids (Bergström). Prostaglandin originates in the seminal vesicles where it is found in an inactive form (Eliason, 1959). *Vesiglandin*, of unknown structure, has also been described by von Euler.

Human semen which has been stored for 2–3 days contains colourless crystals of *spermin* (Vauquelin, 1791):

$$CH_2\text{---}CH_2\text{---}CH_2 \qquad CH_2\text{---}CH_2\text{---}CH_2$$
$$|\qquad\qquad\qquad | \qquad\qquad\qquad |$$
$$NH_2 \qquad NH(CH_2)_4NH \qquad NH_2$$

Florence (1895) identified *choline* in human semen. Fresh semen contains *glycerylphosphorylcholine* (Kahane, Levy and Diament, 1952). *Phosphorylcholine* has also been identified (Lundquist, 1946).

$$H_2C\text{---}OH$$
$$|$$
$$HC\text{---}OH \quad O$$
$$|\qquad\qquad \|$$
$$H_2C\text{---}O\text{---}P\text{---}O\text{---}CH_2\text{---}CH_2\text{---}N^+\text{---}CH_3$$
$$|\qquad\qquad\qquad\qquad\qquad CH_3$$
$$OH \qquad\qquad\qquad\qquad CH_3$$

Glycerylphosphorylcholine

In rabbit, bull, boar and stallion *all choline is in the form of glycerylphosphorylcholine* while in man, monkey, cock and hedgehog *most of the choline is in the form of phosphorylcholine*.

Inositol has been identified in boar semen (Mann, 1951) and is secreted by the *seminal vesicles*. Seminal vesicle fluid is rich in *ergothioneine* (a sulphur-containing base) (Tanret, 1909) which is also found in liver of rat and in erythrocytes. The reducing properties (SH⁻) of this compound would favour sperm motility.

$$CH\text{---}NH$$
$$\| \qquad\qquad\qquad C\text{---}SH$$
$$C\text{---}N$$
$$|$$
$$CH_2$$
$$|$$
$$CH\text{---}N^+ \equiv (CH_3)_3$$
$$|$$
$$C\text{---}O\text{---}$$
$$\|$$
$$O$$

Ergothioneine

Mammalian semen also contains *citric acid* (Schersten, 1929); G. Humphrey and Mann (1946) showed the presence of *fructose*, an important source of energy. In the epididymis and testes the *spermatozoa are inactive* since, in these regions, fructose is not available except in rat in which fructose is localized in the dorsolateral prostate and coagulation gland (Humphrey and Mann, 1949). The seminal vesicles contain fructose and in this region sperm become motile. *Fructolysis leading to lactic acid formation is the principal source of anaerobic metabolism of mammalian sperm.*

Under aerobic conditions, in the absence of fructose and glucose, ejaculated mammalian sperm are inactive. The sperm of animals with aquatic fertilization are aerobic.

The appearance of fructose in the male accessory glands of young mammals is coincident with the beginning of male hormone secretion (Davies, Mann and Rowson, 1957).

Hers (1959) found that slices of seminal vesicles from sheep irreversibly converted glucose to sorbitol and sorbitol to fructose. The fructose formed from ^{14}C-glucose is labelled in position 1. The conversion of glucose to sorbitol and fructose is inhibited by glucosone and glucuronolactone. The conversion of sorbitol to fructose is inhibited by ethylenediaminetetraacetate which inactivates ketose reductase. These results as well as the presence of sorbitol in the tissues suggest that the conversion of glucose to fructose in the seminal vesicles is by the following pathway:

$$\text{Glucose} + \text{NADPH}_2 + \text{H}^+ \underset{\text{aldose reductase}}{\rightleftharpoons} \text{Sorbitol} + \text{NADP}$$

$$\text{Sorbitol} + \text{NAD} \underset{\text{ketose reductase}}{\rightleftharpoons} \text{Fructose} + \text{NADH}_2 + \text{H}^+$$

Overall reaction

$$\text{Glucose} + \text{NADPH}_2 + \text{NAD} \rightleftharpoons \text{Fructose} + \text{NADP} + \text{NADH}_2$$

Another important property of semen was described by Lindahl and Kihlström in 1952. Ejaculated mammalian semen agglutinates after several hours at room temperature. The mammalian organism produces substances which inhibit this agglutination, the *antaglutins*. In the female, antaglutins are found in the ovarian follicles, the Fallopian tubules and mucous of the cervix. Antaglutins of the male are produced in the *prostate gland*. Male antaglutin is probably a *sulphated ester of a tocopherol derivative* (Lindahl and Kihlström, 1954) and it is noticeable that in rats, vitamin E is the antisterility vitamin. The antaglutins of females contain sugars but no sulphur.

Mammalian testes and sperm contain an enzyme, *hyaluronidase*, which depolymerizes mucins to hyaluronic acid. This is the "diffusion factor" of Duran–Reynals. It has been suggested that hyaluronidase is involved in penetration of the ovum by a spermatozoon but it seems now that the enzyme plays no part in this process.

The rate of migration of sperm in the male genital tract has been measured by the elegant technique of Dawson (1957). Rams which have been trained to use an artificial vagina receive injections of ^{32}P and the rate at which various constituents of semen become labelled is estimated. After administration of ^{32}P the first labelled compound to appear is glyceryl-phosphorylcholine (maximum after 15 days) of seminal fluid, labelled phospholipids and ribonucleic phosphorus appear later (maximum after 25 days) and finally desoxyribonucleic acid of sperm (maximum after 52 days). When the testes and epididymis are isolated surgically, glyceryl-phosphorylcholine of seminal fluid is labelled as normal but epididymal sperm are not labelled except in the acid soluble fraction. These results show that with ejaculation occurring every two to three days, the sperm take a minimum of 11 days to pass from the epididymis before ejaculation. Moreover, the glycerylphosphorylcholine content of semen increases in the epididymis during maturation of sperm.

Aspects of male sex physiology

The work of Etienne Wolff and his co-workers has contributed greatly to our knowledge of the growth and differentiation of gonads cultured *in vitro* in an agar medium. These are *organ* cultures and not tissue cultures. The experiments of P. J. Gaillard (1929, 1942, etc.) provided a basis for these studies.

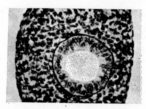 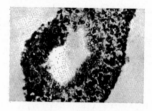

a	*b*

FIG. 141. Action of crystallized male hormone, testosterone propionate, on canals of Müller cultured *in vitro*. (a) Canal of Müller at 7 days: transverse section showing epithelium consisting of tall cylindrical cells surrounded by connective tissue layer. (b) Canal of Müller at 7 days, cultured on a medium containing testosterone propionate. Degeneration of connective tissue, the necrosis is reaching the epithelium of canal which tends to be disorganized (after E. WOLFF and Y. LUTZ-OSTERTAG). (WOLFF, *Experientia*, 1953.)

In collaboration with K. Haffen, E. Wolff (1952) demonstrated that, in an explant of *non-differentiated* gonads of duck (7th day), the genital cords develop normally into ovaries or testes according to sex.

While studying the differentiation of gonads *in vitro* using *synthetic media* in which all the nutritive constituents were known, Wolff and his co-workers observed that two compounds are required for survival and differentiation. These are *cysteine* and *methionine*, two amino acids containing—SH groups. Wolff wrote: "It is remarkable that cysteine alone is required for differentiation of the ovarian cortex of duck embryos but it is not sufficient for differentiation of the testicular medulla of male embryos. *The addition of methionine and arginine to the medium allows normal growth of testes. The gonads thus react to dietary factors as to sex hormones.* We have called this a "hormonomimetic action" although these effects may be due to liberation of small residual amounts or to synthesis of hormones under the influence of the amino acids".

In the invertebrates, Demal (1961) obtained survival of larval gonads of *Drosophila* and *Calliphora* although differentiation of gonads was not seen. Lender and Duveau-Hagege (1963) cultivated *in vitro* gonads from last stage larvae of *Galleria mellonella* (Lepidoptera) on a medium containing salt solution, agar, trehalose, peptone, horse serum and embryonic chick extract (9 days). 67 per cent of the gonads of both sexes survived for at least 7 months. In male gonads, the spermatogonia multiplied, spermatocytes underwent meiosis and developed to spermatozoa (follicles of female gonads differentiated and ovocytes entered previtellogenesis).

Puberty, the threshold of sexual activity (sexual maturation being distinct from sexual differentiation), is initiated by pituitary *gonadotropic hormones* acting on genital receptors, prepared by impregnation with hormones and histological maturation. The hypothalamus does not play a definite role in males. However, in females, *experimental destruction of the mamillary bodies leads to precocious maturation of genital organs.* Hypothalamic implantation to the impubescent rat provides evidence for the existence of hypothalamic factors (GRF = *gonadotropin releasing factor*) which stimulate the secretion of pituitary gonadotropins (Moszkowska and Kordon, 1951).

Injection of *chorionic gonadotropins* also affects sexual maturation. This effect is the basis of a pregnancy test using batracians (Galli-Mainini, 1947). Injection of urine from a pregnant female to male toads or frogs results after several hours in the appearance of some spermatozoa in the cloaca of experimental animals.

The role of the pituitary in sexual maturation has been demonstrated.

A brephoplastic graft of hypophyso-pituitary complex beneath the *tunica albuginea* of the atrophied testis of a hypophysectomized impubic mouse produces spermiogenesis (R. M. May, 1955).

Huignard (1957) examined the fate of intraocular brephoplastic grafts of genital glands of mice. Testicular grafts were the most easily obtained. Spermatogenesis was completed when small volumes of testis were implanted in the anterior chamber of the eye. *It should be noted that if the recipient is a castrated female mouse which has not received hormonal injections, interstitial tissue is rarely seen* (Huignard).

We shall now examine certain facts concerning *castration* and *sexual differentiation*.

Surgical castration, *in utero*, of rabbit embryos, before or during sexual differentiation (Jost, 1947), shows that, *in the absence of genital glands and after sufficiently early castration*, the *genital apparatus acquires feminine characters irrespective of the genetic sex of the embryo*. In males as in castrated females, the ducts of Müller develop into female ducts and the urogenital sinus and external genital organs exhibit feminine characters.

From these results it can be concluded that male hormones inhibit the metabolic processes which determine feminization, a dominant or non-differentiated phenomenon. Thus, the male sex chromosome contains genes which are potential inhibitors of morphogenesis.

Jost *unilaterally castrated* male rabbit foeti. In such a foetus the duct of Müller did not disappear and the duct of Wolff developed only on the side containing the testis. Thus, the androgenic hormone can diffuse only a small distance. The essential fact to remember is that, in rabbit at least, *the mammal is basically feminine in nature*.

This experimental technique has been extended to female foeti receiving a unilateral testicular graft on the side containing the ovary. In these animals the genital ducts *in the region of the testis* are masculinized.

Thus, androgenic activity is limited by the feminine nature of the internal medium.

Jost completed these experiments by determining the role of the pituitary.

Removal of the hypophyso-pituitary complex from a male rabbit foetus by the extreme method of *decapitation* results in abnormal differentiation of the genital apparatus due to *functional insufficiency of the testes*.

The internal structures most closely associated with the testis are masculine, the prostate is rudimentary and the external genital organs are feminine. This state constitutes *masculine pseudohermaphroditism*. These testes are unable to influence these structures although this can occur, in spite of decapitation, when gonadotropic hormones are administered to the foetus (Jost, 1951). From these facts, the importance of impregnation of genital receptors by steroid hormones and gonadotropins in the realization of normal genital development is apparent. Here we have an example of physiological synergism with morphological consequences.

The extent of the female influence on differentiation is demonstrated by the experiments of Wolff and Haffen. Direct parabiosis of two undifferentiated but genetically opposed gonads results in *normal differentiation of the female gonad to an ovary while the male gonad develops into an intermediate structure which more closely resembles an ovary than a testis* (ovotestis). Thus, the ovarian secretion feminizes the male partner. This effect must be attributed to a female hormone. However, this is not just a *qualitative problem* since Hampe, a student of Wolff, showed that injection of low doses of oestrogen benzoate on the eleventh day of incubation stimulated the testicular germinal epithelium of chick embryos. Higher doses produced *cortical* proliferation. Thus, as stated by Wolff; "This is not a question of an ambisexual hormone but of a female hormone which under certain conditions of dose and administration, can stimulate male characters". The action related to hormonal doses had been demonstrated by Butenandt in 1931.

The effects of sex hormones on genital differentiation can be seen more directly. A. Raynaud (1939, 1947) found that injection of *female hormones* to gravid mice females produced *feminization* of male embryos by a direct action on the testis leading to degeneration of interstitial tissue. The development of the male genital tract would be initiated in the normal male embryo by the action of a male hormone secreted by the *embryonic* testis. Androgenic compounds have been extracted from testes of bull foeti (Womack and Koch, 1931) and horse foeti (Catchpole and Cole, 1934). However, experiments of R. Stoll on birds lead to the conclusion that *embryonic male hormones are not exactly identical with known androgens.* Although high doses of androgens such as androsterone and testosterone produce agenesia of the ducts of Müller with the exception of the proximal segment, testicular grafts produce *complete regression.* Evidence for some difference has also been obtained from freemartins. When a cow gives birth to bivitelline twins of opposite sex, the male calf is normal but the female calf shows normal external genital organs but degenerate and sterile ovaries. This female is a freemartin. Lillie (1916), Keller and Tandler (1916) consider that testicular hormones from the male twin influence the sister. In cattle, mice, rats, rabbits, etc. (Jost, Chodkiewicz and Mauléon, 1963), administration of androgens to the gestating mother does not produce a freemartin. There is no ovarian inhibition but complete masculinization of the external genital organs is seen. Thus, the embryonic hormone has a greater effect. However, is the embryonic hormone a different hormone or the nascent form of a known hormone (nascent oxygen, for example, being more active than molecular hormone) acting on organs which are themselves "nascent" or more "adapted" to the conditions imposed?

To recapitulate on the facts concerning the difficult problem of *hormonal* differentiation of sex, one can consider the relatively simple case of lower vertebrates. The facts are summarized in Table 47 (Witschi).

Witschi considers that embryonic genital buds secrete "inductors" which are hormonal compounds but which differ from known sex hormones. The testes develop at the expense of primary proliferation of germinal epithelium which constitutes the *medulla* while the ovary develops from the secondary proliferation of the epithelium, the cortex. Thus, the predominance of the medulla or cortex would depend on secretion of differentiating substances by these regions: *cortexin* producing females and *medullin* producing males.

33

TABLE 47

Masculinizing agents	Feminizing agents
High temperature (Overmaturity of egg) Testicular transplantation Testosterone and androsterone on frogs High doses of oestrone and oestradiol	Low temperature Ovarian transplantation Testosterone in salamander Low doses of oestrone and oestradiol

Reciprocal inhibitory relationships must exist between these substances.

Among species of *Rana* and *Hyla*, daily administration of doses of testosterone as low as $1/500,000,000$ totally and permanently masculinizes the gonads (Gallien, 1937; Witschi).

In *Coris julis* Reinboth (1962, 1963) has described two morphological types of functional males, one of which is similar to females (primary male), whereas the other originates from spontaneous sexual inversion of old females (secondary male). The secondary male shows a distinct colour pattern which may be induced artificially by injection of testosterone to females as well as primary males. Histological examination of several gonads from treated animals, taken a few weeks or several months after the beginning of the experiment, seemed to show no definite signs of an experimental sex-inversion. Even in those animals with ovaries which had almost returned to normal the colour pattern remained the typical one of a secondary male.

Vitamins are important for sexual activity. Delost and Terroine (1955) found that post-natal development of the testes is *retarded* during *biotin* deficiency. Similarly, vitamin T of Goetsch (*termitin* of mushrooms and yeast ingested by insects) activates gametogenesis of insects and vertebrates. This vitamin is found in crude penicillin (Aschkenasy-Lelu, 1953). *Pyridoxine* deficiency retards weight increase of testes of young rats and during the period of slight pyridoxine deficiency secretion of male hormone is almost completely inhibited but spermatogenesis is not inhibited until the final stages of acute deficiency. Testosterone, injected during acute deficiency can restore the weight of the seminal vesicles provided that the duration of treatment is prolonged (Delost and T. Terroine, 1961, 1963). B_1 deficiency in young rats also retards weight increase of testes, the appearance of spermatogenesis and secretion of interstitial glands. Although this inhibition is probably not specific, it cannot be explained entirely in terms of poor nutrition. Ascorbic acid protects against B_1 avitaminosis (Delost, H. Delost and T. Terroine, 1958).

Deficiency in vitamins E, A, B_1, B_2, B_6, B_{12}, C, M, FF and H affects the whole genital tract. In males in particular, A or E deficiency damages the germinal chain while water soluble vitamin deficiency affects the whole male genital apparatus and the interstitial gland (Delost, 1961). It should be noted that wild, nonhibernating mammals exhibit seasonal variations in peripheral endocrine glands: gonads, thyroid and adrenal gland. Temperature, light, population and nutrition (the vitamin content of plants varying with season) are the principal external factors concerned with these endocrine cycles (Delost, 1960). There is also a relationship between nutrition and sexuality in fish. In salmon, ascorbemia falls during spawning to a third of the value at the beginning of migration (Fontaine et al.) and in *Clupea harangus*, fertility is directly related to lipid content (Anokhina, 1959). In general terms, certain fish cease feeding during spawning (the mullet, *Mugil cephalus*), others lose their appetite at this time while others continue to feed (*Cottus gobio*). In mullet, fasting is related to the formation of genital products, and Mislin (1941; reported by Fontaine) called this phase "*synchronous* fasting". Synchronous fasting differs from true fasting in that in *Salmo salar* the muscles of the tail and trunk are equally affected during true fasting but during synchronous fasting the muscles of the tail show little change and the lateral muscles of the body are concerned with autonutrition.

Thus, there is a direct correlation between trophic activity and androgenic glands.

In any way, these phenomena must be explained in terms of enzyme systems and Table 48 indicates the enzymes influenced by androgens.

Finally, in invertebrates removal of gonads does not inhibit the appearance of somatic sexual characters (smaller size of males, bright coloration of males, etc.). In crustaceans, this phenomenon can be explained by the existence of a separate androgenic gland which was discovered by Mme Charniaux-Cotton and in some ways represents a condensed version of the hormonogenic interstitial cells. (In *Maia squinado* this gland can measure 3 cm in length.) The main experiments of Mme Charniaux-Cotton were carried out on the small amphipod *Orchestia gammarella* by ablation and implantation. Masculinized females always remain sterile since, although an efferent duct develops, the duct has no aperture. However, Mme Charniaux-Cotton succeeded in fertilizing eggs with sperm from gonads of females masculinized by androgenic grafts. Thus, from these experiments embryos were obtained from two individuals which were genetically of the same sex. The androgenic gland which secretes hormones which have not been identified (commercial steroids are inactive) is separate from the gonads in the adult but during development is associated with the genital apparatus.

Sexual development is not only associated with the genital organs. There is a masculine "terrain" just as there is a feminine "terrain" (as shown

33*

TABLE 48. *Influence of androgens on "effective enzyme concentration of various tissues"* (R. I. DORFMAN, 1961)

Enzyme system	Species and (conditions)	Influence on tissue concentration	Reference
β-Glucuronidase	mice (male and female)	increase in kidney. No change in liver or spleen	Fishman (1951) Riotton and Fishman (1953)
D-Amino acid oxydase	mice (castration)	decrease in kidney but not other tissues	Clark *et al.* (1943)
	mice (castration + androgen)	increase in kidney	
Arginase	mice (castration + androgen)	increase in kidney but not liver or intestine	Kochakian (1947)
Zymohexase (aldolase)	rat (castration)	decrease in ventral prostate	Butler and Schade (1958)
	rat (castration + testosterone)	up to 25-fold increase in ventral prostate	
Succinic acid dehydrogenase	rat (castration)	decrease in prostate and seminal vesicle but not in perineal musculature	Davis *et al.* (1949) Leonard (1950)
	rat (castration + androgen)	increase in prostate and seminal vesicle but not in perineal musculature	

earlier by Joyet-Lavergne, 1935). Stoll and Maraud (1956) found that from the 16th day of incubation of chick the adrenal glands show sexual dimorphism. There is a relative increase in the chromaffin region of males. This differentiation suggests an action of testicular androgens and can be induced by testosterone administration.

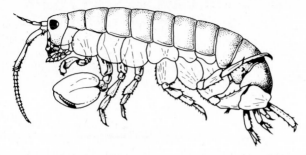

FIG. 142. Male individual of *Orchestia gamarella*. Note the development of the propodite of the 2nd gnathopode which forms the claw. Enlargement: ×7. (CHARNIAUX-COTTON.)

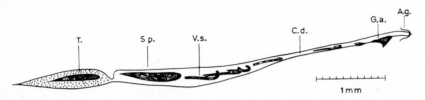

FIG. 143. Schema of the male genital organ of *Orchestia gammarella* with the androgen gland. *T*, testis; *Sp.*, spermatozoids; *V.s.*, seminal vesicles; *C.d.*, deferent canal; *G.a.*, androgen gland; *A.g.*, genital apophysis. (H. CHAR-NIAUX–COTTON.)

Effects of castration

(1) In the *adrenal gland* castration results in retention of zone X.

(2) Castration finally produces *hypothyroidism*. Injection of androgens restores thyroid activity most probably via the hypothalamo-hypophyso-pituitary complex.

(3) Castration is known to produce an increase in the size and number of the *pancreatic* islets of Langerhans (Konevskaïa, 1919) and, according to Champy, the exocrine pancreas degenerates and is invaded by adipose tissue. This latter reaction is delayed, occurring a year after castration. *Specific* and *non-specific stress* leads to polymacronesia of the islets of Langerhans and more specifically, stress of the pelvic sympathetic system results in similar changes to those seen in castrated animals (as shown in guinea pig by Théret in 1953).

(4) The reactions of the *pituitary* are well characterized. The typical structure has been incorrectly termed the "hypophysis of castration" (Fischera, 1905) and should be termed the pituitary of castration. The *basophilic cells hypertrophy* several weeks after castration and contain large vacuoles. At this time, FSH (gonadotropic follicle stimulating hor-

mone) is abundantly secreted. It should be noted that the hormone is not localized in the vacuoles but is synthesized by the thin cytoplasmic layer between the membrane and vacuole. Abolins-Krogis (1952) found that the *maximum activity* of pituitary alkaline phosphatase in castrated guinea pig coincides with the *maximum degranulation of acidophilic cells* and with *marked hyperchromasia of basophilic cells.*

Androgens and senescence

Since the results of Brown-Séquard (1889) concerning sexual rejuvenation induced by testicular extracts, endocrinology of senescence has rapidly developed. Pincus (1955) reported that urinary 17-ketosteroids of females appeared to be derived almost exclusively from adrenal precursors since there was no significant decrease in urine content after ovariectomy. Thus, the difference between the urinary 17-ketosteroid content of females and males of the same age can be considered to provide an estimate of the amount of 17-ketosteroids derived from the testes. The results of Pincus suggest that the urinary ketosteroids of *testicular origin* and from *adrenal precursors* decrease with age. Zone X must be considered in this respect. Pincus wrote: "If the defects in the production of neutral steroid hormones in aged persons are to be rectified it would be necessary to restore the defective anabolic enzyme system". Unfortunately, we have not reached this stage and Pincus has considered a therapeutic substitute designed to increase the excretion of 17-desoxysteroids in aged subjects.

A group of 14 men aged between 70 and 87 years received *per os* a mixture of testosterone, adrenosterone, hydrocortisone and corticosterone. After two months each subject performed a test which consisted of raising two weights of 10 kg to shoulder height at intervals of 5 s. *Those which were least able to raise the weights showed lower excretion of 17-ketosteriods than those subjects which were best able to perform the test.*

This is certainly not the realization of the dream of Brown-Séquard or Faust but these results indicate a possible means of improving the deficiencies induced by changes in steroidogenesis during senescence.

II. Physiology of the female genital system

The *ovary* is the *principal* functional unit of the female genital system. As in the case of the testis, the ovary is both an *exocrine gland* (producing cells, the ova) and an *endocrine gland* (producing oestrogens). The ovary resembles the pituitary in that, in mammals at least, it is a *multiple endocrine gland* since *progesterone* is produced at certain stages.

Anatomy

The early development of the ovary is the same as that of the testis but the medullary cords regress and secondary proliferation of the germinal epithelium produces the cords of Pflüger which become organized into the active female zone, the *cortex*. From this one fact concerning the organogenesis of the ovary we could consider *feminization to be inhibition of masculine potentialities*.

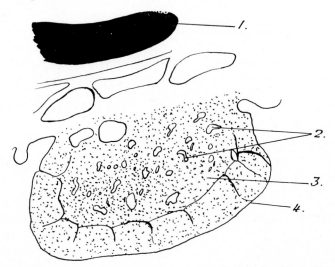

FIG. 144. Section of the left embryonic ovary of chick (18 days). *1*, Vein; *2*, Medullary cavities; *3*, Medulla; *4*, Cortex with sex cords.

The cords of Pflüger fragment to give the *follicles*, each of which contains a germinal cell (ovocyte) surrounded by epithelial cells. Each follicle grows by proliferation of epithelial cells or *follicle cells* which are localized around the edge of the ovocyte in a *thick granulosa* which is resorbed centrally to form the follicle cavity filled with fluid. The whole structure is the *Graafian follicle* which reaches a size of 12 mm in diameter in woman.

The adult ovary contains as many ripe follicles at the surface as there are young per litter (one in woman, sometimes two, rarely three in the two ovaries). After rupture of the follicle the ovocyte passes into the *general cavity* accompanied by several adherent follicle cells. This stage is *ovulation* and results in the formation of a *corpus luteum* which takes the place of the follicle and is produced by multiplication of the cells of the granulosa.

In the embryonic buds of vertebrates there are two pathways associated with the elimination of genital products: a pair of *Wolffian ducts* (which will produce epididymis and vas deferens) and a pair of *Müllerian ducts* (which will produce the infundibulum, oviduct, uterus of mammals and vagina). The Wolffian ducts regress in the female.

The results of Martinovitch (1938) from rat and mouse show that ovaries removed at the 16th day of gestation and cultivated on a medium developed by Fell and Robinson (1929) continue their development. Wolff (1952) transplanted embryonic gonads of mice to the standard medium of Wolff and Haffen and found that these organs differentiated and developed in culture. B. Salzberger (1962) demonstrated that embryonic ovaries of mice removed after the 14th day of gestation and 2 or 3 days after birth

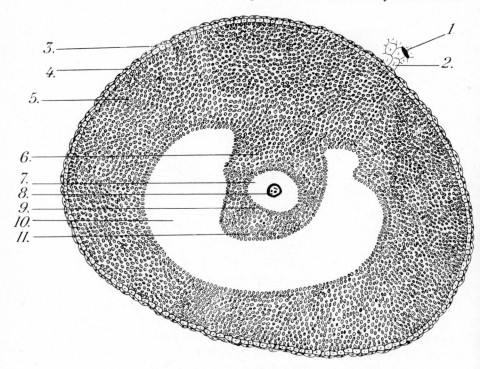

FIG. 145. Section of a ripe follicle of mouse. *1*, vessel; *2*, connective; *3, theca externa*; *4, theca interna*; *5*, granulosa consisting of follicular cells; *6, cumulus oophorus*; *7*, ovocyte; *8*, nucleus; *9, corona radiata*; *10, liquor folliculi* in the *antrum*; *11*, proligerous disc.

can continue normal development in a foreign organism such as the coelom of young chick embryos. In this case the host provides the culture medium. However, reactions can occur between the associated chimeric structures: (a) in the combination of mouse ovary–chick ovary, the two organs are intimately associated to form chimeric organs (the gonad of mouse containing numerous ovocytes and primary follicles occupies the medullary part of the genital gland of chick), (b) in the combination of mouse ovary–chick testis, development of the mouse ovary is inhibited, most germinal cells are destroyed but certain "medullary" cells remain. The *rete ovarii*, appearing as sinuous cords, develop well within chick tissue and form mixed structures with the testicular tubes. Thus, male hormone secreted by the chick testis has an inhibitory action on development of mouse ovary (compare with p. 504).

In some species such as rat, mouse, hamster and mole, the differences between the external genital organs of male and female are less pronounced than in such species as cat, monkey and man (Marshal, 1956). In rat, mouse, hamster and mole, the urinary orifice is at the apex of the genital tubercle which is traversed by the urethra in both sexes. The vaginal orifice is at the base of the clitoris. However, in other animals the urethra does not pass through the clitoris which is little developed. The urinary orifice opens in the same region as the vaginal orifice, in a vestibule (cat, monkey, man) or into an ano-genital depression (guinea pig). In rabbit the vagina opens into the urethra which in the terminal section forms a urethro-genital duct (Jost, 1943).

The genital tubercle of the white rat is formed during the second half of foetal life, the genital tubercle having a similar structure in both sexes up to 18 days. In the penis, the urethra is surrounded by the balano-preputial fold and from the 19th day is found within the genital tubercle. By the 21st day the urethral penis is complete and the urinary orifice is situated at the end of this organ. In the clitoris the urethra is external to the balano-preputial fold. The latter is much less well developed than in the male even in the foetus at term. The urinary orifice remains at the base of the clitoris at birth and does not develop into a clitoral urethra until after birth.

In the female, formation of the urethra, invagination of the balano-preputial fold and fusion of the lower edge of the prepuce occur more slowly than in the male.

In the female foetus, androgens can modify development of the balano-preputial fold and closure of the urethral canal (Dantchakoff, 1938, in guinea pig; Raynaud, 1944, Kerkhof, 1952, in mouse; Greene et al., 1937, in rat). Under these conditions, the external genital organs of the female at birth can have the same appearance as the male penis, if androgen stimulation occurs at an early stage. If stimulation is delayed, until after 19 days, the time of sexual differentiation of the genital tubercle, the clitoral structures no longer respond or respond incompletely to androgens: the organ acquires a structure intermediate between the penis and clitoris. Thus, sexual differentiation of the female genital tubercle into the clitoris is irreversible after foetal life (Lacassagne and Raynaud, 1937; Greene et al., 1938; Turner et al., 1939; Selye, 1940; M. G. Moreau, 1963).

Main aspects of ovarian function

Chemistry of oestrogens

The *follicle hormones* or *oestrogens* are secreted by the cells of the Graafian follicle. *Oestradiol* is the principal hormone produced by these cells. Oestradiol exists in two forms, α and β, according to the position of a hydroxyl group. *The α form is most active and found in urine of pregnant mare, while β oestradiol is found in urine of pregnant woman* (diol, since the molecule contains two alcohol groups). *Oestriol* is the principal oestrogen of urine of pregnant women and of the placenta. Oestriol and almost certainly *oestrone* are formed by oxidation of oestradiol. *Transformation of androgens to oestrogens has been demonstrated in ovary, placenta and adrenal gland.*

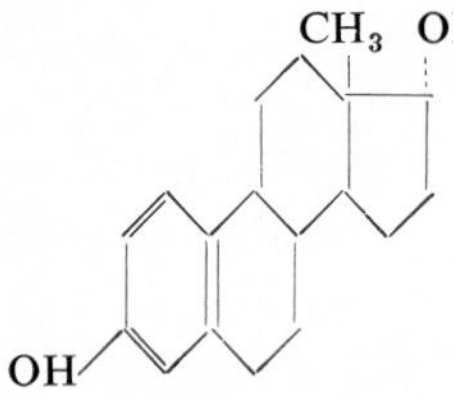

CH$_3$ OH

OH

α-Oestradiol

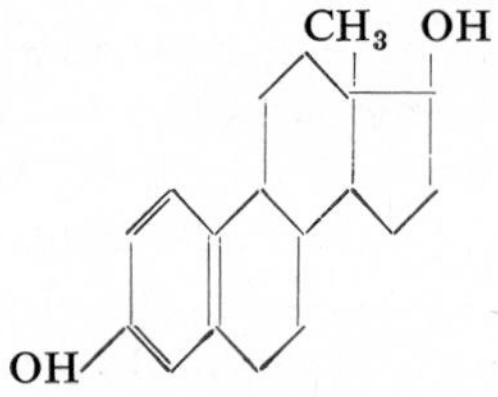

CH$_3$ OH

OH

β-Oestradiol

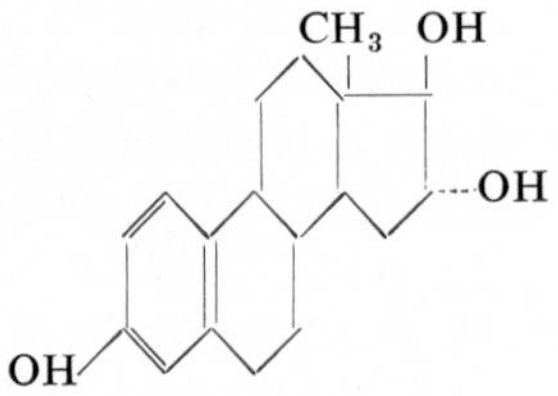

CH$_3$ OH

···OH

OH

Oestriol
(Marrian, 1930)

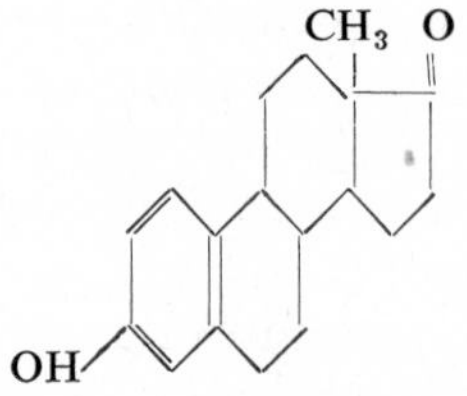

CH$_3$ O

OH

Oestrone
(Butenandt, 1929; Laqueur, 1930)

Oestrone has been found in palm-nuts and oestradiol in female catkins of willow. Thus, oestrogens are not localized only in animals and this fact raises certain problems concerning comparative biochemistry. Furthermore, the Stannius corpuscles of Atlantic salmon contain oestrogens and 17-ketosteroids. Although the oestriol content of the Stannius corpuscle of females is scarcely doubled during the period from the beginning of anadromous migration to spawning and the androgen content remains unaltered, the androgen content increases tenfold in the Stannius corpuscle of males (Cédard and M. Fontaine, 1963). Fontaine considers that these facts confirm the relationship established between the Stannius corpuscles and the adrenal cortex (interrenal gland). It has been suggested that these structures have a common origin, the "coelomoducts" (Willmer, 1960).

Allen and Doisy (1923) were the first to show the presence of oestrogenic activity in the follicular fluid and Doisy (1929) prepared crystalline folliculin from ovaries of the sow.

Other oestrogens have been found. Zondek (1923) demonstrated that urine of gravid mares and of *stallions* contained large quantities of oestrone (1 kg/700,000 kg urine). In 1932, Girard isolated two hormones from the urine of gravid mares. These were *equilin* and *equilenin* which are not present in human urine.

Epuilin has one more double bond than oestrone and equilenin is even more unsaturated. Equilin and equilenin exhibit weak oestrogenic activity.

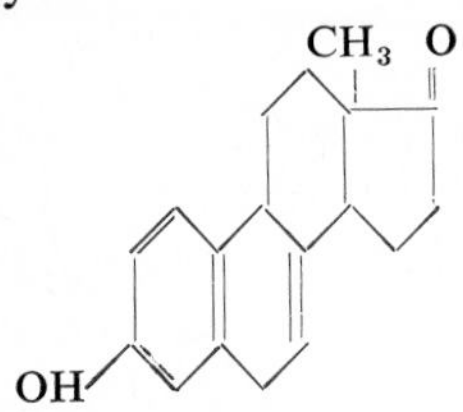

CH_3 O

OH

Equilin

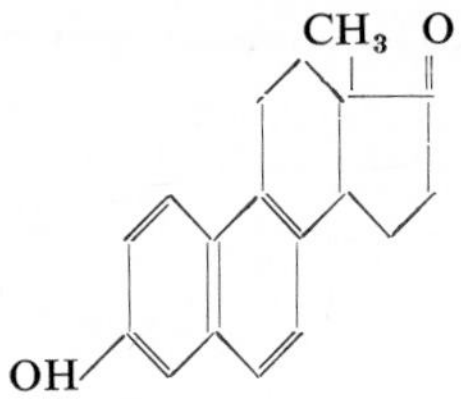

CH_3 O

OH

Equilenin

Among these compounds, oestradiol and oestrone must be considered as the typical ovarian hormone of oestrus: *folliculin = oestradiol + oestrone*. The other compounds are derived from the placenta. But the testis of bull and stallion produce folliculin (in the same way that *androgens* are found in ovary). *Oestrogenic activity is associated with numerous compounds which are chemically unrelated.* Heat can be induced in female rat by vitamin D, diethylstilboestrol, *pp*-dihydroxy-diphenyl, *p*-hydroxy-propenylbenzene. Diethylstilboestrol prevents frigidity in women and females of domestic animals.

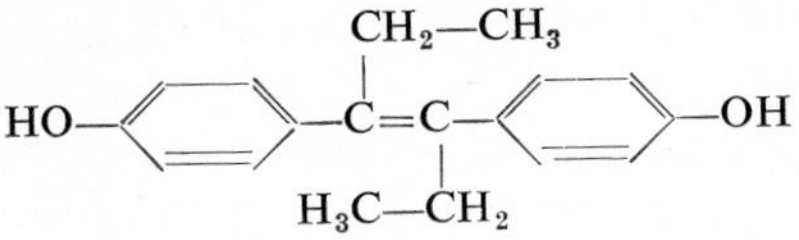

CH_2—CH_3

HO— —C=C— —OH   Diethylstilboestrol
(same activity as oestradiol)

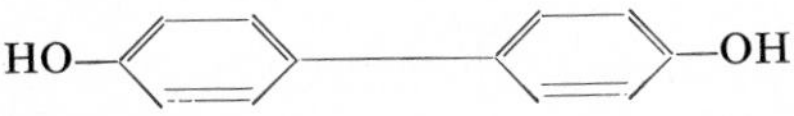

H_3C—CH_2

HO— — —OH   *pp*-Dihydroxy-diphenyl

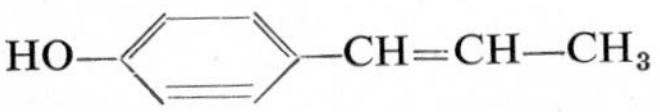

HO— —CH=CH—CH_3 *p*-Hydroxy-propenylbenzene

The similar effects produced by unrelated molecules can be explained in several ways. The compounds may have an *indirect* oestrogenic action or the *receptor* molecules may be more determinant than the excitatory molecules or, in the case of oestrogenic activity, molecular determinism may not be limited and the key-lock linkage may need to be considered to be of the masterkey-lock type. These problems are of particular significance since they are related to that of cancer.

The question of synthetic oestrogens has particular interest in that, when the English chemists Cook and Dodds synthesized oestrogens for the first time, they had, in fact, started out on an entirely different problem.

Cook and Dodds were studying the effects of carcinogenic hydrocarbons on castrated mice. They found that these compounds produced *hypertrophy* of the *mammary gland, vagina* and *involuted uterus*. Although mg quantities of these hydrocarbons were required to produce these effects compared with 1/1000 mg oestradiol, this work led to the synthesis of diethylstilboestrol, a highly active oestrogen.

In 1944 the Swiss chemist, Miescher produced an active oestrogen which, in acknowledgement to Doisy, he called *bisdehydrodoisynolic acid*. In 1949, Courrier, Horeau and Jacques prepared a crystalline compound having all the properties of folliculin but the additional property of being active when administered *orally*. They called this compound *dimethylethyl-allenolic acid* after Allen.

Bisdehydrodoisynolic acid   Dimethylethylallenolic acid

Some physiological actions of oestrogens in the embryo

(1) Chick embryo

(a) During normal ontogenesis, the genital tract of the vertebrate embryo differentiates according to sex which is determined genetically. Differentiation is brought about by the action of inductors of *unknown structure* (cf. p. 505). It has been suggested that steroid hormones related to genital hormones of the adult may be important. Ribouleau (1939) detected oestrogens in the fertilized eggs of chickens. This would appear to indicate that active molecules are formed before morphological differentiation. ("The function creates the organ" or at least, precedes it.) During the first days of incubation the oestrogen content decreases but increases in the male embryo during the 12th to 16th day. Stoll and Maraud (1956) found that from the *beginning of sexual differentiation* at $6\frac{1}{2}$ days the *chick embryo eliminates neutral 17-ketosteroids*. The existence of excretory derivatives implies the existence of circulating steroid hormones although the nature of the oestrogens has yet to be determined.

(b) Androgens, such as testosterone (in the form of propionate) produce regression of Mullerian canals (according to Wolff *et al.*, 1935—1948; Williers *et al.*, 1935; V. Dantchakoff, 1935). Involution of the Mullerian ducts of male chick embryo is induced by male hormones *in situ* and in organ cultures (Wolff, 1936, 1952). Testosterone propionate activates phosphatase of undifferentiated Mullerian ducts of chick embryo but oestrone and oestradiol benzoate also have this effect (Scheib, 1962). De Duve (1955, 1959) found that acid hydrolases of various tissues are localized in specific intracellular organites called lysosomes. Scheib-Pfleger and Wattiaux (1962) consider that, in the Mullerian canals, hydrolases (acid ribonuclease, β-glucuronidase, etc.) are localized in lysosomes and that these enzymes are responsible for the autolysis which characterizes Mullerian regression. However, that androgens induce *regression* of Mullerian ducts (natural process during sexual differentiation) is questionable: acting on chick embryo androsterone and dehydroandrosterone (Gaarenstroom, 1939), testosterone (Stoll 1948–1950; Stoll and Maraud, 1952) produce *agenesia* of Mullerian ducts. Thus the active substance or interacting substances, which lead to normal regression of these ducts seem to be different from testosterone (cf. p. 505).

(2) Embryonic development in sea urchin (Psammechinus miliaris and Paracentrotus lividus)

I. Agrell (1955, 1956) found that oestradiol in concentrations lower than 10^{-6}M *inhibits cellular division of embryos* and stated that oestradiol modifies the synthesis of desoxyribonucleic acid. It is of interest to relate this observation to the relationships between cortisone and desoxyribonucleic acid (cf. p. 431). Testosterone was also found to be inhibitory. This antimitotic action of oestradiol is particularly interesting when one considers other oestrogenic compounds such as the *carcinogenic hydrocarbons* and that, according to Agrell, the action of oestradiol is to potentiate the release of desoxyribonucleic acid from protein by desoxyribonuclease.

Physiological action of oestrogens in adult organs and organisms

Oestrogens change the leucocyte equilibrium. At the start of prolonged oestrogen treatment there is leucocytosis followed after a certain period by leucopenia (Girod and Kehl, 1959).

Action of oestrogens on the ovary and female genital tract

Ovarian insufficiency, either naturally occurring or experimentally induced by X-rays or surgical procedures, leads to regression of genital and mammary structures with uterine atrophy, permanent dioestrus, reduction of mammary glands (genital infantilism). However, if a small part of the gland remains, an almost normal physiological state is maintained by compensatory hypertrophy. This deficiency can be overcome either by administration of female hormones or by ovarian grafts. Ovarian grafts are implanted in renal tissue, in the eye (Markee) or in *mesentery* as practised by Mayer and Soubireau (1948). However, bilateral mesenteric grafts cannot maintain the receptor organs (uterus, vagina, etc.) since, although the transplanted ovaries show hypertrophy, they are situated in a region drained by the portal vein which contains blood of a particular composition. The receptor organs are maintained by *ovarian grafts situated in subcutaneous and muscular tissue.*

The concept of a receptor, which is of great importance in endocrinology, has to be examined: in the chain of events the *inductor organ*, the *chemical messenger* (or messengers) and the *receptor* or *effector organ* (or organs) must be considered.

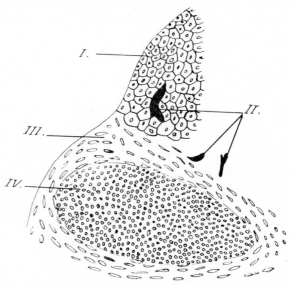

FIG. 146. Section of an ovarian segment of mouse with corpus luteum and follicle. *I*, luteal cells of corpus luteum; *II*, vessels; *III*, connective; *IV*, follicle.

Normal function depends on the integrated reactions of these three systems, a deficiency in one system leading to pathological, dangerous conditions. Hyperfunction of an endocrine gland (such as pituitary) does not produce the usual symptoms when it is coincident with dysfunction of hormonal transport or obliteration of receptors. Oestrogens which counteract the effect of castration, in the normal adult female generally lead to a series of hyperfunctional reactions: hypertrophy and hyperplasia of the uterus which, in the presence of oxytocin, shows regular contractions although oxytocin in the absence of folliculin induces irregular contractions (Brouha and Simonnet, 1927). Oestrogens also induce *congestion of nasal mucosa*, production of *breeding adornment* in certain fish, etc. Through an action on the ovary, oestrogens can lead finally to a form of *lymphoid degeneration.*

In general terms, oestrogens interrupt the sexual cycle by tending to produce permanent oestrus through prolongation of the follicular stage. As a result, oestrogens awaken sexual activity in castrated females or stimulate sexual libido in normal females. Furthermore, oestrogens inhibit the maternal instinct. Thus, oestrogens play an important role in determining behaviour.

In considering the action of oestrogens at the molecular level, Hagerman and Villee (1953) demonstrated that oestradiol increases the rate of utilization of pyruvic acid, an effect which is related to stimulation of isocitrate dehydrogenase. Talaly and Williams-Ashman (1958) found that oestradiol stimulates transhydrogenation between coenzyme I and coenzyme II:

$$NADPH_2 + NAD^+ \rightleftharpoons NADP^+ + NADH_2$$

Two oestradiol dehydrogenases have been described, one of which stimulates the reaction:

$$NAD^+ + oestradiol \rightleftharpoons oestrone + NADH_2 + H^+$$

the other catalyses the reaction:

$$NADP^+ + oestradiol \rightleftharpoons oestrone + NADPH_2 + H^+$$

In addition the uptake of amino acids by slices of placenta incubated *in vitro* is stimulated by oestrogen and the rates of synthesis of protein and nucleic acids by placenta slices is stimulated by oestrogens (Hagerman and Villee, 1960, 1961).

The oestrus cycle

Certain animal species show *seasonal* sexual activity (for example, batracians). However, in primates notably, sexual activity is *continuous* and characterized, in mature females, by a relatively short ovarian cycle.

In woman for example, *from birth* the ovary contains hundreds of thousands of primordial follicles but only about 400 reach maturity during the period from puberty to menopause, follicular maturation occurring once every 28 days.

The corpus luteum takes the place of the follicle and during gestation the corpus luteum is maintained [the functional significance of the corpus luteum will be examined later (p. 549)]. If gestation does not occur the corpus luteum disappears after about 12 days and menstruation follows. In the absence of gestation ovarian cycles follow regularly without interruption until menopause.

Each cycle consists of

(1) *a follicular phase or follicular maturation;*
(2) *ovulation;*
(3) *a luteal phase during which the corpus luteum functions;*
(4) *regression of the corpus luteum.*

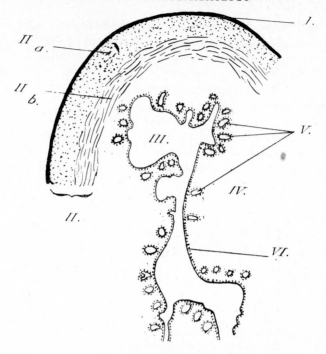

Fig. 147a. Transverse section of rabbit uterus in anoestrus. (Note the small number of endometrial acini). *I*, connective theca; *II*, myometrium; *II$_a$*, layer of longitudinal muscle fibres; *II$_b$*, layer of circular muscle fibres; *III*, lumen; *IV*, chorion; *V*, "glands"; *VI*, prismatic epithelium.

In female rats the cycle lasts only 5 days. Furthermore, numerous follicles mature simultaneously and many ova are released. Since all these ova could be fertilized, *polyembryogenesis* is normal in rat.

In primates the oestrus cycle and menstrual cycle are directly related, menses ceasing after ovariectomy. *The duration of the cycle is variable* (Table 49).

TABLE 49

Primate	Duration of menstrual cycle (in days)
Gorilla	45
Chimpanzee	33–36
	(ovulation on the 18th day; Yerkes)
Orang-Outang	29
Rhesus macaco	28 (as in woman)
Cynocephalic macaco	36

The cycle consists of: *oestrus* or "heat" (heat preceding ovulation), *pro-oestrus* corresponding with the follicular phase, *post-oestrus* corresponding with the progestation phase and finally, *anoestrus*, the period of sexual inactivity (Heape, 1900).

Dioestrus is the resting phase.

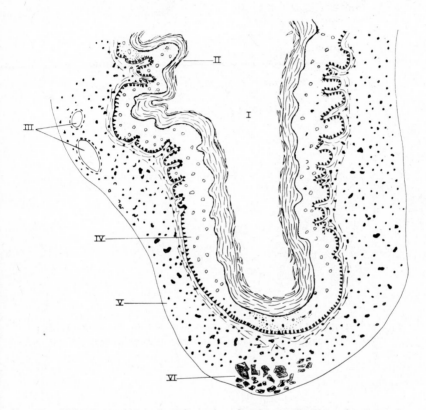

FIG. 147b. Transverse section of a vagina of rodent at the time of oestrus. *I*, lumen; *II*, keratinized lamella cells showing desquamation; *III*, vessels; *IV*, epithelium; *V*, chorion; *VI*, adipose formations.

A single injection of 2·5 mg of testosterone propionate during the first few days after parturition can provide definite sterilization in the female rat. The ovaries remain in the follicular phase and no *corpus luteum* is formed during the period of genital activity. The animals present a *permanent oestrus* and there are important endometrial changes. The lesions are not irreversible: the ovaries can be luteinized, the endometrium and the mammary gland can respond to progesterone and the pituitary of these animals, under the action of reserpine can secrete prolactin. The hypothalamus seems to be involved in the process of hormonal sterilization

34

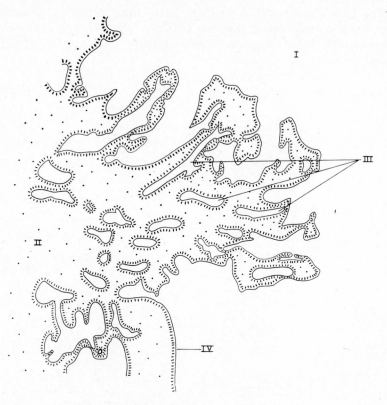

FIG. 147c. Detail of the uterine endometrium (rabbit in oestrus). *I*, lumen; *II*, chorion; *III*, "glands"; *IV*, prismatic epithelium.

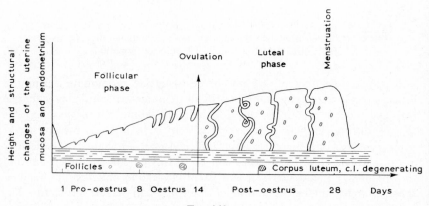

FIG. 148

(Selye, 1940; Bradbury, 1941; Tagasugi, 1952; Barraclough, 1954–1961; school of G. Mayer, 1962–1966).

The cycle is characterized by anatomo-physiological changes which are summarized in Table 50.

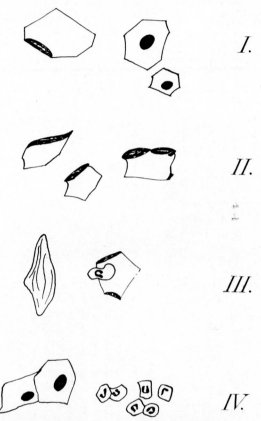

FIG. 149. Vaginal smears (♀rat). *I*, Pro-oestrus: Nucleated malpighian cells and anucleated keratinized cells; *II*, Oestrus: Anucleated keratinized cells showing folds; *III*, Metaoestrus: Keratinized cells with polynuclears; *IV*, Dioestrus: Epithelial cells and polynuclears surrounded by mucus.

The results of Hamilton, Widnell and Tata (1967) show that cyclical variations in uterine cell mass and population is most markedly obvious both in the high content of cytoplasmic RNA and protein and in the nuclear capacity to synthesize RNA at oestrus which then declines, to possibly the lowest levels of the cycle, at dioestrus.

The factors determining menstruation and the significance of menstruation have been studied in the female *Rhesus* monkey in which the duration of the cycle is the same as in woman.

34*

TABLE 50

Events of the oestrus cycle in woman

Phase	pituitary	ovary	uterus	vagina
Pre-ovulation	Secretion of FSH	The ovocyte begins maturation, a cavity develops in the granulosa, the theca interna thickens and thecal cells secrete folliculin	Repair of endometrium desquamated during previous menses; the mucosa thickens, the still rare "glandular tubes" are straight and the arterioles short and slightly sinuous	Rapid increase in the number of acidophilic cells (the total is about 50%)*
Ovulation	Simultaneous secretion of FSH and LH; the equilibrium of this secretion produces ovulation	The Graafian follicle is voluminous, folliculin and simultaneously a small amount of progesterone are secreted. Towards the 13th day the follicle ruptures, this is ovulation; the granulosa is invaded by vessels from the theca interna and transformed to a corpus luteum which begins to secrete large amounts of progesterone and folliculin. According to Gosselin (1957), ovulation is associated with a sharp fall in ovarian cholesterol.	The glandular tubes are hypertrophied, the cells full of glycogen, the arterioles elongate and are coiled	Presence of the maximum total of 60 to 80% of acidophilic cells which are large and flat. Mucus, leucocytes and microbial flora have almost disappeared.

Pre-menstru-ation	Secretion of LH and also of LTH	The corpus luteum synthesizes progesterone and β-oestradiol	The endometrium is thick, tubes are numerous and convoluted and the cells are rich in glyco-gen. Arterioles are numerous and wound round the tubes with which they must not be confused
Menstruation			The spiral arteries undergo more and more prolonged phases of constrictions leading to ische-mia of the surface of the muco-sa; this area breaks away and is eliminated with the blood of menses. The mucosa will finally be reformed from the deeper layers.

* The technique of vaginal smears which is of considerable diagnostic interest and commonly called the technique of Papanicolaou had French origins: from 1889, M. Morau recognized the existence of periodic changes in the epithelium of the vaginal mucosa of small rodents. In 1892, F. Lataste and Retterer confirmed this cycle and associated it with oestrus and ovulation. It was not until 1917 that Stockard and Papanicolaou described the vaginal smear technique for guinea pig and in 1933, Papanicolaou showed the existence of cyclic changes in the cellular structure of vaginal smears of woman and studied with Shorr the effect of various hormones (from pituitary, ovary and testis) on vaginal cytology in woman.

Phelps (1946) grafted uterine endometrial fragments to the eye of female *Rhesus* which had been previously castrated such that observations did not depend on the sexual state of the recipient. Administration of hormones modifies the terrain and typical changes associated with the oestrus cycle can be induced by administration of progesterone and folliculin. Progesterone shows a specific action on endometrial vascularization which would prepare ovum implantation.

Furthermore, after injection of large doses of oestrogens to female *Rhesus*, the uterine epithelium undergoes squamous changes (Kaiser, 1947). However, spiral arteries do not develop but with the *arrest* of oestrogen treatment menstruation occurs. This fact is all the more curious since, according to Schlegel (1949), uterine *arterio-venous shunts* are formed and these would be responsible for ischemia leading to desquamation. Schlegel used a solution of a fluorescent thioflavin compound to show the existence of vascular shunts. Little is known of the mechanism of menstruation. Prostaglandin may be important (Hall and Pickles, 1963).

Menstrual blood, the volume of which varies between 10 and 200 cm^3 in woman, *does not coagulate* even in the presence of thrombin. Champy and his co-workers consider that "menstrual blood is only a variety of serum containing erythrocytes". However, O and G. Smith (1945) stated that menstrual fluid contains a toxin (of the "necrosin" type of Menkin, a substance appearing with destruction of any tissue) and a *fibrinolytic enzyme* which could explain the lack of coagulation of menstrual blood.

The ovary and hypothalamo-hypophyso-pituitary complex

The existence of relationships between the genital glands and hypothalamo-hypophyso-pituitary complex has already been seen when considering the release of puberty and the "hypophysis" of castration (cf. p. 509).

Two facts form the basis of female sexual hormonology in mammals: (1) Allen and Doisy found that follicular fluid induces oestrus in castrated mice or impuberal rats as shown by vaginal smears; (2) Zondek and Ascheim found (by examination of vaginal smears) that pituitary implantation leads to ovarian development in impuberal rats.

The latter fact is of particular interest in this context. It would appear that the quantity of pituitary gonadotropic hormone *before* puberty is sufficient to model the female organs (soma) but insufficient to produce *maturation*. This maturation relates essentially to development of the follicle which implies *maturation of the ovocyte* (germen). Thus, we must consider the significance of this *cytological maturation*. Moricard (1933) showed that the *first mitosis of maturation in rabbit (formation of the first polar region) is released by gonadotropin*. Moricard (1940) confirmed this notable action which demonstrates the influence of the soma on germ cells. The soma represents the medium in which the more or less latent parasitism of the germ cells develops. In this respect, it is useful to recall the facts concerning *pituitary gonadotropins:*

(1) FSH (β cells) produces growth and maturation of follicles but this hormone *is not responsible for secretion of folliculin.*

(2) LH (γ cells) induces *secretion of folliculin* and ovulation with formation of the *non-functional corpus luteum* (no progesterone production). In the male, LH acts on the cells of Leydig. *LH is also secreted by the placenta* (chorionic gonadotropin).

LH and FSH act synergistically: LH can have an action only on the ovary which has been prepared by FSH.

(3) LTH induces secretion of *progesterone* by the corpus luteum which is then functional. LTH is important in milk release in rabbit and guinea pig notably.

FSH and LH are glycoproteins and show the specific histochemical reactions of these compounds such as the reaction of MacManus (1946) or P.A.S. reaction ("periodic acid Schiff"). Using these methods the *large basophilic cells* of the pituitary have been shown to secrete gonadotropins and the *small basophilic cells* to secrete TSH.

By homogenization followed by centrifugal fractionation (Claude, 1940) *acidophilic* pituitary granules can be isolated.

We shall now consider the *function of the gonadotropic system in the hypothalamo-hypophyso-pituitary complex.*

In certain mammals such as rabbit (Friedmann, 1920), cat and ferret *ovulation is released by orgasm.* The pathways of this reflex are neurohumoral and involve the hypothalamus. Everett (1949) stated that in a mammal with *spontaneous* ovulation, such as the rat, ovulation is conditioned by a neurohumoral reflex in which the hypothalamus is stimulated by oestrogens secreted by the ovary. Thus, there are no fundamental differences between spontaneous ovulation and induced ovulation; the differences would depend on the greater sensitivity of the hypothalamo-pituitary or hypothalamo-hypophyso-pituitary system in one case than in the other, that is, on different thresholds of action.

It should be remembered that vascularization of the pituitary differs from that of the hypophysis. The hypophysis is supplied *by the inferior hypophyseal arteries* while the *pituitary receives branches of the superior hypophyseal arteries* from the internal carotid. In the upper part of the pituitary stalk these "superior hypophyseal" arteries form capillary loops which become veins as they pass down towards the pituitary where they branch to form sinuses (portal system). *The neurovascular junction between the hypothalamus and pituitary is in the region of the capillary loops where numerous fibres of the hypothalamo-hypophyseal bundle terminate* (Herlant, 1954). This region has great physiological importance.

Lesions of the pituitary portal veins result in degeneration of the pituitary gonadotropic cells leading to genital atrophy. This degeneration appears to be related to the absence of a hormonal regulator of hypothalamic origin (J. Assenmacher, 1962). It should be noted that according to Assenmacher, the hypothalamus completely controls gonadotropic function but only partially controls thyrotropic and corticotropic function of the pituitary of duck. By systematic lesions in rat, J. Moll and W. de Jong (1963) delimited a hypothalamic thyrotropic region. Lesion of this area resulted in marked inhibition of thyroid activity. In larvae of *Alytes obstetricans*, ablation of the diencephalon produces an immediate loss of sensitivity to thyroxine; the earlier the ablation the greater the inhibition of complete metamorphosis (J. J. Bounhiol, P. Disclos and C. Remy, 1963). Furthermore, thyroid function is related to genital function. M. Fontaine (1961) found that a prolonged series of injections of thyrostimulating pituitary extracts to female eels produces an increase in the gonosomatic ratio, the diameter of the ovary and in the ratio of weight of pituitary to body weight, this increase being more marked at 20°C than at 13 or 16°C.

Kordon and Bachrach (1959) have shown that lesions produced by high frequency electro-coagulation in a region of the hypothalamus limited by the suprachiasmatic and ventromedian nuclei below and the paraventricular and dorsomedian nuclei above lead to ovarian hypertrophy with involution of corpora lutea and predominance of oestrus. In guinea pig, Mazzuca and Barry found that ovulation is suppressed by bilateral electrolytic lesion of the interstitial laterodorsal hypothalamic nucleus (ILDHN), a region which is thought to be involved in control of secretion of gonadostimulin B (LH). Everett (1961) stated that stimulation of the preoptic region of rat on dioestrus day 3 advances ovulation by approximately 24 h. R. D. Lisk (1960) has shown that regulation of FSH and LH secretion is mediated by neural mechanisms located in the arcuate nucleus in the case of males and in both the arcuate nucleus and mamillary body of females.

Thus, recent results support the concept advanced by Hohlweg and Junkmann (1932) of a *sex centre* in the hypothalamus.

Reserpine, an alkaloid of *Rauwolfia*, causes reversible inhibition of the oestrus cycle associated with regressive changes in the ovary and uterus. In the pituitary reserpine induces numerical changes in the basophilic cells which are probably associated with secretion of luteinizing hormone (Tuchmann–Duplessis and Mercier–Parot, 1960). It should be noted that the epiphysis shows marked anti-gonadotropic activity. Serotonin—in which the epiphysis is rich—may be one of the factors acting through the hypothalamus (Moszkowska, 1963).

These questions of inhibition of ovulation are related to the problem of contraception which has considerable demographic and moral significance.

The work of Pincus has led to the development of orally active compounds which inhibit ovulation; 19-norandrosterones are particularly active.

The relationships between the hypothalamo-hypophyso-pituitary complex and genital activity are also demonstrated by certain aspects of the sexual physiology of birds and particularly by *laying*.

First let us examine the essential features of the anatomy of the genital tract of birds.

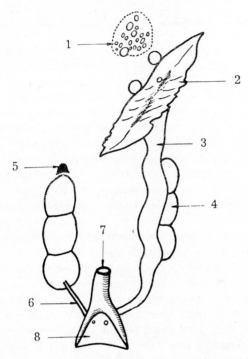

FIG. 150. Female genital organs (pigeon). *1*, Single ovary with ovisacs; *2*, fallopian tube; *3*, oviduct; *4*, kidney; *5*, adrenal capsule; *6*, ureter; *7*, intestine; *8*, cloaca.

Birds contain one ovary, *on the left*; the right ovary (potential testicle, J. Benoit, 1923) atrophies early in life. (In this respect attention should be drawn to the sexuality of the terrain, p. 504.) The left ovary has the appearance of a cluster composed of a large number of ovisacs which are either small and white or voluminous and yellow. The ovisacs, covered with cylindrical epithelium, contain the vitellus of the egg. At the time of maturation the ovisac splits and the eggs pass into the oviduct. The oviduct is capable of marked dilation. The upper region consists of the Fallopian tube and the lower region opens into the cloaca. Once in the oviduct the egg undergoes various changes. The *oviduct* is *not concerned only with the passive extrusion of eggs* but participates in the formation of accessory structures: "albumin", shell membranes, shell.

As in mammals, the pituitary gonadotropic complex of birds controls follicular growth and ovulation. Total hypophysectomy leads to regression of the left ovary and oviduct while *administration of pituitary extracts* of hens leads to secretion of androgens and oestrogens by the ovaries of immature hens. However, the chicken pituitary shows certain peculiarities. The pituitary of hen *is not inhibited by oestrogens* in contrast with the pituitary of mammals of both sexes and of cock. There appears to be continuous secretion of FSH in hen (Nalbandov, 1958). Another notable characteristic is the presence of progesterone (or a progesterone-like compound) in the blood of hens. This is unexpected and demonstrates the complexity of endocrinological problems in which *interconversion of sterols* may be involved. The presence of progesterone in a non-mammal seems to be an evolutionary mistake in the sense that during evolutionary development many trials succeed while many more are abortive; however, *injection of progesterone to hen produces ovulation*, a fact which is even more notable than the presence of progesterone in this animal. Ovulation is not induced by progesterone in totally hypophysectomized hens. Thus, ovulation requires a certain endocrine factor which could be LH (Nalbandov, 1958).

Let us now examine the processes taking place in the oviduct of domestic hen.

The formation of "albumin" is controlled by two hormones: *oestrogen* which regulates anatomical and glandular development of the oviduct and another hormone which leads to formation of "albumin" granules and which may be an *androgen* or *progesterone* which only has a positive action on an oviduct prepared by oestrogens. "Albumin" secretion is induced by the mechanical stimulation of the *presence* within the duct of a solid object: an egg but also a ping-pong ball, etc. The egg remains in the upper oviduct for 3 h and becomes covered with albumin. The egg then passes to the isthmus where the shell membranes are secreted and where the egg remains for $1\frac{1}{2}$ h. From the isthmus the egg passes to the region containing the shell secreting gland, calcification being completed in 22 h. The growth of this gland is dependent on oestrogens which also mobilize Ca^{++} from bones resulting in hypercalcemia. Parathormone is probably important in calcium deposition in the shell. In the domestic fowl, shell formation directly affects the acid–base balance of blood. During calcification there is a decrease in the level of bicarbonate ions in blood leading to compensatory ventilatory alkalosis with a decrease in the p_{CO_2} of blood which maintains the pH at a relatively constant value (P. Mongin and L. Lacassagne, 1964). Thus, the passage of the egg through the various parts of the oviduct occurs at different rates.

The effect of light on genital activity has been demonstrated repeatedly following the work of Bissonette (1930) on spermiogenesis. However, J.

Benoit (1935) has provided the most comprehensive account of this effect.

Artificial white light stimulates gonadal development in immature ducks. This effect is not inhibited by blinding the animals or by sectioning the optic nerves but *is prevented by total hypophysectomy.* These effects can be explained by photostimulation of the hypothalamo-pituitary region through the thin cranium of duck.

By exposing impubescent ducks to light passing through coloured filters, Benoit and Ott (1938) found that the *maximum gonadotropic effect was obtained with a red-orange filter.*

This problem has been considered by many authors and the effect of light has been confirmed in mammals, notably in ferret, rat, etc. Le Gros Clark, McKoewn and Zuckerman (1939) consider the primary receptor to be the *retina* which is of diencephalic origin. Ferrets *in which the optic nerve or visual cortex has been destroyed still show light sensitivity with respect to sexual activity.* For this reason, Le Gros Clark and his co-workers stated, in 1939, that the normal response of the pituitary to retinal stimulation depends on information passing to the hypothalamus via the accessory nerves of the optic tract. Light would be involved in liberation of GRF factor.

Sexual stimulation is not limited to illumination and can be produced also by a more integrated optical stimulus, the visual stimulation. Harper (1904) found that an *isolated pigeon does not ovulate* but ovulation occurs in pigeons placed side by side. Mathews (1939) examined this problem and demonstrated that an opto-genital (psychosomatic) reaction is involved by the following experimental combinations:

(1) male and female together in cage ················→ ovulation

(2) male and female in same cage but separated by
 a glass window ·····································→ ovulation

(3) two females together in cage ·····················→ ovulation

(4) one female in a cage placed before a
 mirror ····················→ ovulation

(5) female isolated in cage without a
 mirror ····················→ no ovulation

This is an example of stimulation by the presence of a similar animal, a particular case of the group effect. This stimulation is dependent on the *hypothalamo-hypophyso-pituitary pathway.* The hypothalamus is linked with the higher centres: the thalamus and cerebral cortex and this link determines the importance of psychic stimuli on hypothalamo-hypophyso-pituitary function. Benoit found photosensitive zones along the optic-hypothalamic pathways and pathways linking the rhinencephalon with the hypothalamus. Benoit and Assenmacher (1953) detected Bargmann-Gomori positive granules in the superficial part of the median eminence of duck in contact with capillary beds. These authors stated that:

(1) section of the superficial part of the median eminence, that is, the upper part of the hypothalmo-hypophyseal tract, results in gonad atrophy;

(2) section of the lower region of the stalk does not modify the gonads.

However, the male house sparrow *(Passer domesticus)* shows testicular development even in the total absence of light (M. and L. Vaugien, 1961). This would appear to result from elongation of the period of wakefulness and feeding which is no longer regulated by the diurnal period of days. Sexual maturation of young male parrots *(Melopsittacus undulatus)* is even more rapid in complete darkness (L. Vaugien, 1953). These results which are in opposition to those obtained from photostimulation of nest builders of temperate countries, could be explained in terms of the tropical origins of this parrot (desynchronization of nest building birds of Europe during autumn and winter; Baker and Ranson, 1938).

In the human species hypothalamo-genital correlations have been demonstrated particularly with respect to *cerebral functions and menstruation.*

In 1946, Reifenstein described, under the name of "hypothalamic amenorrhea" a type of amenorrhea with the following characteristics:

(1) originating from psychological trauma (emotion);

(2) *normal* level of FSH in urine;

(3) absence of oestrogens and thus of oestrogenic action of genital receptors as shown by vaginal smears and uterine biopsy.

This amenorrhea would be due to the absence of secretion of luteostimulin through inhibition of hypothalamic origin. *Luteostimulin (LH) is the hormone of ovulation.* Ovulation can be induced in rabbit by weak electrical stimulation of the tuberal region (Harris, 1950). However, Brooks (1940) produced amenorrhea in she-monkey by destruction of the pituitary stalk or by lesion of the lower part of the stalk, procedures which do not inhibit FSH secretion.

Table 51, modified from Weill and Bernfeld (1954), summarizes the present day concept of hypothalamo-hypophyso-pituitary regulation of genital activity.

This takes account of the complete cycle. The *hypothalamus* has an *indispensible* role in the oestrus cycle. A pituitary graft to a totally hypophysectomized guinea pig produced permanent oestrus but no oestrus cycle (Schweitzer, Charipper and Haterius, 1937).

It must be remembered that oestrogenic hormones undergo inactivation. For example:

$$\text{oestradiol} \rightarrow \text{oestrone} \rightarrow \text{oestriol}$$

in the liver (Schiller and Pincus; Jayle). In the vaginal keratinization test, if oestradiol is considered to have an activity of 1000, oestrone has an

TABLE 51

Hypothalamic stimulation

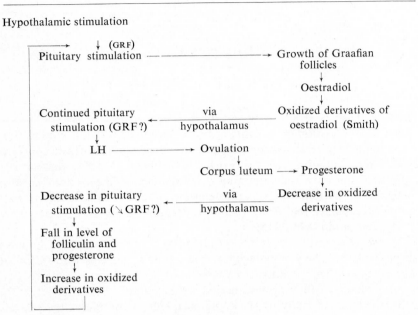

activity of 100 and oestriol of 20. Thus, the state of body systems, particularly of the liver, can modify the endocrine balance.

The hypothalamo-hypophyso-pituitary system regulates other stages of genital life with the exception of gestation in which the endocrine function of the placenta replaces hypothalamo-pituitary function, at least during the major part of pregnancy. Thus, the ♀ mammal contains two endocrine functions more than the ♂, those associated with the corpus luteum and those associated with the placenta.

CHAPTER 23

FERTILIZATION

THE role of nutrition in sexuality is a recognized fact not only with respect to vitamin requirements, for example vitamin E, but also as it concerns sexual differentiation (cf. the work of Wolff and Wolff). Nutrition is important not only in determining *the conditions* of fertilization but also in fertilization itself, that is, in the aetiology of fertilization. In this respect we must consider *the theory of the conjugation of gametes* proposed by P. A. Dangeard (1898–1915).

Dangeard states that "sexual reproduction . . . is not a primitive property of protoplasm"; he postulates that "the union of gametes with fusion of nuclei resulted from a nutritional constraint arising from successive divisions without an intermediate period of nutrition". Thus, gametes are cells which nutritionally are weak and incapable of continuing their development alone. Obviously, this idea of Dangeard needs to be amended and made more precise but the basic concept is good. Following work on the physiology of sea-urchin spermatozoa (1954), I gave support to the idea of the *trophic deficiency of gametes*. In the case of spermatozoa, this deficiency implies a form of *parasitism:* the *male element is a parasite with a life cycle in two hosts* as shown by the following scheme:

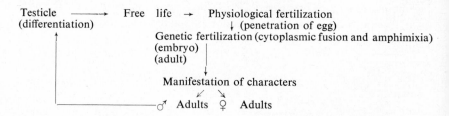

Testicular spermatic biology passes through a stage of *meiosis*. Thus, the male gamete is not a cell but, from the nuclear aspects, a demi-cell, which, because of gene → enzyme interrelationships, could explain the trophic deficiency and the dependence of male gametes.

The coming together of the fertilizing sperm and the ovum is conditioned by a number of physico-chemical and biological factors related to the morphological structure of gametes. In short, this constitutes *the ecology of fertilization.* This can be illustrated by examples from lower animals but, as we shall see, the concept of the ecology of fertilization is equally valid in mammals.

Since the demonstration by Pfeffer (1884) that malic acid has *chemotactile* properties on the antherozoids of ferns, the question of the possible "secretion" by ovules of compounds acting on the physiology of spermatozoids and thus on fertilization mechanisms has been of considerable interest. This is a question of preparation for *fertilization* which, in the strict sense, is consummated when amphimixia is realized. The remaining processes are differentiation, display, trap, secondary adaptation.

In considering the processes of fertilization in nereids, F. R. Lillie proposed the name *fertilisin* for the substance which produces agglutination of specific spermatozoids. Furthermore, seawater which has contained ripe ovules of sea-urchins acquires particular properties towards specific spermatozoids. The seawater *agglutinates spermatozoids* (Derbès, 1847) and *activates flagella movements* (F. R. Lillie). Fertilisin is extracted from sea-urchins by hydrochloric acid (Vasseur), trichloracetic acid (Rybak) or acetic acid (Tyler). Fertilisin is a *sulphated mucopolysaccharide*, the structure of which was elucidated by Vasseur (1954). *In the presence of fertilisin* (or gynogamone = "hormone of female gametes", Hartmann, 1939) *specific spermatozoids undergo morphological changes* (Popa, J. C. Dan): the acrosome, which has a *secretory* significance, liberates a substance which can clearly be seen with the electron microscope (Fig. 151). In the presence of fertilisin, the spermatozoids agglutinate in the region of the acrosomes but after secretion of this substance they become free. These facts explain the agglutination and spontaneous deglutination which are characteristic of this phenomenon. In collaboration with Burstein, I demonstrated (1960) that fertilisin of sea-urchins precipitates with β-lipoproteins in the presence of Ca^{++}, an ion which is found in seawater, and it would appear that this type of combination with spermatic β-lipoprotein could explain agglutination by fertilisin. The agglutination reaction of fertilisin and spermatozoids has been compared to that associated with erythrocyte substance and influenza virus (Burnet, 1951).

The nature of the agent which activates flagella movement is still unknown. The existence of fertilisin would not appear to be restricted to lower animals. Bishop and Tyler (1956) found that an agglutinin was secreted from the *zona pellucida* of mammalian eggs (rabbit, mouse, cow). Furthermore, Thibault and his co-workers (1950–1960) stated that "fertilizins" in the uterus—and then *in presence of uterine secretions*—rendered mammalian sperm more suitable for fertilization. Substances produced by female insects and attracting males can be compared with fertilisins in that they favour the encounter of gametes from opposite sexes. Carlson and Lüscher called these compounds *pheromones*. In the case of the silkworm, an aromatic substance is produced by the scent glands of the female and the "call" is detected by the male chemoreceptors. This compound is a dienic alcohol (Butenandt, Hecker and Stamm).

These facts pose important problems concerning *artificial insemination*. One of the problems is to obtain adequate survival of sperm. Male semen

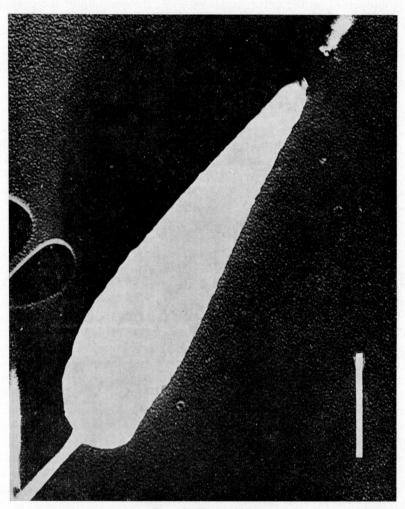

FIG. 151 (DAN, 1952)

can be preserved by slowing the metabolic processes of gametes and contagious germs by *cooling* (Spallanzani introduced this technique); a temperature of $+2°C$ is commonly used but a much lower temperature, that of liquid nitrogen at $-196°C$, can be used (Parkes, 1945). It should be noted that the presence of glycol (Luyet, 1941) or glycerol (J. Rostand,

1946) preserves the integrity of spermatic locomotory mechanism (cf. p. 203).

Cold is not the only factor in the preservation of *fertilizing* spermatozoids. For the purposes of artificial insemination, *preservation and dilution media* are based on sodium citrate (or sulphate or phosphate) + glucose + peptone (Milovanov, 1934) or *egg yolk* (Lardy and Philips, 1939), the latter being one of the most widely used methods. Jacquet (1957) used a medium based on skimmed milk which was equally successful.

Saturation with CO_2 favours preservation of fertilizing ability (Cohn, 1917; Van Demark, Du Mesnil du Buisson, 1957, 1958). In the case of sea-urchin spermatozoids I was able to obtain (1954) long survival by placing sperm in seawater containing penicillin, at laboratory temperature and in specified containers. Sperm lost their fertilizing ability only after 162 h. This would appear to be the extreme limit for such kind of cells characterized by *constant catabolism*.

It may be noted that although fertilization can be followed easily in animals with aquatic fertilization, this is not so in the case of internal fertilization (within female genital tract) particularly in mammals.

Now, fertilization can be considered to comprise:

(1) penetration of the ovum by a spermatozoid, a *general phenomenon;*
(2) the act which leads to favourable conditions for contact between ovum and spermatozoid;
(3) amphimixia.

In mammals, the region in which ovum and spermatozoid come into contact is the upper *opening of the oviduct* and penetration of the egg takes place at this site. Using *radioactive tracers*, an attempt has been made to follow the passage of spermatozoids in the female genital tract but, as yet, it has been possible only to define the correlations between the nucleus and cytoplasm in the egg (Sirlin and Edwards, 1957). The technique consists of peritoneal injection to males either of ^{14}C-*formate* or labelled *adenine* which accumulates in desoxyribonucleic acid of sperm. The labelled groups are detected *by autoradiography of histological sections* (biological extension of the principal observation of Becquerel concerning radioactivity). By this method the development of the male pronucleus in the egg can be followed. These facts concern the cytophysiology of fertilization. We shall examine the macroscopic phenomena of fertilization in greater detail.

Neuro-humoral relationships associated with fertilization—role of gonadotropins

We have already seen that extrusion of the egg is the result of follicle maturation and relates to the double action of FSH and LH. The reactions take the form of a chain of stimulator–effector reactions:

Hypothalamus → pituitary (hypothalamic effector I) → *ovarian follicle* (hypo-thalamic effector II or hypothalamo-pituitary effector) → *uterine endo-metrium* (hypothalamo-pituitary-ovarian effector or hypothalamic effector III).

This chain of reactions gives rise to certain *practical considerations.* Here are two examples:

Can the action of FSH be dissociated from that of LH?

Implantation of a pituitary graft beneath the skin of impuberal rabbit results in the appearance of hemorrhagic follicles, repeated ovulation and the formation of numerous corpora lutea. Precocious puberty can be induced in an impuberal female by administration of large doses of gona-dotropic hormone with both LH and FSH activity but *no effect is obtained following injection of FSH* or progesterone or even oestradiol. The ovary is important in this reaction since no response is seen during ovarian dys-function. Fevold demonstrated that *purified follicle stimulating hormone is not capable of inducing ovarian secretion of oestrogens unless traces of LH are present.* In menstrual disorders associated with absence of follicle maturation—as in the majority of cases of hormonal sterility—the absence of LH is considered to be important and administration of LH in the first stage of the menstrual cycle leads to follicle maturation (Levrier, 1956).

The role of gonadotropins is evident from the work I carried out with T. Gustafson (1952) on *Xenopus laevis*, an aquatic animal in which it was possible to obtain artificial fertilization without sacrificing the male. The aim was to obtain, *at any given time, a large quantity of embryos at the same stage, without killing the male for economic and genetic reasons* (possibility of back-cross), for the purpose of studying the chemical embryology of vertebrates. The method of obtaining fertilizable eggs by injection of gonadotropins to *Xenopus* was known (Shapiro, 1946). In the case of the male *Xenopus*, the only method available until then for obtain-ing large amounts of spermatozoids was removal of the testes. We pretreat-ed males with mare *chorionic* gonadotropin (injection to dorsal lymph sacs) and after 24 h obtained sperm by injecting a large dose of adrenaline which has three effects: (1) general constriction with closure of the cloaca and urine accumulation, (2) massive release of sperm which accumulate in urine, (3) torpor of the animal which favours handling. After 3 h urine together with sperm is removed from the cloaca with a pipette and the ovules previously obtained are inseminated immediately. The spermatozoids ob-tained in this way are extremely *motile.* Fertilization is enhanced by the addition of glycine to urine, glycine probably acting by dissociating pro-teins from the ovular membrane. Under these conditions 100 per cent of embryos can be obtained. Pretreatment of males with gonadotropin can be replaced by ultraviolet illumination (at 3600 A) for 2–3 days; irradiated males tend to exhibit amplexus and to croak. When males are transferred from relatively cold water (12°C) to warm water (20°C) and subjected to irradiation, which activates the hypothalamo-hypophyso-pituitary complex, the testes are to some extent in the same condition as in spring.

Evidence for endocrine interrelationships in fertilization have also been obtained in *invertebrates*. For example:

(1) *Results of* Durchon (1952): certain species of nereids reproduce in the *atoquous state*, that is, without undergoing morphological changes (such as in *Nereis diversicola*); other nereids, at the time of genital maturation, exhibit a form of partial metamorphosis and develop from the atoquous to the epitoquous state (mainly in males). *Nereis* is transformed into *Heteronereis:* the eyes enlarge, the body segments become compressed one against the other, the nereid setae are lost and replaced by pelagic setae *(Nereis succina)*. These morphological changes are related to external causes (temperature and diet) and internal genetic and humoral causes. In epitoquous species, the heteronereidian transformation is regulated by two factors, one *endocrine* factor of *cerebral* origin which is *inhibitory*, the other genital with an excitatory effect.

(2) *Results of* Lubet (1956): while studying the factors inducing release of gametes in mussel, Lubet found that emission of gametes is augmented by removal of internal inhibition; *the disappearance of a neurosecretory product of the cerebral ganglia would appear to be necessary for initiation of egg release.*

Thus, in one way or another, either by facilitation or inhibition, central hormones in particular have an essential role in the physiology of reproduction both in vertebrates and invertebrates.

Physiology of copulation (or coitus)

In animals with internal fertilization, the introduction of sperm to the vagina requires erection of the penis and ejection of seminal fluid.

Erection results from accumulation of blood in the cavernous tissue; this hyperemia is due to local arterial vasodilation. Excitation of the peripheral end of the erector nerve of Eckhard, from the *sacral plexus*, produces vasodilation of erection. Erection is possible in a spinal animal but normally stimuli with cortical or sub-cortical integration are involved in excitation. The nervous pathway for excitation is the *pelvic parasympathetic nerve*. Local sympathetic vasoconstriction leads to flaccidity of erectile organs in male and female. J. W. Watson (1964) called attention to the fact that the penis of bull and ram is adapted to attain erection with minimum transfer of blood. The entire organ is enclosed in a thick dense *tunica albuginea* which is virtually non-distensible; the *corpus cavernosum* is isolated except at the crura and much of the internal volume has been reduced in terms of blood space by an increase in the amount of fibroelastic tissue.

Ejaculation at the moment of orgasm is a reflex reaction, the centres of which are situated in the lumbar region of the spinal cord. The muscles of the epididymis, vas deferens, seminal vesicles and prostate contract and sperm accompanied by glandular secretions are discharged into the posterior urethra between the external and internal sphincters of the bladder. Sperm are ejected by rhythmic contractions of bulbo- and ischio-cavernosus muscles. Sperm penetrate the median female genital tract within 3 mn when there is female orgasm and in 1 h when there is no orgasm. To some extent this explains the sterility of frigid females.

Stimulation of the hypogastric nerves produces ejaculation while bilateral sympathectomy beneath L_2 or section of presacral nerves abolishes ejaculation while preserving the possibility of erection and sensation. Lesion of sacral nerves below S_1 abolishes erection and sensitivity of the penis.

The time of survival of human sperm in the vagina does not exceed 2 days and only sperm in the cervix remain motile for this period of time. Spermatozoids progress in the female genital tract at a rate of about 2 mm/mn; (in ewe and cow passage of spermatozoids from the vagina to the upper part of the Fallopian tube takes about 8 h; Dauzier, 1958). Contractions of the vagina, cervix and uterus augment the speed of ascension of spermatozoids to the Fallopian tubes where fertilization takes place. Fertilization requires a certain concentration of spermatozoids, about 100 million per ml in man. Thus, *the fertilizing spermatozoid is selected randomly*. Spermatic hyaluronidase is thought to be important in penetration of the ovum.

CHAPTER 24

GESTATION

Now that the egg has become a zygote it will develop into an embryo. In many animals with aquatic fertilization the zygote remains free (as in the case of sea-urchin except incubating sea-urchins) or becomes attached either by tendrils or in *nests* (in the case of fish). *Nidification* is a well-known process in birds and *nidation* in mammals is biologically perfected to the highest degree. Let us examine the concept of nidus in the case of fish by considering the classification established by L. Bertin (1952); since the nest represents a *protective area* for the conceptus, the evolution of the nest is of great importance.

Inert nests

(1) Prepared nests

These are fairly common among littoral fish (blennies, labrum, gobies etc.) which select a fissure or even a fragment of immersed pottery. Sometimes the preparation consists only of cleaning while on other occasions camouflage is prepared.

(2) Hollowed nests

These are made by excavating sand or mud. Bertin stated that a pair o, catfish are capable of carrying, pebble by pebble, in a single night, more than 5 litres of material and of hollowing an area of a quarter of a square metre.

(3) Woven nests

The nests of sticklebacks (P. Coste, 1846) are equivalent in perfection to those of birds. The male three-spined stickleback, a freshwater species of France, chooses green material for the walls of the nest while red material is used preferentially around the opening (Wunder).

(4) Floating nests

These are made of plant material and filled with air bubbles by the male (foam nests); examples of this type are seen among fighting fish (Carbonnier, 1872), ornamental fish from Indo-malayan marshes.

Living nests

(1) Heterospecific incubation

The bitterling, a western freshwater species, incubates its eggs in the gills of Anodonta (oxygenation?) (Bresse, 1950).

(2) Buccal incubation

In the Mediterranean *Apogon* (Garnaud, 1950), the nuptial ritual being completed by fertilization, the male swallows the eggs, throws them out, swallows them again and repeats this several times until the eggs are situated in the pharynx where they remain free.

(3) Cutaneous incubation

The female of the marine needle-fish lays threads of 50–800 eggs, according to the species, each thread measuring several centimetres in length. The threads become attached to the ventral region of the male where they remain after being fertilized (phoresia).

(4) Marsupial incubation

The male *Hippocampus* possesses a marsupial pouch in which the eggs are fertilized and incubated; the young will be nourished in the pouch and one can consider this to be a form of gestation; the male will also be responsible for parturition. Incubation is not modified by castration but total hypophysectomy performed during incubation produces premature parturition and malformation of embryos and alevin (Boisseau, 1964).

(5) Ovarian incubation

In the previous cases the nidus has been exterior to the female and viviparity has been associated with the paternal organism. However, in xiphophores (ornamental fish) a change is seen: the eggs are fertilized and develop within the ovary and the embryos show oophagia.

This stage of sexual evolution is of great importance since from this stage onwards the male introduces semen into the female organism with the aid of a penis, priape or gonopod, differentiated from anal or ventral fins.

The walls of the follicle and ovary in the gestating female show hyperemia which augments metabolic exchange between the mother and conceptus; sometimes parietal and embryonic papillae are present which augment the surface for exchange.

Bertin wrote: "In many species of fish we can speak of *viviparity, gestation* and finally *parturition* exactly as in mammals. The only difference is that there is no *placenta* to provide a closed link between the embryo and mother. The *viviparity* in this case is *aplacental*".

(6) Uterine incubation

In sharks a *placenta* exists. In the two divided uterine pouches a foetus is linked to the mother by an umbilical cord and a placenta of uterine and foetal origin. Anatomically this placenta does not resemble that of mammals but functionally the "invention", according to Cuénot, is acquired.

It can be seen that there would need to be only one more improvement —the presence of mammary glands—to pass from fish to mammals and, curiously, from a selachian, a fish considered to be primitive, to mammals. In the fish, primitive vertebrates, one meets all the stages from oviparity to viviparity passing through ovoviviparity. This is a good example of an evolutionary sequence which is explained by the branched character of evolution, that is to say, by a series of trials of which we know only the successes and which lead, as a result, to continual improvement. Moreover, this development is an example of the principle which I have previously stated of spatial vegetal economy and spatial development of central nervous structures. It appears that all these changes contribute to the liberation of creatures, depending on their evolutionary position, from contingencies, vegetal and genital limitations, allowing them the space and time to realize their independence (increasing capacity of information) and, at the human level, allowing the integration of nature which favours propagation and duration of life. All this is determinism not finalism.

Thus, gestation represents a significant evolutionary acquisition. There appears to me to be two essential advantages:

(1) From the quantitative point of view, gestation allows economy of living material; in species in which eggs are free, they are the prey of numerous other animals and a relatively low percentage of embryos will produce adults. When the eggs are laid the mother protects them against destruction. This arises from the fact that the mother is an adult animal capable of defending itself and of locomotion.

(2) The ethology of lower species is characterized by that which can be termed instinctive preoccupations, these being slavery, limitations on freedom or at least on liberation. The activity of nest building takes up time which cannot be used for other activities. Moreover, an important behavioural point, this embryogenetic stage has become through gestation mainly unconscious (vegetal activity give rises to the psychism of alienation) and constituting, potentially at least, an advantage for the work of higher nervous functions.

These remarks have led us to note the balance between that which it is convenient to call the creation of flesh and creation of the spirit. It is necessary to know the way in which these processes are determined by "hazardous" factors. Thus, in *Proteus*, a living fossil urodele of Balkan caves, when the ambient temperature is higher than 15°C there is oviparity and viviparity below 15°C (Kammerer); in this poikilotherm, at least, viviparity is accidental, being ecologically determined.

Gestation in mammals

The structure of the uterus shows anatomical variations according to the type of mammal being considered. Once again the concentration of vegetal organs as a function of evolution will be noted (Fig. 152).

The mammalian embryo develops in structures of this type. In this respect, the uterus is prepared by hormones which, by analogy with the term oestrogens, can be called *gestagens* (Zander, 1959) and particularly by progesterone, the hormone of the *corpus luteum* but also synthesized by the placenta which secretes the hormone during the second phase of gestation. The suprarenal cortex also synthesizes progesterone which serves as a precursor of corticoids. Luteal progesterone would appear after rupture of the follicle and conditions an important development of the endometrium which will favour implantation. (Postulated by A. Prenant, 1898, demonstrated by Bouin and Ancel, 1909, the hormone being isolated by Corner and Allen, 1934).

Thus, gestation is prepared by a progestative phase characterized by modification of the uterus for the fertilized egg. This progestative state depends on sex hormones but also on the general endocrine state. Thyroid hormones act synergistically with sex hormones just as certain corticosteroids such as 11-desoxycorticosterone which augment uterine permeability and facilitate the action of genital hormones (in contrast, cortisone decreases uterine permeability).

The gestative state is characterized by simultaneous secretion of LH (thus of oestrogen) and of prolactin—LTH—(and therefore, progesterone). The blastocysts pass some time lying freely in the uterine lumen before their attachment and subsequent implantation occurs. In the rat and mouse, the blastocyst attachment is initiated by oestrogen; during the first days of the intra-uterine life of the blastocyst the uterine epithelium is dominated first by progesterone (pre-attachment stage), then by progesterone and oestrogen (attachment stage). The pre-attachment stage is characterized by epithelial cells with many apical vesicles and a cell surface with regularly arranged microvilli intermingled with protrusions of the apical cytoplasm. The attachment stage is dominated by epithelial cells with few apical vesicles and a cell surface, where the regular microvilli have changed into irregular projections (O. Nilsson 1966). The structure-function relationships (cells and hormones) are well illustrated by such facts. It is to be noted that actinomycin D inhibits the decidual reaction (Burin and Sartor, 1965). G. Mayer suggests that the states of gestation and lactation (in which LTH has a net action; cf. Chapter 26) differ by varying degrees of neurodepression. Effectively, in rat the hormonal state associated with lactation can be induced by administration of phenothiazine derivatives:

(a) Administration of low doses (0·5 mg/day from oestrus) induces the appearance of a hormonal state in which progesterone acts synergistically

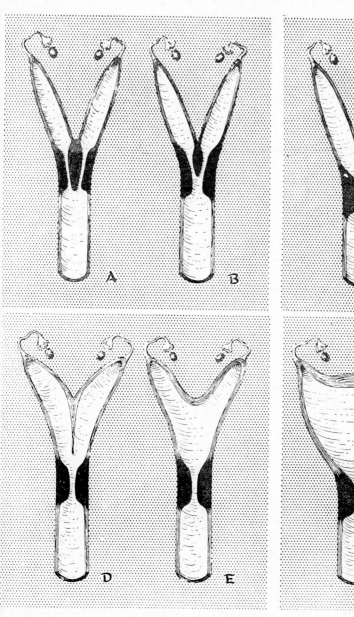

FIG. 152.

Double uterus: 2 cervices,
horns completely separate.
   (A) rat, mouse, rabbit.
   (B) guinea pig.
Two-horned uterus: 1 cervix.
   (C) pig, insectivores.

Bipartitioned uterus: 1 cervix.
   (D) cat, dog, cow, sheep.
   (E) mare.
Simple uterus: 1 cervix.
   (F) primates.
     (NALABANDOV, 1958.)

with oestrogen [(1) the vaginal mucosa carries a stratified, mucous-bearing epithelium as is found during pregnancy, (2) the lobulo-alveolar system of the mammary gland is developed but does not secrete]. In this case the pituitary seems to liberate both LTH and FSH–LH complex which are responsible for secretion of oestrogens.

(b) Administration of high doses (2·5 mg/day) of phenothiazine leads to the appearance of a hormonal state of the lactation type in which oestrogen is absent and progesterone acts alone (the mammary gland is little developed since oestrogen–progesterone synergism is necessary for the normal development of parenchyma but the acini are open and secrete abundantly). The pituitary liberates only LTH which manifests its luteotropic activity in the maintenance of *corpora lutea*.

Progesterone suppresses oestrus as well as ovulation under certain conditions but it stimulates the mammary gland. When the gravid state appears in the female, the *corpus luteum* is maintained and menstruation — as ovulation — ceases; if there is no gestation a new menstrual cycle commences.

The transport of progesterone has been followed with ^{14}C-labelled progesterone: about 75 per cent of the hormone passes in the bile and is eliminated through the faeces.

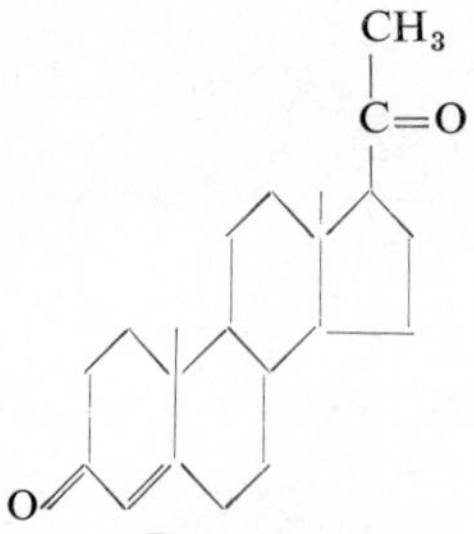

Progesterone 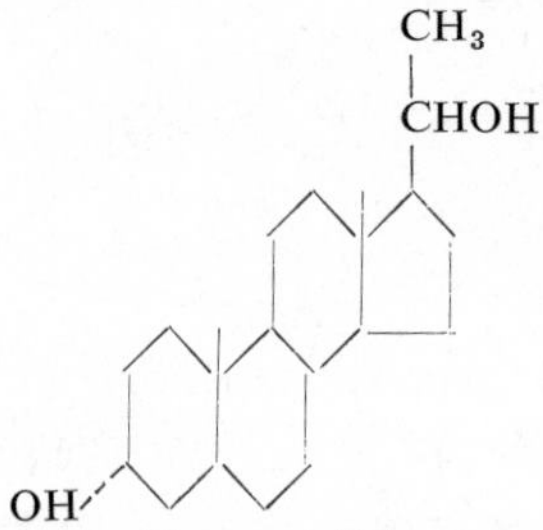   Pregnanediol

Two other progestative hormones are known:

pregnanediol, principal excretory product of progesterone; it is found in urine when the endometrium is in the progestative state;

5-pregnenolone which is characterized by its ability to inhibit uterine motility.

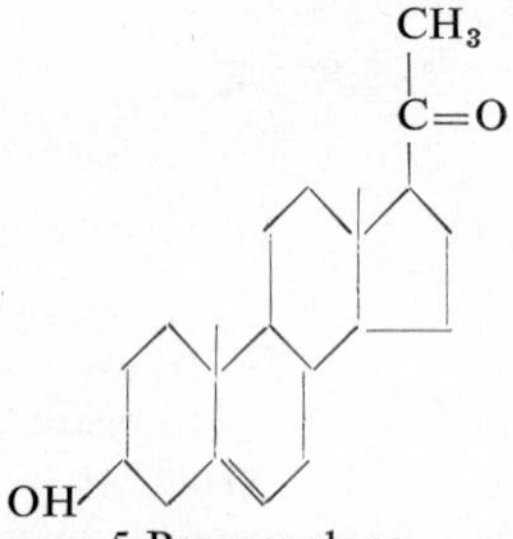

5-Pregnenolone

Furthermore, from the very early stages of gestation urine contains gonadotropic hormone of chorionic origin but also produced by the placenta except in rat (Canivenc, 1956). Injection of urine from a pregnant female to immature female mice leads to the appearance of ovarian hemorrhagic areas. These reactions which can also be obtained in virgin rabbit provide a diagnostic test of pregnancy (tests of Ascheim and Zondek, Friedman and A. Brouha). This important diagnosis can be performed in another way: the urine to be tested is injected to a male frog or toad; when gonadotropin is present, small but characteristic amounts of spermatozoa are found in the cloaca (test of Galli—Mainini). Finally, to avoid the problems associated with biological material, a rapid chemical test for pregnancy in woman has been developed and consists of measuring the *galactose* released by hydrolysis of urinary gonadotropins (Rybak and Levrier, 1959).

Mode of regulation of the gestative equilibrium

In the ♀ rat, the cycle lasts only 4–5 days (here we find the concept of *specific biotic time*, the temporal unit having different values in different animals). Alloiteau and Vignal (1958) repeatedly injected prolactin to animals during the 4 days following oestrus and produced *pseudo-gestation*, that is to say, a dioestrus lasting 12–14 days and interposed between two regular cycles. They noted that *although prolactin stimulates the secretion of progesterone by the corpus luteum, progesterone also augments the secretion of prolactin.*

The role of the ovary varies *according to the animal species* being considered; *two main categories* can be distinguished:

(a) those in which the ovaries are *indispensable* throughout pregnancy;

(b) those in which *castration is compatible, at least after a certain stage,* with development of gestation.

In the rabbit, bilateral castration on the 20th day produces abortion. In guinea pig, if the surgical trauma is insufficient to produce abortion (in other words, if there is no stress) castration after the 15th day does not interrupt gestation.

In most species the progestative body is transformed into a gestative body. This change is determined by the *gravid uterus* (Klein, 1938): for example, removal of the gravid horn from hamster during the second phase of gestation leads to involution of the *corpora lutea* and resumption of the ovarian cycle. Uterine distension does not appear to be involved, no more than the presence of foetus. However, the *trophoblast* and the *placenta* exert a fundamental influence: *removal of the placenta is followed by rapid breakdown of gestative structures* (Klein, 1938). (Cf. p. 549, the role of the placenta as an endocrine gland.)

The *hypothalamus* also appears to be involved in maintenance of the corpus luteum but little information is available in this respect. However, Stutinsky (1953) studied the hypophysis of rat during gestation. He distinguished three phases in the development of the hypophysis during gestation: (1) up to the 15th day; (2) to the end of gestation and associated with an increase in neurosecretion; (3) 48 h *after* gestation; this latter point demonstrating that gestative life does not end abruptly with parturition.

In addition, it should be noted that the hormonal processes of pregnancy depend on definite trophic conditions. For example, inanition or diets deficient in protein determine hormonal disorders similar to those produced by total hypophysectomy: secretion and excretion of gonadotropins are arrested but the administration of progesterone under these conditions maintains pregnancy (Nelson and Evans). In protein deficiency, embryogenesis requires the intervention of glucocorticoids: an adrenalectomized rat provided with a low protein diet aborts systematically in spite of progesterone administration but in 80 per cent of cases, gestation proceeds normally if cortisone is administered with progesterone (Aschkenasy, 1957). However, it is to be noted that in 95 per cent of pregnancies a physiological anaemia develops in which there is marked hypoglobulemia until the last quarter of pregnancy in women (Boycott, 1936; Toulet *et al.*, 1964). All these points raise important problems which will again be met when considering mother–foetus interrelationships (cf. p. 551).

Progress of gestation

I shall take the monkeys as an example.

Pregnancy in the chimpanzee lasts 235 days but only 168 days in *Rhesus* (6 times the duration of the oestrus cycle). In the macaco the chronology of segmentation of the zygote has been followed by Lewis and Hartman in 1933 (Table 52).

TABLE 52

Number of blastomeres	Time in h after ovulation
2	24–34
4	36–48
8	48–72
(arrival in uterus)	
9–16	72–96

Migration of the zygote is brought about by movements of oviduct muscles and ciliary lining (controlled by folliculin). Nidation occurs towards the 9th day after ovulation.

The level of oestrogens in blood and urine of gravid Rhesus augments during gestation to reach a maximum at the time of parturition. The progesterone level would appear to follow cyclic variations, the maximum level appearing every 28 days. Thus, there are *endocrine tides*. The hyperlipemia and parallel hypercholesterolemia would be induced by marked secretion of oestrogens at the end of pregnancy.

Synopsis

Hisaw (1944) distinguished *three phases in gestation of monkey:*

(1) a luteal phase during which the uterus is controlled by the ovary and the *corpus luteum* by LTH;

(2) after zygote implantation, *trophoblastic* LTH appears to be contributing to the maintenance of the corpus luteum: *the placenta* assumes a supplementary or *vicarious pituitary function* to maintain the *corpus luteum*, in fact total hypophysectomy does not lead during this phase to abortion in monkeys;

(3) until the end of pregnancy it appears that the *corpus luteum* is no longer capable of secreting progesterone. *At this stage it is the placenta which assumes the ovarian and pituitary functions.*

The important properties assumed by the placenta must be considered in more detail.

Placenta

The placenta is as rich in enzymes as the liver and contains, for example, lipases, phosphatases, nucleases, amylase, urease, arginase, cytochrome-oxidase, succinic acid dehydrogenase. In the woman the placenta is the richest organ in histaminase of which the functional significance is unknown. It is necessary to state that in foetal circulation the placenta assumes the functions of nutrition, detoxication, excretion. A *placental barrier* is present, that is, the regions of the choriallantoic placenta in which the circulating foetal blood most closely approximates to the maternal blood. Basically, the cat placenta consists of a series of more or less parallel trophoblastic plates or lamellae, separated from one another by regions containing foetal connective tissue traversed by the slender foetal capillaries. Within the lamellae, surrounded by the trophoblast, run the tortuous, sinusoidal maternal capillaries (Wislocki and Dempsey, 1946; Amoroso, 1952). The trophoblast also encloses "decidual" giant cells, lodged between the maternal capillaries; these cells are derived from fibroblasts of the original uterine mucosa which are transformed into giant cells during the invasion of the endometrium by the foetal trophoblast (Wislocki and Dempsey, 1946). Both the giant cells and the maternal capillaries are surrounded by a variable, wide zone of amorphous substance. Lipid droplets are frequently present in the trophoblast which consists of an outer syncytial and an inner cellular layer (Dempsey and Wislocki, 1956).

We shall take as an example the placenta of rat. This is a discoid placenta of the *hemo-chorial type*, that is, the maternal blood is directly in contact with the chorial villi.

There are *two very different regions:* one, peripheral with giant cells, the other, underlying with small cells. The latter are strongly *basophilic* due to the presence of ribonucleic acid as well as secretory cytological structures *(ergastoplasm)* visible with the electron microscope. These *small* trophoblastic cells would secrete prolactin (Mayer, Blanquet, Canivenc, Capot, 1953).

There are two points to be made at this stage, one concerning nucleic acids of the placenta, the other concerning prolactin.

(1) Placental nucleic acids

Mohri (1954) traced chronologically the following events in the placenta of rat (thus, again the time must be estimated in "rat-units" and not in anthropocentric terms):

(a) In the first stage of gestation (between the 10th and 13th day), the level of total desoxyribonucleic acid (which is nuclear and provides an expression of the cellular level), total ribonucleic acid and total nitrogen are almost constant and low;

(b) in the second phase (13th–17th day) there is a sudden increase in these three substances (marked biosynthesis of proteins?);

(c) in the third phase (17th–20th day) the level of total ribonucleic acid and nitrogen continues to increase while that of desoxyribonucleic acid remains almost constant;

(d) in the fourth phase (20th–22nd day, just before parturition), the levels of total ribonucleic acid and nitrogen fall sharply and that of desoxyribonucleic acid remains unchanged.

Thus, there are *critical phases in placental activity* corresponding undoubtedly to the cycles of hormonal tides.

(2) Placental prolactin

Two experimental luteotropic factors are known in rat, oestradiol and prolactin. However, the placenta of rat does not appear to be capable of producing oestrogens (Canivenc and Mayer, 1951) as in the case of dog and cat while the presence of oestrogens has been reported for other animals (woman, ewe, cow). On the other hand, *the placenta of rat contains a prolactin-like substance* (Canivenc, 1952): implantation above the crop of pigeon of a placenta removed from gravid rat on the 12th day results in development, in the region with which the implant is in contact, of abundant proliferation of crop epithelium, analogous to that obtained by prolactin administration in the classical test of Lyons and Page (1935).

Other hormones have been localized in the placenta, for example, *intermedin* (in woman, mare and cow). The concentration of intermedin in the chorion increases during gestation. The role of this compound is not evident but it can be suggested that intermedin would be involved as a hypoglycemic factor (cf. p. 449). Extracts of rat placenta do not contain FSH or LH (or ICSH) (Astwood and Geep, 1938). In considering the steroids the endocrine unit is the foeto-placenta rather than the placenta (Diczfalusy *et al.*).

Thus, for hormonal reasons and other reasons which we have not considered, the gestative state is characterized by a new homoiostasis, by a new functional state in which the placenta, notably through its hormonal polyvalence, is dominant. We have just considered that steroid hormones are detectable, for example, in the placenta of mare in which, at the 200th day of pregnancy, the placenta is the organ containing the highest level of oestrogens (Catchpole and Cole, 1934). Hence, are these steroids secreted by the placenta or are they concentrated in this region? In addition, during gestation the adrenal cortex is hypertrophied and Canivenc has provided arguments which seem to show that the state of the adrenal cortex during pregnancy is dependent on the placenta which would act via the pituitary (ACTH).

Interrelationships of the foetus and maternal organism

We shall give some idea of these interrelationships in mammal by considering several facts which show that gestation is a particular type of *parabiosis*.

(1) Placental and foetal glycogen

In 1859 Claude Bernard showed the presence of glycogen in the placenta of rabbit. This glycogen is localized both in the maternal decidual layer and in the foetal villi. The maximum concentration is found in rabbit during the 21st day of foetal life, after which the total decreases until term. Placental glycogen is not augmented by maternal hyperglycemia or decreased by maternal hypoglycemia; it appears not to be sensitive to maternal secretions of insulin and adrenaline. Jost and Jacquot (1958), examining the question of foetal glycogen, provided evidence for the following facts:

In the rabbit foetus ablation of the hypophyso-pituitary complex, before the time when the liver commences to accumulate glycogen (20th day of gestation), inhibits the accumulation of glycogen (Jost and Hatey, 1949). This inhibition is removed by injection of corticostimulin.

In contrast, decapitation of *rat* foeti only completely inhibits the accumulation of hepatic glycogen when the *mother is adrenalectomized*. Thus, maternal corticoids would act on the foetal liver. Moreover, the rat would contain a *placental* factor sensitizing the foetal liver to corticoids. At the end of pregnancy *in rat* there is an endocrine secretion from the foetal adrenals under pituitary control which is concerned with regulation of the glycogen level of foetal liver (R. Jacquot, 1958). In the foetal liver of rat, uridine-diphosphoglucose-glycogen transglucosylase activity increases during the last days of gestation and this parallels the increase in hepatic glycogen. However, glucose-6-phosphate dehydrogenase shows a net decrease in activity between 17 to 21 days *post-coitus* and glucose-6-phosphatase activity only becomes significant just before birth, at a time when the accumulation of glycogen has been realized. In rat embryos decapitated *in utero* on the 18th day post-coitus, the uridine-diphosphoglucose-glycogen transglucosylase activity is lower than in normal embryos of the same age; glucose-6-phosphate dehydrogenase activity is higher than in normal embryos while glucose-6-phosphatase activity remains very low (Jacquot and Kretschmer, 1963). Thus, there seems to be a relationship between pituitary activity and foetal hepatic metabolism. One point appears to need consideration; it concerns foetal kaliemia. Ashmore, Hastings and their co-workers (1957) have shown that in the liver of adult rat the synthesis of glycogen is much greater in a medium rich in K^+ than in a medium rich in Na^+, Na^+ augmenting phosphorylase activity (Cahill *et al.*, 1957). Now, J. Maniey (1961) states that the maternal kaliemia of rat remains constant from the $18\,^1/_2$ day until term while that of the foetus decreases markedly during the same period but it should be noted that in rat foetus at term the plasma potassium level (5·7 m-equiv./l) is slightly higher than in adults (4·7 m-equiv./l). On the other hand foetal hepatic metabolism of sugars implies protein and lipid metabolic changes.

Measurements of nucleic acids and phospholipids in the foetal liver of rat (Lafarge and Frayssinet, 1964) have shown that the ratios of cytoplasmic RNA to DNA and phospholipids to DNA increase regularly from the 18th day of foetal life and these authors characterize three phases in the hepatic development of rat:

from the 15th to 18th day of foetal life: phase of growth by hyperplasia without hypertrophy;

from the 19th day of foetal life to the 3rd day of post-natal life: phase of cellular repair;

from the 5th day of post-natal life: phase of growth by hyperplasia and hypertrophy.

It should be noted that total hypophysectomy by decapitation of the rat foetus produces hypercholesterolemia and an increase in cholesterol level of liver, a phenomenon which is more marked when the pregnant female is adrenalectomized. This augmentation of foetal hepatic cholesterol can be reduced by injection of cortisone to the foetus and suppressed by ACTH (L. O. Picon, 1961, 1962). In the rat foetus, the ratio of lipids to proteins in the whole body is constant but after birth, this ratio increases rapidly. In the rabbit foetus, the ratio of lipids to proteins in the body increases during the latter stages of gestation, particularly in the liver. Decapitation of the foetus augments the level of lipids in the body of rabbit but not of rat (Picon and Jost, 1963). Let us state in this respect that the level of unsaturated fatty acids in maternal blood is higher than in foetal blood, this difference mainly affecting linoleic acid (Degrelle-Cheymol, 1964; Furuhjelm).

In considering carbohydrate metabolism more particularly, the presence of *fructose* in embryonic fluid of calf foetus, detected by Claude Bernard in 1855, must be reported. Embryonic fructose has been found in many mammals but not in primates. Goodwin (1954) considers that this fructose which does not pass into the maternal organism would constitute a carbohydrate reserve for the foetus.

In my opinion, the fructose, as in mammalian sperm, represents an *economic energy source*. If a comparative examination is made of the pathways of fructose and glucose utilization, it can be stated that to reach the stage of fructose-1-6-diphosphate from fructose two enzymic stages are required (1), while to reach the same stage from glucose, three stages are required (2). Thus, it is a functional advantage to start with fructose in the case of an organism requiring a rapidly mobilizable energy source.

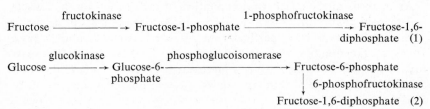

Hers (1959) demonstrated the presence of aldose-reductase in the placenta of ewe, ketose-reductase in foetal liver and sorbitol in foetal blood. This suggests that fructose is formed here as in the seminal vesicles (cf. p. 501), glucose being transformed into sorbitol in the placenta and sorbitol into fructose in the liver.

(2) Pituitary, thyroid and adrenal hormones in the foetus

The facts reported above lead to an examination of the function of a certain number of fundamental endocrine glands of the mammalian foetus.

The role of the pituitary in development of embryos of homoiothermic amniote vertebrates (birds and mammals) has been examined by deprivation methods: (a) X-ray irradiation (Wolff and Stoll, 1937, in chicken; Raynaud and Friley, 1947, in mouse); (b) surgery (Fugo and Witschi, 1938, in birds; Jost, 1947, and Wells, 1947, in rodents). Total hypophysectomy by decapitation produces testicular deficiency in male rabbit foetus, a deficiency which is corrected by administration of gonadotropic hormone. However, decapitation in rat or mouse does not retard differentiation of the genital tract. In the absence of the pituitary the thyroid of rabbit foetus fixes less radioactive iodine (Jost, Morel and Marois). Foetal decapitation also leads to a reduction in the volume of the adrenal cortex at birth. In the rat foetus, the pituitary-adrenal links seem to be established between the 16th and 17th days and the action of the pituitary on the adrenal cortex appears particularly intense between the 18th and 20th days (A. Cohen, 1959). Furthermore, the ascorbic acid content of the adrenals of decapitated rat embryos is lower than normal except in embryos sacrificed at the 18th day. This suggests that the action of the foetal pituitary on adrenal ascorbic acid level commences at 18 days (A. Cohen, 1959); corticostimulin injected to decapitated foeti, restores the normal adrenal ascorbic acid level. The thyroid of rat foetus begins to concentrate iodides towards $17\frac{1}{2}$–$18\frac{1}{2}$ days and at 18 days and 14 h thyroxine can regularly be found in plasma of foeti from thyrodectomized mothers (J. P. Geloso, 1961). The circulating form of thyroxine in foetal blood has been studied by Myant and Osorio (1959–1960): in rabbit foetus, the protein linking with thyroxine is different from that in the mother. Thyroid hormones are not necessary for growth in weight of foeti but are indispensable for development of the normal newborn.

An analysis of the effects of decapitation of rabbit foeti on the testicle and thyroid has shown (Jost and Gonse, 1953) that the role of the foetal pituitary appears to be of particular importance between 22 and 24 days. However, the number of cells staining with the MacManus-Hotchkiss reaction ("periodic acid – Schiff") increases between 19 and 23 days, indicating that the foetal pituitary shows variable phases of activity (Jost, 1962).

(3) Alimentary competition between mother and foetus

Fasting established at the beginning of gestation leads to death of embryos while fasting in the second half of pregnancy produces premature delivery with undersized progeny (Barry in rat, 1920; Jonen in rabbit, 1931).

Protein deficiency affects the mother but has only a slight effect on the foetus; similarly, in those factors concerned with phosphorus and calcium metabolism the *foetus has priority* over the mother.

In the case of vitamin A, vitamin B_1, pantothenic acid and vitamin B_2, the reverse is true, that is, the mother satisfies her requirements before those of the foetus. Jacquot (1952) concludes from this: "The mammalian embryo is in no way a privileged parasite which lives compulsorily to the detriment of the maternal organism". There is competition in one direction or another depending on the type of substrate considered.

(4) Teratology and uterine circulation during gestation

The passage of substances from the mother to the foetus was demonstrated for the first time by Flourens (1861) with the aid of madder dye. Flourens wrote in the XXth lesson of his *Ontologie naturelle:* "It concerns the bones of a foetus, in which all the bones had become red, a very bright red, solely due to the fact that the mother has been subjected to a diet mixed with madder during the last 45 days of gestation. And not only were all the bones red but the teeth became red also".

By a similar process Canivenc was able to impregnate the yolk of chicken eggs with Sudan black in order to increase the contrast between the embryo and its environment for cinematographic purposes (Rybak, Cortot and Canivenc, 1955).

The link across the placenta can result in embryopathy and/or placental disorders when the maternal organism receives certain compounds as, for example, during therapeutic treatment.

In this way, injection of hydrocortisone to the pregnant rat reduces the size of the placenta (Gunberg, 1957). Sex hormones of the adult (androgens and oestrogens) can interrupt sexual differentiation of the foetus and lead to intersexuality (Dantchakoff; Greene; Raynaud, 1936), the malformations mainly affecting the mammary gland (Raynaud, 1938–1956). In rat, injection of 10 mg of oestradiol to the mother on the 14th day of gestation produced, in 100 per cent of cases, total suppression of the mammary gland associated with premature development and hypertrophy of the mammillae (Delost and Jean, 1962, 1963). Pregneninolone (17-ethinyltestosterone) which maintains gestation of the castrated rabbit (Courrier and Jost, 1939) causes masculinization of the female foetus (Courrier and Jost, 1942). The route of administration can be of great significance: allyllestrenol (a synthetic progesterone-like compound) injected *subcutaneously* (20 mg per day) to castrated rats maintains gestation without inducing masculinization of the foetus but, given *orally* to intact rats, it produces disorders of female foeti, notably persistance of the Wolffian ducts (Jost and Moreau, 1963). Injection of cortisone to gestating female mice produces foetal abnormalities and notably palatine fissure (Frazer and Fainstatt, 1951). In rat embryos, injection of high doses of corticostimulin results in a fall in the adrenal ascorbic acid content from $17\frac{1}{2}$ days (A. Cohen, 1958). Rats born to mothers treated with cortisone during gestation and suckled by normal mothers only show a delay of a few hours (in rat-units!) in the onset of puberty and will reproduce normally. However, delays in sexual development are seen in normal rats suckled by mothers treated with cortisone during lactation or gestation (Mercier-Parot, 1957).

36*

The thyroid effects are also to be noted. Injection of a single dose of 50 μg thyroxine to the mother (mouse) on the 18th day of gestation produces a decrease in the rate of postnatal growth of progeny (Carteret and Delost, 1961, 1963). An antithyroid, propylthiouracil, administered to the mother (rat) 5 days before parturition (50 mg/day), produces: (a) an increase in foetal calcemia without a parallel change in maternal calcemia, (2) a decrease in the rate of growth of the parathyroids; these changes would correspond to endocrine hyperactivity of the foetal parathyroids (Pic, 1963; Jost, Pic, Maniey and Legrand, 1963). One will note the action of an antithyroid on the parathyroids, an action seen in both foetus and adult (Malcolm et al., 1949; Manunta and Rowinski, 1953; Jost, 1957; Santini, 1959). This does not prevent the antithyroids administered to the pregnant female from acting on the foetal thyroid to produce hypertrophy and hyperplasia (A.M. Hughes, 1944) but hyperplasia is only seen in a foetus containing an intact pituitary (Jost, 1956–1957, 1959). The antithyroid (propylthiouracil) decreases the affinity of the foetal thyroid for [131]I (Noumura, 1959; M. Maillard, 1959).

Even vasopressin or adrenaline can have serious morphogenetic consequences: when administered to rat foeti during the sensitive phase (15 to 18 days) they produce hemorrhages in the extremities with formation of blisters, followed by necrosis and congenital amputation of affected parts (Jost, 1953).

In female rats made diabetic by pancreatectomy before pregnancy, the effects are similar to those produced by alloxan: difficulty in obtaining pregnancy; marked hyperglycemia of foetus and mother, no systematic increase in foetal weight in the case of marked hyperglycemia but intense reaction of the foetal islets of Langerhans (S. Kozma-De Bokay, Jacquot and Jost, 1961).

Anaesthesia in pregnant females also produces certain changes. Maniey (1963) states that in rat under ether anaesthesia kaliemia of the mother is lower and that of the foetus greater than under nembutal anaesthesia.

The shape and dimensions of the foetus have considerable importance in local blood circulation. In these studies the circulation time is measured. Reynolds (1952) described a technique using very small quantities of *sodium cyanide* injected to the uterus: *the time taken for this toxic substance to reach the carotid sinus* where it produces panting in the mother (chemoreceptors) is measured, this value expressing the circulation time. By this method it is found that circulation decreases up to the 22nd day of gestation in rabbit; from the 24th day circulation increases, that is, the circulation time becomes shorter. These changes are thought to be related to geometric modifications in the foeti during growth. A demonstration of this can be given by *injecting a dye to the maternal aorta* when up to the 16th day of gestation in rabbit the whole uterus is deeply stained. At the 20th day, the dye does not penetrate the most distended parts of the uterus and this penetration is minimum at 22 days. But on the 24th day when the foetus pass from a more or less spherical form to an oblong form, the uterine distribution of dye becomes more extensive and increases still more until the 28th day (this irrigation could favour the work of the uterus at parturition).

Delayed nidation

Courrier and Colonge (1951) have shown that *cortisone administered in the second half of pregnancy is harmful to gestation in rabbit but much less harmful in rat.* On the other hand, Robson (1934) demonstrated that the hypophyso-pituitary complex is indispensable for the maintenance of gestation in rabbit. However, in rat (Selye *et al.*, 1935), total hypophysectomy is harmful only when performed during progestation and hypophysectomy is compatible with development of gestation when carried out after the 12th day. This led Mayer and his co-workers (1955) to compare the action of cortisone on the hormonal equilibrium of progestation and gestation in rat. They found that 10 mg of cortisone daily from the first day of gestation produces disorders of nidation (absence of ovo-implantation or involution of eggs a short time after nidation). In contrast, during gestation, pregnancy is resistant to very high doses of cortisone. The interpretation of such facts is still not clear but let us say that the progestative state is regulated solely by pituito–ovarian relationships while the gestative state is dominated by relationships between the placenta and ovary. This leads one to think that the endocrine ability of the placenta would be greater than that of the pituitary when the organism has to support the surcharge of an additional organism showing intensive development and metabolism.

During the last few years, G. Mayer and his co-workers, and notably Canivenc, have carried out research of great interest concerning the phenomenon of delayed nidation.

The stages associated with the fertilized egg of mammals can be summarized as follows: (1) tubular transit; (2) intra-uterine migration (*stage of free life*); (3) the blastocyst develops a trophoblast which links the germ cell with the uterus *(stage of fixed life)*. Canivenc and Mayer provided evidence for the following facts:

(1) *In gravid rats* castrated on the 4th day of pregnancy and injected daily with 10 mg of *progesterone,* the *fertilized eggs remain free* but the addition of a low dose of *oestradiol* leads to implantation. Thus, by an unknown mechanism, *oestrogens show an economic synergism with progesterone;* in other words, *a low dose of progesterone associated with a certain quantity of oestrogen produces the same effect as larger doses of progesterone administered alone.*

(2) Pincus (1936) observed that when pregnancy is established in a rat *nursing* numerous young (more than 5), ovo-implantation, which under normal conditions takes place 6 days later, is delayed and the delay of nidation is proportional to the number of young suckled. However, local administration of progesterone to the wall of one of the two uterine

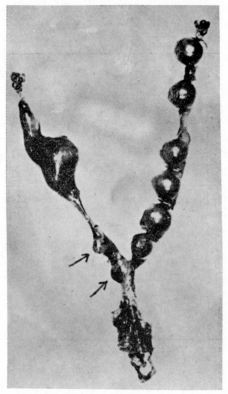

FIG. 153. Size: ×1. Macroscopic image of the uterine horns of a post-partum pregnant female suckling 12 offspring. On the left, treated uterine horn. On the right, untreated uterine horn. The arrows indicate the involuting decidua. (CANIVENC and MAYER, *Annales d'Endocrinologie*, 1955.)

horns (and not in the lumen where it would be toxic to the blastocyst) of rats in lactation on the 3rd day of gestation, allows one to obtain the growth of two types of embryos with different development in the one animal. The eggs implant at the normal time in the treated region of the uterus while in the other horn there is a delay of nidation which varies with the number of young nursed (about 1 day per young above 5). Thus, one mother can have three litters simultaneously: one in lactation, a second embryonic litter and a third embryonic litter but of a different age.

(3) *Stress can produce a delay of nidation.* In rat the normal duration of pregnancy is 21 days and during the first 6 days the fertilized egg is free in the genital tract, then it implants. Canivenc and Mayer (1955) burned a certain area of the skin surface of the mother on the 4th day of pregnancy and delayed the nidation by 25 days (Lataste, 1887, reported a similar effect).

Experiments of Mayer and Thévenot-Duluc have shown that rats deprived of all food during the first 4 days of pregnancy do not exhibit ovo-implantation. Administration of reserpine to rat produces a denutrition phenomenon and inhibits ovo-implantation (Mayer and Meunier, 1959). In this case a hormonal state is created which is characterized by the presence in the organism of progesterone and the absence of oestrogens, such that the administration of low doses of oestrogens before the time of normal nidation results in ovo-implantation at the usual time but later administration of oestradiol results in delayed nidation. This demonstrates that the eggs remain free in the uterus in an inactive form for a variable length of time. When the delay in nidation is produced by burns, progesterone (in a daily dose of 10 mg) inhibits the delay. It seems that under these conditions, each "stress" can be overcome by administration either of oestrogens or of progesterone depending on the extent to which a new hormonal regime is established with the presence of progesterone and absence of oestrogens or presence of oestrogens and absence of progesterone. Nidation takes place only when there is a certain relative level of oestrogens and gestagens. A disturbance of pituitary gonodatropic activity must be responsible for delayed nidation (cf. p. 228).

Furthermore, the adrenal gland is important in "stress" but the delay of nidation is inhibited by *bilateral adrenalectomy.* Histamine is involved in "stress" (cf. p. 226) and is known to stimulate mammalian smooth muscle and notably the uterus (Magnus, 1904). During gestation the histidine decarboxylase activity of the foetus increases markedly and the maternal kidney has a high capacity for histamine formation during gestation (Kahlson *et al.*, 1958, 1959; Rosengren, 1963). It has been suggested (Shelesnyak) that histamine may be important in the formation of the deciduoma but this was not verified by Finn and Keen (1962). Thus, the mechanisms concerned with the processes of nidation depend on more complex phenomena. It should be stressed that uterine nidation, in the rat at least, is the result of a precise hormonal sequence and, in particular, of the presence of a permanent level of progesterone and, on the 4th day of progestation, of a certain level of oestrogen; this equilibrium of progestative hormones can be modified by intervening in the ovary (castration followed by the administration of progesterone) or in the hypothalamo-hypophyso-pituitary system (transplantation of the pituitary complex in the kidney or administration of neuro-depressors) and even by implantation into the kidney of a fragment of non-mature testicle of rat (school of G. Mayer, 1966).

(4) Finally, delayed nidation occurs *naturally* in certain animals. This was first observed by Bischoff in 1854 in roedeer. In 1931 Fisher described this phenomenon in the European badger. In these animals, oestrus and fertilizing coitus take place during July–August but the egg does not develop beyond the blastocyst stage; it remains free, floating in the uterine cavity and does not implant until six months later, in January. The stage of fixed life extends until February–March, the time of parturition.

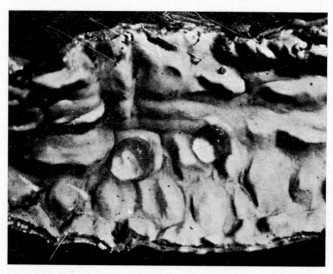

Fig. 154. Magnification: ×3·5. Type II blastocysts found in December in a uterus ligatured in April. (Canivenc, *Annales d'Endocrinologie*, 1957.)

It should be noted that in species showing placentation of the decidual type, birth is accompanied by detachment of a part of the maternal uterine mucosa; this produces a hemorrhage, traces of which remain and indicate the existence of post ovular implantations. These traces remain visible for long periods as yellow-brown areas in species with a long cycle. These placental scars are found in European badger and the pigment is hemosiderin (Canivenc *et al.*, 1962).

The work of Neal and of Canivenc demonstrates the *existence in 80 per cent of cases of a post-partum oestrus* in February or March. Free blastocyst are found in the uterine cavity from March 5th in the south-west of France, in other words, well before the time usually associated with heat in badger. Thus, there is sexual intercourse very shortly after parturition. It is noted (Canivenc, 1965) that the luteal deficiency is common to the different species showing differing ovo-implantation, the cytological signs of inactivity at the level of the corpus luteum seem to be the essential proof of this. The addition of ovarian hormones does not modify the ovoimplantation chronology. In the badger, the pituitary cytology reveals a lack of the LH factor, but the experiments show that, if it is possible to provoke the appearance of new corpora lutea by means of gonadotropins, the same appearance of inactivity is still observed. It remains to be known what determines this delay in ovo-implantation and whether it depends naturally in the badger, as experimentally in rat, on an absolute or relative deficiency in progesterone.

CHAPTER 25

PARTURITION

W<small>E SHALL</small> first examine the uterine processes which, to a great extent, condition the phenomena of delivery.

Physiological chemistry of the uterus

The work of Csapŏ (1950) is fundamental to this problem.

Csapŏ was able to extract *myosin* and *actomyosin* from the uterus of different animals (cf. Volume II). Two significant differences were apparent between striated muscle and uterine muscle:

(1) uterine muscle contains less myosin + actomyosin, in particular, *less actomyosin*;

(2) the threads of uterine myosin *contract more slowly* in the presence of ATP (syneresis) than myosin threads of skeletal striated muscle (cf. p. 186 and Volume II).

However, the quantity of myosin + actomyosin and the percentage of actomyosin increases during gestation.

The contraction intensifies and reaches a maximum in uterine muscle during labour. In the ♀ rat, the level of actomyosin + myosin decreases significantly after parturition but primiparous uteri (first gestation) and particularly multiparous uteri are generally very muscular and show spontaneous contractions which are not seen in the uterus of virgin females.

Csapŏ also showed the presence of adenosine triphosphatase (ATPase) activity in uterine actomyosin, the activity being augmented by Ca^{++} and inhibited by Mg^{++}. A study of the sedimentation diagram from the Svedberg ultracentrifuge shows that *uterine myosin is different from myosin of skeletal striated muscle.*

One fact of interest is that *the quantity of actomyosin is a function of the concentration of oestrogens.* After ovariectomy actomyosin almost completely disappears from the uterus but it can be reconstituted by the injection of oestrogens. The mechanical tension developed by the uterus is a direct measure of the quantity of actomyosin and thus of oestrogens. Chronic stretch not only prevents involution and castration atrophy but, like oestrogen, initiates uterine growth; stretch-induced myometrial hypertrophy is independent of uterine activity, thus *a change in the confi-*

guration of the myometrial cell profoundly alters its biology (A. Csapŏ *et al.*, 1965; cf. also p. 243). Once again, the explanation is found in mechano-molecular physiology and is apparent in the occurrence of a hormone–enzyme link.

The effect of mineral ions on the uterus is still a controversial subject. In general terms it can be said that in considering smooth muscle and in particular the uterus the situation becomes complex.

Grinkraut and Sawaya (1957) studied the problem of action of ions on uterine mechanics in rat. The principal conclusions are as follows:

(a) absence of Na^+ blocks contractions;

(b) absence of K^+ also blocks contractions but a *contracture* develops; in rat this contracture is decreased or inhibited by 1–10 $\mu g/ml$ adrenaline (Daniel, 1964);

(c) absence of Ca^{++} leads to complete relaxation of the uterus;

(d) absence of Mg^{++} produces no clear response.

Contraction of the uterus can also be produced by progesterone *in vitro*; the contraction is blocked by electrical shocks, anoxia or oxytocin (Csapŏ and Corner, 1951).

The causes of uterine contraction and particularly the influence of steroid hormones are not elucidated. With the aid of techniques by which open hollow organs are maintained for long periods, I was able to study directly the interior of *virgin, primiparous, multiparous* and *pregnant* uteri of rats *in vitro*. The opening is made along the ante-mesenteric region such that circular myometrial fibres are sectioned. The uteri show longitudinal contractions *even when they contain embryos* but, *in vivo*, a pregnant uterus (at 21st day) shows no comparable contractions. Progesterone may have an inhibitory action but it is possible that *uterine innervation* by the hypo-gastric nerves can be involved *in vivo*.

It should be noted that uterine contraction is relatively slow compared with that of the heart, but uterine muscle of sow contains about seven times less sarcosomes (mitochondria of muscle) than cardiac muscle and respiration is also less (Gautheron and Gaudemer, 1960). These factors indicate that mitochondria may be involved in determining the intensity of contractile metabolism.

Macroscopic phenomena of parturition

During the last month of gestation in Rhesus, for example, the pubic symphysis relaxes while connective tissue regresses even in the absence of ovaries (Hartmann and Straus, 1939). During labour, uterine contraction is characterized by a series of waves of synchronous contractions passing from the upper part of the uterus to the cervix which has remained closed throughout pregnancy and which now opens. The uterus expels in this way the embryo and associated structures.

Endocrinology of parturition

(1) Sterols

The uterus contracts rhythmically in the presence of oestrogens but shows contractures with progesterone; it is noticeable that the quantity of oestrogens and of progesterone increases until parturition.

Gestation can be prolonged by progesterone and parturition occurs when these injections cease. Thus, it would appear that a *fall* in the level of progesterone determines parturition but if the placenta represents the essential source of progesterone at the end of gestation, there would be a contradiction here. However, if the level of progesterone does not decrease in *absolute value* but the action of progesterone is inhibited by increasing concentrations of oestrogens, the relative deficiency in progesterone could explain parturition. The inhibitory action of oestrogens would determine, at a certain time, an equilibrium which would represent the hormonal threshold of parturition and there would appear to be a decrease in progesterone level.

(2) Let us consider the role of the *pineal gland*: the mammalian epiphysis has the unique ability to synthesize methoxyindoles such as melatonin − 5 methoxy–N–acetyltryptamine (Lerner *et al.*, 1958; Axelrod *et al.*, 1961–1963). Chlorpromazine enhances the accumulation of tritiated melatonin by brain, heart and ovary but neither serotonin nor methoxytryptamine; pretreatment altered the accumulation of tritiated melatonin in tissue although both compounds are similar to melatonin in chemical structures (Wurtman and Axelrod, 1966), while melatonin, according to Hertz-Eshel and Rahamimoff (1965), causes an inhibition of spontaneous and serotonin-induced contractions of the oestrogenized rat uterus.

(3) Relaxin

The pelvic girdle constitutes the basal support of the visceral cavity; the morphology varies with sex. In the female, the cavity *is larger than in the male* and appears to be proportional to the dimensions of the foetal cranium (hypothesis of Schultz).

At the time of parturition, the pubic symphysis opens by relaxation of the sacro-iliac ligaments. This phenomenon can be compared with laying in chickens in which the pubis opens several millimetres for the passage of eggs (Engeler, 1928).

There are two points to be considered:

(a) *sexual dimorphism of the pubic symphysis;*
(b) *relaxation of the pubic symphysis.*

(a) Intratesticular or subcutaneous grafts of ovarian tissue to ♂ mice result in feminization of the pubic symphysis, particularly in castrated mice. In both sexes, oestrogens can induce the appearance of the adult virgin symphysis or that of gestating mothers depending on the dose. Decalcification of the pubis by oestrogens is inhibited by testosterone. The role of oestrogens in regulation of calcemia and particularly the level of oestrogens in birds at the time of laying will be remembered (cf. p. 530). It is possible that parathormone would be mobilized to prepare for parturition.

(b) Hisaw (1929) recognized the hormonal influence on relaxation of the pubic symphysis in guinea pig (discovery of *relaxin*). The test used today is that of Marois, Nataf and Marois (1950); this is a radiological technique for measuring the pubic space in guinea pigs. The chemical nature of relaxin is unknown. According to Frieden and Hisaw (1950) this hormone, extracted by ethanol, may be a low molecular weight protein (about 9000); it contains reducing sugars and 2·7 per cent of cystine. It may be produced by the ovary and can be detected in ovaries of elasmobranchs and, surprisingly, in the testes of birds (Steinetz, Beach and Kroc, 1959).

(4) Hypophyso-pituitary hormones

The fact that *parturition is possible in totally hypophysectomized females* is not an argument against the *necessary* intervention of oxytocin since the hormone may continue to be secreted by the hypothalamus (Cross, 1958). Oxytocin does appear to be *the factor initiating uterine contraction at the time of parturition*.

Collin and Racadot (1953) reported a fall in the level of Bargmann-Gomori positive material in the hypophysis during *post-partum* in guinea pig. But, curiously, they attribute this solely to augmentation in the consumption, not of oxytocin, but of antidiuretic hormone.

Oxytocin can be inactivated by *oxytocinase* (Page) which is found in liver. This enzyme is activated by cysteine or glutathione.

Oxytocin produces contractions of smooth muscles and of uterus specially and is assayed using the uterus of guinea pig (Dale and Laidlaw, 1912) or rat (Suzor *et al.*, 1952). However, since a number of compounds other than oxytocin, such as *histamine, can act on uterine muscle*, certain precautions must be taken in this assay and the use of antihistamines is essential.

The responses of the uterus and cervix are different: in the non-gestating cow, the sensitivity of the cervix to oxytocin is low during the luteal phase but increases during the follicular phase or following administration of oestrogens. The increase in sensitivity of the uterus is less pronounced (Fitzpatrick, 1956). It is interesting to consider this change as a function of the ratio progesterone/oestrogens at parturition, oxytocin activating movements of the uterus dominated by oestrogens and thus facilitating evacuation. As the cervix is more reactive than the uterus, expulsion of the foetus is favoured.

The role of the hypothalamus in parturition is open to debate. However, many workers have found that electrolytic destruction of the pituitary stalk results in prolonged labour, often leading to death of the foetus in rodents, monkeys, cat. Furthermore, Hild and Zetler consider that oxytocin originates in the hypothalamus. The following scheme indicates the present concept of the endocrinology of parturition:

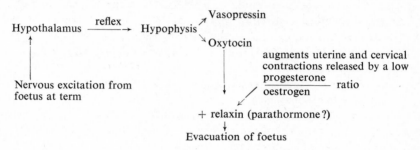

CHAPTER 26

LACTATION

The mammary glands represent the characteristic structures of mammals.

Schultz (1948) found a correlation between the number of mammary glands and offspring in prosimians. There is normally a single pair of glands in monkeys and man but additional pairs can appear and these are situated below the normal pair which *correspond to the lowest pair of pectoral mammary glands of prosimians*. The localization of mammary glands is *thoracic* in hominids and proboscidians and *inguinal* in ungulates and cetaceans.

The mammary glands are derived from sebaceous glands. They are found in both sexes but are normally functional only in the female. *However*, the mammary glands of the male can become functional and numerous experiments on lactation have been carried out on males subjected to appropriate treatments. The spontaneous development of mammary glands in the male is called *gynecomastia*. In the adult the basic structure of the mammary gland is that of a typical *exocrine* gland and consists of *alveolar cells* which secrete milk and *excretory ducts* which lead to the external surface of the *nipple*, the *areola;* these *galactophores* traverse bundles of *smooth* muscle.

When the mammary gland becomes active it passes through *two* stages:

the colostrogenic stage, with secretion of colostrum, introduced by a period of cell proliferation;

the lactogenic phase, with milk secretion.

During both stages, the secretory acini are voluminous as well as the subjacent cubic cells in which the *chondriome is well developed*, as in every case in which there is production of secretory material.

Colostrum is a viscous fluid with approximately the same composition as milk, that is, it contains mineral salts, *lactose, casein (phosphated* protein) and fatty droplets (less abundant than in milk), leucocytes, fairly numerous desquamated epithelial cells, *granular bodies* of Donné (or *colostrum bodies*) filled with fatty droplets (Figs. 157 a and b).

Ling, Kon and Porter found colostrum to be rich in several vitamins but the level of each varies according to species. The colostrum of cow contains 150 μg carotene/100 g while that of mare contains 25 μg/100 g and sow only 1 μg. Colostrum contains 140 μg vitamin A/100 g in cow, 33 μg/100g in mare and 95 μg in sow. The milk of cow is particularly rich in vitamin

567

B_2 (500 μg/100 g) while mare's milk contains large amounts of pantothenic acid (750 μg/100 g). The levels of ascorbic acid, measured in mg, are: 2·5/100 g in cow, 6/100 g in mare and 30/100 g in sow. Colostrum of cows contains more calcium than does milk of cows.

S.J. Folley and his co-workers(1947–1950) have shown that the mammary glands of ruminants are able to oxidize acetate (cf. p. 97) more readily than glucose, while the opposite is true for other mammals, the terrestial

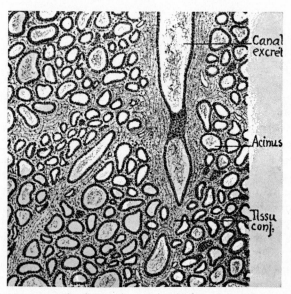

FIG. 155. Mammary gland. Female at the end of gestation (Bouin, hematein and eosin. Low magnification.) Acini with large lumen and cubic epithelium grouped in lobules. Excretory canals. Colostrogenic phase. The general arrangement does not change with the lactogenic phase. (DUBREUIL and BAUDRIMONT, 1959.)

ones at least. It should be noted that, in cow, the introduction of silage (aqueous plant material preserved by fermentation in a silo) or acetic acid to a normal equilibrated diet results in a significant increase in the quantity of milk and total fatty material secreted by the mammary gland (Zelter, 1953).

Milk contains three essential proteins (casein, β-lactoglobulin and α-lactoglobulin) and a disaccharide, *lactose*, containing glucose and galactose.

M. Polonovski and Lespagnol (1928) demonstrated the presence in human milk of a carbohydrate which they called *gynolactose*. The faeces of infants fed on human milk are rich in *Lactobacillus bifidus* but this is absent from infants receiving cow's milk (Tissier, 1900; Moro, 1905).

This bacillus converts lactose to lactic acid and acetic acid (Orla-Jensen, 1943) such that the intra-intestinal pH decreases. This explains the beneficial effect of human milk in resistance of infants to typhoid bacilli which are sensitive to acid. The variety *pennsylvanicus* of *L. bifidus* can proliferate only in the presence of human milk (Giorgy, 1953). The "*bifidus* factor" is gynolactose. Gynolactose is a mixture of at least 14 sugars, certain of which contain nitrogen (R. Kuhn; J. Montreuil). Sialic acid has been shown to be present in human milk (Giorgy). The biosynthesis of lactose requires ATP and UDP, uridine diphosphate (Leloir and Cardini, 1955):

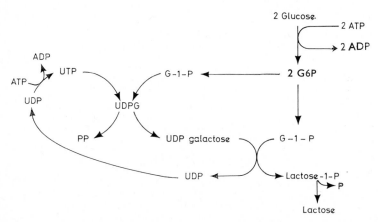

Lactoplasma contains numerous fatty globules, cellular and nuclear fragments, crystals of calcium phosphate but practically no leucocytes. In the presence of rennet, lactoplasma separates into *lactoserum* (lactose+mineral salts+water) and coagulated *casein*. Milk also contains urea, vitamins (mainly A, D, E, K, B_1 and B_2), enzymes (lactase, amylase, lipase, phosphomonoesterases, xanthine oxidase, etc.). The milk of ewes is particularly rich in calcium. Cow's milk also contains 1 mg per cent of rhodanate, a goitrogenic substance, but, according to Virtanen, the consumption of large amounts of milk is not harmful.

We shall now examine the *development* of the *mammary gland* during the different stages of the natural history of the female by modifying to some extent the classification of Dubreuil and Baudrimont:

(1) *Foetal stage.* From the *mammary crest* (ectodermal thickening) only the *lacteal points* (pectoral in humans) remain and these will develop into pilosebaceous buds and galactophoric buds.

(2) *Post-natal stage.* For several days the mammary region of the newborn secretes "witch's milk" which resembles colostrum.

(3) *Infant stage.* The "witch's milk" disappears but mammary activity does not cease: the gland proliferates and the ducts subdivide. This is the terminal stage of mammary growth in the male.

37

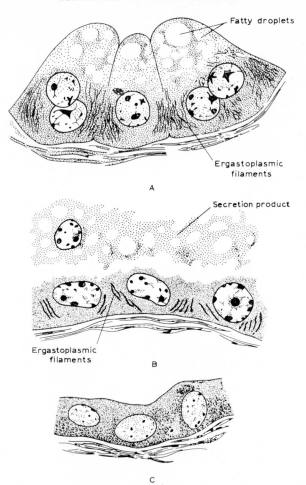

FIG. 156. Secretory cycle of milk secretion. After LIMON. Type of holo-merocrine secretion (by decapitation of cells). (A) Phase of accumulation: the apical pole of secretory cells contains fatty droplets, certain cells contain two nuclei; basal ergastoplasmic filaments. (B) Phase of exocellular excretion: Note the cell nuclei and droplets of fat in the protoplasmic mass produced by the decapitation of secretory cells and constituting the product of secretion. (C) Phase of rest, during which the cells, reduced to their basal part, are reconstituted. (DUBREUIL and BAUDRIMONT, 1959.)

(4) *Pubescent stage. Since the mammary gland is a sex receptor* (target organ) it develops in the girl; in particular, the galactophores branch extensively.

(5) *Stage of active genital life.*

(a) *In the absence of gestation,* the mammary gland shows cyclic histophysiological changes which follow cyclic ovarian secretion (endocrine tides). The follicular phase is one of relative inactivity but the folliculo-progestational phase is marked by multiplication of acini while the mammary secretion passes to the apical regions but does not reach the exterior. The connective tissue is then invaded by leucocytes and at menstruation parts of the gland are inactive.

It is important to note that the various parts of the gland are not at the same stage of activity *simultaneously*, indicating some functional alternation.

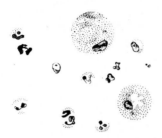

FIG. 157a. Colostrum 24 h after parturition. Globules of colostrum and polynuclear leucocytes. (JOLY, 1923.)

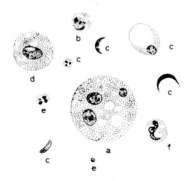

FIG. 157b. Colostrum 48 h; *a*, large globule of colostrum with 3 nuclei, *b*, small globule of colostrum; *c*, crescent shaped body formed from globules of colostrum; *d*, characteristic globule of colostrum; *e*, leucocyte with pycnotic nucleus; *f*, granular leucocyte containing fat globules. (JOLY.)

(b) *During gestation*, in woman for example, during the first five months the ducts and acini proliferate actively. From the sixth month until confinement colostrum is secreted.

(6) *Nursing stage*. Post-partum is characterized by the rapid appearance of milk secretion and the disappearance of Donné bodies and leucocytes.

(7) *Weaning*. With the cessation of suckling, a recurrent *colostrum phase* appears followed by involution of numerous acini, regression of glands and return of normal genital cycles.

Finally, the stages of inactive genital life are reached.

37*

(8) *Menopause*. The acini disappear and only excretory ducts remain but by a process of substitution the connective tissue becomes rich in fats leading in many cases to an increase in the volume of the post-menopause mammary gland.

(9) *Senescence*. Fats disappear and the connective tissue changes to a dense fibrous system; in addition, galactophoric branching resembles that in man.

This series of reactions shows a symmetry of growth and involution (galactophores → acini → galactophores).

What factors determine mammary gland activity?

Endrocrinology of lactation

We are dealing here with a particular case, that in which an endocrine system acts on an exocrine system.

Peripheral systems

Early bilateral ovariectomy inhibits development of mammary glands; after puberty, ovariectomy causes regression of glands. Unilateral castration does not produce these changes. Thus, *mammogenesis is dependent on the ovary*.

Folliculin administered in weekly doses of 2000 to 8000 IU to *male* and *impubescent* castrated female monkeys leads to development of mammary ducts and acini in both sexes while, in castrated *adult* monkeys, growth of acini requires progesterone to be injected.

Progesterone administered *after or at the same time* as folliculin induces proliferation without secretion in castrated impubescent females. This leads us to consider the hormonal equilibrium of pregnancy and lactogenesis in more detail.

I shall again refer principally to the work of Gaston Mayer's school (from 1951):

(1) *Bilateral ovariectomy during gestation in rat produces lactogenesis*. Normally during pregnancy there is *inhibition of milk production* and the ovary appears to be involved in this inhibition (through the pituitary or hypothalamo-hypophyso-pituitary complex?). When the female rat is castrated 2 to 3 days before normal term, the most critical phases for the young appear to correspond with delivery, acquisition of maternal instinct and the beginning of suckling. If the young survive until birth and the mother takes charge of them, they will develop normally (Jost, 1959). In the rat, lactogenesis at the time of delivery is due, not to the disappearance of uterine distension, but to hormonal changes, in particular, the disappearance of progesterone produced by cessation of placental luteotropic function (G. Mayer, 1956). It would be interesting to possess detailed

information on the endocrinology of gestation and parturition in selachians in which there is no lactation which implies a complex endocrine process.

Although the hormonal equilibria of pregnancy and lactation in rat are different, these two states have certain factors in common which are characterized by persistence in the ovary of functional *corpora lutea* and by *hypertrophied adrenal glands. But what is the effect of adrenalectomy on the gravid rat? The response varies according to whether pregnancy has been maintained:*

(1) *Rats in which gestation is maintained* deliver living young at the normal time but the young die within 1 or 2 days after parturition. At delivery the mothers show histologically normal milk production but in general they die 10 days after parturition.

(2) *Rats in which gestation is interrupted* after adrenalectomy die within 2 to 3 days following abortion.

Thus, the question of the *vicarious role of corpora lutea* in adrenalectomized rats during pregnancy and lactation — and perhaps of the placenta — is raised.

Before passing to an examination of the role of the hypothalamo-hypophyso-pituitary complex in lactation, let us recapitulate:

The development of the mammary gland is divided into two distinct phases: that of morphological development (mammogenesis) which involves the structures of the lobulo-alveolar system and that of functional activity (lactation).

Morphological development of the mammary gland begins with the embryonic stage but optimum development is attained during pregnancy. The latter begins with progestation and continues during gestation. Two hormones are involved: progesterone and oestradiol.

The phase of lactation consists of the period of lactogenesis (milk secretion) and that of *galactopoiesis* (or *milk maintenance). The essential factor in these two mechanisms is the liberation of pituitary prolactin.*

This leads us to examine the role in the endocrinology of lactation of the following:

Central endocrine systems

Firstly, I shall refer to the very informative experiments of Li and his co-workers (1955) which indicate the labile nature of genetic sexual terrain in the case in which all structures producing sex hormones are eliminated; in the *immature male rat,* totally *hypophysectomized* and castrated in addition as a precaution and, in certain experiments, *adrenalectomized,* the mammary ducts grow following injections of oestrone and *pituitary growth hormone.* Li and his co-workers were able to direct mammary events at will; thus:

(1) *limited* lobulo-alveolar growth is obtained by injection of oestrone, progesterone and lactogenic hormone, while

(2) *total* lobulo-alveolar growth is obtained by injection of oestrone, progesterone, lactogenic hormone + *growth hormone*.

Furthermore, in glands of the latter type, milk *secretion* is obtained by supplementary administration of *hydrocortisone acetate*.

These experiments provide an impressive summary of the hormonal factors of lactation. The main idea of their work, of great methodological importance, is the substitution of endocrine glands by their hormones.

The need for growth hormone has not passed unnoticed. Since the discovery by Stricker and Grueter (1928) of the existence in the pituitary of a lactogenic principle, an attempt has been made to relate all the effects of the pituitary on the mammary gland solely to the action of this hormone, *prolactin*, isolated by Riddle, Bates and Dykshorn (1932). However, the situation is more complex:

(1) Prolactin (LTH) does not act only on the mammary gland but also on the ovary. G. Mayer has shown that in rat prolactin would act as a *luteotropic factor* for the cyclic *corpus luteum* with prolongation of the cycle and functional activation of the *corpus luteum*. Prolactin also prolongs the duration of the *post-partum* cyclic *corpus luteum* in non-nursing rat and the gestative state of the ovary beyond its term (example of hormonal polyorganotropism).

These responses can be followed in woman using the *technique of vaginal smears*. During pregnancy and lactation vaginal smears show a characteristic appearance:

(a) *During pregnancy*. The intermediate layer of the vaginal epithelium shows hyperplasia while the superficial layer is decreased by marked desquamation. This increase in the intermediate layer results in the appearance in smears of large boat-shaped cells, called *navicular cells of pregnancy* (Fig. 158). Eight weeks after the inhibition of menses, the presence of these characteristic cells allows the pregnant state to be confirmed; thus vaginal smears cannot be used as an early test of pregnancy.

FIG. 158. Navicular cells of pregnancy.

(b) *During post-partum.* The vaginal epithelium involutes but after the tenth day development of the intermediate layer is resumed. Desquamation is associated with relatively large (25–45 μ), flat, cells with globular nuclei and generally basophilic cytoplasm (but certain cells are acidophilic). A large number of histiocytes, leucocytes and erythrocytes are found—as in the case *immediately after post-partum*; this is the characteristic appearance of smears taken during nursing.

In certain experiments of G. Mayer, in cases where prolaction prolongs the period of survival of the *post-partum corpus luteum* of non-nursing rats, the vaginal mucosa is either of the lactation type or of the pregnancy type. However, *in the first case the mammary acini are open and in the latter case they are closed.*

(2) At the present time, it is questioned *whether the essential role in galactopoietic activity is not due to growth hormone* (Folley, 1955). This is obviously of great economic importance (dairy industry). Purified growth hormone injected to cow temporarily increases milk production while prolactin does not have this effect and ACTH decreases the amount of milk.

This question of the galactopoietic ability of growth hormone is not exhausted for at least two reasons:

(a) TSH also shows galactopoietic activity, less pronounced than growth hormone but significant; administration of thyroxine increases milk production.

(b) Growth hormone is diabetogenic in cat (Young, cf. p. 491). The extent to which the galactopoietic activity is related to the ability to augment total blood glucose is a question that must be asked.

Other hormones—other endocrine glands—appear to be involved and particularly oxytocin (liberated at the time of parturition). Oxytocin may initiate lactation by inducing contraction of the mammary gland.

After the initiation of lactation the quantity of milk secreted increases rapidly to reach a maximum after which the amount decreases until the young are weaned.

Neurohormonal relationships during lactation

(1) Section of the *spinal cord* at the level of the eleventh thoracic vertebra in rat (Ingelbrecht, 1935; Eayrs and Baddley, 1954) does not modify the progress of pregnancy or delivery but *milk does not occur.* This fact has not been explained fully, associated total sympathectomy having no effect on the process (Denamur and Martinet, 1959).

(2) Ligature of galactophores does not inhibit milk secretion in rat. However, if *suckling is prevented* by ablation of the nipples, milk secretion is inhibited. If the galactophores are ligatured on one side and the nipples destroyed on the other, suckling of intact nipples is sufficient to maintain

milk secretion even on the injured side. Thus, suckling plays an essential role in milk secretion (Selye), a role which can only be explained by a reflex nervous or neurohormonal action (role of spinal cord?). According to Desclin (1962), the reflex is hormonal and implies liberation of prolactin. This reflex is both lactogenic and luteotropic but the lactogenic phenomenon in rat requires the presence of corticoids and the reflex also must

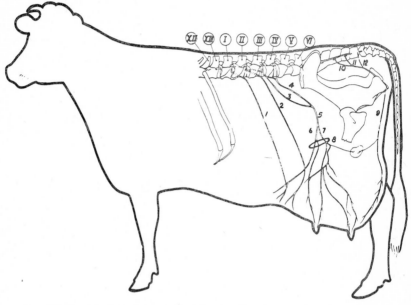

FIG. 159. Innervation of the nipple (after ST. CLAIR). *1*, ventral branch of first lumbar nerve; *2*, ventral branch of second lumbar nerve; *3*, ventral derivation of inguinal nerve; *4*, dorsal derivation of inguinal nerve; *5*, inguinal nerve; *6*, anterior inguinal nerve; *7*, posterior inguinal nerve; *8*, external inguinal ring *9*, perineal nerve; *10*, second sacral origin of pudic nerve; *11*, third sacral origin of pudic nerve; *12*, fourth sacral origin of pudic nerve. (From R. V. SMITH, 1960.)

induce the liberation of ACTH. In mice the processes are similar but it should be noted that prolactin is luteotropic in this case and suckling produces pseudo-gestation.

(3) Lefébure and then Kiefer showed the abundant innervation of the nipple. However, despite many experiments, the pathways of the suckling reflex have not been clearly established (reticular?). The indispensable nature of alveolar emptying in assuring continuation of milk secretion suggests a definite centre, the hypophysis (Folley, Petersen). This directs our study towards the hypothalamo-hypophyso-pituitary complex. A number of facts are available:

Suckling or manual treatment produces liberation of vasopressin in woman, bitch, cow and rabbit.

Andersson has established that mechanical stimulation of the vaginal region of the uterine cervix in mare induces milk ejection and Pickles (1953) has stated that coitus produces milk ejection in nursing mothers. These processes are associated with reflexes involving the hypothalamo-hypophyso-pituitary complex and it is to be noted that Andersson found that milk ejection can be provoked in mare by stimulation of the *central* end of the vagus.

In cow and in ewe and mare, vaginal distension induces an antidiuretic effect (Debackere and Peeters, 1960–1963). In these animals, stimulation of the nipple also can produce inhibition of diuresis. These effects are attributed to reflex liberation of antidiuretic hormone and summation is apparent, since the antidiuretic effect is greater with the second stimulus than with the first effected 30 mn previously. The amount of vasopressin liberated with vaginal distension in hydrated mare and ewe is in the region of 3 mU; in ewe, the antidiuretic action of oxytocin is more than 100 times less than that of vasopressin (Peeters and Debackere).

During lactation there is a *decrease* in the volume of hypophyseal neuro-secretion and in the ganglionic cells of diencephalic secretory nuclei there is a decrease in Nissl material and vacuolization. It is interesting to relate these facts to those concerning migration of fish (cf. p. 465). Furthermore:

(1) *Section of the supra-optico-hypophyseal tract* of nursing females is accompanied by discontinuation of milk secretion; on the other hand:

(2) *Electrical stimulation* of the hypothalamus near the supra-optic nucleus in mare activates milk secretion (Andersson).

(3) *Milk secretion is associated with inhibition of diuresis* (Cross and Harris). This emphasizes the question of the role of the functional link between oxytocin and antidiuretic hormone, particularly in parturition.

(4) Benson and Folley (1956) have shown that repeated injection of oxytocin to nursing mothers increases the production of milk by bringing about the production of large amounts of galactopoietic factors from the pituitary.

(5) Collin and Racadot (1950) stated that the neurohypophyseal reactions which accompany the phenomenon of lactation in guinea pig consist of degeneration of vesicles of isolated cells and groups of cells in the *pars intermedia* (effect on glycemia?).

(6) The mechanisms of *milk ejection* involve a complex reflex arc: the nervous pathways associated with this process pass through the dorsal column of the spinal cord and terminate, according to Tsakhaev, in the nuclei of Goll and Burdach (cf. Volume II). If the fibres of the band of Reil are stimulated immediately before the inferior olive (Andersson, 1951), milk ejection occurs in both the denervated and intact nipple. Thus, the hormonal process is controlled by a nervous process (typical neurohor-

monal process). The fibres pass through the cerebral trunk and reach the *anterior hypothalamus* which is, as we shall see (Volume II), the parasympathetic centre. It is the supra-optic nucleus which is important as shown by Andersson using chronic implantation of stimulating electrodes (excitation produces milk flow). It is of interest that the area immediately next to the supra-optic nucleus initiates *rumination*. Thus, there is a relationship between feeding of the mother and lactation and this represents a very remarkable case of physiological regulation.

These facts demonstrate the direct interrelationships which exist between lactation and the hypothalamo-hypophyso-pituitary complex.

From this study of somatic and sexual endocrinology two points emerge which seem to me to unify the individual processes by integrating them in the organic context. Firstly, the importance of the hypothalamo-hypophyso-pituitary complex arises from the many interrelationships in which it is involved in discrimination and regulation. The hypothalamus, in particular, benefits from its topographical position which is of exceptional importance, being in direct connection with visceral regions and, as we shall see (Volume II), with the cerebral cortex and thus, the possible role of emotional and even intellectual processes in the regulation of the internal environment is conceived (psychosomatic role). Finally, the concept of a *neurohormonal reflex* has gradually been expounded during our study. In considering the integrative systems this will assume a new significance.

INDEX

MY

COCK